Undergraduate Texts in Mathematics

Editors

S. Axler
F.W. Gehring
K.A. Ribet

Springer
New York
Berlin
Heidelberg
Hong Kong
London
Milan
Paris
Tokyo

Undergraduate Texts in Mathematics

(continued after index)

Donald Estep

Practical Analysis
in One Variable

With 211 Illustrations

 Springer

Donald Estep
Department of Mathematics
Colorado State University
Fort Collins, CO 80523
USA
estep@math.colostate.edu

Mathematics Subject Classification (2000): 26-02, 26Axx, 34Axx, 41A05 65H05 65L20

Library of Congress Cataloging-in-Publication Data
Estep, Donald J., 1959–
     Practical analysis in one variable / Donald Estep.
         p. cm. — (Undergraduate texts in mathematics)
     Includes bibliographical references and index.
     ISBN 0-387-95484-8 (alk. paper)
     1. Mathematical analysis. I. Title. II. Series.
QA300 .E785 2002
515—dc21   2002020943

ISBN 0-387-95484-8   Printed on acid-free paper.

Printed in the United States of America.

9 8 7 6 5 4 3 2 1   SPIN 10875295

Typesetting: Pages created by the author using a Springer T$_E$X macro package.

www.springer-ny.com

Springer-Verlag   New York Berlin Heidelberg
A member of BertelsmannSpringer Science+Business Media GmbH

*Dedicated to Lipman Bers, who first showed me
the beauty in mathematics, and to Patty Somers,
who helped me find it again.*

derivatives and whose solutions are functions. Mou...
functions motivates the introduction of the derivative,...
the simplest differential models motivates the introduc...
We investigate the properties of these operations thorough,...
practical application, we derive and analyze the basic transce...
tions as solutions of some classic differential equations. This par...
with a discussion of Newton's method for solving root problems. V...
basic material about numbers and functions in hand, the book tur...
more detailed analysis of functions, including investigations of continui...
sequences of functions, and approximation theory. The book concludes by
discussing the solution of nonlinear differential equations by means of the
essentially important Contraction Mapping Principle and Arzela's theorem
about equicontinuous functions.

While these are classic topics, the material in this book is not arranged
in the usual order found in most real analysis texts. There are two reasons.
One of the few tenets of teaching I have managed to hold after twenty
years is to introduce only one new concept at a time and only introduce a
concept when it is needed. Consequently, material in this book is introduced
in an order motivated by the practical problem of solving models rather
than by the formal style of building the subject from the ground up. Three
important examples are the introduction and use of Lipschitz continuity
well before other notions of continuity, the introduction of differentiation
via the linearization of a function, and the introduction of integration as
an approximation method for solving a differential equation rather than as
a way of computing the area under a curve. Each of these choices yields
distinct pedagogical benefits in terms of motivating ideas and teaching
students how to *do* analysis.

The order of the material in this book is also dictated by the goal of presenting constructive arguments. For example, assuming Lipschitz continuity makes it much easier to give constructive proofs for several fundamental
results like the Mean Value Theorem. Hence, the most general notion of
continuity and general versions of some fundamental results are not presented until the final third of the book, where the discussion becomes more
abstract and sophisticated as well as less constructive.

This book is aimed at two kinds of courses. First, there is the honors calculus sequence typically taken by freshman planning on a technical major.
These students often have advanced placement credit in calculus. Second,
there is the introductory course in real analysis offered to mathematics ma-
jors that have completed calculus. This book has been used successfully for
both kinds of courses at Georgia Tech and Colorado State University. Much
of this material has also been successfully tested at Chalmers University of
Technology in Sweden.

To use this book for such courses, it is necessary to be selective on the
material covered. For a freshman honors calculus course, I lecture on material in Chapters 1–4, 5–7 (briefly), 8–15, and finally calculus material

Preface

Background

I was an eighteen-year-old freshman when I began studying analysis. I
had arrived at Columbia University ready to major in physics or perhaps
engineering. But my seduction into mathematics began immediately with
Lipman Bers' calculus course, which stood supreme in a year of exciting
classes. Then after the course was over, Professor Bers called me into his
office and handed me a small blue book called *Principles of Mathematical
Analysis* by W. Rudin. He told me that if I could read this book over the
summer, understand most of it, and prove it by doing most of the problems,
then I might have a career as a mathematician. So began twenty years of
struggle to master the ideas in "Little Rudin."

I began because of a challenge to my ego but this shallow reason was
quickly forgotten as I learned about the beauty and the power of analysis
that summer. Anyone who recalls taking a "serious" mathematics course
for the first time will empathize with my feelings about this new world
into which I fell. In school, I restlessly wandered through complex analysis,
analytic number theory, and partial differential equations, before eventually
settling in numerical analysis. But underlying all of this indecision was
an ever-present and ever-growing appreciation of analysis. An appreciation
that still sustains my intellect even in the often cynical world of the modern
academic professional.

But developing this appreciation did not come easy to me, and the presentation in this book is motivated by my struggles to understand the

most basic concepts of analysis. To paraphrase J. von Neumann, it is not that we understand mathematics, rather mathematics just becomes familiar with practice. We often understand a difficult concept by considering special cases that make the concept concrete. In turn, our understanding of a concept is shaded by the special cases we consider. After learning about mathematics in specific contexts, it is easy to become convinced that this is the natural and best setting in which to teach these ideas.

I think this is especially true of analysis. I view analysis as the art and the science of estimation. That the practice of analysis is an art is understood by anyone who tries to explain an " epsilon-delta" proof of differentiation to a calculus student. At certain points, the natural response to "Why did you do that?" is "It's obvious, don't you see?" By the science of estimation, I refer to the need for the mathematical rigor that guarantees that any estimates obtained are meaningful and that plausible arguments are true.

Neither an art nor a science can be taught effectively in the abstract. Concepts and techniques that are perfectly well motivated in practical settings simply become a "bag of tricks" in the abstract. Moreover, technical difficulties often become overwhelming when there are no concrete examples to motivate the issues or provide a compelling reason to spend time on the complications. Too often, the mind lacks the firepower to leap past abstract technical mathematics to imagine how the underlying ideas might be used.

Consequently, I present the basic ideas of real analysis in the context of a fundamental problem of applied mathematics, which is approximating solutions of physical models. This approach is natural to me because of my research interests in numerical analysis and applied mathematics. I am a numerical analyst because my first reaction to being faced with a difficult analytic concept is to compute examples. I believe this "experimental" approach to understanding mathematics is natural for many people. So as much as practicable, I present analysis from a constructive point of view. Many major theorems are proved using constructive arguments that can be implemented on a computer and verified by computation. The theorems themselves are motivated in the context of solving models of physical situations that beg for computational solution. I believe that students who implement these proofs and solve the practical problems in this book will develop a "hands-on" understanding of analysis that will serve them well in the future.

Motivation

I have three overt reasons for writing this book, and one covert reason.

First whenever I teach numerical analysis, I am annoyed by the amount of time I spend on topics from basic calculus. From the point of view of

ists and engineers, modern calculus is very unsatisfactory. Stu d much of their time practicing skills that are rarely used and are 1 ght some fundamental ideas that come up repeatedly. One consequ that students studying science and engineering spend a large portic heir time in upper-level mathematics courses on elementary topics at expense of sophisticated material for which they truly need a mathem cian's help.

Second, teaching a modern calculus course is a frustrating experie for many analysts. Calculus should be a course in real analysis becau that is what it is. But the current trend in teaching calculus is to av anything to do with analysis and instead concentrate on solving practica unimportant "exact answer" problems. The conventional wisdom is th analysis is too hard (or put cynically, students are too dumb to learn re mathematics). But having met many bright students over the years, I hav found this rationale increasingly questionable. Rather, this trend migh originate in the observation that teaching rigorous mathematics to youn students requires significant effort and ingenuity from the instructor.

Third, teaching introductory real analysis using a modern abstract approach, even from a beautiful book like Rudin's, is far from optimal. As noted, I have serious doubts as to the effectivness of an abstract approach to teaching analysis. Moreover, this approach has some serious consequences. First of all, it perpetuates the faulty notion that there is some difference between "pure" analysis and the "dirty" topics important to numerical analysis and applied mathematics. This seeds the prejudices of pure and applied mathematicians that are so unfortunate for mathematics. Moreover, it makes the typical introductory real analysis course unattractive to the brightest students in science and engineering, who could benefit from taking such a course.

This book attempts to place the basic ideas of real analysis and numerical analysis together in an applied setting that is both accessible and motivational to young students of all technical persuasions.

This goal reflects my covert reason for writing this book. Namely, this book is a personal statement about how I believe people learn mathematics and how mathematics should therefore be taught.

Usage

This book begins by considering the solution of algebraic models with numeric roots. The discussion leads naturally from the integers through rational numbers and induction to the construction of the real numbers. Interwoven is a thorough discussion of functions, and the high point of this part of the book is the theory of the fixed point iteration for solving nonlinear equations. The next part of the book is concerned with models that involve proper in Chapters 16–30 and 35. I conclude by covering selected material in Chapters 31 and 36–38. A calculus course that follows this syllabus certainly omits several topics covered in a standard course, like a detailed discussion of integration techniques and various standard "applications." I have not found that my students suffer from this. For an advanced calculus/introductory real analysis course, I lecture on material in Chapters 3, 4, 8–15, 16, 18–23, 25–27, 28 and 29 very briefly, 32–35. I then lecture on a selection of material from Chapters 36–41.

The material is supplemented by exercises that range from simple computations to estimates to computational projects. When I teach this material, I assign a mixture of course work, including in-class exams testing basic understanding, take-home problem sets covering the more difficult analytic problems, and "laboratory" projects performed using a computer and requiring a written report.

Acknowledgments

As a student, I was fortunate to have taken analysis courses from a number of fine mathematicians, including Lipman Bers, Jacob Sturm, Hugh Montgomery, Joel Smoller, Jeff Rauch, Ridgway Scott, Claes Johnson, and Stig Larsson. Though I was an indifferent student, they still managed to show me brief glimpses of the beautiful view of analysis they routinely hold in their minds.

This project was originally conceived during conversations with my good friend and colleague Claes Johnson. Both agreement and disagreement with Claes are always tremendously stimulating. Finding the energy to finish the project was due in no small part to my students in honors calculus at Georgia Tech in 1997/8. Their enthusiasm, patience, curiosity, and good nature was unbounded and teaching them was truly a life-altering experience.

I thank Luca Dieci, Sean Eastman, Kenneth Eriksson, Claes Johnson, Rick Miranda, Patty Somers, Jeff Steif, and Simon Tavener for their comments and corrections, which led to substantial improvements. I thank Lars Wahlbin for helping out with several points of mathematical history.

I thank the Division of Mathematical Sciences at the National Science Foundation for many years of support. Specifically, this material is based upon work supported by the National Science Foundation under grants DMS-9506519, DMS-9805748, and DMS-0107832.

Finally, I thank Patty Somers for her support and patience. Being married to an academic mathematician is bad enough, not to mention having to deal with book writing on top of that. Patty does both while convincing me that it is (almost always) fun.

Fort Collins, Colorado   Donald Estep

Contents

II   Differential and Integral Calculus   215

Introduction

Analysis. Thinking about this word evokes a surprising range of emotions. Now, in the prime of my mathematical career, I suppose I feel about analysis the way a master woodworker feels about her woodworking tools. I take professional pride in my skill and the things I have created using my tools, and I have a professional interest in how others use these tools. I am always on the lookout to hone and improve my skills. But I can also remember back to when I was a student first learning analysis. I remember days of angry frustration trying to read a page or two of a paper or a book or trying to do some problem. There were also a few moments of beautiful epiphany with feelings like those I get when hiking through endless trees in my native Appalachians and suddenly coming upon a gap that reveals the beauty of those old mountains.

But I am getting ahead of myself. What is analysis? Analysis presents two faces, depending on the user's orientation. To "applied" mathematicians, analysis means approximation and estimation. The quest of science and engineering is to describe the physical world, explain how it works, and then make predictions for future behavior. Most often, our descriptions of the physical world are mathematical; truly, mathematics is the language of science and engineering. But even though we must often make gross simplifications to obtain a mathematical description, mathematical descriptions of physical situations are often too complicated to be understood directly. Thus enters analysis in the form of estimates on the description, simplifications of the description, and approximations of the description.

To "pure" mathematicians, working far away from applications to the physical sciences, analysis is the study of the limiting behavior of infinite processes. Many mathematical objects, such as the derivative and the integral and even numbers themselves, for example, are best defined as the limit of an infinite process. Dealing with infinity and infinite processes in a rational and mathematically rigorous way is what distinguishes modern mathematics, beginning around the time of Newton and Leibniz, from "classical" mathematics as developed by the ancient Greeks, for example. The difficulty is understanding what it means to find such a limit, since obviously we can never go to the "end" of an infinite process. Modern analysis has placed this issue on a firm mathematical foundation.

But these two viewpoints of analysis describe the same activity. Approximation and estimation imply some notion of a limit, that is, the possibility of obtaining full accuracy as the limit of the approximation process. On the other hand, the concept of a limit also implies an approximation that can be made as accurate as desired. Indeed, many analysts are primarily interested in infinite processes closely associated to mathematical descriptions of physical phenomena.

Well, enough vague words about analysis. This book not only introduces analysis but also defines analysis, or at least the fundamental elements of analysis. That it takes 600 pages to define these elements is not surprising considering that rigorous mathematical analysis is one of the finest and most important intellectual achievements of mankind. This text describes the path created by a number of geniuses who worked in mathematics, science, and engineering.

This book is organized in three parts, which divide the material by subject and difficulty. The first part, *Numbers and Functions, Sequences and Limits*, discusses the basic properties of numbers and functions and introduces the fundamental concept of the limit and its use for solving mathematical models. The second part, *Differential and Integral Calculus*, introduces the derivative and the integral as well as modeling with and the solution of differential equations. The third part, *You Want Analysis? We've Got Your Analysis Right Here,* digs deeper into properties of functions and the solution of differential equations. The three parts cover successively more difficult topics. The three parts are also written in successively more sophisticated styles. In particular, the material in the first part is bound tightly to the solution of models, while the material in the last part is often presented on its own accord in the abstract.

The ideal preparation for reading this book is previous exposure to calculus, such as an advanced placement course in high school or an introductory course in college. The minimum requirements are a course in trigonometry and exposure to analytic geometry.

By the way, I should have included frustration in my description of my feelings about analysis at present as well as in the past. I still get mighty bouts of frustration when I try to learn new analysis or solve problems in my research. The great professional cyclist Greg LeMond said about riding a bike, "It doesn't get any easier, you just get faster." I suppose the same is true of mathematics. Now some analysis seems easy in its familiarity and I struggle with more complicated ideas. But the struggle to understand analysis is really neverending for me. So if it is any consolation to the reader, analysis may never become easy, but at least you will struggle with harder and harder ideas.[1]

If I had not ended up as a mathematician, I would likely have become an engineer or a scientist for the simple reason that I am driven by the same underlying urge as many engineers, mathematicians, and scientists: namely, the urge to *understand*. As long as I can remember, I have hated not knowing *why* something is true. This desire is the primary motivation for the approach to analysis I have taken in this book.

The book is not written in the "theorem–proof" style common in rigorous mathematics textbooks. With a few exceptions, the discussions in this book are not aimed at merely proving that a fact is true; rather, they attempt to explain *why* certain facts are true. Hence, the explanations and discussions are given the primary focus and, for the most part, theorems are just used to summarize the discussions. This prejudice toward understanding is reflected in the problems, which largely demand that you explain why things are true rather than just doing rote computations.

Setting out to understand analysis is a difficult undertaking and it is a rare person who will be able to understand everything the first time out. Indeed, most people can expect to take years to understand some of the basic ideas of analysis, as the author ruefully knows from his own experience. Since we cannot delay things for years while waiting for true enlightenment, it is important not to get "hung up" on some tricky point. If you do not understand something after spending some time on it, just go on. Working during the time when the mathematical foundations of calculus were being questioned seriously (ultimately leading to modern analysis), d'Alembert[2] wrote

Proceed and faith will come unto you.

Discussing wider human conflicts, Winston Churchill put this more pithily,

If you're going through hell, keep going.

---

[1] Funny, that doesn't sound as encouraging as it is meant to be!

[2] The French mathematician Jean Le Rond d'Alembert (1717–1783) was a highly influential scientist and mathematician. His major mathematical results were in differential equations and mechanics. d'Alembert first defined the derivative of a function as the limit of the ratio of small increments and argued that the concept of the limit should be placed on a firm mathematical foundation.

Part I

Numbers and Functions, Sequences and Limits

1
Mathematical Modeling

The first stop on our journey into analysis is the subject of mathematical modeling. In this book, we explain analysis in the context of understanding mathematical models of the physical world. This is not a book on mathematical modeling, however. That is a subject unto itself, requiring not only a mastery of mathematics, but also of particular scientific and engineering fields. Indeed, a large part of the curricula in science and engineering is devoted precisely to creating mathematical models.

Nevertheless, since we intend to analyze mathematical models of the physical world, it is important to understand how mathematical models are created, what they are intended to model, and what kind of information is expected from models. We start by giving two simple examples of the use of mathematics for describing practical situations. The first example is a problem in economics and the second is a problem in surveying. Both have been important fields of application for mathematics since the time of the Babylonians.[1] While the models are very simple, they illustrate fundamental ideas that reoccur repeatedly.

---

[1] The term Babylonian describes various groups of peoples that lived in Mesopotamia in the region around the Tigris and Euphrates rivers roughly in the period of 4000–1000 B.C. Babylonian mathematics included tables of roots of numbers (exact and approximate), solutions of algebra problems, formulas for long sums, and rudimentary geometry. The Babylonians relied on mathematics extensively to organize their daily business.

1.1   The Dinner Soup Model

We are making soup for dinner, and, following a recipe, we ask our room-mate to go to the grocery store and buy 10 dollars worth of potatoes, carrots, and beef according to the proportions 3:2:1 by weight. In other words, so that by weight there is three times as much potatoes as beef and two times as much carrots as beef. At the grocery store, our roommate finds that potatoes are 1 dollar per pound, carrots are 2 dollars per pound, and beef is 8 dollars per pound. Our roommate thus faces the problem of figuring out how much of each ingredient to buy to use up the 10 dollars.

One way to solve the problem is by trial-and-error. Our roommate could take quantities of the ingredients to the cash register in the proportions of 3:2:1 and let the clerk check the price, repeating until a total of 10 dollars is reached. Of course, both our roommate and the clerk could probably think of better ways to spend the afternoon. Another possibility is to describe the problem mathematically on a piece of paper, or make a **mathematical model** of the problem, and then find the correct amounts by doing some computations. If they are simple enough, our roommate might be able to do the computations in his/her head. Otherwise, he/she might use a piece of paper and a pen or a calculator. In any case, the idea is to use brains (and pen and paper or calculator) instead of brute physical work.

The mathematical model may be set up as follows: we note that it suffices to determine the amount of beef, since we'll buy twice as much carrots as beef and three times as much potatoes as beef. We give a name to the quantity to determine; namely, we let x denote the amount of meat in pounds to buy. The symbol x here represents an unknown quantity, or **unknown**, that we are seeking to determine by using available information.

If the amount of meat is x pounds, then the price of the meat to buy is $8x$ dollars by the simple computation

$$\text{cost of meat in dollars} = x \, \text{pounds} \times 8 \, \frac{\text{dollars}}{\text{pound}}.$$

Since there should be three times as much potatoes as meat by weight, the amount of potatoes in pounds is $3x$ and the cost of the potatoes is $3x$ dollars since the price of potatoes is one dollar per pound. Finally, the amount of carrots to buy is $2x$ and the cost is 2 times $2x = 4x$ dollars, since the price is 2 dollars per pound. The total cost of meat, potatoes and carrots is found by summing up the cost of each

$$8x + 3x + 4x = 15x.$$

Since we assume that we have 10 dollars to spend, we get the relation

$$15x = 10, \tag{1.1}$$

which expresses the equality of total cost and available money. This is an **equation** involving the unknown x and data determined by the physical situation. From this equation, our roommate can figure out how much of each ingredient to buy. This is done by dividing both sides of (1.1) by 15, which gives $x = 10/15 = 2/3 \approx 0.667$. So, our roommate should buy 2/3 of a pound of meat, and therefore $2 \times 2/3 = 4/3$ pounds of carrots, and finally $3 \times 2/3 = 2$ pounds of potatoes.

The mathematical model for this situation is (1.1), i.e. $15x = 10$, where x is the amount of meat, $15x$ is the total cost, and 10 is the available money. The modeling consisted in expressing the total cost of the ingredients $15x$ in terms of the amount of beef x. Note that in this model, we only take into account what is essential for the current purpose of buying potatoes, carrots, and meat for the Dinner Soup, and we do not bother to write down the prices of other items, like ice cream or beer.[2] Determining the useful information is an important, and sometimes difficult, part of the mathematical modeling.

Assigning symbols to relevant quantities, known or unknown, is an important step in setting up a mathematical model. The idea of assigning symbols to unknown quantities was introduced by the Babylonians, who used models like the Dinner Soup model to help organize the feeding of the many people working on their irrigation systems.

A nice feature of mathematical models is that they can be reused to describe different situations. For example if we have 15 dollars to spend, then the model is $15x = 15$ with solution $x = 1$. If we have 25 dollars to spend, then the model is $15x = 25$ with solution $x = 25/15 = 5/3$. In general, if the amount of money y is given, then the model is $15x = y$. In this model we use the two symbols x and y, and assume that the amount of money y is given and the amount of beef x is an unknown quantity to be determined from the equation ($15x = y$) of the model. The roles can be reversed and we can think of the amount of beef x as being given and the expenditure y to be determined (according to the formula $y = 15x$). In the first case we would think of the amount of beef x as a function of the expenditure y, and in the second the expenditure y as a function of x.

Before figuring out the mathematical model for the amount of beef to buy, we suggested that our roommate could find the amount using a trial-and-error strategy. We can also carry out such a strategy mathematically. First, we guess $x = 1$, which gives a total cost of ingredients of 15 dollars. Since this is too much, we try a smaller amount of meat, say, $x = .5$, to get a total cost of 7.5 dollars. This is too little, so we increase a little, say, $x = .75$. This gives a total cost of 11.25 dollars. So we try again with a guess between .5 and .75, say, .625. Now we get 9.38 for the total cost (rounding up). Note that we are definitely making progress toward the target cost of 10 dollars with this procedure. We choose an amount between .625 and .75, say, .6875, to get a total cost of ≈ 10.31. Continuing to guess in this

---

[2]No matter how desirable. Intellectual discipline is paramount, after all.

fashion, we find that the guesses tend closer and closer to the correct value $x = 2/3 = 0.66666 \cdots$ that we determined by division.

1.2   The Muddy Yard Model

The author owns a house with a 100×100 m backyard that has the unfortunate tendency to form a muddy lake every time it rains. We show a perspective of the field on the left in Fig. 1.1. Because of the grading in the

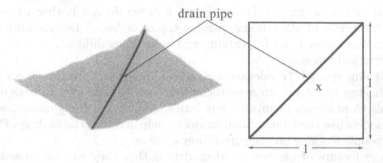

FIGURE 1.1. Perspective of a field with poor drainage and a model describing the dimensions.

yard, he believes the flooding can by stopped by digging a shallow ditch down the diagonal of the yard, laying some perforated plastic drain pipe, then covering the pipe back up. He is faced with the problem of determining the amount of pipe that he needs to purchase. Since a survey of the property only provides the outside dimensions and the locations of corners, and physically measuring the diagonal is not easy to do, he has decided to try to compute the distance using mathematics.

It is quite difficult to describe a yard exactly, so we create a simple model by assuming that the yard is perfectly square and level. In this model, we change to units of 100 m, so the field is 1×1. We plot the model on the right in Fig. 1.1. If we denote the length of the diagonal of the model field by x, then Pythagoras' theorem says that $x^2 = 1^2 + 1^2 = 2$. Therefore, to find the length x of the drain pipe, we have to solve the equation

$$x^2 = 2. \tag{1.2}$$

We call this equation the Muddy Yard model. In general, if we are talking about a field with sides of length y, the model is $y = x^2$.

Solving (1.2) may seem to be deceptively simple at first; the positive solution is just $x = \sqrt{2}$, after all. But this leaves the pertinent question, "What is $\sqrt{2}$?" Walking into a store and asking for $\sqrt{2}$ units of pipe is not going to get a positive response. Chances are that precut pipes do not come in lengths calibrated by $\sqrt{2}$ and a clerk is going to need more concrete information than the symbol "$\sqrt{2}$" to measure out a piece of pipe.

We can try to pin down the value of $\sqrt{2}$ by using a trial-and-error strategy as we did for the Dinner Soup model. We can check easily that $1^2 = 1 < 2$ while $2^2 = 4 > 2$. So $\sqrt{2}$, whatever it is, is between 1 and 2. Next we can check $1.1^2 = 1.21$, $1.2^2 = 1.44$, $1.3^2 = 1.69$, $1.4^2 = 1.96$, $1.5^2 = 2.25$, $1.6^2 = 2.56$, $1.7^2 = 2.89$, $1.8^2 = 3.24$, $1.9^2 = 3.61$. Apparently $\sqrt{2}$ is between 1.4 and 1.5. Next we can try to fix the third decimal. Now we find that $1.41^2 = 1.9881$, while $1.42^2 = 2.0164$. So apparently $\sqrt{2}$ is between 1.41 and 1.42 and likely closer to 1.41. Proceeding in this way, we can apparently determine as many decimal places of $\sqrt{2}$ as we like.

It turns out that we are going to meet many models that have to be solved by using some variation of a trial-and-error strategy. In fact, most mathematical equations cannot be solved exactly by some algebraic manipulations as we can do in the case of the Dinner Soup model (1.1). Consequently, the trial-and-error approach to solving mathematical equations is fundamentally important in mathematics. We shall see that trying to solve equations such as $x^2 = 2$ carries us directly into the very heart of analysis.

1.3 Mathematical Modeling

Based on these examples, we can describe mathematical modeling as a process with three components:

1. formulating the model in mathematical terms

2. understanding the model mathematically

3. solving for the solution of the model

We began this chapter by describing the physical situations in the Dinner Soup model and Muddy Yard model in terms of mathematical equations, which is the first component of mathematical modeling. Note that this aspect is not just mathematical in nature. Formulating equations that describe a physical situation certainly involves mathematics, but it also requires knowledge of physics, engineering, economics, history, psychology, and any other subjects that are relevant to describing the physical setting.

As a second step, we have to determine if the model makes "mathematical sense." "Does it have a solution and is the solution unique?" are important questions, for example. "What are the properties of a solution and do they make sense in terms of the physical situation being modeled?" is another important question. To understand a model mathematically, we have to understand the mathematical components making up the model and the properties of the solution.

The equations we obtained in the Dinner Soup model and the Muddy Yard model, namely, $15x = y$ and $x^2 = y$, are examples of **algebraic equations** in which the data y and the unknown x are both numbers. The models themselves are functions. So we begin our study of analysis by considering the properties of numbers and functions.

As we consider more complicated situations, we encounter models in which the data and the unknown quantities are functions. For example if we want to describe the motion of a satellite, then the description is a function giving the position and perhaps velocity versus time. Such models typically contain derivatives and integrals and are referred to as differential equations or integral equations. Calculus is nothing more (or less) than the science of formulating and solving differential and integral equations.

The last component of modeling is solving the equations in the model in order to determine some new information about the situation being modeled. In the case of the Dinner Soup model, we can solve the model equation explicitly, writing down a formula for the number that satisfies the equation. But, we cannot write down the solution of the Muddy Yard model explicitly and we resort to an iterative "trial-and-error" strategy to compute some digits of the decimal expansion of the solution. So it goes in general; once in a while, we can write down a solution of a model equation explicitly, but most often we can only approximate a solution by means of some iterative computational process.

In this book, we mainly take a **constructive** approach to the problem of analyzing and solving equations in which we seek **algorithms**, or mathematical procedures, through which solutions may be determined or computed as accurately as desired by increasing amounts of work. In this approach, we try to combine components 2 and 3 above.

Chapter 1   Problems

1.1. Suppose that the grocery store sells potatoes for 40 cents per pound, carrots for 80 cents per pound, and beef for 40 cents per *ounce*. Determine the model relation for the total price.

1.2. Suppose that you change the soup recipe to have equal amounts of carrots and potatoes, while the weight of these combined should be six times the weight of beef. Determine the model relation for the total price.

1.3. Suppose you go all out and add onions to the soup recipe in the proportion of 2:1 to the amount of beef, while keeping the proportions of the other ingredients the same. The price of onions in the store is 1 dollar per pound. Determine the model relation for the total price.

1.4. While flying directly over the airport in a holding pattern at an altitude of 1 mile, you see your high-rise condominium from the window. Knowing that the airport is 4 miles from your condominium and pretending that the condominium has height 0, how far are you from home and a cold beer?

1.5. Devise a model giving the length of drain pipe placed between opposing corners of a field that is 100×200 m in size and find an approximate solution using a trial-and-error strategy.

1.6. Devise a model of the draining of a yard that has three sides of approximately the same length 2 assuming that we drain the yard by laying a pipe from one corner to the midpoint of the opposite side. What quantity of pipe do we need?

1.7. A father and his child are playing with a teeter-totter which has a seat board 12 feet long. If the father weighs 170 pounds and the child weighs 45 pounds, construct a model for the location of the pivot point on the board in order for the teeter-totter to be in perfect balance? *Hint:* Recall the principle of the lever which says that the products of the distances from the fulcrum to the masses on each end of a lever must be equal for the lever to be in equilibrium.

1.8. A rectangular plot of land is to be fenced off from a large field that lies along a straight river so that the river forms one side of plot. If fencing is 35 dollars/meter and the rectangular plot is to consist of 100 m^2, find a formula for the cost of the fence in terms of the length of one side of the plot. Note: There are two possible forms for the answer.

2
Natural Numbers Just Aren't Enough

Numbers are a key component of mathematical modeling and therefore we need to develop a deep understanding of the construction and properties of numbers. In this chapter, we begin by considering the natural numbers. These are the numbers $1, 2, 3, \cdots$, which we first learn about as children and encounter most frequently in our daily life. While their properties are familiar to us, it is still worthwhile recalling their properties and how these fit our intuition about counting.

We also recall that the natural numbers alone do not suffice for our daily arithmetic needs. Even simple models involving only natural numbers quickly lead to the integers and then the rational numbers. These numbers extend the natural numbers in the sense that they include the natural numbers as well as numbers that are not natural. We explain how the properties of such extensions are "inherited" from the natural numbers. This is an important idea that we use later on in calculus with great effect.

2.1 The Natural Numbers

The **natural numbers** are familiar from our experience with counting, where we start with 1 and repeatedly add 1 to get the rest; $2 = 1 + 1$, $3 = 2+1 = 1+1+1, 4 = 3+1 = 1+1+1+1, 5 = 4+1 = 1+1+1+1+1$, and so on. Counting is a pervasive activity in human society: we count minutes waiting for the bus to come and the years of our life; the clerk counts change in the store, the teacher counts exam points, Robinson Crusoe counted the days by making cuts on a log. In each of these cases, the unit 1 represents something different; minutes and years, cents, exam points, days; but the operation of addition is the same for all the cases.

Encompassed in this counting experience are the basic rules learned in school about natural numbers. For example, if n and m are natural numbers, then $n + m$ is a natural number and both the **commutative rule for addition**,

$$m + n = n + m,$$

and the **associative rule for addition**,

$$m + (n + p) = (m + n) + p,$$

where m, n, and p are natural numbers, are familiar. The commutative rule $2 + 3 = 3 + 2$ comes from the realization that

$$(1 + 1) + (1 + 1 + 1) = 1 + 1 + 1 + 1 + 1 = (1 + 1 + 1) + (1 + 1).$$

This can be explained in words by observing that if we have 5 donuts in a box, then we can consume them by first eating 2 donuts and then 3 donuts or, equally well, by first eating 3 donuts and then 2 donuts.

In a similar way, we define **multiplication** of two natural numbers m and n denoted by $m \times n = mn$ as the result of adding n to itself m times to get the natural number mn. The **commutative rule for multiplication**,

$$m \times n = n \times m,$$

expresses the fact that adding n to itself m times is equal to adding m to itself n times. This can be pictured by making a square array of dots with m rows and n columns and counting the total number of dots $m \times n$ in two ways: first by summing the m dots in each column and then summing over the n columns and second by summing the n dots in each row and then summing over the m rows (see Fig. 2.1). Other familiar facts are the **associative rule for multiplication**,

$$m \times (n \times p) = (m \times n) \times p,$$

as well as the **distributive rule** combining addition and multiplication,

$$m \times (n + p) = m \times n + m \times p,$$

that hold for any natural numbers m, n, and p. These rules can be verified by similar manipulation of arrays of dots. Since addition and multiplication of natural numbers always produces another natural number, we say the natural numbers are **closed** under the operations of addition and multiplication.

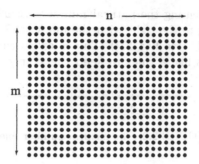

FIGURE 2.1. Illustration of the commutative rule for multiplication $m \times n = n \times m$. The same sum is obtained if the dots are counted across the rows or down the columns.

We define **powers** of natural numbers through repeated multiplication. If n and p are natural numbers, then

$$n^p = \underbrace{n \times n \times \cdots \times n}_{p \text{ multiplications}}.$$

We use the notation of the three dots "$\cdots$" to indicate "continue in the same way." The well-known properties, such as

$$\left(n^p\right)^q = n^{pq}$$
$$n^p \times n^q = n^{p+q}$$
$$n^p \times m^p = (nm)^p,$$

and so on, follow from this definition. By the way, it is useful to remember the formulas

$$(n+m)^2 = n^2 + 2nm + m^2$$
$$(n+m)^3 = n^3 + 3n^2m + 3nm^2 + m^3. \tag{2.1}$$

We also have a clear idea of ranking natural numbers according to size, or **ordering** the numbers. We consider m to be larger than n, $m > n$, if m can be obtained by adding 1 repeatedly to n. The inequality relation satisfies its own set of rules, including

$$m < n \text{ and } n < p \text{ implies } m < p$$
$$m < n \text{ implies } m + p < n + p$$
$$m < n \text{ implies } p \times m < p \times n$$
$$m < n \text{ and } p < q \text{ implies } m + p < n + q,$$

which hold for natural numbers n, m, p, and q. But it is equally important to note that some rules do *not* hold, such as $m < n$ and $p < q$ does not say anything about the relative sizes of $m + q$ and $n + p$. (Why not?)

One way of representing the natural numbers visually is to use a horizontal line extending to the right with the marks 1, 2, 3, spaced at a unit distance consecutively (see Fig. 2.2). This is called the **natural number line**. The line serves like a ruler to keep the points lined up. All of the

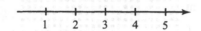

FIGURE 2.2. The natural number line.

arithmetic operations can be interpreted using the number line. For example, adding 1 to a natural number n means shifting from the position of n one unit to the right to $n + 1$ and likewise adding p means shifting p units to the right.

Representing natural numbers as sums of ones like $1+1+1+1+1$, cuts on a log, or beads on a thread quickly becomes impractical, and the positional number system using base 10 is a great improvement. In this system, we use the digits 0, 1, 2, 3, 4, 5, 6, 7, 8, and 9 together with position to represent any natural number efficiently. For example,

$$4711 = 4 \times 10 \times 10 \times 10 + 7 \times 10 \times 10 + 1 \times 10 + 1$$

The choice of the base 10 is of course connected to counting using our fingers.

2.2   Infinity or Is There a Largest Natural Number?

The insight that if n is a natural number, then $n + 1$ is a natural number is an important observation. One consequence is that there cannot be a largest natural number. For given any natural number, we can always find a larger natural number by adding 1 to it.

The principle that there cannot be a largest natural number is expressed by the word infinity, denoted by ∞. We say that there are **infinitely** many natural numbers, or that the set of natural numbers is **infinite**, as a way of saying that in principle it is possible to add 1 to any natural number and thereby obtain a larger natural number. Don't spend too much time philosophizing about the meaning of infinity; it just means that we never stop counting. Infinitely many steps means there is always the potential to take yet another step independent of the number of steps already taken. To have infinitely many donuts means that we can always take yet another donut independent of how many we have already eaten. This potential seems more realistic (and pleasant) than actually eating infinitely many donuts.[1]

2.3   A Controversy About the Set of Natural Numbers

We can easily understand the set $\{1, 2, 3, 4, 5\}$ of the first 5 natural numbers. We can simply write down the numbers $1, 2, 3, 4$, and 5 on a piece of paper and then view the numbers as one entity like a telephone number. Likewise, we can grasp the set $\{1, 2, ..., 100\}$ of the first 100 natural numbers. We can also grasp very large individual numbers. For example, we can understand the numbers $1000 and $10,000 because we buy things that cost that much during our life. We can even comprehend $10,000,000 by imagining what the rest of our life would be like with that much money. But despite this understanding, defining the set of natural numbers is not completely straightforward.

The usual definition of the set of natural numbers, which is denoted by N, is the set of *all* natural numbers. This is a universal concept and turns out to be a useful notion. But there is a catch. Any attempt to write down this set on a piece of paper is rudely interrupted by reality. We simply cannot write down all the members of a set with an infinite number of members.

For this reason, a small group of mathematicians called constructivists and the closely related intuitionists[2] object to definitions involving infinite sets. The constructivists believe that any valid mathematical argument can be reduced to a finite number of operations on numbers with a finite number of digits. This belief has far-reaching consequences that are not at all obvious. A constructivist certainly could not accept the first definition of N, but might accept the definition of N as the set of *possible* natural numbers that can potentially be computed by adding 1 over and over. In this view, N is always under construction and can never actually be completed. The difference between the two definitions of N is subtle, but

---

[1]Of course, this discussion is related to some kind of unlimited thought experiment. In the real world there are limits on how long we can add numbers. Eventually, Robinson Crusoe's log will be filled with cuts. Likewise, a natural number with say 10^{50} digits is impossible to store in a computer since this is about the total number of atoms estimated to be in the Universe. Therefore while in principle there is no largest natural number, in reality there are practical limits on the size of natural numbers we can use. It is important to distinguish what is true in principle and what is true in practice as we study mathematics.

[2]We refer to both groups as constructivists.

leads to major differences in the ways one goes about proving facts about $\mathbb{N}$.[3]

The idea of treating infinity and infinite sets as definite quantities originated with the work of Cantor.[4] Cantor's work on sets has had a great influence on the view of infinity in mathematics, and a majority of mathematicians today believe that the set of all natural numbers is a well-defined entity denoted by $\mathbb{N}$. But Cantor's ideas were controversial at the time, with much of the initial opposition being led by Kronecker,[5] who was the first constructivist. Other constructivists and intuitionists include such important mathematicians as Poincaré,[6] Brouwer,[7] and Weyl.[8]

One important consequence of the constructivist point of view is that a statement about a mathematical object is considered to be meaningful only if it contains an algorithm for computing the object. The expression "a constructive proof" is used to distinguish proofs that reach their conclusions through a constructive algorithm. It is natural to favor such proofs in the context of studying mathematical models because of the ultimate goal of solving the models. Hence for the most part, we provide constructive proofs in this book. On the other hand, there is no denying that the road of strict constructivism is rocky and steep. Thus, we resort to non-constructivist

---

[3]The discussion of the controversies associated with the foundations of analysis in this book is cursory at best. Please see Kline [16] and Davis and Hersh [8] for more detailed and coherent presentations.

[4]Georg Ferdinand Ludwig Philipp Cantor (1845–1915) worked in Germany on number theory and analysis before developing an original theory for infinite sets and infinite numbers. Cantor constructed a description of the real numbers that is closely related to the approach adopted in this book. The controversy surrounding his ideas about sets had a negative impact on his career and contributed to the terrible depression that affected him during his last few years.

[5]Leopold Kronecker (1823–1891) worked in Germany on algebra and number theory. A well known quote of Kronecker's is "God made the positive integers, everything else is the handiwork of man." Kronecker was so opposed to Cantor's ideas that he attempted to prevent an early paper of Cantor's from being published and later argued publicly against his ideas.

[6]The French mathematician Jules Henri Poincaré (1857–1912) created or transformed several areas of mathematics and mathematical physics and derived a theory of special relativity independently of Einstein. He also wrote some very popular books about science.

[7]The Dutch mathematician Luitzen Egbertus Jan Brouwer (1881–1967) made fundamental contributions to topology, which include some early fixed point theorems, before turning to the foundations of mathematics and founding the intuitionists. Interestingly, his most important results rely on the kind of non-constructive arguments that he eventually rejected.

[8]Hermann Klaus Hugo Weyl (1885–1955) worked in Germany before coming to the United States when the Nazis came to power. He made important contributions in analysis, group theory, mathematical physics, number theory, and topology. An interesting self-evaluation by Weyl was "My work always tried to unite the truth with the beautiful, but when I had to choose one or the other, I usually chose the beautiful."

ideas when it is convenient or the alternative is too technical.[9] It is probably fair to say that many mathematicians adopt the same attitude.

2.4 Subtraction and the Integers

Along with our experience of addition, we have an intuitive understanding of the **inverse** operation of **subtraction**. For example, if we have 12 donuts at home in Atlanta, cycle out to Paulding County and back, and then eat 7 donuts, we know there are 5 left. We originally got the 12 donuts by adding individual donuts into a box and we may take away donuts, or subtract them, by taking them back out of the box. Mathematically, we write this as $12 - 7 = 5$.

The need for subtraction also arises if we formulate models using addition and natural numbers. Suppose we wish to determine the number of donuts that add to seven to get twelve, i.e. we want to solve $7 + x = 12$. The answer is given by subtraction, $x = 12 - 7 = 5$.

But this leads immediately to a complication because we can formulate an equation using natural numbers and addition that does not have a natural number as a solution. For example, suppose we have the equation

$$x + 12 = 7.$$

This kind of equation arises if we want to eat 12 donuts but only have 7 available. The solution, which we know is $x = 7 - 12$, can not be a natural number. For if we add any natural number to 12, the resulting natural number is certainly larger than 7. Of course, similar situations arise frequently.

EXAMPLE 2.1. Suppose that we want to buy a titanium bike frame for $2500, while we only have $1500 in the bank. We immediately understand this means that we have to borrow $1000 to buy the frame. This $1000 is a debt and does not represent a positive amount in our savings account. It is not a natural number.

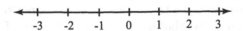

FIGURE 2.3. The integer number line.

Another way to describe this difficulty is to say that the natural numbers are not closed under subtraction, To handle such situations, we **extend** the

---

[9] If this is an intellectual contradiction, we can only be thankful for the tremendous capacity of the human mind to accept concrete statements about objects that are vaguely defined, like the world, soul, love, jazz music, ego, happiness, and N.

natural numbers $\{1, 2, 3, \cdots\}$ by adjoining the negative numbers -1, -2, -3, $\cdots$ together with 0. The result is the set of **integers**

$$\mathbb{Z} = \{\cdots, -3, -2, -1, 0, 1, 2, 3, \cdots\} = \{0, \pm 1, \pm 2, \pm 3, \cdots\}.$$

We say that 1, 2, 3, $\cdots$, are the **positive integers**, while -1, -2. -3, $\cdots$, are the **negative integers**. Graphically, we think of extending the natural number line to the left and then marking the point that is one unit distance to the left of 1 by 0, the point two units to the left of 1 by -1, and so on. The resulting line is called the **integer number line** (see Fig. 2.3). We call the 0 point the **origin** of the number line. The origin is often treated as the "center" of the number line.

Whenever we extend a set of numbers, we need to extend the definitions of the arithmetic operations so they are defined on the new expanded set of numbers as well. We know how to add two natural numbers, but how do we add integers? One consideration for finding the right definitions of the operations is that the "new" arithmetic operations on the integers should agree with the "old" operations when the integers happen to be natural numbers.

An easy way to visualize the extension of the operations is to use the number line. We define the sum of two integers m and n as follows. If n and m are both natural numbers, or positive integers, then $n + m$ is obtained the usual way: starting at 0, we move n units to the right followed by m more units to the right. If n is positive and m is negative, then we get $n + m$ by starting at 0, moving n units to the right, then m units back to the left. Likewise if n is negative and m is positive, then we obtain $n + m$ by starting at 0, moving n units to the left and then m units to the right. Finally if both n and m are negative, then we obtain $n + m$ by starting at 0, moving n units to the left and then m more units to the left. Since the number 0 is neither positive nor negative, when we add 0 we should move neither right nor left. In other words, $n + 0 = n$ for all integers n. It is straightforward to verify that this definition of addition for the integers has all the same properties as addition on the natural numbers.

We can now define the inverse operation of subtraction. First, for an integer n, we define $-n$ to be the integer with the opposite sign. It is clear that these definitions mean that $n + (-n) = (-n) + n = 0$ for any integer n. Finally, we define subtraction of two integers as $n - m = n + (-m)$, and $-n + m = (-n) + m$. Working with this definition, we can now verify that all the properties that we expect from addition, subtraction, and multiplication hold true. By the way, recall that even though subtraction is related to addition, it does not have the same properties. For example, while addition is commutative, subtraction is not in general: $n - m \neq m - n$ unless $n = m$.

We recall that inequalities and the inverse operations can be a little tricky. For example, multiplying by a negative number reverses an inequality, so that $m < n$ implies $-m > -n$.

2.5   Division and the Rational Numbers

We now investigate the inverse operation to multiplication, which is **division**. Actually, we already used division to solve the equation for the Dinner Soup model $15x = 10$ to get $x = 10/15 = 2/3$. Even though 15 and 10 are integers, the answer is not another integer. This motivates the extension of the integers to the **rational numbers**, which is the set $\mathbb{Q}$ of solutions of all equations of the form $nx = m$, where m and n are integers with $n \neq 0$. To make this definition useful, we have to figure out how to compute such solutions of course.

We begin by defining **division with remainder** of a natural number n by another natural number $m \neq 0$ as the process of computing natural numbers p and r such that $n = pm + r$ with $r < m$. This computation can be done mechanically by repeated subtraction.

EXAMPLE 2.2. Suppose that $n = 64$ and $m = 15$. Then

$$64 = 15 + 49$$
$$64 = 15 + 15 + 34 = 2 \times 15 + 34$$
$$64 = 3 \times 15 + 19$$
$$64 = 4 \times 15 + 4.$$

A more efficient method, or **algorithm**, for doing division is the **long division** algorithm in which we systematically divide the denominator into groups of digits of the numerator, working from left to right. We illustrate first by dividing 64 by 15 on the left in Fig. 2.4. Since 15 does not go into 6, we begin the division by considering the first two digits of 64. Since $15 \times 4 = 60$, so we put 4 in line with the 4 in 64 and put the result 60 underneath 64. Subtracting, we get the remaining part of the numerator. In this case, the remaining part is 4 and no further divisions can be carried out. On the right we divide 2418610 by 127. When the remainder of n

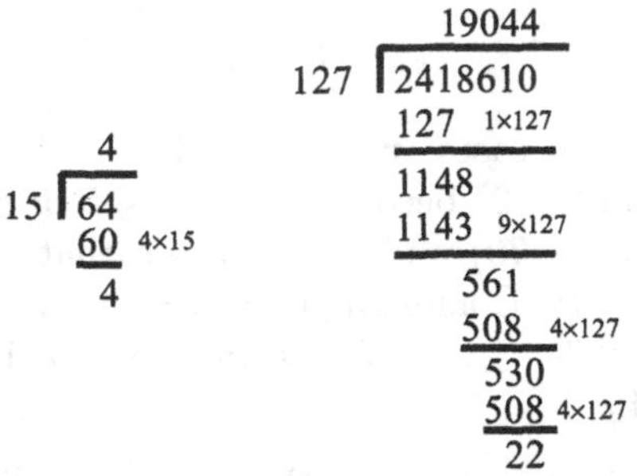

$$
\begin{array}{r}
19044 \\
127\,\overline{)2418610} \\
127 \quad {\scriptstyle 1\times127} \\
\hline
1148 \\
1143 \quad {\scriptstyle 9\times127} \\
\hline
561 \\
508 \quad {\scriptstyle 4\times127} \\
\hline
530 \\
508 \quad {\scriptstyle 4\times127} \\
\hline
22
\end{array}
$$

$$
\begin{array}{r}
4 \\
15\,\overline{)64} \\
60 \quad {\scriptstyle 4\times15} \\
\hline
4
\end{array}
$$

FIGURE 2.4. Two examples of long division.

divided by m is zero, then we obtain a **factorization** $n = pm$ of n as a product of the factors p and m. In this case, we define $n/m = p$ and say that p is the **quotient** of n and m. Likewise $n/p = m$ and m is the quotient of n and p. But, when the remainder of n divided by m is not zero, we have to extend the integers to the **rational numbers**, which are numbers of the form n/m for integers n and $m \neq 0$, in order to define division of all integers. We discuss the properties of rational numbers in Chapter 4.

2.6   Distance and the Absolute Value

Sometimes it is convenient to talk about the distance between two numbers. For example, suppose we have to buy a piece of molding for a doorway and when using a tape measure we position one side of the door frame at 2 inches and the opposite side at 32 inches. We would not go to the store and ask the person for a piece of molding that begins at 2 inches and ends at 32 inches. Instead, we would tell the clerk that we need $32 - 2 = 30$ inches. In this case, 30 is the distance between 32 and 2. We define the **distance** between two integers p and q as $|p - q|$ where the **absolute value** $|\ \ |$ is defined by

$$|p| = \begin{cases} p, & p \geq 0, \\ -p, & p < 0. \end{cases}$$

For example, $|3| = 3$ and $|-3| = 3$. By using the absolute value, we insure that the distance between p and q is the same as the distance between q and p. For example, $|5 - 2| = |2 - 5|$.

In this book, we frequently deal with inequalities combined with the absolute value. We give an example close to every student's heart

EXAMPLE 2.3. Suppose the scores on an exam that are within 5 of 78 out of 100 get a grade of B and we want to write down the list of scores that get a B. This includes all scores x that are a distance of at most 5 from 79, which can be written

$$|x - 79| \leq 5. \tag{2.2}$$

There are two possible cases: $x < 79$ and $x \geq 79$. If $x \geq 79$, then $|x - 79| = x - 79$ and (2.2) becomes $x - 79 \leq 5$ or $x \leq 84$. If $x < 79$, then $|x - 79| = -(x - 79)$ and (2.2) means that $-(x - 79) \leq 5$ or $(x - 79) \geq -5$ or $x \geq 74$. Combining these results, we have $79 \leq x \leq 84$ as one possibility or $74 \leq x < 79$ as another possibility, or in other words, $74 \leq x \leq 84$.

In general if $|x| < b$, then we have the two possibilities $-b < x < 0$ or $0 \leq x < b$, which means that $-b < x < b$. We can actually solve both cases at one time.

EXAMPLE 2.4. $|x - 79| \leq 5$ means that

$$-5 \leq x - 79 \leq 5$$
$$74 \leq x \leq 84 .$$

To solve $|4 - x| \leq 18$, we write

$$-18 \leq 4 - x \leq 18$$
$$18 \geq x - 4 \geq -18 \text{ (Note the changes!)}$$
$$22 \geq x \geq -14.$$

The other direction of inequality is handled differently.

EXAMPLE 2.5. Suppose we want to solve

$$|x - 79| \geq 5. \tag{2.3}$$

Now if $x \geq 79$, then (2.3) becomes $x - 79 \geq 5$ or $x \geq 84$. If $x \leq 79$, then (2.3) becomes $-(x - 79) \geq 5$ or $(x - 79) \leq -5$ or $x \leq -74$. So the answer is all x with $x \geq 84$ or $x \leq -74$. We can write this

$$-(x - 79) \geq 5 \quad \text{or} \quad x - 79 \geq 5$$
$$(x - 79) \leq -5 \quad \text{or} \quad x - 79 \geq 5$$
$$x \leq -74 \quad \text{or} \quad x \geq 84 .$$

Lastly, we mention an important property of $|\ \ |$ called the **triangle inequality**,

$$|a + b| \leq |a| + |b| \tag{2.4}$$

that holds for all rational numbers a and b. We ask you to prove this in Problem 2.12.

2.7   Computer Representation of Integers

Since we use the computer throughout this course, we point out some properties of computer arithmetic from time to time. In particular, we distinguish arithmetic carried out on a computer from the "theoretical" arithmetic described so far.

The fundamental issue that arises when using a computer stems from the physical limitation on memory. A computer must store numbers on a physical device which cannot be "infinite." Hence, a computer can only represent a *finite* number of numbers. Every computer language has a finite limit on the numbers it can represent. It is quite common for a language to have *INTEGER* and *LONG INTEGER* types of variables, where an INTEGER variable is an integer in the range of $\{-32768, -32767, ..., 32767\}$, which are the numbers that take two bytes of storage, and a long integer variable is an integer in the range $\{-2147483648, -2147483647, ..., 2147483647\}$, which are the integers requiring four bytes of storage. This can have some serious consequences, as anyone who programs a loop using an integer index that goes above the permitted largest value discovers. In particular, we cannot check whether some fact is true for all integers using a computer to test each case.

Chapter 2   Problems

2.1. Identify five ways in your life in which you count and the unit "1" for each case.

2.2. Use the natural number line representation to to interpret and verify the equalities (a) $x + y = y + x$ and (b) $x + (y + z) = (x + y) + z$ that hold for any natural numbers x, y, and z.

2.3. Use an array of dots to interpret and verify the distributive rule for multiplication $m \times (n + p) = m \times n + m \times p$.

2.4. Use the definition of n^p for natural numbers n and p to verify that (a) $\left(n^p\right)^q = n^{pq}$ and (b) $n^p \times n^q = n^{p+q}$ for natural numbers n, p, q.

2.5. Verify that (2.1) is true.

2.6. Use the integer number line to illustrate the four possible cases in the definition of $n + m$ for integers n and m.

2.7. Divide (a) 102 by 18, (b) -4301 by 63, and (c) 650,912 by 309 using long division.

2.8. (a) Find all the natural numbers that divide into 40 with zero remainder. (b) Do the same for 80.

Problem 2.9 is an abstract version of long division. We discuss this kind of division in detail later in Chapter 7.

2.9. Use long division to show that

$$\frac{a^3 + 3a^2b + 3ab^2 + b^3}{a + b} = a^2 + 2ab + b^2.$$

2.10. Pick out the *invalid* rules from the following list

$$a < b \text{ implies } a - c < b - c$$
$$(a + b)^2 = a^2 + b^2$$
$$\left(c(a + b)\right)^2 = c^2(a + b)^2$$
$$ab < bc \text{ implies } b < c$$
$$a - b < c \text{ implies } a < c + b$$
$$a + bc = (a + b)c.$$

In each case, find numbers that show the rule is invalid.

2.11. Solve the following inequalities:

(a) $|2x - 18| \leq 22$ (b) $|14 - x| < 6$

(c) $|x - 6| > 19$ (d) $|2 - x| \geq 1$.

2.12. Prove (2.4). *Hint:* Consider different cases of signs of a and b.

2.13. Write a little program in the computer language of your choice that finds the largest integer that the language can represent. *Hint:* Usually one of two things happen if you try to use set an integer variable to a value that is too large: either you get an error message or the computer gives the variable a negative value.

3
Infinity and Mathematical Induction

We claimed that the set of all natural numbers $\mathbb{N}$ is a useful concept. One reason is that it makes it easier to talk about properties of all natural numbers. But this raises the issue of how we should go about proving a property for all natural numbers. If we are asked to prove a property of a few numbers, we could just check the property for each number. In principle, this can be done for any finite set of numbers, though very large sets might cause difficulties. But we can not explicitly check a property for each number in an infinite set of numbers, like $\mathbb{N}$. Mathematical induction is a tool for proving properties of infinite sets of numbers.

3.1   The Need for Induction

We motivate the need for induction using a story about the mathematician Gauss[1] when he was 10. His old-fashioned arithmetic teacher liked to show

---

[1] Carl Friedrich Gauss (1777-1885), sometimes called the Prince of Mathematics, was one of the greatest mathematicians of all times. In addition to an incredible ability to compute (especially important in the 1800s) and an unsurpassed talent for mathematical proof, Gauss had an inventive imagination and a restless interest in nature. He made important discoveries in a staggering range of pure and applied mathematics, as well as important discoveries in science. As part of his investigative method, Gauss relied heavily on "experimental" computation. Unfortunately, Gauss wrote about his work sparingly and several mathematicians that followed him labored to discover theorems that he already knew. Gauss' interest in non-Euclidean geometry gives a good picture of how his mind worked. When Gauss was sixteen, he began seriously to question Euclidean

off to his students by asking them to add a large number of sequential numbers by hand, which the teacher knew from a book could be done quickly using the formula

$$1 + 2 + 3 + \cdots + (n-1) + n = \frac{n(n+1)}{2}. \tag{3.1}$$

Note that the "$\cdots$" indicates that we add all the natural numbers between 1 and n. This formula makes it possible to replace the $n-1$ additions on the left by a multiplication and a division, which is a considerable reduction in work when using a piece of chalk and a slate to do the sums.

By the way, long sums of numbers arise in integration and in models such as computing compound interest on a savings account or adding up populations of animals. Addition formulas like (3.1) are therefore practically useful, which is why we are interested in them.

The teacher posed the sum $1 + 2 + \cdots + 99$ to the class, and almost immediately Gauss came up and laid his slate down on the desk with the correct answer, 4950, while the rest of the class still struggled. How did young Gauss manage to compute the sum so quickly? He did not know the formula (3.1), he just derived it using the following clever argument. To sum $1 + 2 + \cdots + 99$, we group the numbers two by two as follows:

$$\begin{aligned}
1 &+ \cdots + 99 \\
&= (1+99) + (2+98) + (3+97) + \cdots (49+51) + 50 \\
&= 49 \times 100 + 50 = 49 \times 2 \times 50 + 50 = 99 \times 50.
\end{aligned}$$

This agrees with the formula (3.1) with $n = 99$. In Problem 3.9, we ask you to show that this argument can be used to show that (3.1) holds for any natural number n.

Gauss had a special ability to see patterns in numbers that allowed him to discover formulas like (3.1) easily. Most of us do not have that talent. Someone might claim that a formula like (3.1) is true and if we want to use the formula for something important, for example, if our grade depended on adding n numbers, then we would be motivated to try to check if the formula is true.

---

geometry. At the time that Gauss lived, Euclidean geometry had obtained a holy status and was held to be a higher truth that could never be questioned. Yet, Gauss was bothered by the fact that Euclidean geometry rested on postulates that apparently could not be proved, e.g., two parallel lines cannot meet. He went on to develop a theory of non-Euclidean geometry in which parallel lines can meet and this theory seemed to be as good as Euclidean geometry for describing the world. Gauss did not publish his theory, fearing too much controversy, but he decided that it should be tested. In Euclidean geometry, the sum of the angles in a triangle add up to 180° while in the non-Euclidean geometry this is not true. So centuries before the age of modern physics, Gauss conducted an experiment to see if the universe is "curved" by measuring the angles in the triangle made up by three mountain peaks. Unfortunately, the accuracy of his instruments was not sufficient to settle the question.

3.2   The Principle of Mathematical Induction

So the problem is to show that the formula (3.1) is true for *any* natural number n. It is easy enough to verify that it is true for $n = 1$: $1 = 1 \times 2/2$; for $n = 2$: $1 + 2 = 3 = 2 \times 3/2$; and for $n = 3$: $1 + 2 + 3 = 6 = 3 \times 4/2$. Checking the validity in this way for any natural number, one at a time, up to, say, $n = 1000$, would be very tiring but possible. Of course we can use a computer to go further, but even a computer gets stuck when n is very large. We also know in the back of our minds that no matter how many natural numbers n we check, there are always natural numbers that we have not checked.

Instead, we use a technique called the Principle of Mathematical Induction to show that (3.1) is true. The first step is to check that the formula is valid for $n = 1$, which we did above. The second step, which is called the **inductive step**, is to show that if the formula holds for a given natural number then it also holds for the next natural number. The **Principle of Mathematical Induction** states that the formula must therefore hold for any natural number n. This is a fairly intuitive claim. We know that the formula holds for $n = 1$. By the inductive step, it therefore holds for the next number, $n = 2$. But then the inductive step implies it also holds for $n = 3$, and then for $n = 4$, and so on. Since we eventually reach any natural number this way, it is fair to say that the formula holds for any natural number. Of course the Principle of Mathematical Induction is based on the conviction that we eventually reach any natural number if we start at 1 and then add 1 sufficiently many times. We take this as a defining property, or axiom, of the natural numbers.

Now we try to prove (3.1) by showing that the inductive step holds. Therefore, we *assume* that the formula (3.1) is valid for $n = m - 1$, where $m \geq 2$ is a natural number. In other words, we assume that

$$1 + 2 + 3 + \cdots + m - 1 = \frac{(m-1)m}{2}. \tag{3.2}$$

We now want to *prove* that the formula holds for the next natural number $n = m$. To do this for (3.1), we add m to both sides of (3.2) to get

$$1 + 2 + 3 + \cdots + m - 1 + m = \frac{(m-1)m}{2} + m$$
$$= \frac{m^2 - m}{2} + \frac{2m}{2} = \frac{m^2 - m + 2m}{2}$$
$$= \frac{m(m+1)}{2},$$

which shows the validity of the formula for $n = m$. We have verified that if (3.1) is true for any natural number $n = m - 1$ then it is true for the next natural number $n = m$. By induction, we see that (3.1) holds for any natural number n.[2]

We can also write out the proof without introducing the natural number m. In the reformulation, we assume that (3.1) holds with n replaced by $n - 1$, that is, we assume that

$$1 + 2 + 3 + \cdots + n - 1 = \frac{(n-1)n}{2}.$$

Adding n to both sides, we get

$$1 + \cdots + n - 1 + n = \frac{(n-1)n}{2} + n = \frac{n(n+1)}{2},$$

which is (3.1) as desired.

3.3 Using Induction

We emphasize that the method of induction is useful for showing the validity of a given formula, but induction does not say how to get the formula in the first place. In fact, there is no technique for systematically deriving formulas like (3.1). Finding such formulas may require some good intuition, trial and error, or some other insight like the clever idea of Gauss. However, experience with such formulas can make it easier to guess at them. For example, we could argue that the average size of the numbers 1 to n is $n/2$ and since there are n numbers to add, their sum should be something like $n\frac{n}{2}$, which is pretty close to the correct $(n+1)\frac{n}{2}$.

We give two more examples illustrating the use of mathematical induction.

EXAMPLE 3.1. First we show a formula for the **geometric sum** with n terms for the fixed natural number $p > 1$,

$$1 + p + p^2 + p^3 + \cdots + p^n = \frac{1 - p^{n+1}}{1 - p}, \tag{3.3}$$

which holds for any natural number n. Here we think of p as being fixed and the induction is on n. The formula (3.3) holds for $n = 1$, since

$$1 + p = \frac{1 - p^2}{1 - p} = \frac{(1-p)(1+p)}{1 - p} = 1 + p,$$

---

[2]In our experience, most students find that verifying a property like (3.1) for specific n like 1 or 2 or 100 is not so difficult to do. But the general inductive step, i.e., assuming the formula is true for an unspecified natural number and showing that it is true for the next natural number, is troublesome. This sort of abstract argument feels strange at first, but try the problems anyway. Working out some induction proofs is good practice for some of the arguments that we encounter later.

where we use the formula $a^2 - b^2 = (a-b)(a+b)$. Assuming it is true with n replaced by $n-1$, we have

$$1 + p + p^2 + p^3 + \cdots + p^{n-1} = \frac{1-p^n}{1-p}.$$

We add p^n to both sides to get

$$\begin{aligned} 1 + p + p^2 + p^3 + \cdots + p^{n-1} + p^n &= \frac{1-p^n}{1-p} + p^n \\ &= \frac{1-p^n}{1-p} + \frac{p^n(1-p)}{1-p} \\ &= \frac{1-p^{n+1}}{1-p}, \end{aligned}$$

which shows the inductive step.

EXAMPLE 3.2. Induction can also be used to show properties that do not involve sums. For example, we show an inequality that is useful. For any fixed natural number p,

$$(1+p)^n \geq 1 + np \tag{3.4}$$

for any natural number n. The inequality (3.4) is certainly valid for $n = 1$, since $(1+p)^1 = 1 + 1 \times p$. Now assume it holds for $n-1$,

$$(1+p)^{n-1} \geq 1 + (n-1)p.$$

We multiply both sides by the positive number $1 + p$,

$$\begin{aligned} (1+p)^n = (1+p)^{n-1}(1+p) &\geq (1 + (n-1)p)(1+p) \\ &\geq 1 + (n-1)p + p + (n-1)p^2 \\ &\geq 1 + np + (n-1)p^2. \end{aligned}$$

Since $(n-1)p^2$ is nonnegative, we can take it away from the right-hand side and then obtain (3.4).

3.4   A Model of an Insect Population

Induction is often used to derive models as well. We present an example involving the population growth in insects.

We consider a simplified situation of an insect population in which all of the adults breed at one specific time during the first summer they are alive, then die before the following summer. In general, there are many factors that affect the rate of reproduction: the food supply, the weather, pesticides, and even the population itself. But the first time around, we simplify all of this by assuming that the number of offspring that survive to the next breeding season is simply proportional to the number of adults alive during the breeding time. Experimentally this is often a valid assumption if the population is not too large.

The goal of the model is to determine if and when the population of insects reaches a critical size. This is an important for instance if the insects carry disease or consume farm crops.

Because we are describing populations of the insects during different years, we need to introduce a notation that makes it easy to associate variable names with different years. We use the **index notation** to do this. We let P_0 denote the current or **initial** population and P_1, P_2, $\cdots$, P_n, $\cdots$ denote the populations during subsequent years numbered 1, 2, $\cdots$, n, $\cdots$, respectively. The **index** or **subscript** on P_n is a convenient way to denote the year. Our modeling assumptions mean that P_n is proportional to P_{n-1}. We use R to denote the constant of proportionality, so

$$P_n = RP_{n-1}. \tag{3.5}$$

Assuming the initial population P_0 is known, the problem is to figure out when the population reaches a specific level M. In other words, find the first n such that $P_n \geq M$.

To do this, we find a formula expressing the dependence of P_n on n using induction on (3.5). Since (3.5) also holds for $n - 1$, i.e. $P_{n-1} = RP_{n-2}$. Substituting, we find

$$P_n = RP_{n-1} = R(RP_{n-2}) = R^2 P_{n-2}.$$

Now we substitute for $P_{n-2} = RP_{n-3}$, $P_{n-3} = RP_{n-4}$, and so on. After $n - 2$ more substitutions, we find

$$P_n = R^n P_0. \tag{3.6}$$

Since R and P_0 are known, this gives an explicit formula for P_n in terms of n. Note that the way we use induction in this example is different than the previous examples. But the difference is only superficial. To make the induction argument look the same as for the previous examples, we can assume that (3.6) holds for $n-1$ and then use (3.5) to show that it therefore holds for n.

Returning to the question of finding n such that $P_n \geq M$, the model problem is to find n such that

$$R^n \geq M/P_0. \tag{3.7}$$

As long as $R > 1$, R^n eventually grows large enough to do this. For example, if $R = 2$, then P_n grows quickly with n. If $P_0 = 1000$, then $P_1 = 2000$, $P_4 = 32,000$, and $P_9 = 1,024,000$.

Chapter 3 Problems

3.1. Prove the formulas

$$\text{(a)} \quad 1^2 + 2^2 + 3^2 + \cdots + n^2 = \frac{n(n+1)(2n+1)}{6} \qquad (3.8)$$

and

$$\text{(b)} \quad 1^3 + 2^3 + 3^3 + \cdots + n^3 = \left(\frac{n(n+1)}{2}\right)^2 \qquad (3.9)$$

hold for all natural numbers n by using induction.

3.2. Using induction, show the following formula holds for all natural numbers n:

$$\frac{1}{1 \times 2} + \frac{1}{2 \times 3} + \frac{1}{3 \times 4} + \cdots + \frac{1}{n(n+1)} = \frac{n}{n+1}.$$

3.3. Using induction, show the following inequalities hold for all natural numbers n:

$$\text{(a) } 3n^2 \geq 2n + 1 \qquad \text{(b) } 4^n \geq n^2 .$$

Problems 3.4 and 3.5 are different applications of induction than those given in the chapter.

3.4. Prove that $7^n - 4^n$ is a multiple of 3 for all natural numbers n. *Hint: If a is a multiple of 3 then $a = 3b$ for some natural number b.*

3.5. Find a formula for the sum of the odd natural numbers from 1 to n, $1 + 3 + 5 + \cdots + n$, where n is odd, and prove the formula is correct using induction. *Hint: To find the formula, compute the sum for the first few odd numbers.*

Problems 3.6–3.8 have to do with modeling insect populations. Problem 3.8 is considerably harder than Problems 3.6 and 3.7.

3.6. The problem is to model the population of a species of insects that has a single breeding season during the summer. The adults breed during the first summer they are alive, then die before the following summer. Assuming that the number of offspring that survive to the next breeding season is proportional to the square of the number of adults, express the population of the insects as a function of the year.

3.7. The problem is to model the population of a species of insects that has a single breeding season during the summer. The adults breed during the first summer they are alive, then die before the following summer and moreover the adults kill and eat some of their offspring. Assuming that the number of offspring born each breeding season is proportional to the number of adults and that a number of offspring are killed by the adults is proportional to the square of the number of adults and there is no other cause of death, derive an equation relating the population of the insects in one year to the population in the previous year.

3.8. The problem is to model the population of a species of insects that has a single breeding season during the summer. The adults breed during the first and second summers they are alive, then die before their third summer. Assuming that the number of offspring born each breeding season is proportional to the number of adults that are alive and that all insects live out their full lifespan, derive an equation relating the population of the insects in any year past the first to the population in the previous two years.

3.9. Derive the formula (3.1) by verifying the sum

$$
\begin{array}{ccccccccc}
 & 1 & + & 2 & + & \cdots & + & n\text{-}1 & + & n \\
+ & n & + & n\text{-}1 & + & \cdots & + & 2 & + & 1 \\
\hline
 & n{+}1 & + & n{+}1 & + & \cdots & + & n{+}1 & + & n{+}1
\end{array}
$$

and showing that this means that $2(1 + 2 + \cdots + n) = n \times (n + 1)$.

4
Rational Numbers

One motivation for considering rational numbers is that we want to solve
equations of the form

$$qx = p, \tag{4.1}$$

where p and $q \neq 0$ are integers. In particular, we encountered the equation
$15x = 10$ in the Dinner Soup model, which cannot be solved if x is restricted
to be an integer. We need to extend the integers to allow fractions. In school,
we learn the definition of a **rational number** r as any number of the form
$r = p/q$, where p and q are integers with $q \neq 0$. We call p the **numerator**
and q the **denominator** of the **fraction** or **ratio**.

Another motivation for introducing rational numbers is the practical
problem of measuring quantities for which integers alone are too crude an
instrument. When a set of standards for measuring quantities such as the
metric system is created, some arbitrary amount is designated as the unit
measurement. For example, the meter for distance, the gram for weight, and
the minute for time. We measure everything in reference to these units. But
rarely does a quantity measure out to be an exact number of units, and so
we are forced to deal with fractions of the units. We even give names to some
particular units of fractions; centimeters are 1/100 of meters, millimeters
are 1/1000 of a meter, and so on.

Actually, these two motivations are really one and the same. For example, consider our thought process when we are faced with the problem of
dividing a pie into seven equal pieces. As we make the first cut, we try to
guess at the angle of the piece that gives a whole pie if we add it to six

more pieces of the same size. In other words, we try to solve the equation $7x = 1$.

4.1   Operating with Rational Numbers

Recall that when we construct the integers by extending the natural numbers, we need to extend the definitions of the arithmetic operations as well. We do this in a way that preserves the arithmetic properties we know for natural numbers. This is necessary because the integers include the natural numbers.

When we extend the integers to get the rational numbers, we are faced with the same problem. Certainly the rational numbers include the integers, since we understand that $p/1$ should equal p, so again we want the operations on rational numbers to agree with the operations on integers. With this mandate, we can define the arithmetic operations for rational numbers in a unique way.

It is a useful, and even interesting, exercise to work through the definitions of the arithmetic operations for rational numbers. Suppose for example that our roommate has never heard of rational numbers and has asked for an explanation. Fresh from reading the previous chapters, we decide to explain the rational numbers from the point of view of solving equation (4.1). To avoid any confusion with our roommate's previous experience with fractions and the symbols "-" and "/", we use a more abstract notation in the beginning and describe a rational number as that "thing" that solves (4.1), which we denote as an **ordered pair** $x = (p, q)$ where the **first component** p denotes the right-hand side of the equation and the **second component** q denotes the left-hand side of the equation $qx = p$. Whatever notation we use, we need some way to designate the two numbers p and q and also the role of these numbers in the equation (4.1). To orient our roommate, we point out that some of these pairs identify naturally with the familiar integers, namely, $(p, 1)$ is the same as p since the solution of $1x = p$ is $x = p$. Therefore, the rational numbers are an extension of the integers. We also point out that there is no reason to expect that $(p, q) = (q, p)$ in general.

Now we build up the rules for performing arithmetic on the set of such ordered pairs of numbers based on the properties of integers. For example, suppose we want to figure out how to multiply the rational number $x = (p, q)$ by the rational number $y = (r, s)$. We start from the defining equations $qx = p$ and $sy = r$. Multiplying both sides, and naturally assuming that $xs = sx$ should be true since multiplication of integers is commutative, we find

$$qxsy = qsxy = qs(xy) = pr.$$

We conclude that
$$xy = (pr, qs),$$
since $z = xy$ solves the equation $qsz = pr$. We are thus led to the familiar rule
$$(p, q)(r, s) = (pr, qs).$$
Written in the standard notation for rational numbers, this is
$$\frac{p}{q} \times \frac{r}{s} = \frac{pr}{qs}. \tag{4.2}$$

Similarly, to figure out how to add two rational numbers $x = (p, q)$ and $y = (r, s)$, we again start with the defining equations $qx = p$ and $sy = r$. Multiplying both sides of $qx = p$ by s, and both sides of $sy = r$ by q, we find $qsx = ps$ and $qsy = qr$. Adding, and naturally assuming that $qsx + qsy = qs(x+y)$ should be true since addition of integers is associative, we find that
$$qs(x + y) = ps + qr,$$
which suggests that
$$(p, q) + (r, s) = (ps + qr, qs).$$
Written using the standard notation for rational numbers, this is
$$\frac{p}{q} + \frac{r}{s} = \frac{ps + qr}{qs}. \tag{4.3}$$

This says that we add two rational numbers by finding a common denominator.

Inspired by these computations, we define rational numbers to be ordered pairs of integers (p, q) with $q \neq 0$, where the operations of multiplication $\times$ and addition $+$ are defined by (4.2) and (4.3). We can now verify that all the familiar rules for arithmetic with integers hold for rational numbers as well. For example, to show $(p, p) = 1$ for $p \neq 0$, we use (4.3) and (4.2) to get
$$(p, p) + (r, 1) = (p + pr, p) = (p(r + 1), p) = (r + 1, 1)(p, p),$$
so
$$(r, 1) = (p, p)(r + 1 - 1, 1)(p, p) = (r, 1)(p, p),$$
which shows that (p, p) acts just like 1. In standard notation, we get $p/p = 1$ when $p \neq 0$. In a similar way, $(pr, ps) = (r, s)$ for $p \neq 0$, or
$$\frac{pr}{ps} = \frac{r}{s}.$$

Arguing in the same way, we define division $(p, q)/(r, s)$ of the rational number (p, q) by the rational number (r, s) with $r \neq 0$, as
$$(p, q)/(r, s) = (ps, qr)$$

since (ps, qr) solves the equation $(r, s)x = (p, q)$. In the standard notation,

$$\frac{\frac{p}{q}}{\frac{r}{s}} = \frac{ps}{qr}.$$

We can now check the validity of the expected properties of multiplication and division, such as

$$\frac{p}{q} = p \times \frac{1}{q}.$$

Of course, we use x^n with x rational and n a natural number to denote the number x multiplied with itself n times. Following this, we have

$$x^{-n} = \frac{1}{x^n}$$

for natural numbers n and $x \neq 0$. As with integers, we define $x^0 = 1$ for x rational.

It is important to note that these definitions of the arithmetic operations for rational numbers mean that the sum, product, difference, or quotient of any two rational numbers is always another rational number (when defined). In other words, the rational numbers are **closed** under the arithmetic operations. Recall that the initial motivation for defining rational numbers is the observation that we cannot always find integer solutions for (4.1). But we can find a rational number that solves the equation, namely, $x = (p, q)$ since $(q, 1)(p, q) = (pq, q) = (p, 1)$. Moreover, we can find a rational solution of any equation of the form $ax = b$, where a and b are any rational numbers with $a \neq 0$, namely, if $a = (p, q)$ and $b = (r, s)$, then $x = ((r, s), (p, q))\ (= b/a)$.

As a final step, we introduce our roommate to the notation

$$(p, q) = \frac{p}{q} = p/q.$$

The idea of constructing the integers and the rational numbers from the natural numbers and a list of properties, or axioms, that govern arithmetic on the natural numbers originates with Peano.[1]

4.2   Decimal Expansions of Rational Numbers

The most useful way to represent a rational number is in the form of a decimal expansion, such as $1/2 = 0.5$, $5/2 = 2.5$, and $5/4 = 1.25$. In

---

[1]Giuseppe Peano (1858–1932) was an Italian mathematician. He proved his version of Theorem 41.5 and also discovered the method of successive approximation relatively early in his career. Peano's most important work was in mathematical logic and the foundations of analysis.

general, a **finite decimal expansion** is a number of the form

$$\pm p_m p_{m-1} \cdots p_2 p_1 p_0 . q_1 q_2 \cdots q_n, \tag{4.4}$$

where m and n are natural numbers and the **digits** p_m, p_{m-1}, $\cdots$, p_0, q_0, $\cdots$, q_n are equal to one of the natural numbers $\{0, 1, \cdots, 9\}$. We use the "$\cdots$" to indicate the digits that are not written out. The integer part of the decimal number is $p_m p_{m-1} ... p_1 p_0$, while the decimal or fractional part is $0.q_1 q_2 \cdots q_n$. For example, $432.576 = 432 + 0.576$.

The decimal expansion is computed by continuing the long division algorithm "past" the decimal point rather than stopping when the remainder is found. We illustrate in Fig. 4.1.

$$
\begin{array}{r}
47.55 \\
40 \overline{)1902.000} \\
160 \\
\hline
302 \\
280 \\
\hline
22.0 \\
20.0 \\
\hline
2.00 \\
2.00 \\
\hline
.00
\end{array}
$$

FIGURE 4.1. Using long division to obtain a decimal expansion.

It is useful to remember that the decimal expansion (4.4) is actually just a shorthand notation for the number

$$\pm\, p_m 10^m + p_{m-1} 10^{m-1} + \cdots + p_1 10 + p_0$$
$$+\, q_1 10^{-1} + \cdots + q_{n-1} 10^{-(n-1)} + q_n 10^{-n}.$$

EXAMPLE 4.1.

$$432.576 = 4 \times 10^2 + 3 \times 10^1$$
$$+ 2 \times 10^0 + 5 \times 10^{-1} + 7 \times 10^{-2} + 6 \times 10^{-3}.$$

A finite decimal expansion is necessarily a rational number because it is a sum of rational numbers. This can also be understood by writing $p_{m-1} \cdots p_1 . q_1 q_2 \cdots q_n$ as the quotient of the integers $p_m p_{m-1} \cdots p_1 q_1 q_2 \cdots q_n$ and 10^n:

$$p_{m-1} \cdots p_1 . q_1 q_2 \cdots q_n = \frac{p_{m-1} \cdots p_1 q_1 q_2 \cdots q_n}{10^n},$$

like $432.576 = 432576/10^3$.

Computing decimal expansions of rational numbers using long division leads immediately to an interesting observation: some decimal expansions do not "stop." In other words, some decimal expansions contain a never-ending, or **infinite** number, of nonzero digits. For example, the solution of the dinner soup model is $2/3 = .666\cdots$ while $10/9 = 1.11111\cdots$. We carry out the division in Fig. 4.2. As usual, we use "infinite" because we are

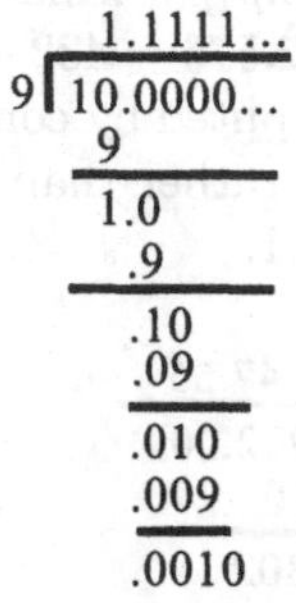

$$
\begin{array}{r}
1.1111... \\
9\overline{)10.0000...} \\
\underline{9} \\
1.0 \\
\underline{.9} \\
.10 \\
\underline{.09} \\
.010 \\
\underline{.009} \\
.0010
\end{array}
$$

FIGURE 4.2. The decimal expansion of 10/9 never stops.

discussing something that continues without stopping. We can find many examples of infinite decimal expansions:

$$\frac{1}{3} = .3333333333\cdots$$

$$\frac{2}{11} = .18181818181818\cdots$$

$$\frac{4}{7} = .571428571428571428571428\cdots$$

Note that all of these examples have the property that the digits in the decimal expansion begin to repeat after some point. The digits in 10/9 and 1/3 repeat in each entry, the digits in 2/11 repeat after every two entries, and the digits in 4/7 repeat after every six entries. We say that decimal expansions with this property are **periodic**. We use the word periodic to describe anything that repeats at a regular interval. In fact, if we consider the process of long division in computing the decimal expansion of p/q, then we realize that *the decimal expansion of any rational number must either be finite or periodic*. If the expansion is not finite, then at every stage in the division process there is a nonzero remainder, which is an integer between 0 and $q - 1$. In other words, there are at most $q - 1$ possibilities for remainders at each step. This means that after at most q divisions, a particular remainder must turn up again. But after that, the subsequent remainders repeat in the same order as before.

The fact that many rational numbers have infinite decimal expansions leads to the same kind of uncertainty encountered when talking about the set of natural numbers. While we know everything about the rational number $5/4 = 1.25$ with a finite decimal expansion, there is an inherent uncertainty to the meaning of the decimal expansion of $10/9$ since we can never write down all the digits. In other words, what is the meaning of the "$\cdots$" in the decimal expansion of $10/9$?

It turns out that we can use the formula for the geometric sum (3.3) to make this more clear. Consider the decimal expansion obtained from $10/9$ by stopping after $n + 1$ divisions, which we write as $1.1 \cdots 1_n$, with n decimals equal to 1 after the point. In long form, this number is

$$1.11 \cdots 11_n = 1 + 10^{-1} + 10^{-2} + \cdots + 10^{-n+1} + 10^{-n},$$

which by (3.3) is

$$1.11 \cdots 11_n = \frac{1 - 10^{-n-1}}{1 - 0.1} = \frac{10}{9}(1 - 10^{-n-1}); \tag{4.5}$$

that is

$$\frac{10}{9} = 1.11 \cdots 11_n + \frac{10^{-n}}{9}. \tag{4.6}$$

The term $10^{-n}/9$ decreases steadily toward 0 as n increases. Therefore, we can make $1.11 \cdots 11_n$ as close as desired to $10/9$ simply by making n sufficiently large. This leads to interpreting

$$\frac{10}{9} = 1.11111111 \cdots$$

as meaning that we can make the numbers $1.111 \cdots 1_n$ as close as desired to $10/9$ by taking n large. Taking sufficiently many decimals in the never-ending decimal expansion of $10/9$ makes the error smaller than any given positive number.

We give another example before considering the general case.

EXAMPLE 4.2. Computing, we find that $2/11 = .1818181818 \cdots$. Taking the first m pairs of the digits 18, we get

$$
\begin{aligned}
.1818 \cdots 18_m &= \frac{18}{100} + \frac{18}{10000} + \frac{18}{1000000} + \cdots + \frac{18}{10^{2m}} \\
&= \frac{18}{100}\left(1 + \frac{1}{100} + \frac{1}{100^2} + \cdots + \frac{1}{100^{m-1}}\right) \\
&= \frac{18}{100}\frac{1 - (100^{-1})^m}{1 - 100^{-1}} = \frac{18}{100}\frac{100}{99}(1 - 100^{-m-1}) \\
&= \frac{2}{11}(1 - 100^{-m}),
\end{aligned}
$$

that is,

$$\frac{2}{11} = 0.1818 \cdots 18_m + \frac{2}{11} 100^{-m}.$$

We thus interpret $2/11 = .1818181818 \cdots$ as meaning that we can make the numbers $.1818 \cdots 18_m$ as close as desired to $2/11$ by making m sufficiently large.

We now consider the general case of an infinite periodic decimal expansion of the form

$$p = .q_1 q_2 \cdots q_n q_1 q_2 \cdots q_n q_1 q_2 \cdots q_n \cdots ,$$

where each period consists of the n digits $q_1 \cdots q_n$. Truncating the decimal expansion after m periods, we use (3.3) to get

$$\begin{aligned}
p_m &= \frac{q_1 q_2 \cdots q_n}{10^n} + \frac{q_1 q_2 \cdots q_n}{10^{n2}} + \cdots + \frac{q_1 q_2 \cdots q_n}{10^{nm}} \\
&= \frac{q_1 q_2 \cdots q_n}{10^n} \left(1 + \frac{1}{10^n} + \frac{1}{(10^n)^2} + \cdots + \frac{1}{(10^n)^{m-1}} \right) \\
&= \frac{q_1 q_2 \cdots q_n}{10^n} \frac{1 - (10^{-n})^m}{1 - 10^{-n}} = \frac{q_1 q_2 \cdots q_n}{10^n - 1} \left(1 - (10^{-n})^m \right).
\end{aligned}$$

That is,

$$\frac{q_1 q_2 \cdots q_n}{10^n - 1} = p_m + \frac{q_1 q_2 \cdots q_n}{10^n - 1} 10^{-nm}.$$

We conclude that

$$p = \frac{q_1 q_2 \cdots q_n}{10^n - 1}$$

in the sense that the difference between the truncated decimal expansion p_m of p and $q_1 q_2 \cdots q_n/(10^n - 1)$ can be made arbitrarily small by increasing the number of periods m, that is, by taking more digits of p into account. Thus, $p = q_1 q_2 \cdots q_n/(10^n - 1)$. We conclude that *every infinite periodic decimal expansion is equal to some rational number.*

We summarize this discussion as the following fundamental result, originally formulated by Wallis.[2]

Theorem 4.1 *A rational number has either a finite or infinite periodic decimal expansion, and vice versa, each finite or infinite periodic decimal expansion represents a rational number.*

---

[2] John Wallis (1616–1703) was an English mathematician who originally studied divinity at Cambridge University because there was no one to advise mathematics students. Wallis made fundamental contributions to the foundations of calculus, especially using analytic techniques to establish important integration formulas that were later used by Newton. Wallis also introduced ∞ to represent infinity and the expression "proof by induction."

4.3   The Set of Rational Numbers

The definition of the set of rational numbers is even more prone to controversy than the definition of the integers. First of all, given any rational number we can always write down a larger rational number. But in addition to this, between any two distinct rational numbers, we can always find another rational number. The common convention is to define the set of rational number denoted $\mathbb{Q}$, to be the set of *all* rational numbers, or

$$\mathbb{Q} = \left\{ x = \frac{p}{q} : p, q \text{ in } \mathbb{Z}, \ q \neq 0 \right\}.$$

Alternatively, we could also define $\mathbb{Q}$ to be the set of *possible* numbers x of the form $x = p/q$, where p and $q \neq 0$ are integers.

4.4   The Verhulst Model of Populations

In next couple of sections, we present models that require rational numbers.

Certain bacteria cannot produce some of the amino acids they need for the production of proteins and cell reproduction. When such bacteria are cultured in growth media containing sufficient amino acids, then the population doubles in size at a regular time interval, say on the order of an hour. If P_0 is the initial population at the current time and P_n is the population after n hours, then we have

$$P_n = 2P_{n-1} \tag{4.7}$$

for $n \geq 1$. This model is similar to the model (3.5) used to describe the insect population in Section 3.4. If the bacteria can keep growing in this way, then we know from that model that $P_n = 2^n P_0$. However if there is a limited amount of amino acid, then the bacteria begin to compete for the resource. As a result, the population can no longer double every hour. The question is what happens to the bacteria population as time increases? For example, does it keep increasing, does it decrease to zero (die out), or does it tend to some constant value?

To model this, we allow the proportionality factor 2 in (4.7) to vary with the population in such a way that it decreases as the population increases. For example, we assume there is a constant $K > 0$ that the population at hour n satisfies

$$P_n = \frac{2}{1 + P_{n-1}/K} P_{n-1}. \tag{4.8}$$

With this choice, the proportionality factor $2/(1 + P_{n-1}/K)$ is always less than 2 and clearly decreases as P_{n-1} increases. We emphasize that there are many other functions that have this behavior. The right choice is the one that gives results that match experimental data from the laboratory. It turns out that the choice made here does fit experimental data well and (4.8) has been used as a model not only for bacteria but also for animals, humans, and insects.

The choice of a mechanism to decrease the population growth rate as the population increases in (4.8) is originally due to Verhulst.[3] We discuss a version of Verhulst's model involving differential equations in Chapter 39.

We now seek a formula expressing how P_n depends on n. We define $Q_n = 1/P_n$, then (4.8) implies that

$$Q_n = \frac{Q_{n-1}}{2} + \frac{1}{2K}.$$

Using induction in the same way used for the insect model in Section 3.4, we get

$$
\begin{aligned}
Q_n &= \frac{1}{2}Q_{n-1} + \frac{1}{2K} \\
&= \frac{1}{2^2}Q_{n-2} + \frac{1}{2K} + \frac{1}{4K} \\
&= \frac{1}{2^3}Q_{n-3} + \frac{1}{2K} + \frac{1}{4K}\frac{1}{8K} \\
&\quad\vdots \\
&= \frac{1}{2^n}Q_0 + \frac{1}{2K}\left(1 + \frac{1}{2} + \cdots + \frac{1}{2^{n-1}}\right).
\end{aligned}
$$

With each hour that passes, we add another term onto the sum, giving R_n. The goal is to figure out what happens to R_n as n increases. Using the formula for the sum of the geometric series (3.3), which obviously holds for rational numbers as well as for integers, we find

$$P_n = \frac{1}{Q_n} = \frac{1}{\frac{1}{2^n}Q_0 + \frac{1}{K}\left(1 - \frac{1}{2^n}\right)}. \tag{4.9}$$

4.5   A Model of Chemical Equilibrium

The solubility of ionic precipitates is an important issue in analytical chemistry. For the equilibrium

$$\mathrm{A}_x\mathrm{B}_y \rightleftharpoons x\,\mathrm{A}^{y+} + y\,\mathrm{B}^{x-} \tag{4.10}$$

---

[3]The Belgian mathematician Pierre Francois Verhulst (1804–1849) worked in mathematics, physics, and social statistics. His most notable achievements were in the study of population dynamics.

for a saturated solution of slightly soluble strong electrolytes, the solubility product constant is given by

$$K_{sp} = [\,A^{y+}\,]^x [\,B^{x-}\,]^y. \tag{4.11}$$

The solubility product constant is useful for predicting whether or not a precipitate can form in a given set of conditions or the solubility of an electrolyte.

For example, we use it to determine the solubility of $Ba(IO_3)_2$ in a .020 mole/liter solution of KIO_3 :

$$Ba(IO_3)_2 \rightleftharpoons Ba^{2+} + 2IO_3^-$$

given that the K_{sp} for $Ba(IO_3)_2$ is 1.57×10^{-9}. We let S denote the solubility of $Ba(IO_3)_2$. By the conservation of mass, $S = [\,Ba^{2+}\,]$ while iodate ions come from both the KIO_3 and the $Ba(IO_3)_2$. The total iodate concentration is the sum of these contributions,

$$[\,IO_3^-\,] = (.02 + 2S).$$

Substituting these into (4.11), we obtain

$$S\,(.02 + 2S)^2 = 1.57 \times 10^{-9}. \tag{4.12}$$

4.6   The Rational Number Line

Recall that we represent the integers using the integer number line, which consists of a line with regularly spaced points. We can also use a line to represent the rational numbers. We begin with the integer number line and then add the rational numbers that have one decimal place:

$$-\cdots, -1, -.9, -.8, \cdots, -.1, 0, .1, .2, \cdots, .9, 1, \cdots.$$

Then we add the rational numbers that have two decimal places:

$$-\cdots, -.99, -.98, \cdots, -.01, 0, .01, .02, \cdots .98, .99, 1, \cdots.$$

Then on to the rational numbers with 3, 4, and more decimal places. We illustrate in Fig. 4.3. The rational line quickly begins to look solid. A solid line would mean that every number is rational, something we discuss later. But in any case, a drawing of a number line appears solid. We call this the **rational number line**.

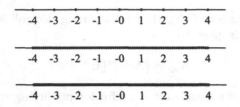

FIGURE 4.3. Filling in the rational number line between -4 and 4 starting with integers, rational numbers with one digit, and rational numbers with two digits, and so on.

Given rational numbers a and b with $a < b$, we say that the rational numbers x such that $a \leq x \leq b$ is a **closed interval**[4] and we denote the interval by $[a, b]$. We also write

$$[a, b] = \{x \text{ in } \mathbb{Q} : a \leq x \leq b\}.$$

The points a and b are called the **endpoints** of the interval. Similarly we define **open** (a, b) and **half-open** intervals $[a, b)$ and $(a, b]$ by

$$(a, b) = \{x \text{ in } \mathbb{Q} : a < x < b\}, \quad [a, b) = \{x \text{ in } \mathbb{Q} : a \leq x < b\},$$

$$[a, b) = \{x \text{ in } \mathbb{Q} : a \leq x < b\}, \text{ and } (a, b] = \{x \text{ in } \mathbb{Q} : a < x \leq b\}.$$

In an analogous way, we write all the rational numbers larger than a number a as

$$(a, \infty) = \{x \text{ in } \mathbb{Q} : a < x\} \text{ and } [a, \infty) = \{x \text{ in } \mathbb{Q} : a \leq x\}.$$

We use ∞ *symbolically* to indicate that there is no right-hand endpoint for the set of numbers larger than a. We write the set of numbers less than a in a similar way. We also represent intervals graphically by marking the points on the rational line segment, as we show in Fig. 4.4. Note how we use an open circle or a closed circle to mark the endpoints of open and closed intervals.

---

[4]To be accurate, we should call this the closed *rational* interval. The term closed has two meanings. One is an interval that contains its endpoints and the second is a more general notion that we avoid describing at the moment. These ideas are the same for intervals of real numbers, but are not the same for intervals of rational numbers. This could lead to confusion, except that we rarely talk about intervals of rational numbers after we discuss real numbers in Chapter 11.

Chapter 4 Problems

4.1. Using the definitions of multiplication and addition of rational numbers, show that if r, s and t are rational numbers, then (a) $r(s+t) = rs + rt$ and (b) $r/(s/t) = rt/s$.

In Problems 4.2 and 4.3, we ask you to find the consequences of making alternate definitions of basic arithmetic operations for rational numbers.

4.2. Suppose that the rational numbers are defined to be ordered pairs $(p, q) = p/q$ of integers where we define the product of two rational numbers (p, q) and (m, n) by $(p, q)(m, n) = (pm, qn)$ as usual and addition as

$$(p, q) + (m, n) = (p + q + m + n, q + n).$$

Find at least one of the arithmetic properties that fails.

4.3. Suppose that the rational numbers are defined to be ordered pairs $(p, q) = p/q$ of integers where we define the product of two rational numbers (p, q) and (m, n) by $(p, q)(m, n) = (pm, qn)$ as usual and addition as

$$(p, q) + (m, n) = (p + m, q + n).$$

Find at least one of the standard arithmetic properties that fails.

4.4. A person running on a large ship runs 8.8 feet/second heading toward the bow while the ship is moving at 16 miles/hour. What is the speed of the runner relative to a stationary observer? Interpret the computation giving the solution as finding a common denominator.

4.5. Compute decimal expansions for (a) 432/125 and (b) 47.8/80.

4.6. Compute decimal expansions for (a) 3/7, (b) 2/13, and (c) 5/17.

4.7. Find rational numbers corresponding to the decimal expansions (a) .4242 4242$\cdots$, (b) .881188118811$\cdots$, and (c) .4290542905$\cdots$.

4.8. Find an equation for the number of milligrams of $Ba(IO_3)_2$ that can be dissolved in 150 ml of water at 25° C with $K_{sp} = 1.57 \times 10^{-9}$ mol^2/L^3. The reaction is

$$Ba(IO_3)_2 \rightleftharpoons Ba^{2+} + 2IO_3^-$$

FIGURE 4.4. Various rational line intervals.

4.9. We invest some money in a bond that yields 9% interest each year. Assuming that we invest any money we make from interest in more bonds for an initial investment of $\$C_0$, write down a model giving the amount of money we have after n years.

4.10. Solve the following inequalities:

(a) $|3x - 4| \leq 1$ (b) $|2 - 5x| < 6$

(c) $|14x - 6| > 7$ (d) $|2 - 8x| \geq 3$.

5
Functions

We turn now to investigate functions, another key component of mathematical modeling. For example, in the Dinner Soup model, the total cost of the purchased food is $15x$ (dollars) where x is the amount of beef in pounds. In other words, for every amount of beef x, there is a corresponding total cost $15x$. We say that the total cost $15x$ is a function of, or depends on, the amount of beef x.

5.1 Functions

In the modern definition of a function, which is due to Dirichlet,[1] f is a **function** of x if for each choice of x in a prescribed set, there is assigned a *unique* value $f(x)$. A general notion of a function was first used by Leibniz,[2] who defined a function to be a quantity that varies along a curve and

---

[1]Johann Peter Gustav Lejeune Dirichlet (1805–1859) worked in Germany. He produced important results in the equilibrium of systems; fluid mechanics; number theory, including the founding of analytic number theory; potential theory; and the theory of Fourier series.

[2]The German mathematician Gottfried Wilhelm von Leibniz (1646–1716) was an important mathematician and philosopher, who also worked as a diplomat, economist, geologist, historian, linguist, lawyer, and theologian. Leibniz was a true interdisciplinary scientist, which is a very rare and exalted state. Leibniz and Newton are credited with inventing calculus independently and roughly at the same time. However, Leibniz developed a better notation for calculus, which we use today. In particular, Leibniz first used the notation dy/dx for the derivative and $\int$ for the integral. In addition to function,

used the expression "a function of." Before Leibniz's time, mathematicians occasionally used the idea of a "relation" between quantities, though defined rather vaguely, and knew about specific functions like the logarithm. The modern notation for a function is due to Euler.[3]

In the Dinner Soup model, the function is $f(x) = 15x$. It is helpful to think of x as the **input**, while $f(x)$ is the corresponding **output**. Correspondingly, we sometimes write $x \rightarrow f(x)$, which visibly represents the idea that the input x is "sent" to the value $f(x)$.

We refer to the input x of a function as a **variable** since x can vary in value. We also call x the **argument** of the function $f(x)$. The prescribed set from which the input to a function f is selected is called the **domain** of the function f and is denoted by $D(f)$. The set of values $f(x)$ corresponding to the choices of x in the domain $D(f)$ is called the **range** $R(f)$ of f.

EXAMPLE 5.1. In the Dinner Soup model with $f(x) = 15x$, we may choose $D(f) = [0,1]$ if we decide that the amount of beef x can vary in the interval $[0,1]$, in which case $R(f) = [0,15]$. We may also choose the domain $D(f)$ to be some other set of possible values of the amount of beef x such as $D(f) = [a,b]$, where a and b are nonnegative rational numbers, with the corresponding range $R(f) = [15a, 15b]$. We might also choose $D(f) = \{x \text{ in } \mathbb{Q}, \ x > 0\}$ with the corresponding $R(f) = \{x \text{ in } \mathbb{Q}, \ x > 0\}$.

In daily life, we stumble over functions right and left.

EXAMPLE 5.2. A car dealer assigns a price $f(x)$, which is a number, to each car x in the lot. The domain $D(f)$ is the set of cars on the lot and the range $R(f)$ is the set of all different prices of these cars.

EXAMPLE 5.3. When we tell time by looking at a clock, we are assigning a numeric value $f(x)$ to the set of angles x made by clock hands. The domain $D(f)$ is the set of angles in a circle that is $D(f) = [0, 360]$

---

he also introduced the terms algorithm, constant, parameter, and variable. Leibniz also designed an early "computing machine."

[3]Leonhard Euler (1707–1783) was born in Switzerland but spent most of his career in Germany and Russia. He worked in nearly all areas of mathematics and was the most prolific mathematician of all time as he published over 850 works. However quality did not suffer with quantity, as it so often does, and Euler made fundamentally important contributions in geometry, calculus and number theory. He synthesized the calculus of Leibniz and Newton into the first definitive calculus textbook, which has heavily influenced almost all subsequent calculus texts. Euler studied differential equations, continuum mechanics, lunar theory, the three body problem, elasticity, acoustics, the wave theory of light, hydraulics, music, and laid the foundation of analytical mechanics. He invented the notation $f(x)$ for a function of x, e for the base of the natural logarithm, i for the square root of -1, π, the Σ notation for sums, and the Δ notation for finite differences, among other things. Euler continued to work during the last seventeen years of his life despite being totally blind.

if we use degrees, and the range $R(f)$ is the set of time instants from 0 to 24 hours.

EXAMPLE 5.4. When the government makes out a tax bill, it is assigning one number $f(x)$, representing the amount owed, to another number x, representing the corresponding salary. The domain $D(f)$ and range $R(f)$ change a lot depending on the political winds.

It is useful to assign a variable name to the output of a function; for example we may write $y = f(x)$. Thus, the value of the variable y is given by the value $f(x)$ assigned to x. We call x the **independent variable** and y the **dependent variable**. x takes on values in the domain $D(f)$ while y takes on values in the range $R(f)$. The names we use for the independent variable and the dependent variables are chosen simply for convenience. The names x and y are common but there is nothing special about these letters. For example, $z = f(u)$ denotes the same function if we do not change f, i.e., the function $y = 15x$ can just as well be written $z = 15u$.

EXAMPLE 5.5. In the used car sales business, we may know that one of the older cars is a lemon, but we nevertheless call it "a creampuff previously owned by a little old lady from Des Moines who only drove it to go to church" simply because that name has some fortunate connotations, while the price we want to get for the car is the same whether it is called a lemon or a creampuff.

5.2   Functions and Sets

So far, we have thought of the input and output of a function as single values. But sometimes we also need to use functions of sets, that is, where the input and output are sets.

EXAMPLE 5.6. When we buy a pad of 200 sheets of paper, we pay a price for the set of sheets and do not compute 200 times the price of an individual sheet.

EXAMPLE 5.7. Though a car dealership sells cars for an individual price, if the dealership goes bankrupt, then the lot of cars may be priced as a collection.

The root of the need to deal with sets is the difficulty we have in thinking about two things at one time. To get out of this limitation, we have to group things together into sets and therefore naturally we have to consider functions on these sets. In this context, we say that the function f defines a **transformation** or **map** of the domain $D(f)$ **onto** the range $R(f)$. We write this symbolically as $f : D(f) \to R(f)$.

EXAMPLE 5.8. For the Dinner Soup model with $f(x) = 15x$ and domain $D(f) = [0,1]$ we write $f : [0,1] \to [0,15]$. If we instead let the domain be $\mathbb{Q}$, then we have $f : \mathbb{Q} \to \mathbb{Q}$. In this case of course the function $f(x) = 15x$ is no longer directly connected to the Dinner Soup model since we allow x to be negative.

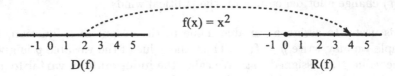

FIGURE 5.1. Illustration of $f : \mathbb{Q} \to \{x$ in $\mathbb{Q}, x \geq 0\}$ with $f(x) = x^2$.

EXAMPLE 5.9. We use the function $f(x) = x^2$ in the Muddy Yard model. In that model, $D(f) = \{x$ in $\mathbb{Q}, x > 0\}$ and $f : \{x$ in $\mathbb{Q}, x > 0\} \to \{x$ in $\mathbb{Q}, x > 0\}$. If we take $D(f) = \mathbb{Q}$, we get $f : \mathbb{Q} \to \{x$ in $\mathbb{Q}, x \geq 0\}$. We illustrate in Fig. 5.1. If we take $D(f) = \mathbb{Z}$, then $f : \mathbb{Z} \to \{0, \pm 1, \pm 2, \pm 3, \pm 4, ...\}$.

EXAMPLE 5.10. The function $f(z) = z+3$ satisfies $f : \mathbb{N} \to \{4,5,6,\cdots\}$ but $f : \mathbb{Z} \to \mathbb{Z}$.

EXAMPLE 5.11. The function $f(n) = 2^{-n}$ satisfies $f : \mathbb{N} \to \{\frac{1}{2}, \frac{1}{4}, \frac{1}{8}, \cdots\}$.

EXAMPLE 5.12. The function $f(x) = 1/x : \{x$ in $\mathbb{Q}, x > 0\} \to \{x$ in $\mathbb{Q}, x > 0\}$. Note that this means that given any x in $\{x$ in $\mathbb{Q}, x > 0\}$, $1/x$ is in $\{x$ in $\mathbb{Q}, x > 0\}$. Vice versa, given any y in $\{x$ in $\mathbb{Q}, x > 0\}$ there is an x in $\{x$ in $\mathbb{Q}, x > 0\}$ with $y = 1/x$.

It is very often tedious or difficult to determine exactly the range of a function f corresponding to the domain. So we often write $f : D(f) \to B$, meaning that for each x in $D(f)$ there is an assigned value $f(x)$ that is in the set B. The range $R(f)$ is included in B, but the set B may be bigger than $R(f)$. In this way, we avoid having to figure out the range $R(f)$ exactly. We then say that f maps $D(f)$ **into** B.

EXAMPLE 5.13. The function $f(x) = x^2$ satisfies both $f : \mathbb{Q} \to \{x$ in $\mathbb{Q}, x > 0\}$ and $f : \mathbb{Q} \to \mathbb{Q}$. This can be seen clearly in Fig. 5.1.

EXAMPLE 5.14. The function

$$f(x) = \frac{x^3 - 4x^2 + 1}{(x-4)(x-2)(x+3)}$$

is defined for all rational numbers $x \neq 4, 2, -3$, so it is natural to define $D(f) = \{x$ in $\mathbb{Q}, x \neq 4, x \neq 2, x \neq 3\}$. It is often the case that we take the domain to be the largest set of numbers for which a function is defined. The range is hard to compute, but certainly we have $f : D(f) \to \mathbb{Q}$.

In the first few examples, we defined functions of sets by first defining the function on members of the set. But it often happens that it is more natural to consider the sets first and then what happens to individual members of the sets second.

EXAMPLE 5.15. A movie consists of a sequence of pictures that are displayed at the rate of 16 pictures per second. We usually watch a movie from the first to the last picture. Afterward we might talk about different scenes in the movie, which corresponds to subsets of the totality of pictures. A very few people, like the film editor and director, might consider the movie on the level of the individual elements in the domain, that is the pictures on the film. When editing the movie, they number the picture frames $1, 2, 3, \cdots, N$ and view the movie as a function f with $D(f) = \{1, 2, \cdots, N\}$ associating to each number n in $D(f)$ the picture frame $f(n)$ with number n.

EXAMPLE 5.16. A telephone directory of the people living in a city like Fort Collins is simply a printed version of the function that assigns a telephone number $f(x)$ to each person x in Fort Collins with a listed number. For example, if $x = $ E. Merckx, then $f(x) = 4631123456$ which is the number of E. Merckx. If we have to find a telephone listing, we first pick up the telephone book, that is the printed representation of the entire domain and range of the function f, and then determine the image, i.e. telephone number, of an individual in the domain.

5.3   Graphing Functions of Integers

So far we have described a function both by listing all its values in a table and by giving a formula like $f(n) = n^2$ and indicating the domain. It is also useful to have a picture, or graph or plot, of a function. A graph of a function is a way of describing the behavior of a function "globally." For example, we can describe the function as increasing in this region and decreasing in another region and thus give an idea of how it behaves without listing specific values.

We begin by describing the graph of a function $f : \mathbb{Z} \to \mathbb{Z}$. Recall that integers are represented geometrically using the integer line. To describe the input and output for $f : \mathbb{Z} \to \mathbb{Z}$, we need two number lines so that we can mark the points in $D(f)$ on one and the points in $R(f)$ on the other. A convenient way to arrange these two number lines is to place them orthogonal to each other and intersecting at the respective origins as

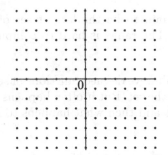

FIGURE 5.2. The integer coordinate plane.

n	f(n)
0	0
1	1
-1	1
2	4
-2	4
3	9
-3	9
4	16
-4	16
5	25
-5	25
6	36
-6	36

FIGURE 5.3. A tabular listing of $f(n) = n^2$ and a graph of the points associated with the function $f(n) = n^2$ with domain equal to the integers.

shown in Fig. 5.2. If we mark the points obtained by intersecting vertical lines through integer points on the horizontal axis with the horizontal lines through integer points on the vertical axis, we get a grid of points as shown in Fig. 5.2. This is called the **integer coordinate plane**. Each number line is called an **axis** of the coordinate plane while the intersection point of the two number lines is called the **origin** and is denoted by 0.

As we saw, a function $f : \mathbb{Z} \to \mathbb{Z}$ can be represented by making a list with the inputs placed side-by-side with the corresponding outputs. We show such a table for $f(n) = n^2$ in Fig. 5.3. We can represent such a table in the rational coordinate plane by marking only those points corresponding to an entry in the table, i.e., marking each intersection point of the line rising vertically from the input and the line extending horizontally from the corresponding output. We draw the plot corresponding to $f(n) = n^2$ in Fig. 5.3.

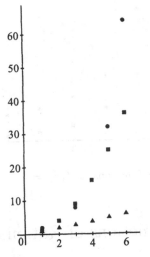

FIGURE 5.4. Plots of the functions ▲ $f(n) = n$, ■ $f(n) = n^2$, and ● $f(n) = 2^n$.

EXAMPLE 5.17. In Fig. 5.4, we plot n, n^2, and 2^n along the vertical axis with $n = 1, 2, 3, .., 6$ along the horizontal axis. The plot suggests 2^n grows more quickly than both n and n^2 as n increases.

EXAMPLE 5.18. In Fig. 5.5, we plot n^{-1}, n^{-2}, and 2^{-n} with $n = 1, 2, .., 6$. These plots indicate that 2^{-n} decreases most rapidly and n^{-1} least rapidly.

We can represent a point on the integer plane as an **ordered pair** of numbers. To each point in the plane located at the intersection of a vertical line passing through n on the horizontal axis and a horizontal line passing through m on the vertical axis, we associate the pair of numbers (n, m). These are the **coordinates** of the point. Using this notation, we can describe the function $f(n) = n^2$ as the set of ordered pairs

$$\{(0, 0), (1, 1), (-1, 1), (2, 4), (-2, 4), (3, 9), (-3, 9), \cdots\}.$$

By arbitrary convention, we always associate the first number in the ordered pair with the horizontal location of the point and the second number with the vertical location.

We can illustrate the idea of a function giving a transformation of its domain into its range nicely using its graph. Consider Fig. 5.3. We start at a point in the domain on the horizontal axis and follow a line straight up to the point on the graph of the function. From this point, we follow a line horizontally to the vertical axis. In other words, we can find the output associated to a given input by tracing first a vertical line and then a horizontal line. By tracing lots of points, we can see that $\mathbb{Z}$ is mapped to $\{x \text{ in } \mathbb{Z}, \ x > 0\}$.

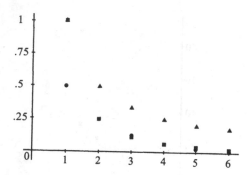

FIGURE 5.5. Plots of the functions ▲ $f(n) = n^{-1}$, ■ $f(n) = n^{-2}$, and •
$f(n) = 2^{-n}$.

5.4   Graphing Functions of Rational Numbers

Now we consider the graph of a function $f : \mathbb{Q} \to \mathbb{Q}$. Following the example
of functions of integers, we plot functions of rational numbers on the **rational coordinate plane** constructed by placing two rational number lines
at right angles and meeting at the origins and then marking every point
that has rational number coordinates. Of course, considering Fig. 4.3, such
a plane appears to be solid, even if it is not solid. We avoid plotting an
example!

   If we try to plot a function $f : \mathbb{Q} \to \mathbb{Q}$ by writing down a list of values, we
realize immediately that graphing a function of rational numbers is more
complicated than graphing a function of integers. When we compute values
of a function of integers, we cannot compute *all* the values because there are
infinitely many integers. Instead, we choose a smallest and largest integer
and compute the values of the functions for those integers in between. For
the same reason, we can not compute all the values of a function defined
on the rational numbers. But now we have to cut off the list in two ways.
We have to choose a smallest and largest number for making the list, but
we also have to decide how many points to use in between the low and
high values. In other words, we cannot compute the values of the function
at all the rational numbers in between two rational numbers. This means
that a list of values of a function of rational numbers always has "gaps" in
between the points where the function is evaluated. We give an example to
make this clear.

EXAMPLE 5.19.   We list some values of the function $f(x) = \frac{1}{2}x + \frac{1}{2}$
defined on the rational numbers in Fig. 5.6 and then plot the function
values in Fig. 5.7.

   The values we list for this example suggest strongly that we should draw
a straight line through the indicated points in order to plot the function. In
other words, we are guessing the values of the function in between the points

x	$\frac{1}{2}x + \frac{1}{2}$		x	$\frac{1}{2}x + \frac{1}{2}$
-5	-2		$-.6$	$.2$
-2.8	$-.9$		$.2$	$.6$
-2	$-.5$		1	1
-1.2	$-.1$		3	2
-1	0		5	3

FIGURE 5.6. A list of some function values of $f(x) = \frac{1}{2}x + \frac{1}{2}$.

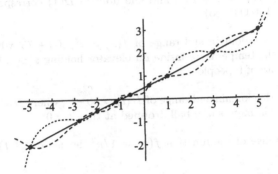

FIGURE 5.7. A plot of some of the points given by $f(x) = \frac{1}{2}x + \frac{1}{2}$ and several functions that pass through those points.

computed by assuming that the function does not do anything strange in between. But there are many functions that agree with $\frac{1}{2}x + \frac{1}{2}$ at the points listed as shown in Fig. 5.7. Deciding whether or not we have evaluated a function defined on the rational numbers enough times to be able to guess its behavior is an interesting and important problem. [4]

In fact, we can use calculus to help with this decision. For now, *we assume that plotted functions vary smoothly between sample points*, which is largely true for the functions we consider in this book.

---

[4] All software packages like *MATLAB* © that plot functions have to make this decision, and not infrequently such software can fail to make a good plot.

Chapter 5   Problems

5.1. Identify four functions you encounter in your daily life and determine the domain and range for each.

5.2. For the function $f(x) = 4x - 2$, determine the range corresponding to (a) $D(f) = (-2, 4]$, (b) $D(f) = (3, \infty)$, and (c) $D(f) = \{-3, 2, 6, 8\}$.

5.3. Given that $f(x) = 2 - 13x$, find the domain $D(f)$ corresponding to the range $R(f) = [-1, 1] \cup (2, \infty)$.

5.4. Determine the domain and range of $f(x) = x^3/100 + 75$, where $f(x)$ is a function giving the temperature inside an elevator holding x people and with a maximum capacity of 9 people.

5.5. Determine the domain and range of $H(t) = 50 - t^2$, where $H(t)$ is a function giving the height in meters of a ball dropped at time $t = 0$.

5.6. Find the range of the function $f(n) = 1/n^2$ defined on $D(f) = \{n \text{ in } \mathbb{N} : n \geq 1\}$.

5.7. Find the domain and a set B containing the range of the function $f(x) = 1/(1 + x^2)$.

5.8. Find the domain of the functions

$$\text{(a)} \ \frac{2 - x}{(x + 2)x(x - 4)(x - 5)} \quad \text{(b)} \ \frac{x}{4 - x^2} \quad \text{(c)} \ \frac{1}{2x + 1} + \frac{x^2}{x - 8} \ .$$

5.9. Consider the function f defined on the natural numbers, where $f(n)$ is the remainder obtained by dividing n by 5 using long division. So, for example, $f(1) = 1$, $f(6) = 1$, $f(12) = 2$, etc. Determine $R(f)$.

5.10. Illustrate the map $f : \mathbb{N} \to \mathbb{Q}$ using two intervals where $f(n) = 2^{-n}$.

5.11. Plot the following functions $f : \mathbb{N} \to \mathbb{N}$ after making a list of at least 8 values: (a) $f(n) = 4 - n$, (b) $f(n) = n - n^2$, (c) $f(n) = (n + 1)^3$.

5.12. Draw three different curves that pass through the points $(-3, -2.5)$, $(-2, -1)$, $(-1, -.5)$, $(0, .25)$, $(1, 1.5)$, $(2, 2)$, $(3, 4)$.

5.13. Plot the functions (a) 2^{-n}, (b) 5^{-n}, and (c) 10^{-n} defined on the natural numbers n. Compare the plots.

5.14. Plot the function $f(n) = \frac{10}{9}(1 - 10^{-n-1})$ defined on the natural numbers.

5.15. Plot the function $f : \mathbb{Q} \to \mathbb{Q}$ with $f(x) = x^3$ after making a table of values.

6
Polynomials

Before further exploring the properties of general functions, we consider the basic example of polynomials. In a concrete way, explained in Chapter 36, polynomials are the "building blocks" for many of the functions encountered in mathematical modeling. Consequently, polynomials crop up repeatedly in analysis.

In this chapter, we develop arithmetic for general polynomials using a convenient notation for sums. Recall that when rational numbers are added, subtracted, or multiplied, the result is another rational number. We show that the analogous property holds for polynomials.

6.1 Polynomials

A **polynomial function**, or **polynomial**, f is a function of the form

$$f(x) = a_0 + a_1 x + a_2 x^2 + a_3 x^3 + \cdots + a_n x^n, \qquad (6.1)$$

where $a_0, a_1, \cdots, a_n$, are given numbers called the **coefficients**. Note how we use the "$\cdots$" in this case to indicate the sum includes the "missing" terms. The domain of a polynomial can be taken to be the set of rational numbers; however, it is difficult to determine the range. Certainly if x is any rational number and the coefficients are rational, then $f(x)$ is another rational number. So the range of a polynomial with rational coefficients contains rational numbers. The question is whether it contains all the rational numbers. It turns out that it may not in general, which we show in Chapter 10.

If n denotes the largest subscript with $a_n \neq 0$, we say that the **degree** of f is n. If all the coefficients a_i are zero, then $f(x) = 0$ for all x and we say that f is the **zero polynomial**. The simplest polynomial beyond the zero polynomial is the **constant** polynomial $f(x) = a_0$ of degree 0. The next simplest cases are the **linear** polynomials $f(x) = a_0 + a_1 x$ of degree 1 and **quadratic** polynomials $f(x) = a_0 + a_1 x + a_2 x^2$ of degree 2 (assuming $a_1 \neq 0$ respectively $a_2 \neq 0$). We use the linear polynomial $f(x) = 15x$ in the Dinner Soup model, the quadratic $f(x) = x^2$ in the Muddy Yard model, and a polynomial of degree 3 to model the solubility of $Ba(IO_3)_2$ in Section 4.5. The polynomials of degree 0, 1, 2, and even 3 are fairly familiar, so we concentrate on developing properties of general polynomials.

6.2   The Σ Notation for Sums

Before exploring arithmetic with polynomials, we introduce a convenient notation for dealing with sums called the **sigma notation** that was invented by Euler. Given any $n + 1$ quantities $\{a_0, a_1, \cdots, a_n\}$ indexed with subscripts, we write the sum

$$a_0 + a_1 + \cdots + a_n = \sum_{i=0}^{n} a_i.$$

The **index** of the sum is i and we assume that it takes on all the integers between the **lower limit**, which is 0 here, and the **upper limit**, which is n here, of the sum.

EXAMPLE 6.1. The finite **harmonic series** of order n is

$$\sum_{i=1}^{n} \frac{1}{i} = 1 + \frac{1}{2} + \frac{1}{3} + \cdots \frac{1}{n}.$$

EXAMPLE 6.2. The **geometric sum** of order n with factor r is

$$1 + r + r^2 + \cdots + r^n = \sum_{i=0}^{n} r^i.$$

The index i is considered to be a **dummy variable** in the sense that it can be renamed or the sum can be rewritten to start at another integer.

EXAMPLE 6.3. The following sums are all the same:

$$\sum_{i=1}^{n} \frac{1}{i} = \sum_{z=1}^{n} \frac{1}{z} = \sum_{i=0}^{n-1} \frac{1}{i+1} = \sum_{i=4}^{n+3} \frac{1}{i-3}.$$

Using the Σ notation, we can write the general polynomial (6.1) in the more condensed form

$$f(x) = \sum_{i=0}^{n} a_i x^i = a_0 + a_1 x^1 + \cdots + a_n x^n.$$

EXAMPLE 6.4. We can write

$$1 + 2x + 4x^2 + 8x^3 + \cdots + 2^{20} x^{20} = \sum_{i=0}^{20} 2^i x^i$$

and

$$1 - x + x^2 - x^3 - x^{99} = \sum_{i=0}^{99} (-1)^{2i-1} x^i$$

since $(-1)^{2i-1} = 1$ if i is odd and $(-1)^{2i-1} = -1$ if i is even.

6.3 Arithmetic with Polynomials

In this section, we work out rules for combining polynomials to get new polynomials. The rules are based on the arithmetic operations for numbers.

We define the sum of two polynomials

$$f(x) = a_0 + a_1 x^1 + a_2 x^2 + \cdots + a_n x^n$$

and

$$g(x) = b_0 + b_1 x^1 + b_2 x^2 + \cdots + b_n x^n$$

as the new polynomial $f + g$ given by

$$(f + g)(x) = (b_0 + a_0) + (b_1 + a_1)x^1 + (b_2 + a_2)x^2 + \cdots (b_n + a_n)x^n;$$

i.e., the sum is the polynomial with coefficients obtained by adding the corresponding coefficients in each summand. Note how we use the parentheses around $f + g$ to indicate that a new polynomial has been constructed. Using the Σ notation,

$$(f + g)(x) = \sum_{i=0}^{n} a_i x^i + \sum_{i=0}^{n} b_i x^i = \sum_{i=0}^{n} (a_i + b_i)x^i.$$

Notice that the value of $f + g$ at the point x can be computed by adding the numbers $f(x)$ and $g(x)$, i.e.,

$$(f + g)(x) = f(x) + g(x).$$

EXAMPLE 6.5. If $f(x) = 1 + x^2 - x^4 + 2x^5$ and $g(x) = 33x + 7x^2 + 2x^5$, then

$$(f + g)(x) = 1 + 33x + 8x^2 - x^4 + 4x^5,$$

where of course we "fill in" the "missing" monomials, i.e., those with coefficients equal to zero in order to use the definition.

EXAMPLE 6.6. If $f(x) = 1 + x^2 - x^4 + 2x^5$ and $g(x) = 33x + 7x^2 + 2x^5$, then

$$\begin{aligned}(f + g)(x) &= (1 + 0x + x^2 + 0x^3 - x^4 + 2x^5) \\ &\quad + (0 + 33x + 7x^2 + 0x^3 + 0x^4 + 2x^5) \\ &= 1 + 33x + 8x^2 + 0x^3 - x^4 + 4x^5 \\ &= 1 + 33x + 8x^2 - x^4 + 4x^5.\end{aligned}$$

In general, to add the polynomials

$$f(x) = \sum_{i=0}^{n} a_i x^i$$

of degree n (assuming that $a_n \neq 0$) and the polynomial

$$g(x) = \sum_{i=0}^{m} b_i x^i$$

of degree m, where we assume that $m \leq n$, we just fill in the "missing" coefficients in g by setting $b_{m+1} = b_{m+2} = \cdots b_n = 0$, and add using the definition.

EXAMPLE 6.7.

$$\sum_{i=0}^{15}(i + 1)x^i + \sum_{i=0}^{30} x^i = \sum_{i=0}^{30} a_i x^i$$

with

$$a_i = \begin{cases} i + 2, & 0 \leq i \leq 15, \\ i + 1, & 16 \leq i \leq 30. \end{cases}$$

In the next step, we define the **product** cf of a constant c and a polynomial

$$f(x) = \sum_{i=0}^{n} a_i x^i,$$

as the polynomial obtained by multiplying each coefficient of the polynomial by c, i.e.,

$$(cf)(x) = \sum_{i=0}^{n} ca_i x^i.$$

This is suggested, of course, by the distributive properties of addition and multiplication of rational numbers.

EXAMPLE 6.8.

$$2.3(1 + 6x - x^7) = 2.3 + 13.8x - 2.3x^7.$$

Note that following this definition, we define the **difference** of two polynomials f and g as $f - g = f + (-g)$.

We can now combine polynomials by multiplying the polynomials by rational numbers and adding the results, and thereby obtain new polynomials. Thus, if f_1, f_2, $\cdots$, f_n are n polynomials and c_1, $\cdots$, c_n are n numbers, then

$$f(x) = \sum_{m=1}^{n} c_m f_m(x)$$

is a new polynomial called the **linear combination** of the polynomials f_1, $\cdots$, f_n. The name is suggested by the fact that a linear function $ax + b$ is defined using the operations of addition and multiplication by a constant. The numbers c_1, $\cdots$, c_n are called the **coefficients** of the linear combination.

EXAMPLE 6.9. The linear combination of $2x^2$ and $4x - 5$ with coefficients 1 and 2 is

$$1(2x^2) + 2(4x - 5) = 2x^2 + 8x - 10.$$

With this definition, a general polynomial

$$f(x) = \sum_{i=0}^{n} a_i x^i$$

can be described as a linear combination of the particular polynomials 1, x, x^2, $\cdots$, x^n, which are called the **monomials**. To make the notation consistent, we set $x^0 = 1$ for all x.

The definitions of the arithmetic operations we have made imply the following theorem:

Theorem 6.1 *A linear combination of polynomials is a polynomial. A general polynomial is a linear combination of monomials.*[1]

As a consequence of these definitions, we obtain a number of rules for linear combinations of polynomials that reflect the corresponding rules for rational numbers. For example, if f, g and h are polynomials and c is rational number, then

$$f + g = g + f, \tag{6.2}$$

$$(f + g) + h = f + (g + h), \tag{6.3}$$

$$c(f + g) = cf + cg. \tag{6.4}$$

---

[1]We discuss linear combinations of polynomials further in Chapter 38.

Finally, we consider the product of general polynomials. We define the product of two monomials x^j and x^i as

$$x^j x^i = x^j \times x^i = x^{j+i},$$

following the same rule that holds for multiplying powers of numbers. Then we define the product of x^j and a polynomial $f(x) = \sum_{i=0}^{n} a_i x^i$ by distributing x^j as follows:

$$x^j f(x) = a_0 x^j + a_1 x^j \times x + a_2 x^j \times x^2 + \cdots + a_n x^j \times x^n$$
$$= a_0 x^j + a_1 x^{1+j} + a_2 x^{2+j} + \cdots + a_n x^{n+j}$$
$$= \sum_{i=0}^{n} a_i x^{i+j},$$

which is a polynomial of degree $n + j$.

EXAMPLE 6.10.

$$x^3 (2 - 3x + x^4 + 19x^8) = 2x^3 - 3x^4 + x^7 + 19x^{11}.$$

Finally, we define the product fg of of two polynomials $f(x) = \sum_{i=0}^{n} a_i x^i$ and $g(x) = \sum_{j=0}^{m} b_j x^j$ by $(fg)(x) = f(x)g(x)$. This means that we can compute $f(x)g(x)$ as follows:

$$(fg)(x) = f(x)g(x) = \left(\sum_{i=0}^{n} a_i x^i \right)\left(\sum_{i=0}^{m} b_i x^i \right)$$
$$= \sum_{i=0}^{n} \left(a_i x^i \sum_{j=0}^{m} b_j x^j \right) = \sum_{i=0}^{n} \left(a_i \sum_{j=0}^{m} b_j x^{i+j} \right)$$
$$= \sum_{i=0}^{n} \sum_{j=0}^{m} a_i b_j x^{i+j}.$$

We consider an example:

EXAMPLE 6.11.

$$(1 + 2x + 3x^2)(x - x^5) = 1(x - x^5) + 2x(x - x^5) + 3x^2(x - x^5)$$
$$= x - x^5 + 2x^2 - 2x^6 + 3x^3 - 3x^7$$
$$= x + 2x^2 + 3x^3 - x^5 - 2x^6 - 3x^7$$

These definitions imply the following theorem:

Theorem 6.2 *The product of a nonzero polynomial of degree n and a nonzero polynomial of degree m is a polynomial of degree $n + m$.*

These definitions also imply the commutative, associative, and distributive properties for polynomials f, g, and h,

$$fg = gf, \tag{6.5}$$
$$(fg)h = f(gh), \tag{6.6}$$
$$(f + g)h = fh + fh, \tag{6.7}$$

hold true.

Products are tedious to compute, but luckily we can use software like $MAPLE^{©}$ to compute them. There are a couple of examples that are good to keep in mind:

$$(x + a)^2 = (x + a)(x + a) = x^2 + 2ax + a^2$$
$$(x + a)(x - a) = x^2 - a^2$$
$$(x + a)^3 = x^3 + 3ax^2 + 3a^2x + a^3.$$

6.4 Equality of Polynomials

We say that two polynomials f and g are equal, $f = g$, if $f(x) = g(x)$ at every point x. Equivalently, $f = g$ if $(f - g)(x)$ is the zero polynomial with all coefficients equal to zero. Note that two polynomials are not necessarily equal because they happen to have the same value at just one point!

EXAMPLE 6.12. $f(x) = x^2 - 4$ and $g(x) = 3x - 6$ are both zero at $x = 2$ but are not equal.

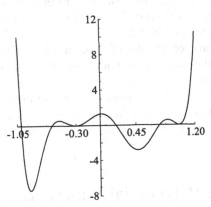

FIGURE 6.1. A plot of $y = 1.296 + 1.296x - 35.496x^2 - 57.384x^3 + 177.457x^4 + 203.889x^5 - 368.554x^6 - 211.266x^7 + 313.197x^8 + 70.965x^9 - 97.9x^{10} - 7.5x^{11} + 10x^{12}$.

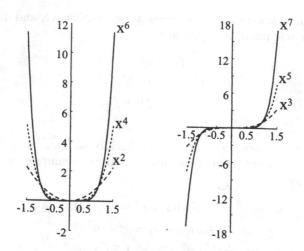

FIGURE 6.2. Plots of some monomials.

6.5   Graphs of Polynomials

A general polynomial of degree greater than 2 or 3 can be quite a complicated function and it is difficult to make meaningful general observations
about their plots. We show an example in Fig. 6.1. When the degree of a
polynomial is large, the tendency is for the plot to have large "wiggles,"
which makes it difficult to plot the function. The value of the polynomial
shown in Fig. 6.1 is 987940.8 at $x = 3$.

On the other hand, we can plot the monomials rather easily. It turns out
that once the degree $n \geq 2$, the plots of the monomials with odd degree
n have a similar shape as do the plots of the monomials with even degree.
We show some samples in Fig. 6.2.

One obvious feature of graphs of monomials is symmetry. When the
degree is even, the plots are symmetric across the y-axis, see Fig. 6.3. This
means that the value of the monomial is the same for x and $-x$, or in
other words $x^m = (-x)^m$ for m even. When the degree is odd, the plots
are symmetric through the origin. In other words, the value of the function
for x is the negative of the value of the function for $-x$ or $(-x)^m = -x^m$
for m odd.

6.6   Piecewise Polynomial Functions

We started this chapter by declaring that polynomials are building blocks
for functions. An important class of functions constructed using polynomials are the **piecewise polynomials**. These are functions that are equal to
polynomials on intervals contained in the domain.

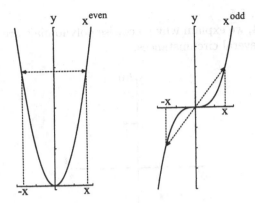

FIGURE 6.3. The symmetries of the monomial functions of even and odd degree.

We already met one example, namely,

$$|x| = \begin{cases} x, & x \geq 0, \\ -x, & x < 0. \end{cases}$$

The function $|x|$ looks like $y = x$ for $x \geq 0$ and $y = -x$ for $x < 0$. We plot it in Fig. 6.4. The most interesting thing to note about the graph of $|x|$ is the sharp corner at $x = 0$, which occurs right at the transition point of this piecewise polynomial.

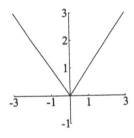

FIGURE 6.4. Plot of $y = |x|$.

Another example is provided by a piecewise constant function that is used to model the current supplied to an electric circuit. Suppose the power is off, so the current is 0, and that we turn the power on, say, with current equal to 1, at time $t = 0$ and then turn it off again at $t = 1$ second. The function $I(t)$ describing the current is called the **step function** and is defined

$$I(t) = \begin{cases} 0, & t < 0, \\ 1, & 0 \leq t \leq 1, \\ 0, & 1 < t. \end{cases} \tag{6.8}$$

We plot I in Fig. 6.5.

In Section 38.4, we explain why piecewise polynomials are preferred over polynomials in several circumstances.

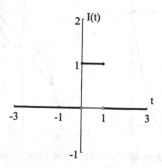

FIGURE 6.5. Plot of the step function $I(t)$.

Chapter 6   Problems

6.1. Write the following finite sums using the summation notation. Be sure to get the starting and ending values for the index correct!

(a) $1 + \frac{1}{4} + \frac{1}{9} + \frac{1}{16} + \cdots + \frac{1}{n^2}$ (b) $-1 + \frac{1}{4} - \frac{1}{9} + \frac{1}{16} - \cdots \pm \frac{1}{n^2}$

(c) $1 + \frac{1}{2 \times 3} + \frac{1}{3 \times 4} + \cdots + \frac{1}{n \times (n+1)}$ (d) $1 + 3 + 5 + 7 + \cdots + 2n + 1$

(e) $x^4 + x^5 + \cdots + x^n$ (f) $1 + x^2 + x^4 + x^6 + \cdots + x^{2n}$.

6.2. Write the finite sum $\sum_{i=1}^{n} i^2$ so that (a) i starts with -1, (b) i starts with 15, (c) the coefficient has the form $(i+4)^2$, and (d) i ends with $n + 7$.

6.3. Rewrite the following polynomials using the summation notation:

(a) $x + 2x^3 + 3x^5 + 4x^7 + \cdots + 10x^{19}$

(b) $2 + 4x + 6x^2 + 8x^3 + \cdots + 24x^{10}$

(c) $1 + x - x^2 + x^3 + x^4 - x^5 + \cdots - x^{17}$.

6.4. Given $f_1(x) = -4 + 6x + 7x^3$, $f_2(x) = 2x^2 - x^3 + 4x^5$ and $f_3(x) = 2 - x^4$, compute the following polynomials:

(a) $f_1 - 4f_2$ (b) $3f_2 - 12f_1$ (c) $f_2 + f_1 + f_3$

(d) $f_2 f_1$ (e) $f_1 f_3$ (f) $f_2 f_3$

(g) $f_1 f_3 - f_2$ (h) $(f_1 + f_2)f_3$ (i) $f_1 f_2 f_3$.

6.5. For a equal to a constant, compute

(a) $(x + a)^2$ (b) $(x + a)^3$ (c) $(x - a)^3$ (d) $(x + a)^4$.

6.6. Compute $f_1 f_2$ where $f_1(x) = \sum_{i=0}^{8} i^2 x^i$ and $f_2(x) = \sum_{j=0}^{11} \frac{1}{j+1} x^j$.

6.7. Plot the function

$$f(x) = 360x - 942x^2 + 949x^3 - 480x^4 + 130x^5 - 18x^6 + x^7$$

using *MATLAB*© or *MAPLE*© . This takes some trial-and-error in choosing a good interval on which to plot. You should make plots on several different intervals, starting with $-.5 \le x \le .5$, then increasing the size.

6.8. Plot the following piecewise polynomials for $-2 \le x \le 2$

(a) $f(x) = \begin{cases} 2, & -2 \le x \le -1, \\ x^2, & -1 < x < 1, \\ x, & 1 \le x \le 2. \end{cases}$ (b) $f(x) = \begin{cases} -1 - x, & -2 \le x \le -1, \\ 1 + x, & -1 < x \le 0, \\ 1 - x, & 0 < x \le 1, \\ -1 + x, & 1 < x \le 2. \end{cases}$

6.9. (a) Show that the monomial x^3 is increasing for all x. (b) Show the monomial x^4 is decreasing for $x < 0$ and increasing for $x > 0$.

7

Functions, Functions, and More Functions

Before continuing the investigation of functions, we describe ways to make complicated functions by combining simpler functions.[1] In fact, this idea was encountered in Chapter 6, where we constructed general polynomials by adding up monomials. We first describe how to add up arbitrary functions, then we consider the operations of multiplication and division. We finish off with composition.

7.1   Linear Combinations of Functions

It is straightforward to generalize the idea of adding functions to produce a new function. The first step is to define the sum $f_1 + f_2$ of two given functions f_1 and f_2 as the function with the value at x given by the sum of the values of $f_1(x)$ and $f_2(x)$, i.e.,

$$(f_1 + f_2)(x) = f_1(x) + f_2(x).$$

---

[1]The idea of combining simple things to get complex ones is fundamental in many different settings. Music is a good example: chords or harmonies are formed by combining single tones, complex rhythmic patterns may be formed by overlaying simple basic rhythmic patterns, single instruments are combined to form an orchestra. Another example is a fancy dinner that is made up of an entree, main dish, dessert, coffee, together with aperitif, wines, and cognac, in endless combinations. Moreover, each dish is formed by combining ingredients like beef, carrots, and potatoes.

For the sum $f_1 + f_2$ to be defined at x, both f_1 and f_2 must be defined at x. Therefore, the domain $D(f_1 + f_2)$ of $f_1 + f_2$ is the intersection $D(f_1) \cap D(f_2)$ of the domains $D(f_1)$ and $D(f_2)$.

EXAMPLE 7.1.   The function $f(x) = x^3 + 1/x$ defined on $D(f) = \{x \text{ in } \mathbb{Q} : x \neq 0\}$ is the sum of the functions $f_1(x) = x^3$ with domain $D(f_1) = \mathbb{Q}$ and $f_2(x) = 1/x$ with domain $D(f_2) = \{x \text{ in } \mathbb{Q} : x \neq 0\}$. The function $f(x) = x^2 + 2^x$ defined on $\mathbb{Z}$ is the sum of x^2 defined on $\mathbb{Q}$ and 2^x defined on $\mathbb{Z}$.

Note this definition implies that $f_1(x) + f_1(x) = 2f_1(x)$ for any valid x. Similarly, we define the product of a given function f and a rational number c, as the new function cf whose value at x is determined by multiplying the value $f(x)$ by c, that is,

$$(cf)(x) = cf(x).$$

Naturally, the domain of cf is $D(cf) = D(f)$.

These definitions are consistent with the definitions we use for polynomials and we get all the familiar commutative, associative, and distributive properties like (6.2)–(6.4).

Combining these operations, we can define a new function $c_1 f_1 + c_2 f_2$ by adding multiples c_1 and c_2 of given functions f_1 and f_2 to form a new function whose domain is the intersection of the domains of f_1 and f_2. This new function is called a **linear combination** of f_1 and f_2. In general, we define the linear combination $a_1 f + \cdots + a_n f_n$ of n functions $f_1, \cdots, f_n$ by

$$(a_1 f + \cdots + a_n f_n)(x) = a_1 f_1(x) + \cdots + a_n f_n(x),$$

where $a_1, \cdots, a_n$ are numbers called the coefficients of the linear combination. Of course the domain of the linear combination $a_1 f + \cdots + a_n f_n$ is the intersection of the domains $D(f_1), \cdots, D(f_n)$.

EXAMPLE 7.2.   The domain of the linear combination of $\left\{ \frac{1}{x}, \frac{x}{1+x}, \frac{1+x}{2+x} \right\}$ given by

$$-\frac{1}{x} + 2\frac{x}{1+x} + 6\frac{1+x}{2+x}$$

is $\{x \text{ in } \mathbb{Q} : x \neq 0, x \neq -1, x \neq -2\}$.

The sigma notation from Section 6.2 is useful for writing general linear combinations.

EXAMPLE 7.3.   The linear combination of $\left\{ \frac{1}{x}, \cdots, \frac{1}{x^n} \right\}$ given by

$$\frac{2}{x} + \frac{4}{x^2} + \frac{8}{x^3} + \cdots + \frac{2^n}{x^n} = \sum_{i=1}^{n} \frac{2^i}{x^i}$$

has domain $\{x \text{ in } \mathbb{Q} : x \neq 0\}$.

Note that there is some uncertainty about linear combinations in the sense that it is generally possible to write a linear combination of a given set of functions in a number of different ways.

EXAMPLE 7.4. A linear combination of the functions $\{1 + x, 1 + x + x^2, x^2\}$ can generally be written in a number of ways. For example,

$$2(1 + x) + (1 + x + x^2) + 3x^2 = (1 + x) + 2(1 + x + x^2) + 2x^2$$
$$= -(1 + x) + 4(1 + x + x^2) + 0x^2.$$

In some situations we want to know that every linear combination of a given set of functions $\{f_1, f_2, \cdots, f_n\}$ can be written in only one way or is **unique**. For example, it is important to know that a polynomial written in the form $a_0 + a_1 x + \cdots + a_n x^n$, which is just a linear combination of the monomials, is a unique polynomial and there is no other linear combination of $\{1, x, \cdots, x^n\}$ that gives the same polynomial.

Whether or not this is true depends on the functions $\{f_1, \cdots, f_n\}$. Suppose that two linear combinations of the functions are equal, say,

$$a_1 f_1(x) + \cdots + a_n f_n(x) = b_1 f_1(x) + \cdots + b_n f_n(x)$$

for all x in the domain. We can rewrite this as

$$(a_1 - b_1) f_1(x) + \cdots + (a_n - b_n) f_n(x) = 0 \qquad (7.1)$$

for all x in the domain. Now if $a_1 = b_1$, $\cdots$, $a_n = b_n$, then the linear combination is unique. So if f_1, $\cdots$, f_n have the property that whenever (7.1) holds for all x in the domain, we have $a_1 - b_1 = \cdots = a_n - b_n = 0$, then every linear combination of f_1, $\cdots$, f_n is unique. We say that the functions $\{f_1, \cdots, f_n\}$ are **linearly independent** on a domain if the only constants $c_1, c_2, \cdots, c_n$ such that

$$c_1 f_1(x) + \cdots + c_n f_n(x) = 0$$

for all x in the domain are $c_1 = \cdots = c_n = 0$. Functions that are not linearly independent are said to be **linearly dependent**.

Theorem 7.1 *If the functions $\{f_1, \cdots, f_n\}$ are linearly independent, then every linear combination of the functions is unique.*

EXAMPLE 7.5. The functions $\{1 + x, 1 + x + x^2, x^2\}$ are linearly dependent since $1(1 + x) - 1(1 + x + x^2) + 1x^2 = 0$ for all x.

EXAMPLE 7.6. The functions $\{x, 1/x\}$ are linearly independent. Suppose that there are constants c_1, c_2 such that

$$c_1 x + c_2 \frac{1}{x} = 0$$

for all $x \neq 0$. Setting $x = 1$ arbitrarily, we get $c_1 + c_2 = 0$ or $c_1 = -c_2$. Setting $x = 2$, we also get $2c_1 + .5c_2 = 0$. But this means $-2c_2 + .5c_2 = -1.5c_2 = 0$ and hence $c_2 = c_1 = 0$.

EXAMPLE 7.7. We can prove that the monomials are linearly independent using induction. First we show that $\{1\}$ is linearly independent, which is easy since $c_0 \times 1 = 0$ implies $c_0 = 0$. Now suppose that $\{1, x, \cdots, x^{n-1}\}$ are linearly independent, i.e., if $a_0 \times 1 + a_1 x + \cdots + a_{n-1} x^{n-1} = 0$ for all x, then $a_0 = a_1 = \cdots = a_{n-1} = 0$. If $c_0 \times 1 + c_1 x + \cdots + c_n x^n = 0$ for all x, then setting $x = 0$, we conclude that $c_0 \times 1 + 0 + \cdots + 0 = 0$ or $c_0 = 0$ and

$$c_0 \times 1 + c_1 x + \cdots + c_n x^n = c_1 x + \cdots + c_n x^n = 0.$$

Factoring, we get

$$c_1 x + \cdots + c_n x^n = x\big(c_1 + c_2 x + \cdots + c_n x^{n-1}\big) = 0$$

for all x. However, this means that

$$c_1 + c_2 x + \cdots + c_n x^{n-1} = 0$$

for all x. By the inductive assumption, this means that $c_1 = \cdots = c_n = 0$ and therefore the monomials $\{1, \cdots, x^n\}$ are linearly independent. By induction, all monomials are linearly independent.

7.2   Multiplication and Division of Functions

We multiply arbitrary functions using the same idea used to multiply polynomials. If f_1 and f_2 are two functions, we define the product $f_1 f_2$ by

$$(f_1 f_2)(x) = f_1(x) f_2(x).$$

EXAMPLE 7.8. The function

$$f(x) = (x^2 - 3)^3 \big(x^6 - \frac{1}{x} - 3\big)$$

with $D(f) = \{x \in \mathbb{Q} : x \neq 0\}$ is the product of the functions $f_1(x) = (x^2 - 3)^3$ and $f_2(x) = x^6 - 1/x - 3$. The function $f(x) = x^2\, 2^x$ is the product of x^2 and 2^x.

The domain of the product of two functions is the intersection of the domains of the two functions. This definition is consistent with the definition we use for polynomials and it implies that the familiar commutative, associative, and distributive properties like (6.5)–(6.7) are true.

Following this lead, given two functions f_1 and f_2, we define the quotient function f_1/f_2 by

$$(f_1/f_2)(x) = \frac{f_1}{f_2}(x) = \frac{f_1(x)}{f_2(x)}$$

as long as $f_2(x) \neq 0$. However, determining the domain requires some additional thought. Not only is it necessary to make sure that both functions are defined, we also have to avoid zeroes in the denominator. Therefore, the domain of the quotient f_1/f_2 of two functions is the intersection of the domains of the two functions taking away points where the denominator f_2 is zero.

EXAMPLE 7.9. The domain of

$$\frac{1 + 1/(x+3)}{2x - 5}$$

is the intersection of $\{x \text{ in } \mathbb{Q} : x \neq -3\}$ and $\{x \text{ in } \mathbb{Q}\}$ excepting $x = 5/2$ or $\{x \text{ in } \mathbb{Q} : x \neq -3, 5/2\}$.

Actually, determining the domain of the quotient of two functions contains a hidden complication. Clearly, we must avoid points at which the denominator is zero and the numerator is nonzero. However, the situation in which both the numerator and denominator are zero at a point is a little less clear.

EXAMPLE 7.10. Consider the quotient

$$\frac{x - 1}{x - 1}$$

with domain $\{x \text{ in } \mathbb{Q} : x \neq 1\}$. Since

$$x - 1 = 1 \times (x - 1) \tag{7.2}$$

for all x, it is natural to "divide" the polynomials to get

$$\frac{x - 1}{x - 1} = 1. \tag{7.3}$$

However, the domain of the constant function 1 is $\mathbb{Q}$ so the left- and right-hand sides of (7.3) have different domains and therefore must represent different functions. We plot the two functions in Fig. 7.1. The

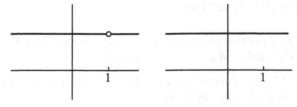

FIGURE 7.1. Plots of $(x-1)/(x-1)$ on the left and 1 on the right.

two functions agree at every point except for the "missing" point $x = 1$.

By the definition, we can only assert that (7.3) holds for $\{x \text{ in } \mathbb{Q} : x \neq 1\}$. On the other hand, (7.3) is just another way of writing (7.2), which holds for all x. So there is a little uncertainty in all of this.

As we said, the quotient f_1/f_2 is defined at all points in the intersection of the domains of f_1 and f_2 taking away points where f_2 is zero. However, if there is a function f such that

$$f_1(x) = f_2(x)f(x)$$

for all x in the intersection of the domains of f_1 and f_2, we often abuse notation and replace f_1/f_2 by f and neglect to mention that we are supposed to avoid those "missing" points where f_2 is zero. In this situation, we say that we have **divided** f_1 by f_2 to get f.

EXAMPLE 7.11. Since $x^2 - 2x - 3 = (x-3)(x+1)$, we write

$$\frac{x^2 - 2x - 3}{x - 3} = x + 1.$$

In most situations, we replace $(x^2 - 2x - 3)/(x - 3)$, which is defined for $\{x \text{ in } \mathbb{Q} : x \neq 3\}$, by $x + 1$, which is defined for all x in $\mathbb{Q}$.

Note that the fact that both f_1 and f_2 are zero at the same point does not mean that we can automatically replace their quotient by a function defined at all points.

EXAMPLE 7.12. Consider the functions

$$\frac{(x-1)^2}{x-1} \quad \text{and} \quad \frac{x-1}{(x-1)^2},$$

both of which are defined for $\{x \text{ in } \mathbb{Q} : x \neq 1\}$. We usually replace the first function by $x - 1$ defined for all x, but dividing the second example gives $1/(x-1)$, which is still undefined for $x = 1$.

In general, it can take a little work to figure out whether the substitution is possible.

We discuss this issue in great detail in Chapter 35.

7.3   Rational Functions

The quotient f_1/f_2 of two polynomials f_1 and f_2 is called a **rational function**. This is the analog of a rational number.

EXAMPLE 7.13. The function $f(x) = 1/x$ is a rational function defined for $\{x \text{ in } \mathbb{Q} : x \neq 0\}$. The function

$$f(x) = \frac{(x^3 - 6x + 1)(x^{11} - 5x^6)}{(x^4 - 1)(x+2)(x-5)}$$

is a rational function defined on $\{x \text{ in } \mathbb{Q} : x \neq 1, -1, -2, 5\}$.

In an example above, we saw that $x - 3$ divides into $x^2 - 2x - 3$ exactly because $x^2 - 2x - 3 = (x - 3)(x + 1)$ so

$$\frac{x^2 - 2x - 3}{x - 3} = x + 1.$$

In the same way, a rational number p/q sometimes simplifies to an integer, in other words q divides into p exactly without a remainder. We can determine if this is true by using long division. It turns out that long division also works for polynomials. Recall that in long division, we match the leading digit of the denominator with the remainder at each stage. When dividing polynomials, we write them as a linear combination of monomials starting with the monomial of highest degree and then match coefficients of the monomials one by one.

EXAMPLE 7.14. We show a couple of examples of polynomial division. In Fig. 7.2, we give an example in which there is no remainder. We

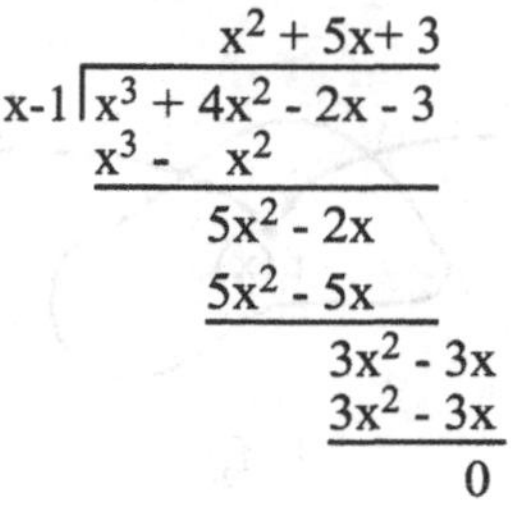

$$
\begin{array}{r}
x^2 + 5x + 3 \\
x-1 \overline{\smash{\big)}\ x^3 + 4x^2 - 2x - 3} \\
\underline{x^3 - x^2} \\
5x^2 - 2x \\
\underline{5x^2 - 5x} \\
3x^2 - 3x \\
\underline{3x^2 - 3x} \\
0
\end{array}
$$

FIGURE 7.2. An example of polynomial division with no remainder.

conclude that

$$\frac{x^3 + 4x^2 - 2x + 3}{x - 1} = x^2 + 5x + 3.$$

In Fig. 7.3, we give an example in which there is a remainder, i.e. we carry out the division to the point where the remaining numerator has lower degree than the denominator. Note that in this example, the numerator is "missing" a term so we fill in the missing term with a zero coefficient to make the division easier. We conclude that

$$\frac{2x^4 + 7x^2 - 8x + 3}{x^2 + x - 3} = 2x^2 - 2x + 15 + \frac{-29x + 48}{x^2 + x - 3}.$$

7.4 Composition of Functions

Given two functions f_1 and f_2, we can define a new function f by first applying f_1 to an input and then applying f_2 to the result: i.e.,

$$f(x) = f_2(f_1(x)).$$

$$\begin{array}{r} 2x^2 - 2x + 15 \\ x^2+x-3\overline{\smash{\big)}\,2x^4 + 0x^3 + 7x^2 - 8x + 3} \\ \underline{2x^4 + 2x^3 - 6x^2} \\ -2x^3+13x^2 - 8x \\ \underline{-2x^3 - 2x^2 + 6x} \\ 15x^2 - 14x + 3 \\ \underline{15x^2 + 15x - 45} \\ -29x + 48 \end{array}$$

FIGURE 7.3. An example of polynomial division with a remainder.

We say that f is the **composition** of f_2 and f_1, and we write $f = f_2 \circ f_1$: that is,

$$(f_2 \circ f_1)(x) = f_2(f_1(x)).$$

We illustrate this operation in Fig. 7.4.

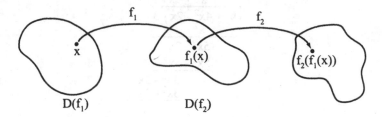

FIGURE 7.4. Illustration of the composition $f_2 \circ f_1$.

EXAMPLE 7.15. If $f_1(x) = x^2$ and $f_2(x) = x + 1$, then $f_1 \circ f_2(x) = f_1(f_2(x)) = (x + 1)^2$, while $f_2 \circ f_1 = f_2(f_1(x)) = x^2 + 1$.

This example illustrates the general fact that $f_2 \circ f_1 \neq f_1 \circ f_2$ most of the time.

Determining the domain of the composition of $f_2 \circ f_1$ can be complicated. Certainly to compute $f_2(f_1(x))$ we have to insure that x is in the domain of f_1 so that $f_1(x)$ is defined. Next we apply f_2 to the result; therefore $f_1(x)$ must have a value that is in the domain of f_2. Therefore the domain of $f_2 \circ f_1$ is the set of points x in $D(f_1)$ such that $f_1(x)$ is in $D(f_2)$.

EXAMPLE 7.16. Let $f_1(x) = 3 + 1/x^2$ and $f_2(x) = 1/(x - 4)$. Then $D(f_1) = \{x$ in $\mathbb{Q} : x \neq 0\}$ while $D(f_2) = \{x$ in $\mathbb{Q} : x \neq 4\}$. Therefore to compute $f_2 \circ f_1$, we must avoid any points where $3 + 1/x^2 = 4$ or $1/x^2 = 1$ or $x = 1$ and $x = -1$. We conclude that $D(f_2 \circ f_1) = \{x$ in $\mathbb{Q} : x \neq 0, 1, -1\}$.

Chapter 7   Problems

7.1. Determine the domains of the following functions:

(a) $3(x-4)^3 + 2x^2 + \dfrac{4x}{3x-1} + \dfrac{6}{(x-1)^2}$

(d) $\dfrac{(2x-3)^{\frac{2}{x}}}{4x+6}$

(b) $2 + \dfrac{4}{x} - \dfrac{6x+4}{(x-2)(2x+1)}$

(e) $\dfrac{6x-1}{(2-3x)(4+x)}$

(c) $x^3\left(1+\dfrac{1}{x}\right)$

(f) $\dfrac{4}{x+2} + \dfrac{6}{x^2+3x+2}$.

7.2. Write the following linear combinations using the sigma notation and determine the domain of the result:

(a) $2x(x-1) + 3x^2(x-1)^2 + 4x^3(x-1)^3 + \cdots + 100x^{101}(x-1)^{101}$

(b) $\dfrac{2}{x-1} + \dfrac{4}{x-2} + \dfrac{8}{x-3} + \cdots + \dfrac{8192}{x-13}$.

7.3. (a) Let $f(x) = ax+b$, where a and b are numbers, and show that $f(x+y) = f(x) + f(y)$ for *all* numbers x and y if and only if $b = 0$. (b) Let $g(x) = x^2$ and show that $g(x+y) \neq g(x) + g(y)$ unless x and y have special values.

7.4. Write the function in Example 7.4 as a linear combination in two new ways.

7.5. Write the following linear combinations in two new ways:

(a) $2x + 3 + 5(x+2)$

(b) $2x^2 + 4x^4 - 2(x^2 + x^4)$

(c) $\dfrac{1}{x} + \dfrac{2}{x-1} + \dfrac{3}{x(x-1)}$.

7.6. Show the following functions are linearly dependent or prove they are linearly independent:

(a) $\{x+1, x-3\}$

(b) $\{x, 2x+1, 4\}$

(c) $\{2x-1, 3x, x^2+x\}$

(d) $\left\{\dfrac{1}{x}, \dfrac{1}{x-1}, \dfrac{1}{x(x-1)}\right\}$

(e) $\{4x+1, 6x^2-3, 2x^2+8x+2\}$

(f) $\left\{\dfrac{1}{x}, \dfrac{1}{x^2}, \dfrac{1}{x^3}\right\}$.

7.7. Prove the functions $\left\{\dfrac{1}{x}, \dfrac{1}{x^2}, \dfrac{1}{x^3}, \cdots\right\}$ are linearly independent using induction.

7.8. Use polynomial division on the following rational functions to show that the denominator divides the numerator exactly or to compute the remainder if not:

(a) $\dfrac{x^2 + 2x - 3}{x - 1}$ (b) $\dfrac{2x^2 - 7x - 4}{2x + 1}$

(c) $\dfrac{4x^2 + 2x - 1}{x + 6}$ (d) $\dfrac{x^3 + 3x^2 + 3x + 2}{x + 2}$

(e) $\dfrac{5x^3 + 6x^2 - 4}{2x^2 + 4x + 1}$ (f) $\dfrac{x^4 - 4x^2 - 5x - 4}{x^2 + x + 1}$

(g) $\dfrac{x^8 - 1}{x^3 - 1}$ (h) $\dfrac{x^n - 1}{x - 1}$, n in $\mathbb{N}$.

7.9. Prove the formula for the geometric sum (3.3) holds by using induction on long division on the expression

$$\frac{p^{n+1} - 1}{p - 1}.$$

Hint: Another way to view this is to note that $(1-p)(1+p+p^2+p^3+\cdots+p^n) = 1+p+p^2+p^3+\cdots+p^n-p-p^2-p^3-\cdots-p^n-p^{n+1}$ and that a lot of terms cancel in this last sum.

7.10. Given $f_1(x) = 3x - 5$, $f_2(x) = 2x^2 + 1$, and $f_3(x) = 4/x$, write out formulas for the following functions:

(a) $f_1 \circ f_2$ (b) $f_2 \circ f_3$ (c) $f_3 \circ f_1$ (d) $f_1 \circ f_2 \circ f_3$.

7.11. With $f_1(x) = 4x + 2$ and $f_2(x) = x/x^2$, show that $f_1 \circ f_2 \neq f_2 \circ f_1$.

7.12. Let $f_1(x) = ax + b$ and $f_2(x) = cx + d$, where a, b, c, and d are rational numbers. Find a condition on the numbers a, b, c, and d that implies that $f_1 \circ f_2 = f_2 \circ f_1$ and produce an example that satisfies the condition.

7.13. For the given functions f_1 and f_2, determine the domain of $f_2 \circ f_1$:

(a) $f_1(x) = 4 - \dfrac{1}{x}$ and $f_2(x) = \dfrac{1}{x^2}$

(b) $f_1(x) = \dfrac{1}{(x-1)^2} - 4$ and $f_2(x) = \dfrac{x+1}{x}$.

8
Lipschitz Continuity

Recall that when we graph functions of rational numbers, we make a leap of faith and assume that the function varies "smoothly" between the sample points. In fact, a basic problem in calculus is determining how much a function changes for a given change in input. In this chapter, we investigate the conditions under which a function varies smoothly as well as making the idea of varying smoothly more precise. As a first step in understanding smooth behavior, we introduce a property of functions called continuity that eliminates sharp changes.

Up to this point, we have been collecting facts about numbers and functions that we need for analysis, but have not been spending much time on the *practice* of analysis, that is on making estimates. Analysis depends on making estimates and it is necessary to learn the craft of estimation in order to do analysis. We start explaining the process of making estimates in this chapter. One way to learn to make estimates is to first study and understand other people's arguments, then apply those same arguments to new problems.

8.1   Continuous Behavior and Linear Functions

A function behaves **continuously**, or is **continuous**, if the change in its values can be made small by making the change in the input small. To make this definition more precise, we consider the behavior of a linear polynomial.[1] The value of a constant polynomial doesn't change when we change the input, so the linear polynomial is the first interesting example to consider. Based on their graphs, linear polynomials certainly behave continuously.

Suppose the linear function is $y = mx + b$ and we set $y_1 = mx_1 + b$ and $y_2 = mx_2 + b$ to be the values at two rational numbers x_1 and x_2. The change in the input is $|x_2 - x_1|$ and the corresponding change in the output of the function is $|y_2 - y_1|$. We can compute a relation between these:

$$|y_2 - y_1| = |(mx_2 + b) - (mx_1 + b)| = |m||x_2 - x_1|. \tag{8.1}$$

In other words, the absolute value of the change in the function values is proportional to the absolute value of the change in the input values with constant of proportionality equal to $|m|$. In particular, this means that we can make the change in the output arbitrarily small by making the change in the input small, which certainly fits our intuition that a linear function varies continuously.

EXAMPLE 8.1.   Let $f(x) = 2x$ give the total number of miles for an "out and back" bicycle ride that is x miles one way. To increase a given ride by a total of 4 miles, we increase the one way distance x by $4/2 = 2$ miles while to increase a ride by a total of .01 miles, we increase the one way distance x by .005 miles.

In contrast, the step function (6.8) is not continuous, or is **discontinuous**, at 0. We recall the graph of $I(t)$ in Fig. 8.1. If we choose $t_1 < 0$ and

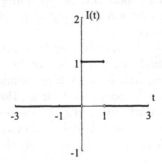

FIGURE 8.1. Plot of the step function $I(t)$.

$t_2 > 0$, then $|I(t_2) - I(t_1)| = 1$ regardless of the size of $|t_2 - t_1|$. Functions like the step function that make sudden changes for small changes in input

---

[1]By the way, we consider linear polynomials because we actually know everything about them. This won't be the last time that we base an investigation of complicated functions on the behavior of linear polynomials!

cause a lot of trouble. Think about what happens to electronic equipment when the power supply abruptly spikes during a thunderstorm.

Note that the slope m of the linear function $f(x) = mx + b$ determines how much the function values change as the input value x changes. The steeper the line, the more the function changes for a given change in input. We illustrate in Fig. 8.2.

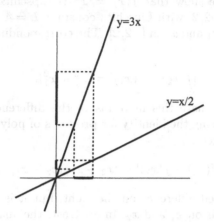

FIGURE 8.2. The outputs of these two lines change a different amount for a given change in input.

EXAMPLE 8.2. Suppose that $f_1(x) = 4x + 1$ while $f_2(x) = 100x - 5$. To increase the value of $f_1(x)$ at x by an amount of .01, we change the value of x by $.01/4 = .0025$. On the other hand, to change the value of $f_2(x)$ at x by an amount of .01, we change the value of x by $.01/100 = .0001$.

8.2 The Definition of Lipschitz Continuity

The idea is to extend the relation between the change in the output of a linear function and the change in the input to general functions. Let $f : I \to \mathbb{Q}$ be a function defined on a set I of rational numbers and taking on rational values $f(x)$ for each rational x.

EXAMPLE 8.3. A typical example of a set I is a rational interval, i.e. $\{x \text{ in } \mathbb{Q} : a \leq x \leq b\}$ for some rational numbers a and b.

In general, if x_1 and x_2 are two numbers in I, then $|x_2 - x_1|$ is the change in the input and $|f(x_2) - f(x_1)|$ is the corresponding change in the output. We say that f is **Lipschitz continuous** with **Lipschitz constant L** on I if there is a nonnegative constant L such that

$$|f(x_1) - f(x_2)| \leq L|x_1 - x_2| \tag{8.2}$$

for all x_1 and x_2 in I.

EXAMPLE 8.4. A linear function $f(x) = mx + b$ is Lipschitz continuous with Lipschitz constant $L = |m|$ on the entire set of rational numbers $\mathbb{Q}$.

EXAMPLE 8.5. We show that $f(x) = x^2$ is Lipschitz continuous on the interval $I = [-2, 2]$ with Lipschitz constant $L = 4$. We choose two rational numbers x_1 and x_2 in $[-2, 2]$. The corresponding change in the function values is

$$|f(x_2) - f(x_1)| = |x_2^2 - x_1^2|.$$

The goal is to estimate this in terms of the difference in the input values $|x_2 - x_1|$. Using the identity for products of polynomials derived in Chapter 6, we get

$$|f(x_2) - f(x_1)| = |x_2 + x_1|\,|x_2 - x_1|. \tag{8.3}$$

We have the desired difference on the right, but it is multiplied by a factor that depends on x_1 and x_2. In contrast, the analogous relationship (8.1) for the linear function has a factor that is constant, namely, $|m|$. At this point, we have to use the fact that x_1 and x_2 are in the interval $[-2, 2]$, which means that

$$|x_2 + x_1| \leq |x_2| + |x_1| \leq 2 + 2 = 4$$

by the triangle inequality. We conclude that

$$|f(x_2) - f(x_1)| \leq 4|x_2 - x_1|$$

for all x_1 and x_2 in $[-2, 2]$.

If there is no such Lipschitz constant, then the function may not behave continuously.

EXAMPLE 8.6. The step function $I(t)$ is *not* Lipschitz continuous on any interval that contains 0, for example, $[-.5, .5]$. If we choose $t_1 < 0$ and $t_2 > 0$ in $[-.5, .5]$, then $I(t_1) = 0$ and $I(t_2) = 1$ and there is no constant L such that

$$|I(t_2) - I(t_1)| = 1 \leq L|t_2 - t_1|$$

for all such t_1 and t_2 since given any value of L, we can make $L|t_2 - t_1|$ arbitrarily small by choosing t_2 close to t_1.

Lipschitz continuity quantifies the idea of continuous behavior using the Lipschitz constant L. If L is moderately sized, then small changes in input yield small changes in the function's output, but a large Lipschitz constant means that the function's values may make a large change when the input values change by a small amount.

However, it is important to note that there is a certain amount of inherent imprecision in the definition of Lipschitz continuity (8.2), and we have to be circumspect about drawing conclusions when a Lipschitz constant is large. The reason is that (8.2) is only an *upper estimate* on how much the function changes, and the actual change might be much smaller than indicated by a Lipschitz constant.[2]

EXAMPLE 8.7. Example 8.5 shows that $f(x) = x^2$ is Lipschitz continuous on $I = [-2, 2]$ with Lipschitz constant $L = 4$. It is also Lipschitz constant on I with Lipschitz constant $L = 121$ since

$$|f(x_2) - f(x_1)| \leq 4|x_2 - x_1| \leq 121|x_2 - x_1|.$$

But the second value of L greatly overestimates the change in f, whereas the value $L = 4$ is just about right when x is near 2, since $2^2 - 1.9^2 = .39 = 3.9 \times (2 - 1.9)$ and $3.9 \approx 4$.

To determine a Lipschitz constant, we have to make some estimates and the result can vary greatly depending on how difficult the estimates are to compute.

Note that if we change the interval, then we expect to get a different Lipschitz constant L.

EXAMPLE 8.8. We show that $f(x) = x^2$ is Lipschitz continuous on the interval $I = [2, 4]$, with Lipschitz constant $L = 8$. Starting with (8.3), for x_1 and x_2 in $[2, 4]$ we have

$$|x_2 + x_1| \leq |x_2| + |x_1| \leq 4 + 4 = 8$$

so

$$|f(x_2) - f(x_1)| \leq 8|x_2 - x_1|$$

for all x_1 and x_2 in $[2, 4]$.

The reason that the Lipschitz constant is bigger in the second example is clear from the graph (see Fig. 8.3), where we show the change in f corresponding to equal changes in x near $x = 2$ and $x = 4$. Because $f(x) = x^2$ is steeper near $x = 4$, f changes more near $x = 4$ for a given change in input.

EXAMPLE 8.9. $f(x) = x^2$ is Lipschitz continuous on $I = [-8, 8]$ with Lipschitz constant $L = 16$ and on $I = [-400, 200]$ with $L = 800$.

---

[2]That is why we avoid talking about a Lipschitz constant as if it is unique.

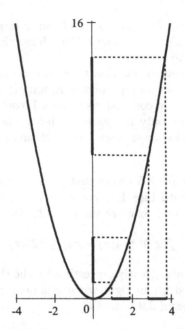

FIGURE 8.3. The change in $f(x) = x^2$ for equal changes in x near $x = 2$ and $x = 4$.

The definition of Lipschitz continuity is due to Lipschitz.[3] It is not the most general notion of continuity. However, the assumption of Lipschitz continuity is frequently encountered in mathematical modeling and analysis in science and engineering. Moreover, the assumption of Lipschitz continuity makes it possible to give constructive proofs of some fundamental results in analysis. So we use Lipschitz continuity until Chapter 32, where we explore other notions of continuity.

8.3   Bounded Sets of Numbers

In all of the examples involving $f(x) = x^2$, we use the fact that the interval under consideration is of finite size. A set of rational numbers I is **bounded** with size less than or equal to $b-a$ if I is contained in the finite interval $[a, b]$. We often try to take $[a, b]$ to be the smallest interval with this property, in

---

[3]Rudolf Otto Sigismund Lipschitz (1832–1903) worked in Germany, first as a high school teacher and later as a university professor. Lipschitz studied number theory, the theory of Bessel functions, Fourier series, differential equations, analytical mechanics, and potential theory. He used the Lipschitz condition to generalize a result of Cauchy on the existence of a solution of an ordinary differential equation.

which case we set $|I| = b - a$ to be the **size** of I.[4] Any finite interval $[a, b]$ is bounded with $|[a, b]| = b - a$.

EXAMPLE 8.10. The set of rational numbers $I = [-1, 500]$ is bounded with $|I| = 501$ but the set of even integers is not bounded.

While linear functions are Lipschitz continuous on the unbounded set $\mathbb{Q}$, functions that are not linear are usually only Lipschitz continuous on bounded sets.

EXAMPLE 8.11. The function $f(x) = x^2$ is *not* Lipschitz continuous on the set $\mathbb{Q}$ of rational numbers. This follows from (8.3) because $|x_1 + x_2|$ can be made arbitrarily large by choosing x_1 and x_2 freely in $\mathbb{Q}$, so it is not possible to find a constant L such that

$$|f(x_2) - f(x_1)| = |x_2 + x_1||x_2 - x_1| \leq L|x_2 - x_1|$$

for all x_1 and x_2 in $\mathbb{Q}$.

8.4 Monomials

Continuing the investigation of continuous functions, we next show that the monomials are Lipschitz continuous on bounded intervals, as we expect based on their graphs.

EXAMPLE 8.12. We show that the function $f(x) = x^4$ is Lipschitz continuous on $I = [-2, 2]$ with Lipschitz constant $L = 32$. We choose x_1 and x_2 in I and we estimate

$$|f(x_2) - f(x_1)| = |x_2^4 - x_1^4|$$

in terms of $|x_2 - x_1|$.

To do this, we first show that

$$x_2^4 - x_1^4 = (x_2 - x_1)(x_2^3 + x_2^2 x_1 + x_2 x_1^2 + x_1^3)$$

by multiplying out

$$(x_2 - x_1)(x_2^3 + x_2^2 x_1 + x_2 x_1^2 + x_1^3)$$
$$= x_2^4 + x_2^3 x_1 + x_2^2 x_1^2 + x_2 x_1^3 - x_2^3 x_1 - x_2^2 x_1^2 - x_2 x_1^3 - x_1^4$$

and then canceling the terms in the middle to get $x_2^4 - x_1^4$.

---

[4]The existence of such a smallest interval is an interesting question that we address in Chapter 32.

This means that

$$|f(x_2) - f(x_1)| = |x_2^3 + x_2^2 x_1 + x_2 x_1^2 + x_1^3| \, |x_2 - x_1|.$$

We have the desired difference $|x_2 - x_1|$ on the right and we just have to bound the factor $|x_2^3 + x_2^2 x_1 + x_2 x_1^2 + x_1^3|$. By the triangle inequality,

$$|x_2^3 + x_2^2 x_1 + x_2 x_1^2 + x_1^3| \leq |x_2|^3 + |x_2|^2 |x_1| + |x_2||x_1|^2 + |x_1|^3.$$

Now because x_1 and x_2 are in I, $|x_1| \leq 2$ and $|x_2| \leq 2$, so

$$|x_2^3 + x_2^2 x_1 + x_2 x_1^2 + x_1^3| \leq 2^3 + 2^2 2 + 2 2^2 + 2^3 = 32$$

and

$$|f(x_2) - f(x_1)| \leq 32|x_2 - x_1|.$$

Recall that a Lipschitz constant of $f(x) = x^2$ on I is $L = 4$. The fact that a Lipschitz constant of x^4 is larger than the constant for x^2 on $[-2, 2]$ is not surprising considering the plots of the two functions, see Fig. 6.3.

We can use the same technique to show that the function $f(x) = x^m$ is Lipschitz continuous where m is any natural number.[5]

EXAMPLE 8.13. The function $f(x) = x^m$ is Lipschitz continuous on any interval $I = [-a, a]$, where a is a positive rational number, with Lipschitz constant $L = ma^{m-1}$. Given x_1 and x_2 in I, we want to estimate

$$|f(x_2) - f(x_1)| = |x_2^m - x_1^m|$$

in terms of $|x_2 - x_1|$. We can do this using the fact that

$$x_2^m - x_1^m = (x_2 - x_1)(x_2^{m-1} + x_2^{m-2} x_1 + \cdots + x_2 x_1^{m-2} + x_1^{m-1})$$

$$= (x_2 - x_1) \sum_{i=0}^{m-1} x_2^{m-1-i} x_1^i.$$

We show this by first multiplying out

$$(x_2 - x_1) \sum_{i=0}^{m-1} x_2^{m-1-i} x_1^i = \sum_{i=0}^{m-1} x_2^{m-i} x_1^i - \sum_{i=0}^{m-1} x_2^{m-1-i} x_1^{i+1}.$$

To see that there is a lot of cancellation among the terms in the middle in the two sums on the right, we separate the first term out of the first

---

[5]This is the first really hard estimate we encounter. When reading a difficult piece of analysis, we have to avoid concentrating so hard on the details that we forget about the goal of the analysis and how we might get there.

sum and the last term in the second sum,

$$(x_2 - x_1) \sum_{i=0}^{m-1} x_2^{m-1-i} x_1^i$$

$$= x_2^m + \sum_{i=1}^{m-1} x_2^{m-i} x_1^i - \sum_{i=0}^{m-2} x_2^{m-1-i} x_1^{i+1} - x_1^m,$$

and then change the index in the second sum to get

$$(x_2 - x_1) \sum_{i=0}^{m-1} x_2^{m-1-i} x_1^i$$

$$= x_2^m + \sum_{i=1}^{m-1} x_2^{m-i} x_1^i - \sum_{i=1}^{m-1} x_2^{m-i} x_1^i - x_1^m = x_2^m - x_1^m.$$

This is tedious, but it is good practice to go through the details and make sure this argument is correct.

This means that

$$|f(x_2) - f(x_1)| = \left| \sum_{i=0}^{m-1} x_2^{m-1-i} x_1^i \right| |x_2 - x_1|.$$

We have the desired difference $|x_2 - x_1|$ on the right and we just have to bound the factor

$$\left| \sum_{i=0}^{m-1} x_2^{m-1-i} x_1^i \right|.$$

By the triangle inequality,

$$\left| \sum_{i=0}^{m-1} x_2^{m-1-i} x_1^i \right| \le \sum_{i=0}^{m-1} |x_2|^{m-1-i} |x_1|^i.$$

Now because x_1 and x_2 are in $[-a, a]$, $|x_1| \le a$ and $|x_2| \le a$. So

$$\left| \sum_{i=0}^{m-1} x_2^{m-1-i} x_1^i \right| \le \sum_{i=0}^{m-1} a^{m-1-i} a^i = \sum_{i=0}^{m-1} a^{m-1} = m a^{m-1}$$

and

$$|f(x_2) - f(x_1)| \le m a^{m-1} |x_2 - x_1|.$$

8.5   Linear Combinations of Functions

Now that we have seen that the monomials are Lipschitz continuous on a given interval, it is a short step to show that any polynomial is Lipschitz continuous on a given interval. But rather than just do this for polynomials, we show that a linear combination of Lipschitz continuous functions is Lipschitz continuous.

Suppose that f_1 is Lipschitz continuous with constant L_1 and f_2 is Lipschitz continuous with constant L_2 on the interval I.[6] Then $f_1 + f_2$ is Lipschitz continuous with constant $L_1 + L_2$ on I, because if we choose two points x and y in I, then

$$
\begin{aligned}
|(f_1 + f_2)(y) - (f_1 + f_2)(x)| &= |(f_1(y) - f_1(x)) + (f_2(y) - f_2(x))| \\
&\leq |f_1(y) - f_1(x)| + |f_2(y) - f_2(x)| \\
&\leq L_1|y - x| + L_2|y - x| \\
&= (L_1 + L_2)|y - x|
\end{aligned}
$$

by the triangle inequality. The same argument shows that $f_2 - f_1$ is Lipschitz continuous with constant $L_1 + L_2$ as well. It is even easier to show that if $f(x)$ is Lipschitz continuous on an interval I with Lipschitz constant L, then $cf(x)$ is Lipschitz continuous on I with Lipschitz constant $|c|L$.

From these two facts, it is a short step to extend the result to any linear combination of Lipschitz continuous functions. Suppose that $f_1, \cdots, f_n$ are Lipschitz continuous on I with Lipschitz constants $L_1, \cdots, L_n$, respectively. We use induction, so we begin by considering the linear combination of two functions. From the remarks above, it follows that $c_1 f_1 + c_2 f_2$ is Lipschitz continuous with constant $|c_1|L_1 + |c_2|L_2$. Next given $i \leq n$, we assume that $c_1 f_1 + \cdots + c_{i-1} f_{i-1}$ is Lipschitz continuous with constant $|c_1|L_1 + \cdots + |c_{i-1}|L_{i-1}$. To prove the result for i, we write

$$
c_1 f_1 + \cdots + c_i f_i = (c_1 f_1 + \cdots + c_{i-1} f_{i-1}) + c_n f_n.
$$

But the assumption on $(c_1 f_1 + \cdots + c_{i-1} f_{i-1})$ means that we have written $c_1 f_1 + \cdots + c_i f_i$ as the sum of two Lipschitz continuous functions, namely $(c_1 f_1 + \cdots + c_{i-1} f_{i-1})$ and $c_n f_n$. The result follows by the result for the linear combination of two functions. By induction, we have proved the following theorem:

Theorem 8.1 *Suppose that* $f_1, \cdots, f_n$ *are Lipschitz continuous on I with Lipschitz constants* $L_1, \cdots, L_n$ *respectively. The linear combination* $c_1 f_1 + \cdots + c_n f_n$ *is Lipschitz continuous on I with Lipschitz constant* $|c_1|L_1 + \cdots + |c_n|L_n$.

---

[6]Likely, f_1 and f_2 are Lipschitz continuous on intervals I_1 and I_2, respectively, and we take $I = I_1 \cap I_2$.

Corollary 8.2 *A polynomial is Lipschitz continuous on any bounded interval.*

EXAMPLE 8.14. We show that the function $f(x) = x^4 - 3x^2$ is Lipschitz continuous on $I = [-2, 2]$, with constant $L = 44$. For x_1 and x_2 in $[-2, 2]$, we have to estimate

$$|f(x_2) - f(x_1)| = |(x_2^4 - 3x_2^2) - (x_1^4 - 3x_1^2)|$$
$$= |(x_2^4 - x_1^4) - (3x_2^2 - 3x_1^2)|$$
$$\leq |x_2^4 - x_1^4| + 3|x_2^2 - x_1^2|.$$

Example 8.13 shows that x^4 is Lipschitz continuous on I with constant 32 while x^2 is Lipschitz continuous on $[-2, 2]$ with constant 4. Therefore

$$|f(x_2) - f(x_1)| \leq 32|x_2 - x_1| + 3 \times 4|x_2 - x_1| = 44|x_2 - x_1|.$$

8.6   Bounded Functions

Lipschitz continuity is related to another important property of a function called boundedness. A function f is **bounded** on a set of rational numbers I if there is a constant M such that

$$|f(x)| \leq M \text{ for all } x \text{ in } I.$$

Note that in every case, the arguments needed to verify the definition of Lipschitz continuity (8.2) involve showing that some function is bounded on a given interval.

EXAMPLE 8.15.   To show that $f(x) = x^2$ is Lipschitz continuous on $[-2, 2]$ in Example 8.5, we proved that $|x_1 + x_2| \leq 4$ for x_1 and x_2 in $[-2, 2]$.

It turns out that a function that is Lipschitz continuous on a bounded domain is automatically bounded on that domain. To be more precise, suppose that a function f is Lipschitz continuous with Lipschitz constant L on a bounded set I with size $|I|$ and choose a point y in I. Then for any other point x in I

$$|f(x) - f(y)| \leq L|x - y|.$$

Now, $|x - y| \leq |I|$. Also, since $|c + d| \leq |e|$ implies $|c| \leq |d| + |e|$ for any numbers c, d, e, we get

$$|f(x)| \leq |f(y)| + L|x - y| \leq |f(y)| + L|I|.$$

Even though we don't know $|f(y)|$, we do know that it is finite. This shows that $|f(x)|$ is bounded by the constant $M = |f(y)| + L|I|$ for any x in I, and we have proved:

Theorem 8.3 *A Lipschitz continuous function on a bounded set I is bounded on I.*

EXAMPLE 8.16. In Example 8.14, we showed that $f(x) = x^4 + 3x^2$ is Lipschitz continuous on $[-2, 2]$ with Lipschitz constant $L = 44$. Using this argument, we find that

$$|f(x)| \le |f(0)| + 44|x - 0| \le 0 + 44 \times 2 = 88$$

for any x in $[-2, 2]$. In fact, since x^4 is increasing for $0 \le x$, $|f(x)| \le |f(2)| = 16$ for any x in $[-2, 2]$. So the estimate on the size of $|f|$ using the Lipschitz constant 44 is not very accurate.

By the way, bounded functions are *not* necessarily Lipschitz continuous.

EXAMPLE 8.17. The step function in Model 6.8 is a bounded function that is not Lipschitz continuous on any set I that contains 0 and points near 0.

8.7   Products and Quotients of Functions

The next step in investigating which functions are Lipschitz continuous is to consider the product of two Lipschitz continuous functions on a bounded interval I. We show that the product is also Lipschitz continuous on I. More precisely, if f_1 is Lipschitz continuous with constant L_1 and f_2 is Lipschitz continuous with constant L_2 on a bounded interval I, then $f_1 f_2$ is Lipschitz continuous on I. We choose two points x and y in I and estimate by using the old trick of adding and subtracting the same quantity

$$
\begin{aligned}
|f_1(y)&f_2(y) - f_1(x)f_2(x)| \\
&= |f_1(y)f_2(y) - f_1(y)f_2(x) + f_1(y)f_2(x) - f_1(x)f_2(x)| \\
&\le |f_1(y)f_2(y) - f_1(y)f_2(x)| + |f_1(y)f_2(x) - f_1(x)f_2(x)| \\
&= |f_1(y)| \, |f_2(y) - f_2(x)| + |f_2(x)| \, |f_1(y) - f_1(x)|.
\end{aligned}
$$

Theorem 8.3, which says that Lipschitz continuous functions are bounded, implies there is some constant M such that $|f_1(y)| \le M$ and $|f_2(x)| \le M$ for $x, y \in I$. Using the Lipschitz continuity of f_1 and f_2 in I, we find

$$
\begin{aligned}
|f_1(y)f_2(y) - f_1(x)f_2(x)| &\le ML_1|y - x| + ML_2|y - x| \\
&= M(L_1 + L_2)|y - x|.
\end{aligned}
$$

We summarize as follows:

Theorem 8.4 *If f_1 and f_2 are Lipschitz continuous on a bounded interval I, then $f_1 f_2$ is Lipschitz continuous on I.*

EXAMPLE 8.18. The function $f(x) = (x^2+5)^{10}$ is Lipschitz continuous on the set $I = [-10, 10]$ because $x^2 + 5$ is Lipschitz continuous on I and therefore $(x^2 + 5)^{10} = (x^2 + 5)(x^2 + 5) \cdots (x^2 + 5)$ is as well by Theorem 8.4.

We can extend this result to unbounded intervals provided we know that f_1 and f_2 are bounded in addition to being Lipschitz continuous.

Continuing the investigation, we consider the quotient of two Lipschitz continuous functions. In this case, however, we require more information about the function in the denominator than just that it is Lipschitz continuous. We also have to know that it does not become too small. To understand this, we first consider an example.

EXAMPLE 8.19. We show that $f(x) = 1/x^2$ is Lipschitz continuous on the interval $I = [1/2, 2]$, with Lipschitz constant $L = 64$. We choose two points x_1 and x_2 in I and we estimate the change

$$|f(x_2) - f(x_1)| = \left| \frac{1}{x_2^2} - \frac{1}{x_1^2} \right|$$

by first doing some algebra

$$\frac{1}{x_2^2} - \frac{1}{x_1^2} = \frac{x_1^2}{x_1^2 x_2^2} - \frac{x_2^2}{x_1^2 x_2^2} = \frac{x_1^2 - x_2^2}{x_1^2 x_2^2} = \frac{(x_1 + x_2)(x_1 - x_2)}{x_1^2 x_2^2}.$$

This means that

$$|f(x_2) - f(x_1)| = \left| \frac{x_1 + x_2}{x_1^2 x_2^2} \right| |x_2 - x_1|.$$

Now we have the looked-for difference on the right, we just have to bound the factor in front. The numerator of the factor is the same as in Example 8.5 and so

$$|x_1 + x_2| \leq 4.$$

Also,

$$x_1 \geq \frac{1}{2} \text{ implies } \frac{1}{x_1} \leq 2 \text{ implies } \frac{1}{x_1^2} \leq 4$$

and likewise $\frac{1}{x_2^2} \leq 4$. So we get

$$|f(x_2) - f(x_1)| \leq 4 \times 4 \times 4 \, |x_2 - x_1| = 64|x_2 - x_1|.$$

Note, we use the fact that the left-hand endpoint of the interval I is $1/2$. The closer the left-hand endpoint is to zero, the larger the Lipschitz constant has to be. In fact, $1/x^2$ is *not* Lipschitz continuous on $(0, 2]$.

We mimic this example in the general case f_1/f_2 by assuming that the denominator f_2 is **bounded below** by a positive constant. We give the proof of the following theorem as Problem 8.16.

Theorem 8.5 *Assume that f_1 and f_2 are Lipschitz continuous on a bounded set I with constants L_1 and L_2 and moreover assume there is a constant $m > 0$ such that $|f_2(x)| \geq m$ for all x in I. Then f_1/f_2 is Lipschitz continuous on I.*

EXAMPLE 8.20. The function $1/x^2$ does not satisfy the assumptions of Theorem 8.5 on the interval $(0, 2]$ and it is not Lipschitz continuous on that interval.

8.8   The Composition of Functions

We conclude the investigation into Lipschitz continuity by considering the composition of Lipschitz continuous functions. This is actually easier than either products or ratios of functions. The only complication is that we have to be careful about the domains and ranges of the functions. Consider the composition $f_2(f_1(x))$. Presumably, we have to restrict x to an interval on which f_1 is Lipschitz continuous and we also have to make sure that the values of f_1 are in a set on which f_2 is Lipschitz continuous.

So we assume that f_1 is Lipschitz continuous on I_1 with constant L_1 and that f_2 is Lipschitz continuous on I_2 with constant L_2. If x and y are points in I_1, then as long as $f_1(x)$ and $f_1(y)$ are in I_2, we have

$$|f_2(f_1(y)) - f_2(f_1(x))| \leq L_2|f_1(y) - f_1(x)| \leq L_1 L_2|y - x|.$$

We summarize as a theorem.

Theorem 8.6 *Let f_1 be Lipschitz continuous on a set I_1 with Lipschitz constant L_1 and f_2 be Lipschitz continuous on I_2 with Lipschitz constant L_2 such that $f_1(I_1) \subset I_2$. Then the composite function $f_2 \circ f_1$ is Lipschitz continuous on I_1 with Lipschitz constant $L_1 L_2$.*

EXAMPLE 8.21. The function $f(x) = (2x - 1)^4$ is Lipschitz continuous on any bounded interval since $f_1(x) = 2x - 1$ and $f_2(x) = x^4$ are Lipschitz continuous on any bounded interval. If we consider the interval $[-.5, 1.5]$, then $f_1(I) \subset [-2, 2]$. From Example 8.12, we know that x^4 is Lipschitz continuous on $[-2, 2]$ with Lipschitz constant 32, while a Lipschitz constant of $2x - 1$ is 2. Therefore, f is Lipschitz continuous on $[-.5, 1.5]$ with constant 64.

EXAMPLE 8.22. The function $1/(x^2 - 4)$ is Lipschitz continuous on any closed interval that does not contain either 2 or -2. This follows because $f_1(x) = x^2 - 4$ is Lipschitz continuous on any bounded interval while $f_2(x) = 1/x$ is Lipschitz continuous on any closed interval that does not contain 0. To avoid zero, we must avoid $x^2 = 4$ or $x = \pm 2$.

Chapter 8   Problems

8.1. Verify the claims in Example 8.9.

In Problems 8.2–8.7, verify the definition of Lipschitz continuity to show the indicated functions are Lipschitz continuous.

8.2. Show that $f(x) = x^2$ is Lipschitz continuous on $[10, 13]$.

8.3. Show that $f(x) = 4x - 2x^2$ is Lipschitz continuous on $[-2, 2]$.

8.4. Show that $f(x) = x^3$ is Lipschitz continuous on $[-2, 2]$.

8.5. Show that $f(x) = |x|$ is Lipschitz continuous on $\mathbb{Q}$.

8.6. Show that $f(x) = 1/x^2$ is Lipschitz continuous on $[1, 2]$.

8.7. Show that $f(x) = 1/(x^2 + 1)$ is Lipschitz continuous on $[-2, 2]$.

8.8. In Example 8.12, we show that x^4 is Lipschitz continuous on $[-2, 2]$ with Lipschitz constant $L = 32$. Explain why this is a reasonable value for a Lipschitz constant.

Problems 8.9–8.12 deal with functions for which verifying the Lipschitz continuity condition is either problematic or impossible.

8.9. Compute a Lipschitz constant of $f(x) = 1/x$ on the intervals (a) $[.1, 1]$, (b) $[.01, 1]$, and $[.001, 1]$.

8.10. Explain why $f(x) = 1/x$ is not Lipschitz continuous on $(0, 1]$.

8.11. (a) Explain why the function

$$f(x) = \begin{cases} 1, & x < 0, \\ x^2, & x \geq 0, \end{cases}$$

is *not* Lipschitz continuous on $[-1, 1]$. (b) Is f Lipschitz continuous on $[1, 4]$?

8.12. Suppose a Lipschitz constant L of a function f is equal to $L = 10^{100}$. Discuss the continuity properties of $f(x)$ and in particular decide if f continuous from a practical point of view.

8.13. Assume that f_1 is Lipschitz continuous with constant L_1 and f_2 is Lipschitz continuous with constant L_2 on a set I, and let c be a number. Show that $f_2 - f_1$ is Lipschitz continuous with constant $L_1 + L_2$ on I and cf_1 is Lipschitz continuous with constant $|c|L_1$ on I.

8.14. Show that a Lipschitz constant of a polynomial $f(x) = \sum_{i=0}^{n} a_i x^i$ on the interval $[-c, c]$ is

$$L = \sum_{i=1}^{n} |a_i| i c^{i-1} = |a_1| + 2c|a_2| + \cdots + nc^{n-1}|a_n|.$$

8.15. Explain why $f(x) = 1/x$ is not bounded on $\{x \text{ in } [-1, 1], \ x \neq 0\}$.

8.16. Prove Theorem 8.5.

8.17. Use the theorems in this chapter to show that the following functions are Lipschitz continuous on the given intervals and compute a Lipschitz constant, or prove they are not Lipschitz continuous:

(a) $f(x) = 2x^4 - 16x^2 + 5x$ on $[-2, 2]$ (b) $\dfrac{1}{x^2 - 1}$ on $\left[-\dfrac{1}{2}, \dfrac{1}{2}\right]$

(c) $\dfrac{1}{x^2 - 2x - 3}$ on $[2, 3)$ (d) $\left(1 + \dfrac{1}{x}\right)^4$ on $[1, 2]$.

8.18. Show the function

$$f(x) = \frac{1}{c_1 x + c_2(1 - x)},$$

where $c_1 > 0$ and $c_2 > 0$, is Lipschitz continuous on $[0, 1]$.

9
Sequences and Limits

With the tools to create complicated functions in hand, we can now model more complicated situations. However, as with the simple Dinner Soup model, solving models almost invariably leads to a lot of work. In fact, usually we cannot solve for the solution of a model in terms of a value that can be written down concretely like an integer. The best we can do in general is approximate a solution to increasing accuracy by increasing amounts of work. This trade-off can be quantified in the mathematical terms of the limit.

The infinite decimal expansion of rational numbers discussed in Chapter 4 is a particular example of a limit. The limit is the fundamental concept of analysis.[1]

9.1 The First Encounter with Sequences and Limits

We begin with the infinite decimal expansion of 10/9, which by (4.6) can be written

$$\frac{10}{9} = 1.11\cdots 11_n + \frac{1}{9}10^{-n}.$$

---

[1]The limit also has the dubious honor of being one of the more confusing topics in analysis and the history of analysis has been a struggle to come to grips with certain evasive aspects of its definition.

Rewriting this equation, we get an **estimate** on the difference between $1.111\cdots 11_n$ and $10/9$,

$$\left|\frac{10}{9} - 1.11\cdots 11_n\right| \le 10^{-n}. \tag{9.1}$$

If we consider $1.11\cdots 11_n$ as an approximation of $10/9$, then (9.1) means that the **error** $|10/9 - 1.11\cdots 11_n|$ can be made arbitrarily small by taking n large. If we want the error to be smaller than 10^{-9}, then we simply choose $n \ge 10$. Computing $1.11\cdots 1_n$ requires more work as n increases, for which we gain increased accuracy. Trading work for accuracy is the idea behind limits.[2]

The concept of the limit applies to the set, or **sequence**, of successive approximations $\{1.1,\ 1.11,\ 1.111,\ \cdots,\ 1.1\cdots 1_n,\ \cdots\}$. The name sequence suggests that set is to be considered as being ordered from left to right. In general, a sequence of numbers is a neverending ordered list of numbers $\{a_1,\ a_2,\ a_3,\ \cdots,\ a_n,\ \cdots\}$ called elements, where the index notation is used to distinguish the elements. We also write

$$\{a_1, a_2, a_3, \cdots, a_n, \cdots\} = \{a_n\}_{n=1}^{\infty}.$$

The symbol ∞ indicates that the list continues forever in the same sense that the natural numbers $1, 2, 3, ...$, continue forever.

EXAMPLE 9.1. The sequence of even natural numbers can be written

$$\{2, 4, 6, \cdots\} = \{2n\}_{n=1}^{\infty}$$

and the odd natural numbers as

$$\{1, 3, 5, 7, \cdots\} = \{2n - 1\}_{n=1}^{\infty}$$

and some other sequences:

$$\left\{1, \frac{1}{2^2}, \frac{1}{3^2}, \frac{1}{4^2}, \cdots\right\} = \left\{\frac{1}{n^2}\right\}_{n=1}^{\infty}$$

$$\left\{\frac{1}{3^2}, \frac{1}{4^2}, \cdots\right\} = \left\{\frac{1}{n^2}\right\}_{n=3}^{\infty}$$

$$\{1, 2, 4, 8, \cdots\} = \{2^i\}_{i=0}^{\infty}$$

$$\{-1, 1, -1, 1, -1 \cdots\} = \{(-1)^j\}_{j=1}^{\infty}$$

$$\{1, 1, 1, \cdots\} = \{1\}_{k=1}^{\infty}.$$

---

[2]An estimate like (9.1) gives a quantitative measurement of how much accuracy is gained for each increase in work, and so such estimates are useful not only to mathematicians but to engineers and scientists.

Note the index of a sequence is a **dummy variable** that can be called anything we like.

EXAMPLE 9.2.

$$\{n + n^2\}_{n=1}^{\infty} = \{j + j^2\}_{j=1}^{\infty} = \{\text{Frodo} + (\text{Frodo})^2\}_{\text{Frodo}=1}^{\infty}.$$

We can also change its value so the sequence begins with a different number by reformulating the coefficients.

EXAMPLE 9.3.

$$\{1 + 1^2, 2 + 2^2, 3 + 3^2, \cdots\} = \{n + n^2\}_{n=1}^{\infty} = \{(j - 2) + (j - 2)^2\}_{j=3}^{\infty}.$$

See Problems 9.1–9.3.

The sequence $\{1.11 \cdots 11_n\}_{n=1}^{\infty}$ has the property that each number in the sequence is a better approximation to $10/9$ than the preceding number and as we move from left to right the numbers approach $10/9$ in value. In contrast, a single element, even one with many digits like 1.1111111111111, has a fixed accuracy. We say that the sequence $\{1.11 \cdots 11_n\}_{n=1}^{\infty}$ **converges** to $10/9$ and that $10/9$ is the **limit** of the sequence $\{1.11 \cdots 11_n\}_{n=1}^{\infty}$ because the difference between $10/9$ and $1.11 \cdots 11_n$ can be made arbitrarily small by taking the index large.

9.2 The Mathematical Definition of a Limit

The estimate (9.1) implies that $10/9$ can be approximated to any specified accuracy by taking terms in the sequence with sufficiently large index. This is the observation that we use to define the convergence of a general sequence. We explain the logic of the definition with an example.

EXAMPLE 9.4. To tighten or loosen a hex bolt with head diameter $2/3$, a mechanic needs to use a socket wrench of a slightly bigger size. The tolerance on the difference between the sizes of the bolt and the wrench depend on the tightness, the material of the bolt and the wrench, and conditions such as whether the bolt threads are lubricated and whether the bolt is rusty or not. If the wrench is too large, then the head of the bolt is in danger of being stripped before the bolt can be tightened or loosened. Two wrenches with different tolerances are shown in Fig. 9.1.

Now suppose that we have an infinite set of wrenches of sizes .7, .67, .667, $\cdots$ that can be represented as a sequence $\{.66 \cdots 667_n\}_{n=1}^{\infty}$. In every case the wrenches are bigger than $2/3$, but not by much. In fact, we ask you to show that

$$\left| .66 \cdots 667_n - \frac{2}{3} \right| < 10^{-n} \qquad (9.2)$$

FIGURE 9.1. Two socket wrenches with different tolerances.

in Problem 9.5. Naturally, (9.2) suggests that $\{.6\cdots 67_n\}_{n=1}^{\infty}$ converges to the limit 2/3.

What does this mean in practical terms for the mechanic? Given any specified tolerance on the size, she can reach into the tool chest and pull out a wrench that meets the criteria. The accuracy tolerance is not up to the mechanic, it is specified by some second party like a bicycle manufacturer. Rather, the mechanic has to meet any specified accuracy to avoid voiding a warranty. The cost of being able to meet any specified tolerance is having to stock an expensive set of wrenches.

In general, we say that a sequence $\{a_n\}_{n=1}^{\infty}$ converges to a limit A if it is possible to make the terms a_n arbitrarily close to A by taking the index n sufficiently large. In other words, the difference $|a_n - A|$ can be made as small as desired by taking n large. When this is true, we write

$$\lim_{n\to\infty} a_n = A.$$

It is convenient to translate this definition into mathematical notation. From the statement, we can guess that there are two quantities involved: a bound on the size of $|a_n - A|$ and a corresponding number that indicates how large the index has to be to achieve the bound.

In the two examples considered so far, the relation between the size of $|a_n - A|$ and n is given by (9.1) and (9.2), both of which guarantee that a_n agrees with A to at least $n-1$ decimal places. But in general we cannot expect to gain a whole digit of accuracy each time n is increased by 1. So the mathematical statement of the convergence has to be more flexible. Therefore, we specify the size of $|a_n - A|$ by using a general variable ϵ rather than specifying a number of digits.[3] The bound on $|a_n - A|$ should

---

[3]Traditionally, ϵ and δ are used to denote small quantities, though not interchangeably. Thinking of δ as "difference" and ϵ as "error" may help clarify their usage.

be satisfied for all sufficiently large n. In other words, given ϵ, there should be a number N such that the bound is satisfied for all n larger than N.[4]

Putting this together, the mathematical statement that a sequence converges to a limit reads

$$\lim_{n \to \infty} a_n = A$$

if for any $\epsilon > 0$ there is a number $N > 0$ such that

$$|a_n - A| \le \epsilon \text{ for all } n \ge N.$$

We emphasize that the value of N depends on ϵ, and in particular, we expect that decreasing ϵ means that N increases.

EXAMPLE 9.5. We verify the definition for $\{1.11 \cdots 11_n\}_{n=1}^{\infty}$. We want to show[5] that given $\epsilon > 0$ there is a $N > 0$ such that

$$\left| 1.11 \cdots 11_n - \frac{10}{9} \right| \le \epsilon$$

for all $n \ge N$. Suppose that ϵ has the decimal expansion

$$\epsilon = 0.000 \cdots 00 p_m p_{m+1} \cdots$$

where the first digit of ϵ that is nonzero is p_m in the mth place. By (9.1), if we choose $n \ge m$, then

$$\left| 1.11 \cdots 11_n - \frac{10}{9} \right| \le 10^{-n} = .00 \cdots 001 \le \epsilon.$$

Therefore given any $\epsilon > 0$, if we choose $N = m$, where the first nonzero digit of ϵ is in the mth place, then $|1.11 \cdots 11_n - 10/9| < \epsilon$ for $n \ge N$.

We next present a couple of less familiar examples.

EXAMPLE 9.6. We show that $\{1/n\}_{n=1}^{\infty}$ converges to 0, i.e.,

$$\lim_{n \to \infty} \frac{1}{n} = 0.$$

This is intuitively obvious since $1/n$ can be made as close to 0 as desired by taking n large. It is also visible in a plot of the elements of the sequence (see Fig. 5.5). But to satisfy the annoying mathematician who specifies an $\epsilon > 0$, we show there is an $N > 0$ such that

$$\left| \frac{1}{n} - 0 \right| \le \epsilon$$

---

[4]What happens if the bound is satisfied for only some of the n larger than N? We address this question in Chapter 32.

[5]It is often helpful to begin these problems by writing out the definition.

for all $n \geq N$. In this case, it is not too hard to determine N since $n \geq 1/\epsilon$ implies that $1/n \leq \epsilon$. Therefore given any $\epsilon > 0$, if $N = 1/\epsilon$, then

$$\left| \frac{1}{n} - 0 \right| = \frac{1}{n} \leq \epsilon \qquad (9.3)$$

for $n \geq N$. Note that if ϵ decreases, then N naturally increases.

In Problem 9.8, we ask you to show that

$$\lim_{n \to \infty} \frac{1}{n^p} = 0$$

where p is any natural number.

EXAMPLE 9.7. We show that

$$\lim_{n \to \infty} \frac{1}{2^n} = 0.$$

This limit is suggested by considering a plot of the elements of the sequence (see Fig. 5.5). By the definition, given $\epsilon > 0$, we show that there is an N such that

$$\left| \frac{1}{2^n} - 0 \right| = \frac{1}{2^n} \leq \epsilon$$

for $n \geq N$. Certainly

$$2^4 = 16 \geq 10$$

so given any natural number m

$$\frac{1}{2^{4m}} \leq \frac{1}{10^m}.$$

So if ϵ has the decimal expansion $\epsilon = 0.000 \cdots 00 p_m p_{m+1} \cdots$, where the first digit of ϵ that is nonzero is p_m in the mth place, then

$$\frac{1}{2^{4m}} \leq \frac{1}{10^m} \leq \epsilon.$$

Therefore, given any $\epsilon > 0$, if $N = 4m$, where the first nonzero digit of ϵ is in the mth place, then $|1/2^n - 0| < \epsilon$ for $n \geq N$.

In general, it is possible to show that

$$\lim_{n \to \infty} r^n = 0$$

when $|r| < 1$ by using a proof similar to the case $r = 1/2$. We ask you to do the case $|r| < 1/2$ in Problem 9.9. We can show the general result easily after introducing the logarithm in Chapter 28.

We continue with a more complicated example.

EXAMPLE 9.8. We show that the limit of the sequence $\{\frac{n}{n+1}\}_{n=1}^{\infty} =$ $\{\frac{1}{2}, \frac{2}{3}, \cdots\}$ equals 1, that is

$$\lim_{n \to \infty} \frac{n}{n+1} = 1. \tag{9.4}$$

This is suggested by a plot of the elements (see Fig. 9.2). We begin by

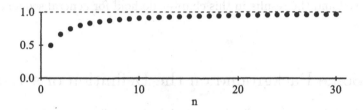

FIGURE 9.2. The elements of the sequence $\{n/(n+1)\}$.

simplifying the difference

$$\left|1 - \frac{n}{n+1}\right| = \left|\frac{n+1-n}{n+1}\right| = \frac{1}{n+1}.$$

Note that this intuitively shows that $\frac{n}{n+1}$ can be made arbitrarily close to 1 by taking n large and so 1 is the limit. To verify the definition, suppose $\epsilon > 0$ is given. By the equality above,

$$\left|1 - \frac{n}{n+1}\right| = \frac{1}{n+1} \le \epsilon$$

provided that $n \ge 1/\epsilon - 1$. Given $\epsilon > 0$, if $N \ge 1/\epsilon - 1$, then

$$\left|1 - \frac{n}{n+1}\right| \le \epsilon$$

for $n \ge N$.

We conclude with an important observation. The examples of convergent sequences presented so far have the property that the elements of the sequence are rational and the limit of the sequence is a rational number. This is important because up to now arithmetic is only defined for rational numbers, and subtraction is used in the definition of the limit. In other words, *the definition of a limit presented so far only makes sense if the sequence consists of rational numbers and the limit of the sequence is a rational number.*

This raises the question: does a convergent sequence of rational numbers always converge to a rational number? The short answer is no, and therein lies a mystery that confounded generations of mathematicians attempting to establish a rigorous foundation for analysis. We explain how this can be in Chapter 10. *In this chapter, we simply assume that all sequences consist of rational numbers, all convergent sequences converge to a rational limit, and the domains and ranges of all functions are contained in* $\mathbb{Q}$. Technically, this means that the discussion in this chapter does *not* apply to all rational sequences that converge. In Chapter 11, we show that this assumption can be removed and the results in this chapter do hold for general sequences of numbers.

9.3   Some Background on the Definition of a Limit

Leibniz and Newton[6] implicitly used the idea of limits in their versions of calculus. Newton in particular formulated a definition in words that is close to the modern definition. But they did not express the idea of a limit in quantitative terms, and this opened up the early versions of calculus to criticism about its rigor. As a consequence, there was increasing recognition of the need for a more precise definition of a limit among mathematicians following Leibniz and Newton. Cauchy[7] was the first person to write down the modern definition of the limit. We owe the notation "lim" to Weier-

---

[6]The English mathematician Sir Isaac Newton (1643–1727) is credited jointly with the discovery of calculus together with Leibniz. The scope of Newton's achievements in mathematics and physics has likely never been surpassed by any other individual. Remarkably, he made some of his most important scientific discoveries, including the composition of white light, calculus, and his law of gravitation, while waiting out the Great Plague of 1644–45 at home. A few years later, Newton became the Lucasian Professor at Cambridge University; a position he held for eighteen productive years. Though active in scientific politics, Newton tended to publish his results many years after they were derived, and only then at the urging of his colleagues. Newton was always prone to depression and nervous irritability, and the strain of publishing his renowned *Principia* caused him to turn away from research in physics and mathematics for the last forty years of his life. During this period, Newton worked successfully as Master of the Mint and acquired a fortune. He also worked in theology and alchemy, producing little that is remembered.

[7]Augustin Louis Cauchy (1789–1857) was born and worked in France. Cauchy was incredibly productive, publishing nearly as many papers as Euler (789) while producing fundamentally important results in nearly all areas of mathematics including real and complex analysis, ordinary and partial differential equations, matrix theory, Fourier theory, elasticity, and the theory of light. Cauchy worried about the foundations of analysis and gave the first rigorous proofs of some well-known results in calculus, including the first general existence result for ordinary differential equations and the Mean Value Theorem. Cauchy wrote down the first "ϵ-δ proof" in proving the Mean Value Theorem. Cauchy was a principled person in political terms and his career path took up and down swings as the political winds changed.

strass,[8] who was also a key figure in clarifying the mathematical meaning of the limit.

9.4   Divergent Sequences

For the sake of comparison, it is a good idea to examine some sequences that do not converge, or **diverge**. There are lots of ways for a sequence to diverge; for example, consider the divergent sequences

$$\{-6, 2, -.4, -.7, 5, 6.1, 2, 9.9, -3, .2, 1, 7, 28, .3, -5.4, \cdots\}, \tag{9.5}$$
$$\{(-1)^n\}_{n=1}^{\infty} = \{-1, 1, -1, 1, \cdots\}, \tag{9.6}$$
$$\{(-n)^n\}_{n=1}^{\infty} = \{-1, 4, -27, \cdots\}, \tag{9.7}$$
$$\{n^2\}_{n=1}^{\infty} = \{1, 4, 9, 16, \cdots\}. \tag{9.8}$$

In each case, there is no single number that the terms in the sequence approach as n increases. In general, we expect that if the elements of a sequence are written down at "random" it is likely that the sequence diverges. The sequences that converge are special.

Also in general, there is relatively little that can be said about a divergent sequence. However, we do distinguish one special case of divergence. No pattern in the elements of (9.5) can be seen, while the elements of (9.6) oscillate between two values, but never get close to one value. The elements of (9.7) also oscillate, but now become bigger as the index increases. Whereas the elements of (9.8) simply become bigger as the index increases, which is predictable behavior. We distinguish this case by saying that a sequence **diverges to infinity**, and write

$$\lim_{n \to \infty} a_n = \infty,$$

whenever the terms grow without bound as the index increases. Mathematically, we say that a sequence diverges to infinity if given any positive M there is a natural number N such that $a_n \geq M$ for $n \geq N$.

---

[8]Karl Theodor Wilhelm Weierstrass (1815–1897) began his career as a high school teacher before bursting on the mathematical scene with a few papers and acquiring a professorship at the University of Berlin. Weierstrass made fundamental contributions to bilinear forms, infinite series and products, the theory of functions, the calculus of variations, and the foundations of real analysis. While Weierstrass published relatively few papers, he was a tremendous lecturer, and his seminars and courses had great impact on mathematics. Much of the modern concern with complete rigor in analysis originated with Weierstrass. Weierstrass introduced the notation lim, | |, and the ϵ-δ definitions of continuity and limit of a function. Weierstrass was also the first mathematician to sponsor a woman for a doctorate degree. This was the talented Russian mathematician Sofia Vasilyevna Kovalevskaya (1850–1891).

EXAMPLE 9.9. We show that $\lim_{n\to\infty} n^2 = \infty$ by verifying the definition. For $n \geq 1$, $n^2 \geq n$. Hence, given any M

$$n^2 \geq n \geq M,$$

provided $n \geq M$. This means we take $N = M$.

Similarly, we say that the sequence diverges to minus infinity, and write

$$\lim_{n\to\infty} a_n = -\infty,$$

if given any negative M we can find a natural number N such that $a_n \leq M$ for $n \geq N$.

9.5   Infinite Series

An important example of sequences is provided by series, or infinite sums. We begin by recalling the geometric sum discussed in Chapter 3.

EXAMPLE 9.10. Recall that the sum

$$1 + r + r^2 + \cdots + r^n = \sum_{i=0}^{n} r^i$$

is called the **geometric sum** of order n with factor r. Both to understand infinite decimal expansions and the Verhulst population model in Section 4.4, we determined the value of this sum as we take more and more terms. If we let

$$s_n = \sum_{i=0}^{n} r^i,$$

then this is the same thing as studying the convergence of the sequence $\{s_n\}_{n=0}^{\infty}$.

To study the convergence, we use the formula

$$s_n = \sum_{i=0}^{n} r^i = \frac{1 - r^{n+1}}{1 - r}$$

that was proved using induction for any $r \neq 1$. If $|r| < 1$, then as n increases r^{n+1} approaches zero in value. Hence it is reasonable to guess that

$$\lim_{n\to\infty} s_n = \frac{1}{1 - r}$$

for $|r| < 1$. This is also suggested by plotting s_n, as in Fig. 9.3.

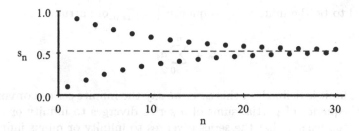

FIGURE 9.3. The elements of the partial sum $\{s_n\}$ of the geometric series with $r = -.9$.

We verify this is true using the definition of the limit. Given $\epsilon > 0$, we show there is an N such that

$$\left| \frac{1 - r^{n+1}}{1 - r} - \frac{1}{1 - r} \right| = \left| \frac{r^{n+1}}{1 - r} \right| \leq \epsilon$$

for all $n \geq N$. Equivalently, given $\epsilon > 0$ we show there is an $N > 0$ such that

$$|r|^{n+1} \leq \epsilon |1 - r| \tag{9.9}$$

for $|r| < 1$. But $\epsilon |1 - r|$ is a fixed number once ϵ is specified, while $|r|^{N+1}$ can be made as small as desired by taking N sufficiently large when $|r| < 1$. Certainly (9.9) holds for any $n \geq N$ once N is determined. This verifies the definition of convergence, albeit rather vaguely. After we introduce the logarithm in Chapter 28, an exact relationship between N and ϵ is easy to determine.

Based on this result, we call the limit of $\{s_n\}_{n=0}^{\infty}$ the **geometric series** and write

$$\lim_{n \to \infty} s_n = \sum_{i=0}^{\infty} r^i = 1 + r + r^2 + \cdots .$$

Since this limit is defined for $|r| < 1$, we say the geometric series converges for $|r| < 1$ and write

$$1 + r + r^2 + \cdots = \frac{1}{1 - r}. \tag{9.10}$$

The sequence $\{s_n\}_{n=0}^{\infty}$ is called the sequence of **partial sums** of the series.

In general, the **infinite series**

$$\sum_{i=0}^{\infty} a_i = a_0 + a_1 + \cdots$$

is defined to be the limit of the sequence $\{s_n\}_{n=0}^{\infty}$ of partial sums

$$s_n = \sum_{i=0}^{n} a_i$$

when the limit is defined. In this case, we say the infinite series **converges**.

If the sequence of partial sums of a series diverges to infinity or minus infinity, then we say that the series diverges to infinity or minus infinity.

EXAMPLE 9.11. The series $\sum_{i=1}^{\infty} i = 1+2+3+\cdots$ diverges to infinity. This follows because the partial sum $s_n = \sum_{i=1}^{n} i$ satisfies $s_n \geq n$ for all n. Therefore the partial sums increase without bound as the index increases.

In this book, infinite series are given much less space than is usual for calculus and analysis texts. Historically, infinite series were crucial to the development of analysis. Indeed, much of early analysis, including differentiation formulas, integration, and the existence of solutions of differential equations, was justified using the properties of infinite series, and many important analysts worked on the properties of infinite series. Nevertheless, the role of infinite series in real analysis, which is analysis in the space of real numbers, has diminished greatly in this century.[9]

There is a fundamental difference between the smoothness of functions encountered in real analysis and those encountered in complex analysis, which is analysis in the space of complex numbers. The so-called analytic functions that lie at the heart of complex analysis are very smooth and for this reason are closely connected to infinite series.[10] For this reason, infinite series remain a central topic of complex analysis. We believe the natural place to learn about infinite series is in a course in complex analysis, and point to Ahlfors [1], for example.

9.6   Limits Are Unique

In the next few sections, we work out some useful properties of convergent sequences. The sequences studied in this book are usually constructed as approximations to a quantity that we want to compute, e.g., $\{1.11\cdots 1_n\}_{n=1}^{\infty}$ is a sequence of approximations to $10/9$. The properties developed in this chapter make it possible to combine approximations of different quantities to form an approximation of a new quantity.

---

[9]This is one reason that the chapter on infinite series in a standard calculus book is one of the least popular and least motivated topics from the students' point of view.

[10]The connection between smoothness of functions and infinite series is discussed further in Chapter 37.

We begin with the observation that the limit of a convergent sequence is unique. It certainly makes sense that it is impossible for the terms in a sequence to become arbitrarily close to two different numbers at the same time. Suppose in fact that a sequence $\{a_n\}_{n=1}^{\infty}$ converges to two numbers A and B. To show that A and B are equal, we show that the distance $|A-B|$ is zero. To estimate this difference, we use a variation of the triangle inequality (2.4) that reads

$$|a - b| \leq |a - c| + |c - b| \quad \text{for all } a, b, c. \tag{9.11}$$

We ask you to prove this in Problem 9.14. Using (9.11) with $a = A$, $b = B$, and $c = a_n$, we get

$$|A - B| \leq |a_n - A| + |a_n - B|$$

for any n. Now because a_n converges to A, both $|a_n - A|$ and $|a_n - B|$ can be made smaller than $|A - B|/4$ by taking n sufficiently large. But this means that $|A - B| \leq |A - B|/2$, which can only hold if $|A - B| = 0$.

Theorem 9.1 *A sequence can have at most one limit.*

On this topic, there is a minor hiccup regarding the uniqueness of infinite decimal expansions. For example, it is straightforward to show (Problem 9.15) that $\lim_{n \to \infty} 0.99 \cdots 99_n = 1$. This means that 1 has two decimal representations, namely, $1.000 \cdots = 0.9999 \cdots$. So we have to decide what we mean when we write $a = b$, where a and b are not written in the same way.

A standard approach is to interpret $a = b$ as meaning that we can show that $|a - b|$ is smaller than any positive number. This is equivalent to writing $a = 0$ if we can show that $|a|$ is smaller than any positive number. Correspondingly, if $|a|$ is bigger than some positive number, then we write $a \neq 0$. With this definition, we can write $.999 \cdots = 1$ without trouble.[11] Another approach is to simply avoid decimal representations ending in repeated digits of 9 by replacing any such expansion by the equivalent expansion ending with all 0 digits. Hence when $.999 \cdots$ occurs, it is replaced by $1.000 \cdots = 1$.

9.7 Arithmetic with Sequences

It turns out that if we perform arithmetic on sequences that converge, we end up with another convergent sequence. For example, suppose that

---

[11]However, this interpretation could trouble a constructivist because verifying that $a = b$ for arbitrary a and b nominally requires showing $|a - b|$ is smaller than an infinite number of positive numbers. Verifying that $|a - b|$ is smaller than any finite number of positive numbers can not settle the issue. This is discussed in more detail in Chapter 11.

$\{a_n\}_{n=1}^\infty$ converges to A and $\{b_n\}_{n=1}^\infty$ converges to B. Then $\{a_n + b_n\}_{n=1}^\infty$, the sequence obtained by adding the terms of each individual sequence, converges to $A+B$. Since we are trying to prove that $\{a_n+b_n\}_{n=1}^\infty$ converges to $A+B$, we estimate the difference $|(a_n + b_n) - (A+B)|$. Inequality (9.11) implies

$$|(a_n + b_n) - (A + B)| = |(a_n - A) + (b_n - B)|$$
$$\leq |a_n - A| + |b_n - B|.$$

Since $|a_n - A|$ and $|b_n - B|$ can be made as small as desired by taking n large, $|(a_n + b_n) - (A + B)|$ can be made as small as desired by taking n large.[12]

Likewise, $\{a_n b_n\}_{n=1}^\infty$ converges to AB. This requires a frequently useful trick of adding and subtracting the same quantity. We have

$$|(a_n b_n) - (AB)| = |a_n b_n - a_n B + a_n B - AB|$$
$$= |a_n(b_n - B) + B(a_n - A)|$$
$$\leq |a_n||b_n - B| + |B||a_n - A|.$$

We also need the fact that the numbers $|a_n|$ are smaller than $|A| + 1$ for n large, which follows because $|a_n - A|$ can be made as small as desired for n large. So for n large,

$$|(a_n b_n) - (AB)| \leq (|A| + 1)|b_n - B| + |B||a_n - A|.$$

Now we can make the differences on the right arbitrarily small by taking n large.

The analogous properties hold for the difference and quotient of two sequences (Problems 9.16 and 9.18). We summarize as a theorem.

Theorem 9.2 *Suppose that $\{a_n\}_{n=1}^\infty$ converges to A and $\{b_n\}_{n=1}^\infty$ converges to B. Then $\{a_n+b_n\}_{n=1}^\infty$ converges to $A+B$, $\{a_n-b_n\}_{n=1}^\infty$ converges to $A - B$, $\{a_n b_n\}_{n=1}^\infty$ converges to AB, and if $b_n \neq 0$ for all n and $B \neq 0$, $\{a_n/b_n\}_{n=1}^\infty$ converges to A/B.*

We say that a sequence $\{a_n\}$ is **bounded** if there is a constant M such that $|a_n| \leq M$ for all indices n. The previous argument also justifies the following theorem We ask you to provide the details in Problem 9.27.

Theorem 9.3 *A convergent sequence is bounded.*

---

[12]This proof introduces a new level of sophistication into discussions involving convergence. We did not write down the formal definition of convergence to a limit with its ϵ and corresponding N. Rather, we use informal language about making certain quantities "as small as desired" by choosing indices "sufficiently large." Once we understand how to argue from the formal definition, then it is convenient to use informal language. But we point out that we can modify this and the following informal arguments to use the definition of convergence. We pose this as Problem 9.19.

This discussion is a bit tedious but it can make computing the limit of a complicated sequence much easier.

EXAMPLE 9.12. Consider $\{2 + 3n^{-4} + (-1)^n n^{-1}\}_{n=1}^{\infty}$.

$$\lim_{n\to\infty} (2 + 3n^{-4} + (-1)^n n^{-1})$$

$$= \lim_{n\to\infty} 2 + 3 \lim_{n\to\infty} n^{-4} + \lim_{n\to\infty} (-1)^n n^{-4}$$

$$= 2 + 3 \times 0 + 0 = 2.$$

EXAMPLE 9.13. We compute the limit of

$$\left\{ 4\frac{1 + n^{-3}}{3 + n^{-2}} \right\}_{n=1}^{\infty}$$

by using Theorem 9.2 to argue

$$\lim_{n\to\infty} 4\frac{1 + n^{-3}}{3 + n^{-2}} = \lim_{n\to\infty} 4\frac{\lim_{n\to\infty}(1 + n^{-3})}{\lim_{n\to\infty}(3 + n^{-2})}$$

$$= 4\frac{\lim_{n\to\infty} 1 + \lim_{n\to\infty} n^{-3}}{\lim_{n\to\infty} 3 + \lim_{n\to\infty} n^{-2}}$$

$$= 4\frac{1 + 0}{3 + 0} = \frac{4}{3}.$$

Each step in the computations in Examples 9.12 and 9.13 is justified because we obtain new limits that are defined after every application of Theorem 9.2. On the other hand, if we attempt to use Theorem 9.2 to manipulate a sequence and end up with limits that are undefined at some point, then the computation is *not* justified by Theorem 9.2.

EXAMPLE 9.14. Clearly,

$$\lim_{n\to\infty} \frac{1+n}{n^2} = \lim_{n\to\infty} \left(\frac{1}{n^2} + \frac{1}{n}\right) = 0.$$

However, if we try to use Theorem 9.2 to compute the limit as

$$\lim_{n\to\infty} \frac{1+n}{n^2} \text{ `` = ''} \frac{\lim_{n\to\infty} 1 + n}{\lim_{n\to\infty} n^2},$$

we get nonsense.

9.8 Functions and Sequences

A common way to make a complicated sequence is to apply a function to each term in a sequence and there by get a new sequence.

EXAMPLE 9.15. In the Verhulst model Model 4.4, we consider the sequence

$$\{P_n\}_{n=1}^{\infty} = \left\{ \frac{1}{\dfrac{1}{2^n}Q_0 + \dfrac{1}{K}\left(1 - \dfrac{1}{2^n}\right)} \right\}_{n=1}^{\infty}.$$

The sequence $\{P_n\}$ is obtained by applying the function

$$f(x) = \frac{1}{Q_0 x + \frac{1}{K}(1 - x)}$$

to the terms in the sequence $\left\{\frac{1}{2^n}\right\}$.

EXAMPLE 9.16. As part of solving the Muddy Yard model in Chapter 10, we need to compute

$$\lim_{n \to \infty} (a_n)^2$$

for a special sequence $\{a_n\}$. Here, we have applied $f(x) = x^2$ to $\{a_n\}$.

Therefore, it is natural to investigate the convergence of a sequence obtained by applying a function to a convergent sequence.

By the way, there are usually several different ways to write a given sequence in terms of functions and other sequences.

EXAMPLE 9.17. Consider

$$\lim_{n \to \infty} \left(n^{-2} + 3\right)^4. \tag{9.12}$$

We can choose $\{a_n\} = \left\{\frac{1}{n}\right\}$ and $f(x) = (x^2 + 3)^4$ so (9.12) can be written

$$\lim_{n \to \infty} f(a_n) = \lim_{n \to \infty} \left((a_n)^2 + 3\right)^4.$$

We can also choose $\{a_n\} = \left\{\frac{1}{n^2}\right\}$ and $f(x) = (x + 3)^4$ so (9.12) can be written

$$\lim_{n \to \infty} f(a_n) = \lim_{n \to \infty} (a_n + 3)^4.$$

Another possibility is $\{a_n\} = \left\{\frac{1}{n^2} + 3\right\}$ and $f(x) = x^4$ so (9.12) can be written

$$\lim_{n \to \infty} f(a_n) = \lim_{n \to \infty} (a_n)^4.$$

See Problem 9.4 for more examples.

The idea behind convergence is to show that the terms in a sequence become close to the limit as the index increases. If we apply a function to a sequence with a limit and the function changes arbitrarily with small changes of input, i.e., the function is not continuous, then we cannot really expect much.

EXAMPLE 9.18. The sequence

$$\left\{-1, \frac{1}{2}, \frac{-1}{3}, \frac{1}{4}, \cdots\right\} = \left\{\frac{(-1)^n}{n}\right\}$$

has the limit

$$\lim_{n\to\infty}\left\{\frac{(-1)^n}{n}\right\} = 0.$$

But the sequence obtained by applying the step function $I(t)$ to this sequence,

$$\left\{I(-1), I\left(\frac{1}{2}\right), I\left(\frac{-1}{3}\right), I\left(\frac{1}{4}\right), \cdots\right\} = \{0, 1, 0, 1, \cdots\},$$

diverges.

In other words, it only makes sense to try to compute the limit in such situations if the function behaves continuously. So *we assume that the function is Lipschitz continuous.*

Now suppose that $\{a_n\}$ converges to the limit A, where all the a_n and A belong to a set I on which f is Lipschitz continuous with Lipschitz constant L. We define the sequence $\{b_n\}$ by $b_n = f(a_n)$ and we show that

$$\lim_{n\to\infty} b_n = f(A).$$

Actually this follows directly from the definitions of a limit and Lipschitz continuity. We want to show that $|b_n - f(A)|$ can be made arbitrarily small by taking n large. But

$$|b_n - f(A)| = |f(a_n) - f(A)| \le L|a_n - A|,$$

since a_n and A are in I. We can make the right-hand side arbitrarily small by taking n sufficiently large since a_n converges to A. We summarize as

Theorem 9.4 *Let $\{a_n\}$ be a sequence with $\lim_{n\to\infty} a_n = A$ and f a Lipschitz continuous function on a set I such that a_n is in I for all n and A is in I. Then*

$$\lim_{n\to\infty} f(a_n) = f\left(\lim_{n\to\infty} a_n\right). \tag{9.13}$$

EXAMPLE 9.19. In the Verhulst model Model 4.4, we need to compute

$$\lim_{n\to\infty} P_n = \lim_{n\to\infty} \frac{1}{\frac{1}{2^n}Q_0 + \frac{1}{K}\left(1 - \frac{1}{2^n}\right)}.$$

The sequence $\{P_n\}$ is obtained by applying the function

$$f(x) = \frac{1}{Q_0 x + \frac{1}{K}(1 - x)}$$

to the terms in the sequence $\left\{\frac{1}{2^n}\right\}$. In this case, f is Lipschitz continuous on any bounded interval, say, $[0, 1]$. Since $1/2^n$ is in $[0, 1]$ for all n as is $\lim_{n \to \infty} 1/2^n = 0$, we compute easily $\lim_{n \to \infty} P_n = f(0) = K$.

EXAMPLE 9.20. The function $f(x) = x^2$ is Lipschitz continuous on bounded intervals, therefore, if $\{a_n\}$ converges to A, then

$$\lim_{n \to \infty} \left(a_n\right)^2 = A^2$$

We can apply this rule to compute more complicated examples as well.

EXAMPLE 9.21. By Corollary 8.2 and Theorem 9.2,

$$\lim_{n \to \infty} \left(\frac{3 + \frac{1}{n}}{4 + \frac{2}{n}}\right)^9 = \left(\lim_{n \to \infty} \frac{3 + \frac{1}{n}}{4 + \frac{2}{n}}\right)^9$$

$$= \left(\frac{\lim_{n \to \infty}(3 + \frac{1}{n})}{\lim_{n \to \infty}(4 + \frac{2}{n})}\right)^9$$

$$= \left(\frac{3}{4}\right)^9.$$

EXAMPLE 9.22. By Corollary 8.2 and Theorem 9.2,

$$\lim_{n \to \infty} \left((2^{-n})^7 + 14(2^{-n})^4 - 3(2^{-n}) + 2\right)$$

$$= 2 \times 0^7 + 14 \times 0^4 - 3 \times 0 + 2 = 2.$$

9.9 Sequences with Rational Elements

We conclude the discussion of computing limits of sequences by considering sequences in which the elements are rational functions of the index. Such examples are common in modeling, and moreover there is a trick that enables such sequences to be analyzed relatively easily.

EXAMPLE 9.23. Consider

$$\left\{\frac{6n^2 + 2}{4n^2 - n + 1000}\right\}_{n=1}^{\infty}.$$

Before computing the limit, we work out what happens when n becomes large. In the numerator, $6n^2$ is much larger than 2 when n is large and likewise in the denominator, $4n^2$ becomes much larger than $-n + 1000$ in size when n is large. So we might guess that for n large,

$$\frac{6n^2 + 2}{4n^2 - n + 1000} \approx \frac{6n^2}{4n^2} = \frac{6}{4}.$$

To see this is a good guess for the limit, we use a trick to put the sequence in a better form to compute the limit,

$$\lim_{n\to\infty} \frac{6n^2 + 2}{4n^2 - n + 1000} = \lim_{n\to\infty} \frac{(6n^2 + 2)n^{-2}}{(4n^2 - n + 1000)n^{-2}}$$

$$= \lim_{n\to\infty} \frac{6 + 2n^{-2}}{4 - n^{-1} + 1000n^{-2}}$$

$$= \frac{6}{4},$$

where we finish the computation as usual.

The trick of multiplying top and bottom of a ratio by a power can also be used to figure out when a sequence converges to zero or diverges to infinity.

EXAMPLE 9.24.

$$\lim_{n\to\infty} \frac{n^3 - 20n^2 + 1}{n^8 + 2n} = \lim_{n\to\infty} \frac{(n^3 - 20n^2 + 1)n^{-3}}{(n^8 + 2n)n^{-3}}$$

$$= \lim_{n\to\infty} \frac{1 - 20n^{-1} + n^{-3}}{n^5 + 2n^{-2}}.$$

We conclude that the numerator converges to 1 while the denominator increases without bound. Therefore,

$$\lim_{n\to\infty} \frac{n^3 - 20n^2 + 1}{n^8 + 2n} = 0.$$

EXAMPLE 9.25.

$$\lim_{n\to\infty} \frac{-n^6 + n + 10}{80n^4 + 7} = \lim_{n\to\infty} \frac{(-n^6 + n + 10)n^{-4}}{(80n^4 + 7)n^{-4}}$$

$$= \lim_{n\to\infty} \frac{-n^2 + n^{-3} + 10n^{-4}}{80 + 7n^{-4}}.$$

We conclude that the numerator grows in the negative direction without bound while the denominator tends toward 80. Therefore,

$$\left\{ \frac{-n^6 + n + 10}{80n^4 + 7} \right\}_{n=1}^{\infty} \quad \text{diverges to } -\infty.$$

9.10 Calculus and Computing Limits

In a standard calculus course, it is easy to get the impression that calculus is about computing limits. Even in this book on analysis, we present many examples and problems on computing limits. However, *rarely are we able* to compute the limits of the sequences that arise in mathematical modeling. In general practice, the best we can do is to first determine that a sequence converges theoretically, and then compute an element of the sequence corresponding to an index that is sufficiently large so that the element is a reasonable approximation of the limit.[13]

9.11   Computer Representation of Rational Numbers

The decimal expansion $\pm p_m p_{m-1} \cdots p_1.q_1 q_2 \cdots q_n$ uses the base 10 system, and consequently each of the digits p_i and q_j may take on one of the 10 values $0, 1, 2, ...9$. Of course, it is possible to use bases other than 10. For example, the Babylonians used the base 60 and thus their digits range between 0 and 59. The computer operates with the base 2 and the two digits 0 and 1. A base 2 number has the form

$$\pm p_m 2^m + p_{m-1} 2^{m-1} + ... + p_2 2^2 + p_1 2 + q_1 2^{-1} + q_2 2^{-2}$$
$$+ ... + q_{n-1} 2^{n-1} + q_n 2^n,$$

which we write as

$$\pm p_{m-1}...p_1.q_1 q_2 ....q_n = p_m p_{m-1}...p_1 + 0.q_1 q_2 ....q_n$$

where n and m are natural numbers, and each p_i and q_j takes the value 0 or 1. For example, in base 2,

$$11.101 = 1 \cdot 2^1 + 1 \cdot 2^0 + 1 \cdot 2^{-1} + 1 \cdot 2^{-3}.$$

In the floating point representation of a computer using the standard 32 bits, which is known as **single precision**, numbers are represented in the form

$$\pm r 2^N,$$

where $0 \le r < 1$ is the **mantissa** and the **exponent** N is an integer. Out of the 32 bits, 23 bits are used to store the base, 2 are used to store the mantissa, 7 bits are used to store the exponent, and finally 1 bit is used to store the sign. Since $2^{10} \approx 10^{-3}$ this gives 6 to 7 decimal digits for the mantissa while the exponent N may range from -126 to 127, implying that the absolute value of numbers stored on the computer may range from approximately 10^{-40} to 10^{40}. Numbers outside these ranges cannot be stored by a computer using 32 bits. Some languages permit the use of

---

[13]Which raises the practical problem of determining an index that is sufficiently large to yield a desired accuracy.

double precision using 64 bits for storage with 11 bits used to store the exponent, giving a range of $-1022 \leq n \leq 1023$, with 52 bits used to store the the mantissa, giving about 15 decimal places.

We point out that the finite storage capability of a computer has two consequences for storing rational numbers. The first consequence occurred for integers: namely, only rational numbers within a finite range can be stored. The second consequence is more subtle, but is actually more serious. This is the fact that only a finite number of digits can be stored. Any rational number that requires more than the finite number of digits in its decimal expansion, which includes any rational number with an infinite periodic expansion, is stored on a computer with an error. So, for example, 2/11 is stored as .1818181 or .1818182, depending on whether the computer rounds or not.

Now, introducing an error in the 7th or 15th digit of a single number would not be so serious except for the fact that such **round-off** errors accumulate when arithmetic operations are performed. For example, if two numbers with a small error are added, the result has a slightly larger possible error.[14] This is a complicated and dry subject, and we won't go into further detail. But we show that the accumulation of errors can have some startling consequences with an example of a divergent series.

EXAMPLE 9.26. To begin with, we show that the **harmonic series**

$$\sum_{i=1}^{\infty} \frac{1}{i}$$

diverges. This means that the sequence $\{s_n\}_{n=1}^{\infty}$ of partial sums

$$s_n = \sum_{i=1}^{n} \frac{1}{i}$$

diverges. To see this, we write a partial sum out for a large n and group the terms as shown:

$$1 + \overline{\frac{1}{2}} + \overline{\frac{1}{3} + \frac{1}{4}} + \overline{\frac{1}{5} + \frac{1}{6} + \frac{1}{7} + \frac{1}{8}} + \overline{\frac{1}{9} + \frac{1}{10} + \cdots + \frac{1}{15} + \frac{1}{16}}$$
$$+ \overline{\frac{1}{17} + \cdots + \frac{1}{32}} + \cdots$$

The first "group" is 1/2. The second group is

$$\frac{1}{3} + \frac{1}{4} \geq \frac{1}{4} + \frac{1}{4} = \frac{1}{2}.$$

---

[14]The accumulation of errors is commonly encountered in science experiments.

The third group is

$$\frac{1}{5} + \frac{1}{6} + \frac{1}{7} + \frac{1}{8} \geq \frac{1}{8} + \frac{1}{8} + \frac{1}{8} + \frac{1}{8} = \frac{1}{2}.$$

The fourth group,

$$\frac{1}{9} + \frac{1}{10} + \frac{1}{11} + \frac{1}{12} + \frac{1}{13} + \frac{1}{14} + \frac{1}{15} + \frac{1}{16},$$

has 8 terms that are larger than $1/16$, so it also gives a sum larger than $8/16 = 1/2$. We can continue in this way, taking the next 16 terms, all of which are larger than $1/32$, then the next 32 terms, all of which are larger than $1/64$, and so on. With each group, we get a contribution to the overall sum that is larger than $1/2$.

When we take n larger and larger, we can combine more and more terms in this way, making the sum larger in increments of $1/2$ each time. The partial sums therefore just become larger and larger as n increases, which means the partial sums diverge to infinity.

Note that by the arithmetic rules, the partial sum s_n should be the same whether the sum is computed in the "forward" direction,

$$s_n = 1 + \frac{1}{2} + \frac{1}{3} + \cdots \frac{1}{n-1} + \frac{1}{n},$$

or the "backward" direction,

$$s_n = \frac{1}{n} + \frac{1}{n-1} + \cdots + \frac{1}{3} + \frac{1}{2} + 1.$$

In Fig. 9.4, we list various partial sums in both the forward and backward directions computed using FORTRAN with single precision variables with about 7 places of accuracy. Note two things about these results. First of all, the partial sums s_n all become equal when n is large enough even though theoretically they should keep increasing as n increases. Second of all, the forward and backward sums do not give the same value! This is all due to the effects of the errors accumulating as the sums are computed.

Chapter 9 Problems

Problems 9.1–9.4 are exercises in using index notation.

9.1. Write the following sequences using the index notation:

(a) $\{1, 3, 9, 27, \cdots\}$

(b) $\{16, 64, 256, \cdots\}$

(c) $\{1, -1, 1, -1, 1, \cdots\}$

(d) $\{4, 7, 10, 13, \cdots\}$

(e) $\{2, 5, 8, 11, \cdots\}$

(f) $\{125, 25, 5, 1, \dfrac{1}{5}, \dfrac{1}{25}, \dfrac{1}{125}, \cdots\}$.

9.2. Determine the number of different sequences there are in the following list and identify the sequences that are equal:

(a) $\left\{ \dfrac{4^{n/2}}{4 + (-1)^n} \right\}_{n=1}^{\infty}$

(b) $\left\{ \dfrac{2^n}{4 + (-1)^n} \right\}_{n=1}^{\infty}$

(c) $\left\{ \dfrac{2^{\text{car}}}{4 + (-1)^{\text{car}}} \right\}_{\text{car}=1}^{\infty}$

(d) $\left\{ \dfrac{2^{n-1}}{4 + (-1)^{n-1}} \right\}_{n=2}^{\infty}$

(e) $\left\{ \dfrac{2^{n+2}}{4 + (-1)^{n+2}} \right\}_{n=0}^{\infty}$

(f) $\left\{ 8\dfrac{2^n}{4 + (-1)^{n+3}} \right\}_{n=-2}^{\infty}$.

9.3. Rewrite the sequence $\left\{ \dfrac{2 + n^2}{9^n} \right\}_{n=1}^{\infty}$ so that (a) the index n runs from -4 to ∞, (b) the index n runs from 3 to ∞, and (c) the index n runs from 2 to $-\infty$.

n	forward sum	backward sum
10000	9.787612915039062	9.787604331970214
100000	12.090850830078120	12.090151786804200
1000000	14.357357978820800	14.392651557922360
10000000	15.403682708740240	16.686031341552740
100000000	15.403682708740240	18.807918548583980
1000000000	15.403682708740240	18.807918548583980

FIGURE 9.4. Forward $1 + \frac{1}{2} + \cdots + \frac{1}{n}$ and backward $\frac{1}{n} + \frac{1}{n-1} + \cdots + \frac{1}{2} + 1$ partial harmonic sums for various n.

9.4. Rewrite the following sequences as a function applied to another sequence three different ways:

(a) $\left\{ \left(\dfrac{n^2 + 2}{n^2 + 1} \right)^3 \right\}_{n=1}^{\infty}$

(b) $\left\{ (n^2)^4 + (n^2)^2 + 1 \right\}_{n=1}^{\infty}$.

Verify the definition of convergence (or divergence) to do Problems 9.5–9.10.

9.5. Prove (9.2).

9.6. Show that $\lim\limits_{n \to \infty} r^n = \infty$ for any r with $|r| \geq 2$.

9.7. Show the following limits hold:

(a) $\lim\limits_{n \to \infty} \dfrac{8}{3n+1} = 0$ (b) $\lim\limits_{n \to \infty} \dfrac{4n+3}{7n-1} = \dfrac{4}{7}$ (c) $\lim\limits_{n \to \infty} \dfrac{n^2}{n^2+1} = 1.$

9.8. Prove that

$$\lim_{n \to \infty} \frac{1}{n^p} = 0,$$

where p is any natural number.

9.9. Show that $\lim\limits_{n \to \infty} r^n = 0$ for any r with $|r| \leq 1/2$.

9.10. Show the following hold:

(a) $\lim\limits_{n \to \infty} -4n + 1 = -\infty$ (b) $\lim\limits_{n \to \infty} n^3 + n^2 = \infty.$

Use the material on geometric series to do Problems 9.11–9.13.

9.11. Find the values of
(a) $1 - .5 + .25 - .125 + \cdots$
(b) $3 + \dfrac{3}{4} + \dfrac{3}{16} + \cdots$
(c) $5^{-2} + 5^{-3} + 5^{-4} + \cdots$

9.12. Find formulas for the sums of the following series, assuming $|r| < 1$:
(a) $1 + r^2 + r^4 + \cdots$
(b) $1 - r + r^2 - r^3 + r^4 - r^5 + \cdots$

9.13. A classic paradox posed by Zeno[15] can be solved using the geometric series. Suppose you are in Paulding county on your bike, 32 miles from your house in Atlanta. You break a spoke, you have no more food and you drank the last of your water, you forgot to bring money, and it starts to rain: the usual activities that make cycling so much fun. While riding home, as you are wont to do, you begin to think about how far you have to ride. Then you have a depressing thought: you can never get home! You think to yourself: first I have to ride 16 miles, then 8 miles after that, then 4 miles, then 2, then 1, then 1/2, then 1/4, and so on. Apparently you always have a little way to go, no matter how close you are, and you have to add up an infinite number of distances to get anywhere! Some of the Greek philosophers did not understand how to interpret a limit of a sequence, so this caused them a great deal of trouble. Explain why there is no paradox involved here using the sum of a geometric series.

---

[15]The Greek philosopher Zeno ($\approx$490 B.C.) is best known for his paradoxes.

Problems 9.14–9.27 have to do with the theoretical results on converging and diverging sequences.

9.14. Show that (9.11) holds using (2.4) and the fact that $a - c + c - b = a - b$.

9.15. Show that
$$\lim_{n \to \infty} 0.99 \cdots 99_n = 1,$$
where $0.99 \cdots 99_n$ contains n decimals all equal to 9.

9.16. Suppose that $\{a_n\}_{n=1}^{\infty}$ converges to A and $\{b_n\}_{n=1}^{\infty}$ converges to B. Show that $\{a_n - b_n\}_{n=1}^{\infty}$ converges to $A - B$.

9.17. Show that if $\lim_{n \to \infty} a_n = A$, then for any constant c (a) $\lim_{n \to \infty} (c + a_n) = c + A$ and $\lim_{n \to \infty} (c a_n) = cA$.

9.18. Suppose that $\{a_n\}_{n=1}^{\infty}$ converges to A and $\{b_n\}_{n=1}^{\infty}$ converges to B. Show that if $b_n \neq 0$ for all n and $B \neq 0$, then $\{a_n/b_n\}_{n=1}^{\infty}$ converges to A/B. *Hint:* Write
$$\frac{a_n}{b_n} - \frac{A}{B} = \frac{a_n}{b_n} + \frac{a_n}{B} - \frac{a_n}{B} - \frac{A}{B},$$
and use the fact that for n sufficiently large, $|b_n| \geq B/2$. Be sure to say why the last fact is true!

9.19. Rewrite the proofs for Theorem 9.2 using the formal definition of convergence.

9.20. Prove Theorem 9.3. *Hint:* Consider the argument for Theorem 9.2.

9.21. Suppose that $\{a_n\}$ is a sequence that converges to the limit A. Prove that $\{a_n^2\}$ converges to A^2 without using Theorem 9.2 or Theorem 9.4.

9.22. Suppose that $\{c_n\}$ is a sequence such that there are numbers a and b with $a \leq c_n \leq b$ for all indices n and $\{c_n\}$ converges to C. Prove that $a \leq C \leq b$.

9.23. Suppose that there are three sequences $\{a_n\}$, $\{b_n\}$, and $\{c_n\}$ such that $a_n \leq c_n \leq b_n$ for all indices n and $\{a_n\}$ and $\{b_n\}$ both converge to the limit A. Prove that $\{c_n\}$ also converges to A.

9.24. Suppose that there are two sequences $\{a_n\}$ and $\{b_n\}$ with $a_n \leq b_n$ for all indices n and $\{a_n\}$ diverges to ∞. Prove that $\{b_n\}$ diverges to ∞.

9.25. Suppose that there are two sequences $\{a_n\}$ and $\{b_n\}$ such that $\{a_n\}$ diverges to ∞ and $\{b_n\}$ is bounded. Prove that $\{a_n + b_n\}$ diverges to ∞.

9.26. Explain why each of the following claims are true or give an example that shows why it is false.

    (a) If $\{a_n\}$ and $\{b_n\}$ are divergent sequences, then $\{a_n + b_n\}$ diverges.

    (b) If $\{a_n\}$ and $\{a_n + b_n\}$ are both sequences that converge, then $\{b_n\}$ converges.

    (c) If $\{a_n\}$ is a convergent sequence with limit A and $a_n > 0$ for all n, then $A > 0$.

9.27. Suppose $\{a_n\}$ is a convergent sequence with limit A such that a_n and A are all in a set I on which the function f is Lipschitz continuous with constant L. Suppose further that $f(a_n)$ and $f(A)$ are all in a set J on which the function g is Lipschitz continuous with constant K. Prove that $\lim_{n\to\infty} g(f(a_n)) = g(f(A))$.

Use the theoretical results about converging sequences to evaluate the limits in Problems 9.28–9.29.

9.28. Compute the following limits:

(a) $\displaystyle\lim_{n\to\infty} \left(\frac{n+3}{2n+8}\right)^{37}$

(b) $\displaystyle\lim_{n\to\infty} \left(\frac{31}{n^2} + \frac{2}{n} + 7\right)^4$

(c) $\displaystyle\lim_{n\to\infty} \frac{1}{\left(2 + \frac{1}{n}\right)^8}$

(d) $\displaystyle\lim_{n\to\infty} \left(\left(\left(\left(1+\frac{2}{n}\right)^2\right)^3\right)^4\right)^5$.

9.29. Compute the limits of the sequences $\{a_n\}_{n=1}^{\infty}$ with the indicated terms or show they diverge:

(a) $a_n = 1 + \dfrac{7}{n}$

(b) $a_n = 4n^2 - 6n$

(c) $a_n = \dfrac{(-1)^n}{n^2}$

(d) $a_n = \dfrac{2n^2 + 9n + 3}{6n^2 + 2}$

(e) $a_n = \dfrac{(-1)^n n^2}{7n^2 + 1}$

(f) $a_n = \left(\dfrac{2}{3}\right)^n + 2$

(g) $a_n = \dfrac{(n-1)^2 - (n+1)^2}{n}$

(h) $a_n = \dfrac{1 - 5n^8}{4 + 51n^3 + 8n^8}$

(i) $a_n = \dfrac{2n^3 + n + 1}{6n^2 - 5}$

(j) $a_n = \dfrac{\left(\frac{7}{8}\right)^n - 1}{\left(\frac{7}{8}\right)^n + 1}$.

Before doing Problem 9.30, consider the warning given before Example 9.14.

9.30. Compute $\displaystyle\lim_{n\to\infty} \left(\sqrt{n^2 + n} - n\right)$. *Hint:* Multiply by $\dfrac{\sqrt{n^2+n+n}}{\sqrt{n^2+n+n}}$ and simplify the numerator.

9.31. Determine the number of digits used to store rational numbers in the programming language that you use and whether the language truncates or rounds.

9.32. The **machine number** μ is the smallest positive number μ stored in a computer that satisfies $1 + \mu > 1$. Note that μ is not zero! For example, explain the fact that in a single precision language $1 + .00000000001 = 1$. Write a little program that computes an approximation of the μ for your computer and programming language. *Hint:* $1 + .5 > 1$ in any programming language on any computer. Also $1 + .25 > 1$. Continue...

10
Solving the Muddy Yard Model

With basic properties of numbers and functions at hand, we can now solve more sophisticated mathematical models. We start by considering the solution of the Muddy Yard model (see Section 1.2),

$$f(x) = x^2 - 2 = 0. \tag{10.1}$$

Recall that to solve the Dinner Soup model $15x = 10$ to get $x = 2/3$, we extended the integers to get the rational numbers. It turns out that the solution of (10.1) is not a rational number and we have to extend the rational numbers to include a new set of numbers called the irrational numbers in order to solve (10.1).

It may seem counterintuitive to worry about solving (10.1) since we "know" the solution is $x = \sqrt{2}$. Of course that is true by definition, but the question remains: what is $\sqrt{2}$? To simply say that it is the solution of (10.1), or "that number" that is equal to 2 when squared, is circular reasoning and not much help when we go to buy the corrugated pipe.

10.1 Rational Numbers Just Aren't Enough

In Section 1.2, we found that $\sqrt{2} \approx 1.41$ using a trial-and-error strategy. But computing $1.41^2 = 1.9881$, we see that $\sqrt{2}$ is not exactly equal to 1.41. A better guess is 1.414, but then we get $1.414^2 = 1.999386$. We use

$MAPLE^©$ to compute the decimal expansion of $\sqrt{2}$ to 415 places:

$x = 1.4142135623730950488016887242096980785696718753$
$76948073176679737990732478462107038850387534 32$
$76415727350138462309122970249248360558 50737212$
$6441214970999358314132226659275055927557999505$
$0115278206057147010955997160597027453459686201$
$4728517418640889198609552329230484308714321450$
$8397626036279952514079896872533965463318088296$
$4062061525835239505474575028775996172983557522$
$0337531857011354374603408498847160386899970699,$

but using $MAPLE^©$ again, we find that

$x^2 = 1.99999999999999999999999999999999999999999999$
$99999999999999999999999999999999999999999999$
$99999999999999999999999999999999999999999999$
$99999999999999999999999999999999999999999999$
$99999999999999999999999999999999999999999999$
$99999999999999999999999999999999999999999999$
$99999999999999999999999999999999999999999999$
$99999999999999999999999999999999999999999999$
$99999999999999999999999999999999999999999999$
$99999999998638103700279039354754492148156 7520$
$7193643367223922486271791890987870158099 60232$
$6405972613126407604056912999503092957478 31888$
$5969500708874056058336501652271573809445 59332$
$0690045817264222173935969533242515158760 23360$
$4272994889141803598971038204956184812333 32162$
$5160160972831371230644994979436534796986 29776$
$6833340665770240318513306002427232125175 27304$
$3547767486608089987807935797774759645877 08250$
$3170068870585486010.$

The number $x = 1.4142\cdots699$ satisfies the equation $x^2 = 2$ to a high degree of precision but not exactly. In fact, it turns out that no matter how many digits we use in a guess with a finite decimal expansion, we never get a number that gives exactly 2 when squared.

We show that $\sqrt{2}$ cannot be a rational number using a proof by contradiction. That is, we show that assuming that $\sqrt{2}$ is a rational number of the form p/q, where p and q are natural numbers, leads to a contradiction or logical impossibility.[1]

To do this, we need some facts about natural numbers. A **factor** of a natural number n is a natural number p that divides into n without leaving a remainder. For example, 2 and 3 are both factors of 6. A natural number n always has factors 1 and n since $1 \times n = n$. A natural number n is called a **prime number** if the only factors of n are 1 and n. The first few prime numbers are $\{2, 3, 5, 7, 11, \cdots\}$. The only even prime number is 2.

Suppose that we take the natural number n and try to find two factors $n = pq$.[2] There are two possibilities:

- The only two factors are 1 and n: i.e., n is prime;

- There are two factors p and q, neither of which is 1 or n.

In the second case, $p \leq n/2$ and $q \leq n/2$, since the smallest possible factor not equal to 1 is 2.

Now we repeat by factoring p and q separately. In each case, either the number is prime or we factor it into a product of smaller natural numbers. Then we continue with the smaller factors. Eventually this process stops since n is finite and the factors at any stage are no larger than half the size of the factors of the previous stage. When the process has stopped, we have **factored** n into a product of prime numbers. It turns out that this factorization is unique except for order.[3]

One consequence of the factorization into prime numbers is the following fact. Suppose that 2 is a factor of n. If $n = pq$ is any factorization of n, it follows that at least one of the factors p and q must have a factor of 2.

Now assume that $\sqrt{2} = p/q$, where all common factors in the natural numbers p and q have been divided out. For example if p and q both have the factor 3, then we replace p by $p/3$ and q by $q/3$, which does not change the quotient p/q. We write this as $\sqrt{2}q = p$, where p and q have no common factors, or squaring both sides, $2q^2 = p^2$. By the fact just mentioned, p must contain the factor 2; therefore p^2 contains two factors of 2, and we can write $p = 2 \times \bar{p}$ with $\bar{p}$ a natural number. Thus $2q^2 = 4 \times \bar{p}^2$, that is, $q^2 = 2 \times \bar{p}^2$. But the same argument implies that q must also contain a factor of 2. This contradicts the original assumption that p and q had no common factors

---

[1]Constructionists and intuitionists do not like proof by contradiction, since it is inherently non-constructive. Likewise we generally avoid proof by contradiction, but this is a pretty argument and well worth an exception. On occasion, we also use proof by contradiction when the alternative is clumsy.

[2]It is straightforward to write a program to search for all the factors of a given natural number n by systematically dividing by all the natural numbers up to n (see Problem 10.2).

[3]First proved by Gauss.

so assuming $\sqrt{2}$ to be rational leads to a contradiction and $\sqrt{2}$ cannot be a rational number.[4]

10.2 Infinite Nonperiodic Decimal Expansions

The decimal expansion of any rational number is either finite or infinite periodic; and vice versa, any decimal expansion that is finite or infinite periodic represents a rational number. The periodic pattern in a decimal expansion of a rational number may take a long time to appear. But it does eventually, and once the pattern is determined, then we know the complete decimal expansion of the rational number in the sense that we no longer have to divide to determine the digits. In fact, we can give the value for any number digit. For example, the 231st digit of $10/9 = 1.111\cdots$ is 1 and the 103rd digit of $.56565656\cdots$ is 5.

But there is no reason to think that all infinite decimal expansions eventually begin to repeat. For example, the decimal expansion of $\sqrt{2}$, if it exists, cannot be finite or infinite periodic. In fact, it is easy (see Problem 10.6) to write down infinite nonperiodic decimal expansions like

$$2.12112111211112111112\cdots, \qquad (10.2)$$

where "$\cdots$" means "continue in the same pattern." This decimal expansion clearly never repeats, so it cannot correspond to a rational number. We call an infinite nonperiodic decimal expansion an **irrational** number because it cannot be a rational number. To solve models with irrational solutions, we have to extend the set of rational numbers to include the irrational numbers.

The irrational number (10.2) is special because there is a distinct pattern to its digits and we "know" this number's decimal expansion in the same way we know the infinite decimal expansion of a rational number. In short, we know all the digits involved and can give the value of any number digit that might be specified (see Problem 10.7). In general, we cannot expect to see such nice patterns in the digits of an infinite nonperiodic decimal expansion. In particular, if we would examine the digits of the decimal expansion of $\sqrt{2}$, we would not be able to discern any pattern whatsoever. It is almost as if the digits occur "randomly."

---

[4]This is a modification of the classic proof that was likely discovered during the fifth century B.C. around the time that the ancient Greek school of philosophy called the Pythagoreans fell into decline. The Pythagorean philosophy revolved around explaining the world in terms of properties of numbers, and one of their principle assumptions was that all numbers could be computed as the ratio of natural numbers. This is contradicted by the irrationality of $\sqrt{2}$, of course. Geometrically, on the other hand, it seems like the diagonal of the unit square must exist. It is tempting to think that this contributed to the rise in the importance of geometry over the following two centuries.

The crux of trying to make sense of irrational numbers is that we would have to give every digit of an irrational number in order to specify it completely. This is practically impossible. In the real world, we can only write down a finite number of digits.

We get around this difficulty by viewing an irrational number as the limit of a sequence of rational numbers that we can use to compute the digits of the irrational number to any desired accuracy. For each irrational number, we specify an algorithm that produces a sequence of increasingly accurate rational approximations. In other words, *we never write down an irrational number, we only specify a procedure for computing it to any desired accuracy.*

10.3   The Bisection Algorithm for the Muddy Yard Model

In order to devise an algorithm for computing an irrational number, we need to have some information about the number.[5] In this case, we know that $\sqrt{2}$ satisfies (10.1) if it exists. We describe an algorithm that generates a sequence of rational numbers that satisfy (10.1) with successively increasing accuracy. The algorithm uses a trial and error strategy that checks whether a given rational number x satisfies $f(x) < 0$ or $f(x) > 0$, i.e., if $x^2 < 2$ or $x^2 > 2$. The algorithm only involves computations with rational numbers, so there is never any uncertainty about how to use it. However, because the numbers produced by the algorithm are rational, none of them can ever actually equal $\sqrt{2}$.

The algorithm actually produces two sequences $\{x_i\}$ and $\{X_i\}$ that are the endpoints of intervals $[x_i, X_i]$ that contain $\sqrt{2}$ and become smaller as i increases. We begin by noting that $f(x) = x^2 - 2$ is a strictly increasing function for rational numbers $x > 0$; that is, $0 < x < y$ implies $f(x) < f(y)$. This follows because $0 < x < y$ means that $x^2 < xy < y^2$.

Now $f(1) < 0$ since $1^2 < 2$ and $f(2) > 0$ since $2^2 > 2$. Therefore, $f(x) < 0$ for all rational $0 < x \leq 1$ and $f(x) > 0$ for all rational $x \geq 2$ (see Fig. 10.1). We therefore naturally look for a solution of (10.1) between 1 and 2. So as the first step of the algorithm, we set $x_0 = 1$ and $X_0 = 2$ as shown in Fig. 10.1.

Next we choose a point between $x_0 = 1$ and $X_0 = 2$ and check the sign of f at that point. For the sake of symmetry, we choose the midpoint $1.5 = (1 + 2)/2$ and find that $f(1.5) > 0$ (see Fig. 10.1). Since this means that $f(x) > 0$ for rational $x \geq 1.5$ and we know that $f(x) < 0$ for $x \leq 1$, we naturally look for a solution of (10.1) between 1 and 1.5. We set $x_1 = 1$ and $X_1 = 1.5$ as shown in Fig. 10.2.

---

[5]This is a subtle point related to the constructivism controversy.

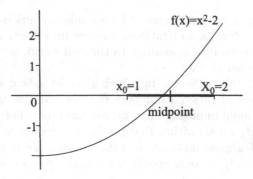

FIGURE 10.1. The first interval computed using the Bisection Algorithm.

Continuing the process, we next check the midpoint 1.25 of $x_1 = 1$ and $X_1 = 1.5$ to find that $f(1.25) < 0$, as shown in Fig. 10.2. We therefore

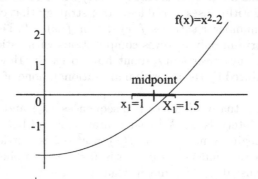

FIGURE 10.2. The second interval computed using the Bisection Algorithm.

naturally look for the solution of (10.1) between 1.25 and 1.5 and set $x_2 = 1.25$ and $X_2 = 1.5$. We then check the sign of f at the midpoint of x_2 and X_2, which is 1.375, and find that $f(1.375) < 0$. As before, we look for a solution of (10.1) between 1.375 and 1.5.

We can continue to search in this way as long as desired, each time determining two rational numbers that apparently "trap" a solution of (10.1). This procedure is called the **Bisection Algorithm**. We list the output for 20 steps from a $MATLAB^{©}$ m-file implementing this algorithm in Fig. 10.3.

i	x_i	X_i
0	1.00000000000000	2.00000000000000
1	1.00000000000000	1.50000000000000
2	1.25000000000000	1.50000000000000
3	1.37500000000000	1.50000000000000
4	1.37500000000000	1.43750000000000
5	1.40625000000000	1.43750000000000
⋮	⋮	⋮
10	1.41406250000000	1.41503906250000
⋮	⋮	⋮
15	1.41418457031250	1.41421508789062
⋮	⋮	⋮
20	1.41421318054199	1.41421413421631

FIGURE 10.3. 20 steps of the Bisection Algorithm for computing an approximate root of (10.1).

10.4 The Bisection Algorithm Converges

Continuing this procedure, we generate two sequences of rational numbers $\{x_i\}_{i=0}^{\infty}$ and $\{X_i\}_{i=0}^{\infty}$ with the property that

$$x_0 \leq x_1 \leq x_2 \leq \cdots \quad \text{and} \quad X_0 \geq X_1 \geq X_2 \geq \cdots$$

In other words, the terms x_i either increase or stay constant while the X_i always decrease or remain constant as i increases. Moreover, by construction, the distance between X_i and x_i strictly decreases as i increases. In fact,

$$0 \leq X_i - x_i \leq 2^{-i} \quad \text{for } i = 0, 1, 2, \cdots \tag{10.3}$$

i.e., the difference between the value x_i for which $f(x_i) < 0$ and the value X_i for which $f(X_i) > 0$ is halved for each increase in i. This means that as i increases, *more and more digits in the decimal expansions of x_i and X_i agree.*

The estimate (10.3) on the difference of $X_i - x_i$ also implies that the terms in the sequence $\{x_i\}_{i=0}^{\infty}$ become closer to each other as the index increases. This follows because $x_i \leq x_j < X_j \leq X_i$ if $j > i$, so (10.3) implies

$$|x_i - x_j| \leq |x_i - X_i| \leq 2^{-i} \quad \text{if } j \geq i. \tag{10.4}$$

We illustrate in Fig. 10.4. We call a sequence in which the terms become closer to each other with increasing index a **Cauchy sequence**.

In particular, this means that when $2^{-i} \leq 10^{-N-1}$, the first N decimals of x_j are the same as the first N decimals in x_i for any $j \geq i$. We conclude that the sequence $\{x_i\}_{i=0}^{\infty}$ determines a specific decimal expansion. To get

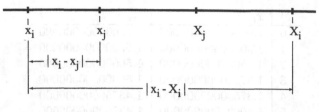

FIGURE 10.4. $|x_i - x_j| \le |X_i - x_i|$.

the first N digits of this expansion we simply take the first N digits of any number x_j in the sequence with $2^{-j} \le 10^{-N-1}$ since (10.4) implies that all such x_j agree in the first N digits.

Considering the idea behind convergence, it is natural to think that the sequence $\{x_i\}_{i=0}^{\infty}$ converges to the decimal expansion determined by its elements. But this decimal expansion does not have to be finite or infinite periodic. In fact, we believe that it corresponds to $\sqrt{2}$, in which case it must be infinite nonperiodic. Since this decimal expansion does not represent a rational number, the previous definition of convergence of a sequence of rational numbers, which assumes the sequence converges to a rational number, does not apply. In fact, we cannot even write down that definition in this case, since that definition uses the limit which is not yet defined! We get out of this dilemma by simply *defining* the limit, $\lim_{i \to \infty} x_i$, of $\{x_i\}_{i=0}^{\infty}$ to be the infinite nonperiodic decimal expansion determined by the x_i.

10.5   ... and the Limit Solves the Muddy Yard Model

Defining a new kind of number and a new kind of convergence as we did in the last section is simple. But showing that the definition is useful is a little more difficult. When we extended the integers to get the rational numbers, we also had to figure out how to compute with rational numbers in ways that are consistent with the rules we know for integers. Likewise in order for this definition to be useful, we have to figure out how to compute with the new number produced by the Bisection Algorithm. In particular, the definition had better lead to the conclusion that $\lim_{i \to \infty} x_i = \sqrt{2}$! In view of (10.1), this means we want

$$f(\lim_{i \to \infty} x_i) = 0.$$

Since we have so far only defined functions of rational numbers, we have to extend the previous definitions so that $f(\lim_{i \to \infty} x_i)$ makes sense.

A way past this obstacle is suggested by the result in Theorem 9.4, which says that for any Lipschitz continuous function f defined on a set of rational numbers,

$$f(\lim_{i\to\infty} x_i) = \lim_{i\to\infty} f(x_i)$$

when x_i is rational for all i and $\lim_{i\to\infty} x_i$ is also rational. In this case, $f(x) = x^2 - 2$ is certainly Lipschitz continuous but $\lim_{i\to\infty} x_i$ is irrational so the theorem does not apply. We skip past this trouble by *defining*

$$f(\lim_{i\to\infty} x_i) = \lim_{i\to\infty} f(x_i) \tag{10.5}$$

if the second limit exists. Since x_i is rational for all i, $f(x_i)$ is always defined. Moreover $f(x_i)$ is also rational for all i so $\{f(x_i)\}$ is a sequence of rational numbers that we believe converges to the rational number 0 and therefore we can use the definition of convergence for rational sequences.

Therefore to show that $\sqrt{2} = \lim_{i\to\infty} x_i$, we have to show that $\lim_{i\to\infty} f(x_i) = 0$ using the standard definition of convergence of a rational sequence. The rational numbers x_i and X_i are always between 0 and 2, and in Example 8.5 we saw that x^2 is Lipschitz continuous with constant $L = 4$ on the rational numbers between 0 and 2. Therefore, (10.3) implies that for any $i \geq 1$,

$$|f(x_i) - f(X_i)| \leq 4|x_i - X_i| \leq 2^{-i}.$$

Since $f(x_i) < 0 < f(X_i)$, we can take away the absolute value signs to get

$$f(X_i) - f(x_i) \leq 2^{-i}.$$

But $f(X_i)$ being positive and $f(x_i)$ being negative means (see Problem 10.10) that

$$|f(x_i)| \leq 2^{-i} \quad \text{and} \quad |f(X_i)| \leq 2^{-i}.$$

Now $f(x_i)$ and $f(X_i)$ are rational for all i, so there is no problem taking limits in the usual way. Therefore,

$$\lim_{i\to\infty} f(x_i) = 0 \quad \text{and} \quad \lim_{i\to\infty} f(X_i) = 0.$$

These limits imply that x_i and X_i become closer to being solutions of (10.1) as i increases. The definition (10.5) therefore implies that $f(\sqrt{2}) = 0$ and

$$\lim_{i\to\infty} x_i = \sqrt{2}$$

as claimed.

Chapter 10 Problems

10.1. Use the *evalf* function in $MAPLE^©$ to compute $\sqrt{2}$ to 1000 places and then square the result and compare to 2.

10.2. (a) Write a $MATLAB^©$ routine that tests a given natural number n to see if it is prime. *Hint:* Systematically divide n by the smaller natural numbers from 2 to $n/2$ to check whether there are factors. Explain why it suffices to check up to $n/2$. (b) Use this routine to write a $MATLAB^©$ routine that finds all the prime numbers less than a given number n. (c) List all the prime numbers less than 1000.

10.3. Factor the following integers into a product of prime numbers: (a) 60, (b) 96, (c) 112, (d) 129.

10.4. Find two natural numbers p and q such that pq contains a factor of 4 but neither p nor q contains a factor of 4. This means that the fact that some natural number m is factor of a product $n = pq$ does not imply that m must be a factor of either p or q. Why doesn't this contradict the fact that if pq contains a factor of 2, then at least one of p or q contains a factor of 2?

10.5. (a) Show that $\sqrt{3}$ (see Problem 1.6) is irrational. *Hint:* Use a powerful mathematical technique: try to copy a proof you already know. (b) Do the same for $\sqrt{a}$ where a is any prime number.

10.6. Specify three different irrational numbers using the digits 3 and 4.

10.7. Determine the 347th digit of (10.2).

10.8. (a) Compute 20 steps of the Bisection Algorithm for computing $\sqrt{2}$ starting with $x_0 = 1$ and $X_0 = 2$. (b) Compute the errors $|x_i - \sqrt{2}|$ and see if there is a pattern to the decrease in error as i increases.

10.9. Show that $\sqrt{2} = \lim_{i \to \infty} X_i$ where $\{X_i\}_{i=0}^{\infty}$ is the sequence produced by the Bisection Algorithm.

10.10. Show that if $a < 0$ and $b > 0$, then $b - a < c$ implies $|b| < c$ and $|a| < c$.

11
Real Numbers

Dealing with the existence of irrational numbers in a mathematically correct way is the hallmark of modern analysis. The difficulty with irrational numbers stems from the fact that it is generally impossible to write down the complete decimal expansion of an irrational number. To understand how this causes trouble, consider the addition of irrational numbers. We add two numbers with finite decimal expansions by starting with the rightmost digit and working left. We add two arbitrary rational numbers by finding a common denominator. But neither of these techniques works for irrational numbers.

To overcome such difficulties, we devise a way of describing an irrational number in terms of the better-understood rational numbers. Recall that in Chapter 10, we showed that the Bisection Algorithm can be used to compute any number of digits of the irrational number $\sqrt{2}$ by means of a sequence of rational numbers. Since this algorithm is the only concrete information we have (so far) about $\sqrt{2}$, it is natural to identify the symbol "$\sqrt{2}$" with the algorithm itself.[1]

The mathematical definition of irrational numbers such as $\sqrt{2}$ lies at the heart of the controversy of constructivism. Using a constructive interpretation, in which $\sqrt{2}$ is an algorithm to compute the decimal expansion of the solution of $x^2 = 2$ to any desired accuracy, while conveying a certain nobility of thought, does raise the need to explain how to compute with

---

[1]Certainly when we need $\sqrt{2}$ in some practical computation, we replace it by a rational approximation.

irrational numbers such as $\sqrt{2}$. Explaining how to compute with irrational numbers, defined by means of sequences of rational numbers, is the goal of this chapter.

11.1   Irrational Numbers

In Chapter 10, we define irrational numbers to be those numbers with infinite nonperiodic decimal expansions. It is easy to write down such numbers, for example

$$.212112111211112\cdots,$$

which we can consider as being determined by the sequence of rational numbers

$$\{.2, .21, .212, .2121, .21211, .212112, \cdots\}.$$

We also saw that $\sqrt{2}$ is irrational, though the digits in its decimal expansion are not so easy to describe. Indeed at present, we can only compute the decimal expansion of $\sqrt{2}$ using the Bisection Algorithm for computing the root of $x^2 - 2 = 0$. This produces a sequence of rational numbers $\{x_i\}$ that defines the digits of $\sqrt{2}$ as $i \to \infty$.

To generalize these examples, we consider sequences of rational numbers that define a unique decimal expansion. However, we cannot invoke the indicated decimal expansion, as in the definition of the limit, to characterize such sequences. The reason is that we have not yet figured out how to compute with such numbers when the decimal expansion is infinite nonperiodic. That is precisely what we do in this chapter.

Instead, we use a condition that guarantees a sequence $\{x_i\}$ defines a unique decimal expansion that does not involve the expansion. We assume that $\{x_i\}$ is a **Cauchy sequence**, which means that for any $\epsilon > 0$ there is an N such that

$$|x_i - x_j| < \epsilon \text{ for } i, j > N.$$

Another way to write this condition is that given $\epsilon > 0$ there is a N such that $i > N$ implies

$$|x_{i+j} - x_i| < \epsilon$$

for all $j > 0$. In particular, by choosing $\epsilon = 10^{-n-1}$ for some natural number i, we can guarantee that the elements x_i agree in the first n digits for all i sufficiently large. In other words, $\{x_i\}$ defines a unique decimal expansion.[2]

EXAMPLE 11.1. Consider a rational or irrational number x with decimal expansion

$$x = \pm p_m \cdots p_0.q_1 q_2 q_3 \cdots,$$

---

[2]Cantor called such sequences "fundamental sequences."

where the digits are specified in some way, such as $.2121121112\cdots$. In this case, it is reasonable to consider the sequence $\{x_i\}$ of rational numbers

$$x_i = \pm p_m \cdots p_0.q_1 \cdots q_i$$

obtained by truncating the decimal expansion of x. If x has a finite decimal expansion itself, then the elements in $\{x_i\}$ just equal x at some point.

With this choice of $\{x_i\}$, we immediately conclude that for all i,

$$|x_j - x_i| \le 10^{-i} \text{ for } j \ge i.$$

EXAMPLE 11.2. In Chapter 10, we show that the Bisection Algorithm for solving $x^2 - 2 = 0$ produces a Cauchy sequence of rational numbers $\{x_i\}$ that define the unique decimal expansion of $\sqrt{2}$. We gain about one digit in the decimal expansion of $\sqrt{2}$ for every three iterates of the Bisection Algorithm.

Given a Cauchy sequence of rational numbers $\{x_i\}$, we **identify** the unique decimal expansion x defined by $\{x_i\}$, whether finite, infinite periodic, or infinite nonperiodic, with $\{x_i\}$ and write $x \sim \{x_i\}$. The word identify and the notation $\sim$ indicate that something subtle is going on. Indeed, we would like to write "$x = \lim_{i\to\infty} x_i$," but we cannot do that yet because we have not defined what "$x =$" means when x has an infinite nonperiodic decimal expansion![3]

Before we get to that, there is an important issue of uniqueness to settle. Namely, a Cauchy sequence of rational numbers defines a unique decimal expansion. However, any given decimal expansion can be identified with many different Cauchy sequences of rational numbers.

EXAMPLE 11.3. We can identify $.212112111211112\cdots$ with $\{.2, .21, .212, \cdots\}$ and $\{.212, .212112, .212112111, \cdots\}$ and ...

EXAMPLE 11.4. In the following chapters, we develop several other algorithms for solving for the root of $x^2 - 2 = 0$ that produce Cauchy sequences of rational numbers that are identified with $\sqrt{2}$ but are not the same as the sequence produced by the Bisection Algorithm.

We would like to characterize the relationship between sequences that are identified with the same decimal expansion. We suppose that $\{x_i\}$ and $\{\tilde{x}_i\}$ are two Cauchy sequences of rational numbers with $x \sim \{x_i\}$ and $x \sim \{\tilde{x}_i\}$. First, it follows that $\{x_i - \tilde{x}_i\}$ is a Cauchy sequence of rational numbers. The triangle inequality means that

$$|(x_i - \tilde{x}_i) - (x_j - \tilde{x}_j)| \le |x_i - x_j| + |\tilde{x}_i - \tilde{x}_j|.$$

---

[3]Of course, if x is rational, then $x \sim \{x_i\}$ means precisely that $x = \lim_{i\to\infty} x_i$.

Given $\epsilon > 0$, there are N and $\tilde{N}$ such that $|x_i - x_j| < \epsilon/2$ for $i, j > N$ and likewise $|\tilde{x}_i - \tilde{x}_j| < \epsilon/2$ for $i, j > \tilde{N}$. It follows that $|(x_i - \tilde{x}_i) - (x_j - \tilde{x}_j)| < \epsilon$ for $i, j > \max\{N, \tilde{N}\}$. Now for any $n > 0$ there is a N such that the first n digits to the right of the decimal point in the decimal expansion of x_i agree with the corresponding digits in the expansion of x for $i > N$. Likewise, there is a $\tilde{N}$ such that the first n digits to the right of the decimal point in the decimal expansion of $\tilde{x}_i$ agree with the corresponding digits of x for $i > \tilde{N}$. Hence, for any $n > 0$, there is a M such that $|x_i - \tilde{x}_i| < 10^{-n}$ for $i > M$. We conclude that $\{x_i - \tilde{x}_i\}$ converges to zero, i.e., $\lim_{i \to \infty} x_i - \tilde{x}_i = 0$.[4]

Likewise, in Problem 11.1, we ask you to show that if $\{x_i\}$ and $\{\tilde{x}_i\}$ are Cauchy sequences of rational numbers such that $\lim_{i \to \infty} x_i - \tilde{x}_i = 0$, then $\{x_i\}$ and $\{\tilde{x}_i\}$ are identified with the same decimal expansion. This proves the following:

Theorem 11.1 *Suppose that $\{x_i\}$ and $\{\tilde{x}_i\}$ are Cauchy sequences of rational numbers. Then $\{x_i\}$ and $\{\tilde{x}_i\}$ are identified with the same decimal expansion if and only if $\lim_{i \to \infty} x_i - \tilde{x}_i = 0$.*[5]

11.2 Arithmetic with Irrational Numbers

We next define the basic arithmetic operations for irrational numbers. To do this, we require some basic facts about Cauchy sequences of rational numbers.

For one thing, if $\{x_i\}$ is a Cauchy sequence of rational numbers, then there is a N such that $|x_j - x_i| < 1$ for $j \geq i > N$. This means that

$$|x_j| \leq |x_{N+1}| + 1 \text{ for } j \geq N,$$

and therefore

$$|x_i| \leq \max\{|x_1|, \cdots, |x_N|, |x_{N+1}| + 1\} \text{ for all } i.$$

We conclude the following:

Theorem 11.2 *A Cauchy sequence of rational numbers is bounded.*

In Theorem 9.2, we show how to do arithmetic with limits. The same kinds of rules hold for Cauchy sequences of rational numbers. For example,

---

[4]Note we can use the idea of a limit defined in Chapter 9 because the limit 0 is rational.

[5]This result is important in practice because it implies that it does not matter which Cauchy sequence of rational numbers identified with a given irrational number is used to define operations with that irrational number in the limit of large index.

we show that the fact that $\{x_i\}$ and $\{y_i\}$ are Cauchy sequences means that $\{x_i + y_i\}$ is a Cauchy sequence. This just uses the triangle inequality,

$$|(x_i + y_i) - (x_j + y_j)| \leq |x_i - x_j| + |y_i - y_j|.$$

Now we can make $|x_i - x_j|$ and $|y_i - y_j|$ as small as we like by taking i and j large, so we can make $|(x_i + y_i) - (x_j + y_j)|$ arbitrarily small as well.[6] In the same way, we show that the fact that $\{x_i\}$ and $\{y_i\}$ are Cauchy sequences means that $\{x_i y_i\}$ is a Cauchy sequence. We estimate using some of the standard tricks and the triangle inequality

$$\begin{aligned}
|x_i y_i - x_j y_j| &= |x_i y_i - x_i y_j + x_i y_j - x_j y_j| \\
&\leq |x_i y_i - x_i y_j| + |x_i y_j - x_j y_j| \\
&= |x_i|\,|y_i - y_j| + |y_j|\,|x_i - x_j|.
\end{aligned}$$

Theorem 11.2 implies that the numbers $|x_i|$ and $|y_i|$ are all bounded by some constant. Calling this C, we get

$$|x_i y_i - x_j y_j| \leq C|y_i - y_j| + C|x_i - x_j|.$$

We can make the right-hand side small by taking i and j large, so this shows that $\{x_i y_i\}$ is a Cauchy sequence.

We ask you to treat the cases of subtraction and division in Problem 11.4. We summarize as a theorem.

Theorem 11.3 *Let $\{x_i\}$ and $\{y_i\}$ be Cauchy sequences of rational numbers. Then $\{x_i + y_i\}$, $\{x_i - y_i\}$, and $\{x_i y_i\}$ are all Cauchy sequences of rational numbers as well. If $y_i \neq 0$ for all i and $\{y_i\}$ is identified with a number that is not 0, then $\{x_i/y_i\}$ is also a Cauchy sequence of rational numbers.*

We let x and y be irrational numbers that are identified with Cauchy sequences of rational numbers $\{x_i\}$ and $\{y_i\}$, respectively. Now consider the definition of $x + y$. If x and y have finite decimal expansions, then we compute their sum by adding digit by digit starting with the first digit on the right, i.e., "at the end." When x and y are irrational, there is no "end," so it is not immediately clear how to compute the sum. But there is no problem computing the sum $x_i + y_i$. Moreover, $\{x_i + y_i\}$ is a Cauchy sequence of rational numbers that is identified with a unique decimal expansion. So we *define* $x + y$ to be the unique decimal expansion

$$x + y \sim \{x_i + y_i\}.$$

---

[6] This is essentially the same argument used to prove Theorem 11.1, which we carried out in gory detail.

EXAMPLE 11.5. We add

$$x = \sqrt{2} = 1.4142135623730950488\cdots$$

and

$$y = \frac{1043}{439} = 2.3758542141230068337\cdots$$

by adding $x_i + y_i$ for $i \geq 1$ in Fig. 11.1.

i	x_i	y_i	$x_i + y_i$
1	1	2	3
2	1.4	2.3	3.7
3	1.41	2.37	3.78
4	1.414	2.375	3.789
5	1.4142	2.3758	3.7900
$\vdots$	$\vdots$	$\vdots$	$\vdots$
10	1.414213562	2.375854214	3.790067776
$\vdots$	$\vdots$	$\vdots$	$\vdots$
15	1.42421356237309	2.37585421412300	3.79006777649609
$\vdots$	$\vdots$	$\vdots$	$\vdots$

FIGURE 11.1. Computing the decimal expansion of $\sqrt{2} + 1043/439$ by using the truncated decimal sequences.

We define the other operations in the same way. For example, we *define xy* to be the unique decimal expansion identified with the Cauchy sequence $\{x_i y_i\}$,

$$xy \sim \{x_i y_i\},$$

and similarly with division and subtraction.

With these definitions, we can easily show that the usual commutative, distributive, and associative rules for these operations hold. For example, addition is commutative since

$$x + y \sim \{x_i + y_i\} = \{y_i + x_i\} \sim y + x.$$

Note we define = to mean that the numbers on both sides of the equality are identified with the same decimal expansion. The point is that once we replace x and y by using the *rational* sequences $\{x_i\}$ and $\{y_i\}$, then we inherit the properties we know and love from the rational numbers. We summarize as a theorem. We ask you to complete the proof in Problem 11.5.

Theorem 11.4 Arithmetic Properties of Numbers *With these definitions of the arithmetic operations, the following properties hold for all rational and irrational numbers x, y, and z:*

- $x + y$ *is either rational or irrational.*

- $x + y = y + x.$

- $x + (y + z) = (x + y) + z.$

- $x + 0 = 0 + x = x$ *and* 0 *is the unique number with this property.*

- *There is a unique rational or irrational number* $-x$ *such that* $x + (-x) = (-x) + x = 0.$

- xy *is either rational or irrational.*

- $xy = yx.$

- $(xy)z = x(yz).$

- $1 \cdot x = x \cdot 1 = x$ *and* 1 *is the unique number with this property.*

- *If* x *is not the rational number* 0, *there is unique a rational or irrational number* x^{-1} *such that* $x \cdot x^{-1} = x^{-1} \cdot x = 1.$

- $x(y + z) = xy + xz.$

11.3   Inequalities for Irrational Numbers

We use the same approach to define inequalities involving irrational numbers. But, we need to be a little careful defining inequalities. For suppose that $\{x_i\}$ and $\{y_i\}$ are Cauchy sequences of rational numbers such that $x_i < y_i$ for all i. It does *not* follow that $\{x_i\}$ and $\{y_i\}$ correspond to different decimal expansions.

EXAMPLE 11.6.   Let $x_i = .999 \cdots 9$ with i digits and $y_i = 1$ for all i. We have

$$x_i < y_i \text{ for all } i \tag{11.1}$$

yet $1 \sim \{x_i\}$ and $1 \sim \{y_i\}$.

The problem is that x_i can approach arbitrarily close to y_i as i increases.

Suppose that $\{x_i\}$ is a Cauchy sequence of rational numbers identified with the decimal expansion x. If there is a constant c such that

$$x_i \leq c < 0 \text{ for all } i \text{ sufficiently large,}$$

then we say that $x < 0$. This condition prevents x_i from approaching arbitrarily close to 0. Likewise if there is a constant c such that

$$x_i \geq c > 0 \text{ for all } i \text{ sufficiently large,}$$

then we say that $x > 0$. When neither of these conditions hold, then $x = 0$.

Now suppose that $\{x_i\}$ and $\{y_i\}$ are Cauchy sequences of rational numbers identified with decimal expansions x and y, respectively. We say that $x < y$ if $x - y < 0$, $x = y$ if $x - y = 0$, and $x > y$ if $x - y > 0$ using the definitions above. Similarly, we define $x \leq y$ if $x < y$ or $x = y$, and so on.[7]

It is easy to check that these definitions follow the rules we expect to hold for inequalities.

EXAMPLE 11.7. For example, $x \leq y$ implies $-x \geq -y$, since $x_i \leq y_i$ for all i implies $-x_i \geq -y_i$ for all i. Moreover, $x \leq y$ and $w \leq z$ implies $x + w \leq y + z$ since $x_i \leq y_i$ and $w_i \leq z_i$ for all i implies $x_i + w_i \leq w_i + z_i$ for all i.

We summarize the crucial properties as a theorem, which we ask you to prove in Problem 11.6.

Theorem 11.5 Order Properties of Numbers *With these definitions of the arithmetic operations and relations $<$, $=$, and $>$, the following properties hold for all rational and irrational numbers x, y, and z:*

- *Exactly one of $x < y$, $x = y$, or $x > y$ holds.*

- *$x < y$ and $y < z$ implies $x < z$.*

- *$x < y$ implies $x + z < y + z$.*

- *If $z > 0$, $x < y$ implies $xz < yz$.*

These definitions allow a nice interpretation of the meaning of $\sim$. If $\{x_i\}$ is a Cauchy sequence of rational numbers and $x \sim \{x_i\}$, then we write $x = \lim_{i \to \infty} x_i$ because given any $\epsilon > 0$ there is a N such that $|x - x_i| < \epsilon$ for $i > N$. Consequently, we drop the use of $\sim$ for the remainder of the book and simply speak about the limit of a Cauchy sequence of rational numbers.

Once we have inequalities for real numbers, we can define intervals with real endpoints in the obvious way. The set of real numbers x between a and b, $\{x : a < x < b\}$, is called the **open interval** between a and b and is denoted by (a, b), where a and b are called the **endpoints** of the interval. An open interval does not contain its endpoints. The **closed interval** $[a, b]$ is a set $\{x : a \leq x \leq b\}$, which does contain its endpoints. Finally, we can

---

[7] A constructivist might object to these definitions on the grounds that they cannot be verified for arbitrary real numbers x and y by finite sequential computation. Suppose that we check and find that $x_i < y_i - c$ for $i = 1, 2, \cdots, N$ for some large N and some constant $c > 0$. We still cannot conclude that $x < y$, since it might be that $x_i = y_i$ for $i > N$. Practically speaking, of course, we can only check the definition for a finite number of terms in the sequences. However, avoiding the use of inequalities raises many complications and in this instance we yield to self-preservation.

have a **half-open interval** with one end open and the other closed, such as $(a, b] = \{x : a < x \leq b\}$.

We also have "infinite" intervals such as $(-\infty, a) = \{x : x < a\}$ and $[b, \infty) = \{x : b \leq x\}$.

11.4   The Real Numbers

In this rest of this book, we discuss rational and irrational numbers simultaneously, so it is convenient to introduce a set that includes both kinds of numbers. A **real** number is what we call any number that is either rational or irrational and the set of real numbers $\mathbb{R}$ is defined to be the set of all rational and irrational numbers. Since each decimal expansion defines a real number, we also could define the set of real numbers $\mathbb{R}$ to be the set of all possible decimal expansions.

This definition is controversial for the same reasons that the definition of $\mathbb{Q}$ as the set of all rational numbers is controversial (see Section 4.3). The set $\mathbb{R}$ contains arbitrarily large numbers, and moreover there are infinitely many real numbers between any two different real numbers. *In addition*, we have not specified any algorithm that determines the digits of an arbitrary real number. Ideally, every time an irrational number arises in a mathematical discussion, we should provide an algorithm for computing its digits to any desired accuracy. But generally devising such an algorithm requires some information about the number in question, such as knowing it is the root of some equation. So not only is it rather tedious to specify algorithms for each number that pops up, it also cumbersome to talk about a general number since we do not have an algorithm that computes all irrational numbers. Therefore to make life simple, we often talk about a general real number *without specifying how the digits are to be computed*.[8]

11.5   Please Oh Please, Let the Real Numbers Be Enough

There is one last subtle question about the construction of the real numbers hanging in the air. We have followed a long path to get to this point. We began with natural numbers and quickly found that the demands of common arithmetic required the extension of natural numbers to the integers and then the rational numbers. We then discovered that the solutions of models involving rational numbers, and indeed the rational numbers themselves, are naturally connected to infinite sequences of rational numbers.

---

[8]And so we stray far from the path of constructivism.

But sequences of rational numbers that converge do not necessarily converge to a rational limit, and so we extended the rational numbers to get the irrational numbers.

It is reasonable at this point to fervently hope that we have constructed all the numbers. In other words, we are faced with the question: do the real numbers contain all numbers or do we need to extend the real numbers to get some new, larger system of numbers? We now show that the real numbers are sufficient.[9]

The definitions of arithmetic for real numbers insure that the result of combining real numbers using arithmetic is another real number. But sequences of real numbers present a more subtle problem. We have to worry about sequences of real numbers because they crop up everywhere in analysis and mathematical modeling.

Our particular concern is to show that a Cauchy sequence of *real* numbers converges to a real limit. If a Cauchy sequence of real numbers did not converge to a real limit, we would be forced to look for a new, larger set of numbers. Here, we use the same definition of convergence and Cauchy sequence as used for sequences of rational numbers. A sequence of real numbers $\{x_i\}$ **converges** to x if for any $\epsilon > 0$ there is a natural number N such that $|x_i - x| < \epsilon$ for $i > N$. If this holds, we write $x = \lim_{i \to \infty} x_i$. Likewise, a sequence of real numbers $\{x_i\}$ is a **Cauchy sequence** if for any $\epsilon > 0$ there is a N such that $|x_i - x_j| < \epsilon$ for $i, j > N$.

Showing that a sequence of real numbers converges using the definition is no different than showing a sequence of rational numbers converges, and we point back to Chapter 9 for examples and problems. To illustrate how to use the definition to show that a sequence of real numbers is a Cauchy sequence, we recall three examples from Chapter 9.

EXAMPLE 11.8. In Example 9.6, we analyze the convergence of $\{1/i\}_{i=1}^{\infty}$. To show this is a Cauchy sequence, we compute

$$\left| \frac{1}{i} - \frac{1}{j} \right| = \left| \frac{j-i}{ij} \right| = \left| \frac{1}{i} \right| \left| \frac{j-i}{j} \right|.$$

Since $j \geq i \geq 1$, $j - i \leq j$ and therefore

$$\left| \frac{j-i}{j} \right| \leq \frac{j}{j} = 1,$$

so

$$\left| \frac{1}{i} - \frac{1}{j} \right| \leq \frac{1}{i}.$$

---

[9]That is, the real numbers are sufficient if we stay out of the domain of complex numbers. We do not have room in this book to motivate the need for extending the real numbers to get the complex numbers. But be assured that extending the real numbers to get the complex numbers is much easier than constructing the real numbers.

In other words if we choose N to be the smallest natural number larger than $1/\epsilon$, then the definition of a Cauchy sequence holds.

EXAMPLE 11.9. In Example 9.8, we study the convergence of $\{i/(i+1)\}_{i=1}^{\infty}$. To show this is a Cauchy sequence, we compute

$$\left|\frac{i}{i+1} - \frac{j}{j+1}\right| = \left|\frac{(j+1)i - (i+1)j}{(i+1)(j+1)}\right| = \left|\frac{i-j}{(i+1)(j+1)}\right|$$
$$= \left|\frac{1}{i+1}\right|\left|\frac{j-i}{j+1}\right|.$$

Since $j \geq i \geq 1$, $j - i \leq j$, while $j + 1 \geq j$, so

$$\left|\frac{j-i}{j+1}\right| \leq \frac{j}{j} = 1$$

and

$$\left|\frac{i}{i+1} - \frac{j}{j+1}\right| \leq \frac{1}{i+1}.$$

In other words if we choose N to be the smallest natural number larger than $1/\epsilon - 1$, then the definition of a Cauchy sequence holds.

EXAMPLE 11.10. Recall that in Example 9.10, we show that the geometric series for r

$$1 + r + r^2 + r^3 + \cdots$$

converges if the sequence of partial sums $\{s_n\}$,

$$s_n = 1 + r + r^2 + \cdots r^n = \frac{1 - r^{n+1}}{1 - r}$$

converges. We show that the sequence of partial sums is a Cauchy sequence when $|r| < 1$. For $m \geq n \geq 1$, we compute

$$|s_n - s_m| = \left|\frac{1 - r^{n+1}}{1 - r} - \frac{1 - r^{m+1}}{1 - r}\right|$$
$$= \left|\frac{(1-r)(1-r^{n+1}) - (1-r)(1-r^{m+1})}{(1-r)^2}\right|$$
$$= \left|\frac{r^{n+1} + r^{n+2} - r^{m+1} - r^{m+2}}{(1-r)^2}\right|$$
$$= \left|\frac{r^{n+1}}{(1-r)^2}\right| \left|1 + r - r^{m-n} - r^{m+1-n}\right|.$$

Since $m \geq n \geq 1$ and $|r| < 1$, $|r|^{m-n} < 1$ and $|r|^{m+1-n} < 1$. Therefore

$$\left|1 + r - r^{m-n} - r^{m+1-n}\right| \leq |1| + |r| + |r|^{m-n} + |r|^{m+1-n} \leq 4$$

and

$$|s_n - s_m| \leq \frac{4}{|1 - r|^2} |r|^{n+1}.$$

Now $4/|1 - r|^2$ is fixed and since $|r| < 1$ we can make $|r|^{n+1}$ as small as we like by taking n large. So the sequence of partial sums is a Cauchy sequence.

We return to the question of whether or not a Cauchy sequence of real numbers has to converge to a real number. Each element x_i of a Cauchy sequence of real numbers $\{x_i\}$ can be approximated to arbitrary accuracy by a Cauchy sequence of rational numbers.[10] We need to distinguish the sequences for different elements so we use a double index. Let $\{x_{ij}\}_{j=1}^{\infty} = \{x_{i1}, x_{i2}, x_{i3}, \cdots\}$ denote a Cauchy sequence of rational numbers that converges to x_i.

EXAMPLE 11.11. If

$$x_i = 4.12112111211112 \cdots ,$$

then

$$x_{i1} = 4.1$$
$$x_{i2} = 4.12$$
$$x_{i3} = 4.121$$
$$x_{i4} = 4.1211$$

$$\vdots \qquad \vdots$$

is one possibility.

We have a lot of sequences, which we write out in a big chart:

$$
\begin{array}{cccccccc}
x_{11} & x_{12} & x_{13} & x_{14} & x_{15} & \cdots & \rightarrow & x_1 \\
x_{21} & x_{22} & x_{23} & x_{24} & x_{25} & \cdots & \rightarrow & x_2 \\
x_{31} & x_{32} & x_{33} & x_{34} & x_{35} & \cdots & \rightarrow & x_3 \\
x_{41} & x_{42} & x_{43} & x_{44} & x_{45} & \cdots & \rightarrow & x_4 \\
\vdots & & & & \ddots & & & \vdots
\end{array}
$$

Since $\{x_i\}$ is a Cauchy sequence, we can make $|x_i - x_j|$ as small as desired by taking $j \geq i$ sufficiently large. This means in particular that $\{x_i\}$ defines a unique decimal expansion x. We want to show that x is the limit of a Cauchy sequence of rational numbers. To construct the sequence, we argue like this. Since $\{x_{1j}\}$ converges to x_1,

$$|x_1 - x_{1j}| < 10^{-1}$$

---

[10]If nothing else, we can use the sequence formed by truncating the decimal expansion of x_i.

for all j sufficiently large. Let m_1 denote the smallest index such that

$$|x_1 - x_{1m_1}| < 10^{-1}.$$

In the same way, let m_2 denote the smallest index such that

$$|x_2 - x_{2m_2}| < 10^{-2}.$$

In general, let m_i denote the smallest index (which is always finite) such that

$$|x_i - x_{im_i}| < 10^{-i}.$$

We claim that $\{x_{im_i}\}$ is a Cauchy sequence of rational numbers that converges to x. This follows from definition in fact. First we estimate

$$\begin{aligned}|x_{im_i} - x_{jm_j}| &= |x_{im_i} - x_i + x_i - x_j + x_j - x_{jm_j}| \\ &\leq |x_{im_i} - x_i| + |x_i - x_j| + |x_j - x_{jm_j}| \\ &\leq 10^{-i} + |x_i - x_j| + 10^{-j}.\end{aligned}$$

Since $\{x_i\}$ is a Cauchy sequence, given $\epsilon > 0$ there is an N such that

$$10^{-i} \leq \epsilon/3, \quad |x_i - x_j| < \epsilon/3, \quad 10^{-j} < \epsilon/3,$$

and hence $|x_{im_i} - x_{jm_j}| < \epsilon$, for $i, j > N$. So $\{x_{im_i}\}$ is a Cauchy sequence. By construction, the first i digits to the right of the decimal point in the decimal expansion of x_{im_i} agree with the corresponding digits of x_i. But this means that $\{x_{im_i}\}$ defines the same decimal expansion as $\{x_i\}$, which by definition implies

$$\lim_{i \to \infty} x_{im_i} = x.$$

On the other hand, if $\{x_i\}$ is a sequence that converges to a limit x, then the estimate

$$|x_i - x_j| \leq |x_i - x| + |x - x_j|$$

implies immediately that $\{x_i\}$ is a Cauchy sequence. We summarize with a theorem.

Theorem 11.6 Cauchy Criterion for Convergence *A Cauchy sequence of real numbers converges to a unique real number and any sequence of real numbers that converges is a Cauchy sequence.*

We describe this result by saying that the real numbers $\mathbb{R}$ are **complete**.

Theorem 11.6 and the preceding discussion imply the following important observation.

Theorem 11.7 Denseness of the Rational Numbers *Any real number can be approximated to arbitrary accuracy by a Cauchy sequence of rational numbers.*

For if a number x is rational, we simply take the sequence with constant values $\{x\}$. If a number is irrational, we can use the sequence of truncated decimal expansions $\{x_n\}$ as in Example 11.1. We describe this result by saying the rational numbers are a **dense** subset of the real numbers.[11]

The definitions we have made for convergence and Cauchy sequence for sequences of real numbers means that all of the usual properties that hold for sequences of rational elements carry over to sequences of real numbers.

In particular, the same arguments used to show Theorems 9.2, 11.2, and 11.3 also imply the following results.

Theorem 11.8 *Suppose that $\{x_n\}_{n=1}^{\infty}$ converges to x and $\{y_n\}_{n=1}^{\infty}$ converges to y. Then $\{x_n + y_n\}_{n=1}^{\infty}$ converges to $x+y$, $\{x_n - y_n\}_{n=1}^{\infty}$ converges to $x - y$, $\{x_n y_n\}_{n=1}^{\infty}$ converges to xy, and if $y_n \neq 0$ for all n and $y \neq 0$, $\{x_n/y_n\}_{n=1}^{\infty}$ converges to x/y.*

Theorem 11.9 *A Cauchy sequence of real numbers is bounded.*

Theorem 11.10 *Let $\{x_i\}$ and $\{y_i\}$ be Cauchy sequences of real numbers. Then $\{x_i + y_i\}$, $\{x_i - y_i\}$, and $\{x_i y_i\}$ are all Cauchy sequences of real numbers as well. If $y_i \neq 0$ for all i and $\lim_{i \to \infty} y_i \neq 0$, then $\{x_i/y_i\}$ is also a Cauchy sequence of real numbers.*

11.6   Some History of the Real Numbers

We can summarize the main result in this chapter as the following description of the essence of the real numbers.

Theorem 11.11 Real Number Theorem *The real numbers are complete and the rational numbers are dense in the real numbers.*

It is surprising to learn that the construction of the real numbers was the final step in the long struggle to put analysis and Calculus on a rigorous mathematical foundation. In hindsight, it seems obvious that it would be difficult to make mathematical sense out of notions like limits, which are inherent to analysis and calculus, without a complete system of numbers like the reals. Otherwise, we would constantly be dealing with convergent sequences of numbers whose limits are not understood. On the other hand,

---

[11]Theorems 11.6 and 11.7 are important in practice as well as in theory. If we want to compute the limit of a sequence of real numbers using a computer, we are faced with the problem that the computer cannot store general irrational elements that occur in the sequence. But these theorems imply that we can replace a Cauchy sequence of real numbers by a Cauchy sequence of rational numbers with the same limit. This means that we can approximate the limit of a sequence of real numbers on a computer to within the computer's accuracy even though we have not used all of the exact elements of the sequence.

the real numbers act just like the rational numbers, so early analysts who relied on intuition based on properties of rational numbers encountered no inconsistencies.

In any case, modern explanations of analysis always present the construction of the real numbers before turning to deeper topics in analysis. There are several possible tacks. The approach we have adopted, i.e., extending the rational numbers to get the irrational numbers, is essentially due to Cantor, who constructed the first rigorous theory for the real numbers.

What might be considered the classic approach for the construction of real numbers was first championed by Hilbert[12] some time after Cantor's work. We would begin by laying out the properties that we expect to hold for a system of real numbers as a set of axioms. The first set of axioms describe how the numbers are to be combined using arithmetic. We assume that the set of numbers R together with the operations of addition $+$, multiplication $\times = \cdot$ satisfy the properties listed in Theorem 11.4. These are called the *Field Axioms*. In addition, we assume there is an ordering operation $<$ on R satisfying the properties in Theorem 11.5, which are called the *Order Axioms*. A set of numbers R satisfying the Field and Order Axioms is called an *ordered field*. The rational numbers are an example of an ordered field.

Finally, the set R is assumed to satisfy the *Axiom of Completeness*:

Any nonempty set of numbers in R that is bounded above has a least upper bound in R.

This axiom is required to insure that the number system R is continuous in the sense that there are no "gaps" between the numbers in R. The Axiom of Completeness is the property that distinguishes the real numbers from the rational numbers.[13] We know that the rational numbers do not represent *all* numbers. The completeness property insures that R does cover all numbers.

The classic existence result for the real numbers shows that the real numbers are an ordered field and the rational numbers are dense in the real numbers. This result is usually proved by approximating an arbitrary real number by rational numbers using the Dedekind cut,[14] which is based

---

[12]David Hilbert (1862–1943) was a very influential German mathematician who proved fundamentally important results in many areas including algebra, algebraic numbers, the calculus of variations, functional analysis, integral equations, and mathematical physics. Hilbert (co-)wrote several influential text books and, in a famous talk given at the Second International Congress of Mathematics, posed 23 outstanding mathematical problems known as Hilbert's problems. Some of these are still unsolved and drive mathematical research today. A solution of one of Hilbert's problems is considered a tremendous achievement among mathematicians. Hilbert attempted to establish a consistent axiomatic and logical description of numbers. While ultimately unsuccessful, Hilbert's approach has heavily influenced the way modern mathematics is presented.

[13]There are several ways to formulate this property.

[14]Named after its inventor, the German mathematician Julius Wihelm Richard Dedekind (1831–1916). Along with a construction of the real numbers, Dedekind made

on the idea that each number divides the other numbers into two sets, those less than and those greater than the number. See Rudin [19] for a presentation of this approach.

However we choose to introduce the irrational numbers, developing their basic properties requires some work. We like Cantor's approach because it is an example of a powerful general technique: namely, approximate an unknown quantity by known quantities and then show the unknown quantity inherits important properties of the approximations. This approach is used with great success in the study of functions, the solution of root problems, the solution of differential equations, and so on. Indeed, we can fairly claim that modern computational science is based philosophically on this idea. Many important mathematical models in science and engineering are too difficult for mathematical analysis and any information about solutions is obtained almost exclusively through approximation made on a computer. We believe that such approximations accurately reveal important properties of the solution, though we can rarely prove that is true.

---

important contributions in the study of mathematical induction, the definition of finite and infinite sets, and algebraic number theory. Dedekind was also a very clear expositor and lecturer, and his style has strongly influenced the way modern mathematics is written down.

Chapter 11 Problems

11.1. Complete the proof of Theorem 11.1.

11.2. Let x be the limit of the sequence of rational numbers $\{x_i\}$ where the first $i - 1$ decimal places of x_i agree with the first $i - 1$ decimal places of $\sqrt{2}$, the ith decimal place is equal to 3, and the rest of the decimal places are zero. Is $x = \sqrt{2}$? Give a reason for your answer.

Problems 11.3–11.10 have to do with arithmetic and inequalities for real numbers.

11.3. Prove that if a Cauchy sequence of rational numbers $\{y_i\}$ is identified with a decimal expansion y that is not the rational number 0, then there is a constant $c > 0$ such that $|y_i| \geq c$ for all i sufficiently large.

11.4. Let $\{x_i\}$ and $\{y_i\}$ be Cauchy sequences of rational numbers. (a) Show that $\{x_i - y_i\}$ is a Cauchy sequence of rational numbers. (b) Assume that $y_i \neq 0$ for all i and that $\{y_i\}$ is identified with a number that is not 0. Prove that $\{x_i/y_i\}$ is a Cauchy sequence. *Hint:* Use Problem 11.3.

11.5. Complete the proof of Theorem 11.4.

11.6. Prove Theorem 11.5. (b) Theorem 11.5 is sufficient to conclude the other usual properties of inequalities hold. For example, prove that $x < y$ implies $-x > -y$.

11.7. Suppose that x and y are two real numbers and $\{x_i\}$ and $\{y_i\}$ are the sequences generated by truncating their decimal expansions. (a) Estimate $|(x + y) - (x_i + y_i)|$. (b) Estimate $|xy - x_i y_i|$. *Hint:* Explain why $|x_i| \leq |x| + 1$ for i sufficiently large.

11.8. Let $x = .37373737 \cdots$ and $y = \sqrt{2}$ and $\{x_i\}$ and $\{y_i\}$ be the sequences generated by truncating their decimal expansions. Compute the first 10 terms of the sequences defining $x + y$ and $y - x$ and the first 5 terms of the sequences defining xy and x/y.

11.9. Let x be the limit of the sequence $\left\{\dfrac{i}{i+1}\right\}$. Show that $\dfrac{i}{i+1} < 1$ for all i. Is $x < 1$?

11.10. If x and y are real numbers and $\{y_i\}$ is any sequence that converges to y, show that $x < y$ implies $x < y_i$ for all i sufficiently large.

Problems 11.11–11.13 have to do with Cauchy sequences of real numbers.

11.11. Show that the following sequences are Cauchy sequences:

$$\text{(a) } \left\{\frac{1}{(i+1)^2}\right\} \qquad \text{(b) } \left\{4 - \frac{1}{2^i}\right\} \qquad \text{(c) } \left\{\frac{i}{3i+1}\right\}.$$

11.12. Show that the sequence $\{i^2\}$ is *not* a Cauchy sequence.

11.13. Let $\{x_i\}$ denote the sequence of real numbers defined by

$$x_1 = .373373337\cdots$$
$$x_2 = .337733377333377\cdots$$
$$x_3 = .333777333377733333777\cdots$$
$$x_4 = .333377773333377773333337777\cdots$$
$$\vdots$$

(a) Show that the sequence is a Cauchy sequence and (b) find $\lim\limits_{i\to\infty} x_i$. This shows that a sequence of irrational numbers can converge to a rational number.

11.14. Prove Theorems 11.8, 11.9, and 11.10.

Problems 11.15 and 11.16 are relatively difficult.

11.15. Let $\{x_i\}$ be an increasing sequence, $x_{i-1} \leq x_i$, which is bounded above; i.e., there is a number c such that $x_i \leq c$ for all i. Prove that $\{x_i\}$ converges. *Hint:* Use a variation of the argument for the convergence of the Bisection Algorithm.

11.16. Explain why there are infinitely many real numbers between any two distinct real numbers by giving a systematic way to write them down.

12
Functions of Real Numbers

The functions we have investigated so far have been defined on sets of rational numbers. Now that we have constructed the real numbers, it is natural to consider functions defined on sets of real numbers.

EXAMPLE 12.1. The constant function $f(x) = \sqrt{2}$ for any rational or irrational x has the domain of real numbers and the range of one irrational number.

Once we have defined functions of real numbers and investigated their properties, we can run full bore into the study of mathematical models and their solution.

12.1 Functions of a Real Variable

Actually, there is no trouble extending the definitions made for functions of rational numbers to functions of real numbers. *Ideas such as the linear combination, product, and quotient of functions are the same.* Likewise, we define a function f to be **Lipschitz continuous** on a set of real numbers I if there is a constant L such that

$$|f(x_2) - f(x_1)| \leq L|x_2 - x_1| \text{ for all } x_1, x_2 \text{ in } I.$$

We ask you to show that the properties of Lipschitz continuous functions contained in Theorems 8.1–8.6 hold for functions defined on real numbers in Problem 12.1. We summarize as a theorem.

Theorem 12.1 *Let f_1 and f_2 be Lipschitz continuous on a set of real numbers I. Then $f_1 + f_2$ and $f_1 - f_2$ are Lipschitz continuous. If the domain I is bounded, then f_1 and f_2 are bounded and $f_1 f_2$ is Lipschitz continuous. If I is bounded and moreover $|f_2(x)| \geq m > 0$ for all x in I, where m is some constant, then f_1/f_2 is Lipschitz continuous.*

If f_1 is Lipschitz continuous on a set of real numbers I_1 with Lipschitz constant L_1 and f_2 is Lipschitz continuous on a set of real numbers I_2 with Lipschitz constant L_2 and $f_1(I_1) \subset I_2$, then $f_2 \circ f_1$ is Lipschitz continuous on I_1 with Lipschitz constant $L_1 L_2$

12.2   Extending Functions of Rational Numbers

Given that, theoretically at least, functions of real numbers behave as we expect, we are still left with the practical issue of producing some interesting examples. The main goal in this section is to show that the functions defined on rational numbers that we have met so far correspond in a natural way to functions defined on real numbers. By natural, we mean that important properties of a function on rational numbers, such as Lipschitz continuity, also hold for the corresponding function on real numbers.

The correspondence between functions of rational and real numbers is based on the same idea (10.5) used to show that the limit of the Bisection Algorithm is $\sqrt{2}$. Namely, for a real number x, which is the limit of the sequence of rational numbers $\{x_i\}$, we define

$$f(x) = \lim_{i \to \infty} f(x_i). \tag{12.1}$$

We say that we have **extended** f from the rational numbers to the real numbers and call f the **extension** of f.[1]

EXAMPLE 12.2.   We evaluate $f(x) = .4x^3 - x$ for $x = \sqrt{2}$ using the truncated decimal sequence $\{x_i\}$ in Fig. 12.1.

With a little thought, we see that this definition makes sense only if f is continuous because we depend on the fact that small changes in the input to f produces small changes in the output of f. In fact, if f is Lipschitz continuous on a set of rational numbers I and $\{x_i\}$ is a Cauchy sequence of rational numbers in I, then $\{f(x_i)\}$ is also a Cauchy sequence since

$$|f(x_i) - f(x_j)| \leq L|x_i - x_j|,$$

and we can make the right-hand side arbitrarily small by taking i and j large. This means that $\{f(x_i)\}$ converges to a limit and it makes sense to

---

[1]Using the same notation for f and its extension is a potential source of confusion. But after this chapter, we deal only with the extensions of the usual functions.

i	x_i	$.4x_i^3 - x_i$
1	1	$-.6$
2	1.4	.0976
3	1.41	.1212884
4	1.414	.1308583776
5	1.4142	.1313383005152
6	1.41421	.1313623002245844
7	1.414213	.1313695002035846388
8	1.4142135	.13137070020305452415
9	1.41421356	.13137084420304793147444064
10	1.414213562	.131370849003047922153528131312
$\vdots$	$\vdots$	$\vdots$

FIGURE 12.1. Computing the decimal expansion of $f(\sqrt{2})$ for $f(x) = .4x^3 - x$ by using the truncated decimal sequence.

talk about $\lim_{i \to \infty} f(x_i)$. In Problem 12.2, we ask you to show this also holds for functions defined on sets of real numbers and Cauchy sequences of real numbers. We summarize as a theorem.

Theorem 12.2 *Suppose that f is Lipschitz continuous on a set of real numbers I with constant L and $\{x_i\}$ is a Cauchy sequence in I. Then $\{f(x_i)\}$ is also a Cauchy sequence.*

EXAMPLE 12.3. We can extend any polynomial to be defined on the real numbers since a polynomial is Lipschitz continuous on any bounded set of rational numbers.

EXAMPLE 12.4. The previous example means that we can extend $f(x) = x^n$ to the real numbers for any integer n. We can also show that $f(x) = x^{-n}$ is Lipschitz continuous on any set of rational numbers that avoids 0. Therefore $f(x) = x^n$ can be extended to the real numbers where n is any integer provided that when $n < 0$, $x \neq 0$.

EXAMPLE 12.5. The function $f(x) = |x|$ is Lipschitz continuous on the set of rational numbers, so it can be extended to the real numbers.

It turns out that if f is Lipschitz continuous on a set of rational numbers, then its extension is also Lipschitz continuous with the same constant. Suppose that f is Lipschitz continuous on the interval of rational numbers $I = (a, b)$ with constant L. We assume that x and y are two real numbers between a and b that are the limits of Cauchy sequences of rational numbers $\{x_i\}$ and $\{y_i\}$, respectively. It follows that x_i and y_i are also contained in (a, b) for all sufficiently large i. We remove any of the finite number of elements from these sequences that do not lie in (a, b) to obtain sequences that converge to x and y and lie entirely in (a, b). We just call these potentially new sequences $\{x_i\}$ and $\{y_i\}$. Now by definition,

$$|f(x) - f(y)| = \left| \lim_{i \to \infty} (f(x_i) - f(y_i)) \right|.$$

By the Lipschitz assumption on f,

$$\begin{aligned} |f(x) - f(y)| &= \lim_{i \to \infty} |f(x_i) - f(y_i)| \\ &\leq L \lim_{i \to \infty} |x_i - y_i| \\ &\leq L|x - y|. \end{aligned}$$

This shows f is Lipschitz continuous as claimed. We ask you to treat the case that either end of the interval is closed or a real number in Problem 12.3. We summarize as a theorem.

Theorem 12.3 *Suppose that f is Lipschitz continuous on the set of rational numbers in (a, b). The extension of f to $[a, b]$ is Lipschitz continuous with the same constant. The same holds if either end of the interval is closed.*

EXAMPLE 12.6. Following Examples 12.3, 12.4, and 12.5, polynomials, $f(x) = x^n$ for integer n, and $f(x) = |x|$ are Lipschitz continuous on suitable domains.

By the way, if f is Lipschitz continuous on the closed interval $[a, b]$, it is immediately Lipschitz continuous on the open interval (a, b). But even if f is defined on $[a, b]$ and Lipschitz continuous on (a, b), it may not be Lipschitz continuous on $[a, b]$ for the simple reason that it may be discontinuous at one of the endpoints a or b.

12.3   Graphing Functions of a Real Variable

We represent real intervals graphically in the same way as rational intervals (see Section 4.6).[2]

In practice, graphing a function of real numbers is no different than graphing a function of the rational numbers. We face the same dilemma: namely, we can only evaluate the function at a finite set of rational numbers.

---

[2]It is interesting, however, to contemplate the difference between intervals of rational and real numbers. Theoretically, intervals of rational numbers appear to be solid but are not, since all the irrational numbers are left out, while intervals of real numbers are solid. Of course, in practice, the computer can only draw intervals of rational numbers and in fact numbers with finite decimal expansions. At least it draws enough points that the intervals look solid, so we can maintain appearances.

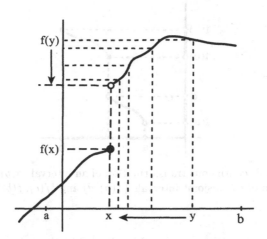

FIGURE 12.2. A function f that is discontinuous at x. As y approaches x from the right, $f(y)$ does not approach $f(x)$. To draw f on the interval $[a, b]$, we must lift the pen at x.

This means that we have to decide both how large a range of points to use and how "dense" a set of points to use. We then draw the graph assuming that the function varies smoothly in between the points where we evaluate the function. For example, this means that we get the same graphs for the polynomial functions whether we consider the polynomials as functions of rational numbers or real numbers.

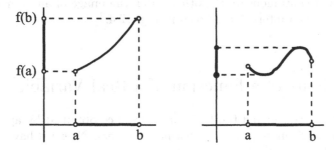

FIGURE 12.3. In the example on the left, f transforms the open interval (a, b) into the open interval $(f(a), f(b))$. In the example on the right, f transforms the open interval (a, b) into the closed interval.

The assumption that the function varies smoothly is critical. Suppose that a function f is Lipschitz continuous with constant L on some set of real numbers and choose a number x in this set. If y is any other real number in the set, then $|f(y) - f(x)| \leq L|y - x|$. If we consider moving y toward x, then this means that the value of $f(y)$ approaches the value of $f(x)$. In

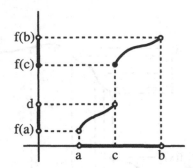

FIGURE 12.4. A discontinuous transformation of an interval (a, b). The image of (a, b) is the union of the disjoint intervals $(f(a), d)$ and $[f(c), f(b))$.

other words, we cannot have any sudden jumps in the value of f from one point to the next near x. This suggests that if f is Lipschitz continuous on an interval, then we can draw its graph on the interval without lifting the pen. Vice versa, if f has a sudden jump in value at a point x, then it cannot be Lipschitz continuous on any interval that contains x. We illustrate in Fig. 12.2. When a function f has a sudden jump in value at a point x, we use solid and open circles to denote the two dangling ends of f at the point x, with the solid circle denoting the value of f at x.

Another way to think about this is to consider functions as transformations. An open interval can be transformed into another open interval or a closed interval or a half-open interval, as we illustrate in Fig. 12.3. If the function is discontinuous on the other hand, the image of an interval does not have to be an interval, as shown in Fig. 12.4.[3]

12.4   Limits of a Function of a Real Variable

Previously, we considered the limit of a sequence obtained by applying a function to a given sequence of distinct numbers. Now we have defined functions of a real variable on an interval I and we are interested in how the function behaves as the inputs in the interval tend to some number c. We say that f **converges** to a limit L as x approaches c in I and write

$$\lim_{x \to c} f(x) = L$$

---

[3]These examples suggest that the image of a real interval under a Lipschitz continuous function is another real interval. We can rephrase this as an equivalent question: Does the image of an interval (a, b) under a Lipschitz continuous function take on every value in between $f(a)$ and $f(b)$? We answer this question in Theorem 13.2.

if for any sequence $\{x_n\}$ with

$$x_n \text{ in } I \text{ for all } n, \quad x_n \neq c \text{ for all } n, \quad \lim_{n\to\infty} x_n = c, \qquad (12.2)$$

we have

$$\lim_{n\to\infty} f(x_n) = L$$

in the usual sense of sequences of numbers. In other words, f converges to L if $\{f(x_n)\}$ converges to L for any sequence $\{x_n\}$ that converges to c but *never actually takes on the value c*.[4]

The fact that we do not allow the sequences actually to reach point c is a subtle but important point. In defining the limit as above, we are interested in the behavior of the function f as the inputs tend to some number c, but *not* necessarily in the value of f at c. In fact, f may not even be defined at the point c, and if defined, its limit may not be the same as its value at c. We illustrate these cases in Fig. 12.5.[5]

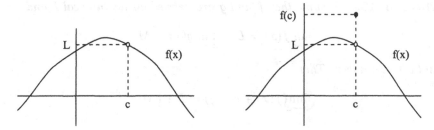

FIGURE 12.5. Both of the functions plotted converge to L as $x \to c$. However, the function on the left is undefined at c and the function on the right has $f(c) \neq L$.

Both of the functions shown in Fig. 12.5 are discontinuous at c. But this is a mild form of discontinuous behavior since the functions have well-defined limits at c. The situation for more "strongly" discontinuous functions is much different. Consider the step function $I(t)$ shown in Fig. 12.6. This function does not have a limit at either 0 or 1, because we can find sequences $\{t_i\}$ that converge to either 0 or 1, for which $\{I(t_i\}$ converges to 1 or 0 or does not even converge at all.

EXAMPLE 12.7. For example, consider applying the step function $I(t)$ to the sequences $\{1/n\}$, $\{-1/n\}$, and $\{(-1)^n/n\}$ for $n \geq 1$. In the first case, $\{I(1/n)\}$ converges to 1 and in the second case $\{I(-1/n)\}$ converges to -1. In the third case, $\{I((-1)^n/n)\} = \{-1, 1, -1, 1, \cdots\}$ does not converge.

---

[4]If I can be understood from the function, it is usually omitted from the definition.

[5]Likewise, when we take the limit of a sequence $\lim_{i\to\infty} a_i$, the index i never actually takes on the value "∞".

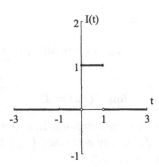

FIGURE 12.6. Plot of the step function $I(t)$.

With this definition, we immediately obtain some useful properties of limits of functions from the properties of limits of sequences of numbers. We ask you to prove the following in Problem 12.12.

Theorem 12.4 *Assume that f and g are defined on an interval I and*

$$\lim_{x \to c} f(x) = L, \quad \lim_{x \to c} g(x) = M$$

and c is a number. Then

$$\lim_{x \to c} (f(x) + cg(x)) = L + cM$$

and

$$\lim_{x \to c} f(x)g(x) = LM.$$

If $M \neq 0$, then

$$\lim_{x \to c} \frac{f(x)}{g(x)} = \frac{L}{M}.$$

Ideally, we have to check this definition of the limit of a function for *every* appropriate sequence, which certainly makes this definition difficult to verify in practice. This motivates seeking another formulation of the definition. The basic idea is that f converges to L at c if we can make $f(x)$ arbitrarily close to L by taking x close to c. We can write this in mathematical terms as given any $\epsilon > 0$ there is a $\delta > 0$ such that

$$|f(x) - L| < \epsilon \text{ for all } x \neq c \text{ in } I \text{ with } |x - c| < \delta.$$

We can show these two definitions are equivalent with a deceptively simple argument. Suppose the second definition holds and let $\{x_n\}$ be a sequence satisfying (12.2). For any $\epsilon > 0$ there is a $\delta > 0$ such that $|f(x) - L| < \epsilon$ for x with $0 < |x - c| < \delta$. Moreover, there is an N such that $|x_n - c| < \delta$ for $n \geq N$. Hence, for $n \geq N$, $|f(x_n) - L| < \epsilon$. So the first definition holds. Now assume the second condition does *not* hold. Then there is an $\epsilon > 0$ such that for every n there is an x_n with $0 < |x_n - c| < 1/n$ but $|f(x_n) - L| \geq \epsilon$. But $\{x_n\}$ converges to c while $\{f(x_n)\}$ does not converge to L, so the first definition cannot hold either.

We summarize as a theorem.

Theorem 12.5 Weierstrass's Characterization of a Limit of a Function *Suppose f is a function defined on an interval I. Then $f(x)$ converges to a number L as x approaches c in I if and only if for every $\epsilon > 0$ there is a $\delta > 0$ such that $|f(x) - L| < \epsilon$ for all x in I with $0 < |x - c| < \delta$.*

We have shown a number of examples of proving a function applied to a given sequence results in a convergent sequence. We finish this section by presenting an example where we verify the alternate formulation.

EXAMPLE 12.8.   We show that $f(x) = x^2$ converges to 1 as x approaches 1. We know this is true from our previous discussion since x^2 is Lipschitz continuous on an interval containing 1, and if $\{x_n\}$ is any sequence converging to 1, then $\lim_{n\to\infty}(x_n)^2 = \left(\lim_{n\to\infty} x_n\right)^2 = 1^2 = 1$.

To verify the alternate formulation, we suppose that $\epsilon > 0$ is given. We want to show that we can make

$$|x^2 - 1| < \epsilon$$

by taking x sufficiently close to 1. For x in $[0, 2]$, $x + 1 < 3$ and therefore $|x^2 - 1| = |x + 1||x - 1| \leq 3|x - 1|$. Hence, if we restrict x in $[0, 2]$ so that $|x - 1| < \epsilon/3$, then

$$|x^2 - 1| = |x - 1||x + 1| \leq |x - 1||x + 1| < \frac{\epsilon}{3}3 = \epsilon.$$

This means the second definition holds with $\delta = \epsilon/3$.

Chapter 12 Problems

12.1. Show Theorem 12.1 is true.

12.2. Complete the proof of Theorem 12.2.

12.3. Complete the proof of Theorem 12.3. *Hint:* It is possible to approximate a real number x by a sequence of rational numbers $\{x_i\}$ that satisfy $x_i \leq x$ or $x_i \geq x$ for all i by using the sequence of truncated decimal expansions obtained by rounding down or up.

12.4. Compute the first 5 terms of the sequence that defines the value of the function $f(x) = \dfrac{x}{x+2}$ at $x = \sqrt{2}$. *Hint:* Follow Fig. 12.1 and use the *evalf* function of $MAPLE^{\copyright}$ in order to determine all the digits.

12.5. Let $\{x_i\}$ be the sequence with $x_i = 3 - \dfrac{2}{i}$ and $f(x) = x^2 - x$. What is the limit of the sequence $\{f(x_i)\}$?

12.6. Show that $|x|$ is Lipschitz continuous on the real numbers $\mathbb{R}$.

12.7. Construct and draw a function that is defined on $[0, 1]$, Lipschitz continuous on $(0, 1)$, but not Lipschitz continuous on $[0, 1]$.

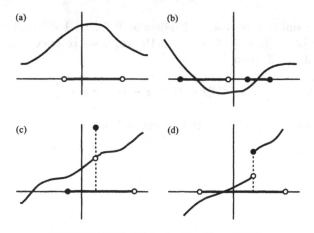

FIGURE 12.7. Plots for Problem 12.11.

12.8. Assume that the two sequences $\{x_i\}_{i=1}^{\infty}$ and $\{y_i\}_{i=1}^{\infty}$ have the same limit

$$\lim_{i \to \infty} x_i = \lim_{i \to \infty} y_i = x$$

and the elements x_i and y_i are all in an interval I containing x. If g is Lipschitz continuous on I, prove that

$$\lim_{i \to \infty} g(x_i) = \lim_{i \to \infty} g(y_i).$$

Problems 12.9–12.11 have to do with plotting a function of a real variable.

12.9. Produce an interval that contains all the points $3 - 10^{-j}$ for $j \geq 0$ but does not contain 3.

12.10. Using $MATLAB^{©}$ or $MAPLE^{©}$, graph the following functions on one graph: $y = 1 \times x$, $y = 1.4 \times x$, $y = 1.41 \times x$, $y = 1.414 \times x$, $y = 1.4142 \times x$, $y = 1.41421 \times x$. Use your results to explain how you could graph the function $y = \sqrt{2} \times x$.

12.11. In the plots shown in Fig. 12.7, draw the images of the indicated sets of real numbers under the indicated functions.

Problems 12.12–12.14 have to do with limits of a function of a real variable.

12.12. Verify Theorem 12.4.

12.13. Show that x^3 converges to 8 as x approaches 2 using the alternate formulation in Theorem 12.5.

12.14. Show that $1/x$ converges to 1 as x approaches 1 using the alternate formulation in Theorem 12.5.

13
The Bisection Algorithm

We turn the focus onto developing a general method for solving mathematical models. It turns out that the Bisection Algorithm used to approximate $\sqrt{2}$ can serve equally well to approximate a root of any Lipschitz continuous function f in a given interval $[a, b]$ provided $[a, b]$ has the property that $f(a)$ has the opposite sign as $f(b)$. In this chapter, we describe how to use the Bisection Algorithm to solve a general root problem and show that it converges. As an application, we use it to solve a difficult chemical model. We also discuss some practical matters that arise when using the Bisection Algorithm.

13.1 The Bisection Algorithm for General Root Problems

Given a function f, the problem is to compute a root

$$f(\bar{x}) = 0 \tag{13.1}$$

in a given interval $[a, b]$. We assume that f is Lipschitz continuous on $[a, b]$ and $f(a)$ and $f(b)$ have opposite signs.[1]

As with the Muddy Yard model, the Bisection Algorithm produces two sequences $\{x_i\}$ and $\{X_i\}$ that are the endpoints of intervals $[x_i, X_i]$ that

---

[1] An efficient way to check this in practice is to verify that $f(a)f(b) < 0$.

contain a root $\bar{x}$ of (13.1) and become smaller as i increases. Since $f(a)$ and $f(b)$ have the opposite signs we set $x_0 = a$ and $X_0 = b$.

In the first step, we check the sign of f at the midpoint $\bar{x}_1 = (x_0 + X_0)/2$. If $f(\bar{x}_1) = 0$, then we have found a root and we stop. Otherwise $f(\bar{x}_1)$ has the opposite sign as one of $f(x_0)$ or $f(X_0)$. We set $x_1 = x_0$ and $X_1 = \bar{x}_1$ if $f(x_0)$ and $f(\bar{x}_1)$ have the opposite sign. Otherwise, we set $x_1 = \bar{x}_1$ and $X_1 = x_1$.

In step 2, we compare the sign of $f(\bar{x}_2)$ at the midpoint $\bar{x}_2 = (x_1 + X_1)/2$ with the signs of $f(x_1)$ and $f(X_1)$. If $f(\bar{x}_2) = 0$ we stop; otherwise we define the new interval $[x_2, X_2]$ using the points $\{x_1, \bar{x}_2, X_1\}$ so that $f(x_2)$ and $f(X_2)$ have opposite signs.

Continuing this process, we produce a sequence of intervals $[x_i, X_i]$ with $f(x_i)$ and $f(X_i)$ having the opposite signs. The algorithm can be described like this.

Algorithm 13.1 Bisection Algorithm

1. Set the initial values $x_0 = a$ and $X_0 = b$ where $f(a)$ and $f(b)$ have opposite signs.

2. Given two rational numbers x_{i-1} and X_{i-1} with the property that $f(x_{i-1})$ and $f(X_{i-1})$ have the opposite signs, set $\bar{x}_i = (x_{i-1} + X_{i-1})/2$.

   - If $f(\bar{x}_i) = 0$, stop.
   - If $f(\bar{x}_i)f(X_{i-1}) < 0$, set $x_i = \bar{x}_i$ and $X_i = X_i$.
   - If $f(\bar{x}_i)f(x_{i-1}) < 0$, set $x_i = x_i$ and $X_i = \bar{x}_i$.

3. Increase i by 1 and go back to step 2 as desired.

13.2   Solving the Model of Chemical Equilibrium

In Section 4.5, we derived a model

$$S(.02 + 2S)^2 - 1.57 \times 10^{-9} = 0 \tag{13.2}$$

that gives the solubility S of $Ba(IO_3)_2$ in a .020 mol/L solution of KIO_3. We use the Bisection Algorithm to solve (13.2).

Unfortunately, as posed, the roots of (13.2) are very small, which makes it difficult to graph the function and find the initial interval $[a, b]$. We show the plot in Fig. 13.1. So we first *rescale* the problem to more convenient variables. Rescaling a problem in order to make it easier to find the roots is often necessary in practice. We rescale the variables in this problem for the same reason we change from kilometers to meters when measuring how far a baby crawls in 5 minutes. We can measure the distance in either unit

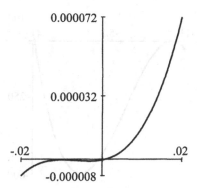

FIGURE 13.1. A plot of the function $S(.02 + 2S)^2 - 1.57 \times 10^{-9}$ in the model (13.2). Apparently there are roots near $-.01$ and 0, but it is hard to see them!

but kilometers produces awkwardly small results, at least if we want to do some boasting.[2]

We first multiply both sides of (13.2) by 10^9 to get

$$10^9 \times S(.02 + 2S)^2 - 1.57 = 0.$$

Next, we write

$$10^9 \times S(.02 + 2S)^2 = 10^3 \times S \times (10^3)^2 \times (.02 + 2S)^2$$
$$= 10^3 \times S \times (10^3 \times (.02 + 2S))^2$$
$$= 10^3 \times S \times (20 + 2 \times 10^3 \times S)^2.$$

If terms of the new variable $x = 10^3 S$, we want to find the roots of

$$f(x) = x(20 + 2x)^2 - 1.57 = 0. \qquad (13.3)$$

If we find a root x of (13.3), then we can find the physical value by taking $S = 10^{-3}x$.

This new function, shown in Fig. 13.2, has much more reasonably sized coefficients and the roots are not nearly as small as in the original formulation. Moreover, f is a polynomial and therefore Lipschitz continuous on any bounded interval. It appears that f might have one root near 0 and another root near -10, but we ignore the negative root, if it exists, because we cannot have "negative" solubility.

Since the positive root of (13.3) is near 0, we choose $x_0 = -.1$ and $X_0 = .1$ and apply Algorithm 13.1 for 20 steps. We show the results in Fig. 13.3. This suggests that the root of (13.3) is $x \approx .00392$ or $S \approx 3.92 \times 10^{-6}$.

---

[2]Unfortunately, there is no real "technique" to rescaling variables. It just takes practice and experience.

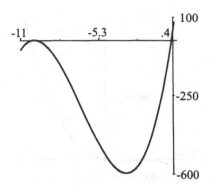

FIGURE 13.2. A plot of the function $f(x) = x(20 + 2x)^2 - 1.57$ in (13.3).

13.3   The Bisection Algorithm Converges

To show that the Bisection Algorithm converges to a root of (13.1), we show that the sequence $\{x_i\}$ is a Cauchy sequence, and hence has a limit, and then show that the limit is a root.

The convergence of the algorithm is the same as for computing $\sqrt{2}$. At step i, either f is zero at the midpoint $\bar{x}_i$ of x_{i-1} and X_{i-1} and we have computed a root, i.e. $f(\bar{x}_i) = 0$, or $f(\bar{x}_i)$ has the opposite sign of either $f(x_{i-1})$ or $f(X_{i-1})$ and we get a new interval $[x_i, X_i]$ which is half the size of the previous interval. After i steps, we conclude that

$$0 \le X_i - x_i \le 2^{-i}(X_0 - x_1) = 2^{-i}(b - a).$$

Arguing as when we computed $\sqrt{2}$, we find that

$$|x_i - x_j| \le 2^{-i}(b - a) \quad \text{if } j \ge i. \tag{13.4}$$

This means that $\{x_i\}$ is a Cauchy sequence and therefore converges to a unique real number $\bar{x}$.

To verify that $\bar{x}$ is a root of f, we use the definition

$$f(\bar{x}) = f(\lim_{i \to \infty} x_i) = \lim_{i \to \infty} f(x_i).$$

This makes sense because f is Lipschitz continuous on $[a, b]$. Now suppose that $f(\bar{x})$ is not zero, say, for example, $f(\bar{x}) > 0$. Since f is Lipschitz continuous in an interval around $\bar{x}$, the values of $f(x)$ are close to $f(\bar{x})$ for all points x close to $\bar{x}$. Since $f(\bar{x}) > 0$, this means that $f(x) > 0$ for x close to $\bar{x}$.

More precisely if we choose $\delta > 0$ sufficiently small, then f is positive for all points in the interval $(\bar{x} - \delta, \bar{x} + \delta)$ (see Fig. 13.4). But if we choose i

i	x_i	X_i
0	-0.10000000000000	0.10000000000000
1	0.00000000000000	0.10000000000000
2	0.00000000000000	0.05000000000000
3	0.00000000000000	0.02500000000000
4	0.00000000000000	0.01250000000000
5	0.00000000000000	0.00625000000000
⋮	⋮	⋮
10	0.00390625000000	0.00410156250000
⋮	⋮	⋮
15	0.00391845703125	0.00392456054688
⋮	⋮	⋮
20	0.00392189025879	0.00392208099365

FIGURE 13.3. 20 steps of the Bisection Algorithm applied to (13.3) using $x_0 = -.1$ and $X_0 = .1$.

so $2^{-i} < \delta$, then x_i and X_i are both within δ of $\bar{x}$ and so $f(x_i)$ and $f(X_i)$ are both positive (see Fig. 13.4). But this contradicts the choice of x_i and X_i in the Bisection Algorithm, since $f(x_i)$ and $f(X_i)$ must be opposite in sign. A similar argument works for $f(\bar{x}) < 0$. Therefore, $f(\bar{x}) = 0$.

We summarize this as a theorem.

Theorem 13.1 Bolzano's Theorem *If f is Lipschitz continuous in an interval $[a, b]$ and $f(a)$ and $f(b)$ have opposite signs, then f has at least one root in (a, b) and the Bisection Algorithm starting with $x_0 = a$ and $X_0 = b$ converges to a root of f in (a, b).*[3]

We name this theorem after Bolzano,[4] who proved an early version.

One consequence of Bolzano's theorem is the following well-known and important theorem, which we ask you to prove in Problem 13.9.

---

[3]There can very well be more than one root of f in (a, b), and if there is more than one root, it is unclear which root is found by the Bisection Algorithm.

[4]Bernard Placidus Johann Nepomuk Bolzano (1781–1848) lived and worked in what is now the Czech Republic. Ordained as a Roman Catholic priest, he held positions as a professor of theology and philosophy while devoting considerable time to mathematics. Bolzano was particularly concerned about the foundations of mathematics. He attempted to place calculus on a more rigorous foundation by eliminating "infinitesimals." He also investigated infinite sets and infinity, anticipating Cantor. Bolzano used a modern notion of continuity of a function and derived both his namesake theorem and the Intermediate Value Theorem. However, his proof was incomplete because he lacked a rigorous theory of real numbers. Bolzano also used the idea of a Cauchy sequence a few years before Cauchy, though Cauchy was probably ignorant of this. Much of Bolzano's work was never published, which diminished the impact of his results.

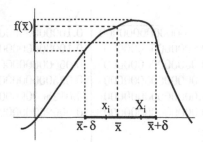

FIGURE 13.4. If f is Lipschitz continuous and $f(\bar{x}) > 0$, then f is positive at all nearby points.

Theorem 13.2 Intermediate Value Theorem *Suppose that f is Lipschitz continuous on an interval $[a, b]$. Then for every d between $f(a)$ and $f(b)$, there is at least one point c between a and b such that $f(c) = d$.*

Regarding the discussion in Section 12.3, this theorem implies that the image of an interval under a Lipschitz continuous function is another interval.

13.4   When to Stop the Bisection Algorithm

Now that we know that $\{x_i\}_{i=0}^\infty$ converges to $\bar{x}$, it would be useful to know how quickly the sequence converges. In other words, we would like to have an estimate of the error of the iteration

$$|x_i - \bar{x}| = \left| x_i - \lim_{j \to \infty} x_j \right| \qquad (13.5)$$

for any i. Remember that we do not know $\bar{x}$, so we cannot simply compute $|\bar{x} - x_i|$! It is important to have an estimate on (13.5), for example, in order to know how many iterations of the Bisection Algorithm to perform in order to determine the value of $\bar{x}$ to a required accuracy.

The difference (13.5) can be made arbitrarily small by taking i large but we want more precise information. Now, if $j \geq i$, then x_j agrees to more decimal places with $\bar{x}$ than x_i. So for large j, x_j is a lot closer to $\bar{x}$ than x_i and $|x_i - x_j|$ is a good approximation of $|x_i - \bar{x}|$. We estimate using the triangle inequality

$$|x_i - \bar{x}| = |(x_i - x_j) + (x_j - \bar{x})|$$
$$\leq |x_i - x_j| + |x_j - \bar{x}|.$$

This estimates the distance between x_i and $\bar{x}$ in terms of the distance between x_i and x_j and the distance between x_j and $\bar{x}$. Now given any $\epsilon > 0$, $|x_j - \bar{x}| \leq \epsilon$ if j is sufficiently large. So (13.4) implies that for any $\epsilon > 0$,

$$|x_i - \bar{x}| \leq 2^{-i}(b - a) + \epsilon.$$

Since ϵ can be arbitrarily small, we conclude

$$|x_i - \bar{x}| \leq 2^{-i}(b - a).$$

EXAMPLE 13.1. Since $2^{-10} \approx 10^{-3}$, we gain approximately 3 decimal places for every 10 successive steps of the Bisection Algorithm. We can see this predicted gain in accuracy in the numbers listed in Fig. 10.3 and Fig. 13.3, for example.

13.5 Power Functions

Now that we have the Bisection Algorithm to compute roots, we can define a^r for any positive real number a and real number r. So far we have only defined a^r when r is an integer.

We first consider the case that r is rational, i.e., $r = p/q$ for integers p and q. For $a > 0$, we define $a^{p/q}$ to be the positive root of

$$f(x) = x^q - a^p = 0. \tag{13.6}$$

Such a root exists by the Intermediate Value Theorem since a^p is a fixed positive number so $x^q > a^p$ for all sufficiently large x and likewise $x^q < a^p$ for $x = 0$. If we define $x_0 = 0$ and choose X_0 sufficiently large, then the Bisection Algorithm started on $[x_0, X_0]$ converges to a root of (13.6).

EXAMPLE 13.2. Recall that we defined the $2^{1/2} = \sqrt{2}$ as the root of $f(x) = x^2 - 2^1$ and computed its value by applying the Bisection Algorithm starting with the interval $[1, 2]$, where $f(1) < 0$ and $f(2) > 0$.

Using this definition, it is possible to show that the properties of exponents, such as $a^r a^s = a^{r+s}$, $a^{-r} = 1/a^r$, and $(a^r)^s = a^{rs}$, that hold for integer powers also hold for rational powers. However, it is difficult to do this now, while it is easy after defining the logarithmic function. So we delay proving these properties until Chapter 28.

Based on the discussion about real numbers, a^r is defined for a real number r by taking the limit of a^{r_i} as $i \to \infty$ where r_i is the truncated decimal expansion of r with i decimal digits. But to show this makes sense, we have to show that a^r is Lipschitz continuous in r and that is not so easy using this definition. So again we delay discussing this definition further until we introduce the logarithmic function, which makes it all much easier.

With the ability to compute the value of a^r for any non-negative a given a real number r, it is natural to define the **power function** with power r,

$$x^r,$$

defined for x in the set of non-negative real numbers. Here we consider r to be a fixed real number. Again we delay discussing the details until we can use the logarithm, but it is possible to show that x^r is Lipschitz continuous on bounded intervals for any $r \geq 1$ and Lipschitz continuous on real intervals $[a, b]$ with $a > 0$ for any $0 \leq r < 1$. We show plots of x^r for these two cases in Fig. 13.5.

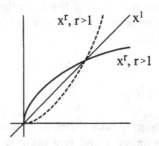

FIGURE 13.5. Plots of x^r for $r < 1$, $r = 1$, and $r > 1$.

EXAMPLE 13.3. We verify that $x^{1/2} = \sqrt{x}$ is Lipschitz continuous on any bounded interval $[a, b]$ with $a > 1$. The properties of exponents imply that

$$\left(x^{1/2} - y^{1/2}\right)\left(x^{1/2} + y^{1/2}\right) = \left((x^{1/2})^2 - (y^{1/2})^2\right) = x - y;$$

so

$$\left|x^{1/2} - y^{1/2}\right| = \frac{x - y}{\left|x^{1/2} - y^{1/2}\right|} \leq \frac{1}{2a^{1/2}}|x - y|,$$

provided that $x \geq a > 0$ and $y \geq a > 0$. Therefore the Lipschitz constant of $\sqrt{x}$ on $[a, b]$ is $1/2\sqrt{a}$. When a is smaller, the constant becomes larger. If we examine the graph of x^r for $r < 1$ in Fig. 13.5, we see that when x is near 0, the function makes large changes for small changes in input.

13.6 Computing Roots by the Decasection Algorithm

It turns out that there are many different ways to compute a root of a function. The choice of the method depends on the circumstances of the problem we have to solve.

To illustrate how a different method can work, we describe a variation of the Bisection Algorithm called the Decasection Algorithm. Like the Bisection Algorithm, the Decasection Algorithm produces a sequence of numbers

$\{x_i\}_{i=0}^{\infty}$ that converges to a root $\bar{x}$. However with the Decasection Algorithm, there is a close connection between the index i of x_i and the number of decimal places x_i and $\bar{x}$ have in common.

The Decasection Algorithm looks the same as the Bisection Algorithm except that at each step the current interval is divided into 10 subintervals instead of 2. We start as before by choosing $x_0 = a$ and $X_0 = b$ so that $f(x_0)f(X_0)) < 0$. Next we compute the value of f at nine equally spaced points between x_0 and X_0. More precisely, we set $\delta_0 = (X_0 - x_0)/10$ and check the signs of f at the points x_0, $x_0 + \delta_0$, $x_0 + 2\delta_0$, $\cdots$, $x_0 + 9\delta_0$, $x_0 + 10\delta_0 = X_0$. There must be two consecutive points at which f has opposite signs, so we set x_1 and X_1 to be two consecutive points where $f(x_1)f(X_1) < 0$.

Now we continue the algorithm by evaluating f at nine equally spaced points $x_1 + \delta_1$, $x_1 + 2\delta_1$, $\cdots$, $x_1 + 9\delta_1$ with $\delta_1 = (X_1 - x_1)/10$. We choose two consecutive numbers x_2 and X_2 from among x_1, $x_1 + \delta_1$, $x_1 + 2\delta_1$, $\cdots$, $x_1 + 9\delta_1$, X_1 with $f(x_2)f(X_2) < 0$. We then proceed to compute $[x_3, X_3]$ and so forth.

By construction

$$|x_i - X_i| \le 10^{-i}(b - a)$$

and the same argument used for the Bisection Algorithm implies that

$$\lim_{i \to \infty} x_i = \bar{x} \quad \text{and} \quad \lim_{i \to \infty} X_i = \bar{x}; \tag{13.7}$$

and, moreover,

$$|x_i - \bar{x}| \le 10^{-i}(b - a).$$

So we gain approximately one digit of accuracy for each step of the Decasection Algorithm.

We show the first 14 steps of this algorithm applied to $f(x) = x^2 - 2$ starting on $[1, 2]$ in Fig. 13.6.

Once we have more than one method to compute a root of a function, it is natural to ask which method is "best." We have to decide what we mean by "best" of course. For this problem, best might mean "most accurate" or "cheapest," for example.

However, for this problem, accuracy is apparently not an issue since both the Decasection and Bisection algorithms can be executed until we get 16 places or whatever number of digits is used for floating point representation. Therefore the way to compare the methods is by the amount of computing time it takes to achieve a given level of accuracy. This computing time is often called the **cost** of the computation, a holdover from the days when computer time was actually purchased by the second.

The cost involved in one of these algorithms can be determined by figuring out the cost per iteration and then multiplying by the total number of iterations we need to reach the desired accuracy. In one step of the Bisection Algorithm, the computer must compute the midpoint between two

i	x_i	X_i
0	1.00000000000000	2.00000000000000
1	1.40000000000000	1.50000000000000
2	1.41000000000000	1.42000000000000
3	1.41400000000000	1.41500000000000
4	1.41420000000000	1.41430000000000
⋮	⋮	⋮
9	1.41421356200000	1.41421356300000
⋮	⋮	⋮
14	1.41421356237309	1.41421356237310

FIGURE 13.6. 14 steps of the Decasection Algorithm applied to $f(x) = x^2 - 2$ implemented in a $MATLAB^{©}$ m-file.

points, evaluate the function f at that point and store the value temporarily, check the sign of the function value, and then store the new x_i and X_i. We assume that the time it takes for the computer to do each of these operations can be measured and we define

$$C_m = \text{cost of computing the midpoint}$$
$$C_f = \text{cost of evaluating } f \text{ at a point}$$
$$C_{\pm} = \text{cost of checking the sign of a variable}$$
$$C_s = \text{cost of storing a variable.}$$

The total cost of one step of the bisection algorithm is

$$C_m + C_f + C_{\pm} + 4C_s,$$

and the cost after N_b steps is

$$N_b(C_m + C_f + C_{\pm} + 4C_s). \tag{13.8}$$

One step of the Decasection Algorithm has a considerably higher cost because there are 9 intermediate points to check. The total cost after N_d steps of the Decasection Algorithm is

$$N_d(9C_m + 9C_f + 9C_{\pm} + 20C_s). \tag{13.9}$$

On the other hand, the difference $|x_i - \bar{x}|$ decreases by a factor of $1/10$ after each step of the Decasection Algorithm as compared to a factor of $1/2$ after each step of the Bisection Algorithm. Since $1/2^3 > 1/10 > 1/2^4$, this means that the Bisection Algorithm requires between 3 and 4 times as many steps as the Decasection Algorithm in order to reduce the initial size $|x_0 - \bar{x}|$ by a given factor. So $N_b \approx 4N_d$. This gives the cost of the Bisection Algorithm as

$$4N_d(C_m + C_f + C_{\pm} + 4C_s) = N_d(4C_m + 4C_f + 4C_{\pm} + 16C_s)$$

as compared to (13.9). This means that the Bisection Algorithm is cheaper to use than the Decasection Algorithm.

Chapter 13 Problems

13.1. Implement Algorithm 13.1 in a program to find a root of a general function f. Test your program by computing a root of $f(x) = x^2 - 2$ starting with the interval $[1, 2]$ and comparing the results to Fig. 10.3.

Problems 13.2–13.5 involve using the Bisection Algorithm to solve a model equation. The program from Problem 13.1 would be useful.

13.2. In the model for the solubility of $Ba(IO_3)_2$, suppose that K_{sp} for $Ba(IO_3)_2$ is 1.8×10^{-5}. Find the solubility S to 10 decimal places.

13.3. In the model for the solubility of $Ba(IO_3)_2$, determine the solubility of $Ba(IO_3)_2$ in a .037 mol/L solution of KIO_3 to 10 decimal places.

13.4. The power P delivered into a load R of a simple class A amplifier of output resistance Q and output voltage E is

$$P = \frac{E^2 R}{(Q + R)^2}.$$

Find all possible solutions R for $P = 1$, $Q = 3$, and $E = 4$ to 10 decimal places.

13.5. Van der Waal's model for one mole of an ideal gas, including the effects of the size of the molecules and the mutual attractive forces, is

$$\left(P + \frac{a}{V^2}\right)(V - b) = RT,$$

where P is the pressure, V is the volume of the gas, T is the temperature, R is the ideal gas constant, a is a constant depending on the size of the molecules and the attractive forces, and b is a constant depending on the volume of all the molecules in one mole. Find all possible volumes V of the gas corresponding to $P = 2$, $T = 15$, $R = 3$, $a = 50$, and $b = .011$ to 10 decimal places.

Problems 13.6–13.8 are concerned with the accuracy of the Bisection Algorithm. The program from Problem 13.1 would be useful.

13.6. (a) Compute 30 steps of the Bisection Algorithm applied to $f(x) = x^2 - 2$ starting with (1) $x_0 = 1$ and $X_0 = 2$; (2) $x_0 = 0$ and $X_0 = 2$; (3) $x_0 = 1$ and $X_0 = 3$; and (4) $x_0 = 1$ and $X_0 = 20$. Compare the errors $|x_i - \sqrt{2}|$ of the results at each step and explain any observed differences in accuracy.

(b) Using the results from (a), plot (1) $|X_i - x_i|$ versus i; (2) $|x_i - x_{i-1}|$ versus i; and (3) $|f(x_i)|$ versus i. In each case, determine if the plotted quantity decreases by a factor of $1/2$ after each step.

13.7. Display the output of 40 steps of the Bisection Algorithm applied to $f(x) = x^2 - 2$ using $x_0 = 1$ and $X_0 = 2$. Describe anything you notice about the last 10 values x_i and X_i and explain what you see. *Hint:* Consider the floating point representation on the computer you use.

13.8. Apply the Bisection Algorithm to the function

$$f(x) = \begin{cases} x^2 - 1, & x < 0 \\ x + 1, & x \geq 0, \end{cases}$$

starting with $x_0 = -.5$ and $X_0 = 1$. Explain the results.

Problems 13.9 and 13.10 have to do with the Intermediate Value Theorem.

13.9. Show that Theorem 13.2 is true.

13.10. Modify Algorithm 13.1 to get a program that computes a point c with $f(c) = d$, where d is any number between $f(a)$ and $f(b)$ and f is Lipschitz continuous on $[a, b]$. Test it by finding the point where $f(x) = x^3$ equals 9, noting that $f(2) = 8$ and $f(3) = 27$.

Problem 13.11 has to do with the power function. Review Example 13.3 before doing the problem.

13.11. (a) Prove that $x^{1/3}$ is Lipschitz continuous on any bounded interval $[a, b]$ with $a > 0$. (b) Prove that $x^{3/2}$ is Lipschitz continuous on any bounded interval $[a, b]$ with $a \geq 0$.

Problems 13.12 and 13.13 have to do with modifications of the Bisection Algorithm.

13.12. (a) Write down an algorithm for the Decasection Algorithm in a form similar to that of Algorithm 13.1. (b) Program the algorithm and then compute $\sqrt{2}$ to 15 places. (c) Show that (13.7) holds. (d) Show that (13.9) is valid.

13.13. (a) Devise a Trisection Algorithm to compute a root $f(\bar{x}) = 0$. (b) Implement the Trisection Algorithm. (c) Compute $\sqrt{2}$ to 15 places. (c) Show that the endpoints produced by the Tridiagonal Algorithm form a Cauchy sequence. (d) Show that limit of the sequence is $\bar{x}$; (e) Estimate on $|x_i - \bar{x}|$. (e) Compute the cost of the Tridiagonal Algorithm from Problem 13.13 and compare to the costs of the Bisection and Decasection Algorithms.

14

Inverse Functions

Possessing the ability to solve general root problems yields a surprising range of benefits. As a particular application, we study the process of "undoing" a given function, by which we mean reversing the action of a function to trace from a given output back to the corresponding input. This is called inverting a function and the function that undoes the action of a given function is called its inverse function. Finding the inverse function to a given function generalizes the problem of solving a root problem for the function. The concept of the inverse function is very powerful, and we use the idea of inverse functions to derive some particular functions, like root functions and later exponential functions and the inverse trigonometric functions.

We investigate inverse functions in two ways. First, we conduct a "geometric" investigation based on graphs. After figuring out what happens using pictures, we go back and re-derive the results analytically. The analytic investigation applies to wider circumstances so it is more generally useful.

14.1 A Geometric Investigation

In Fig. 14.1, we recall the idea that a function "sends" an input point x to an output point $y = f(x)$. More precisely, leaving from the point x on the x axis, we follow a vertical line up to where it intersects the graph of f and then trace a horizontal line over to the y axis. We say that the **image** of x is y.

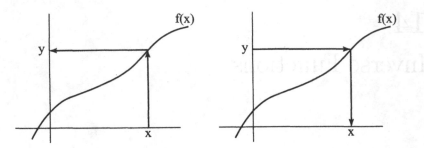

FIGURE 14.1. (Left) A function sends the point x to the point y. (Right) To compute the inverse function, we start at the input y and trace back to find the corresponding input x.

The idea behind the **inverse function** is to reverse this process and start with the output y and find the corresponding input x. Visually, we can think of following a horizontal line over from y to the graph of f and then following a vertical line down to x, as shown on the right in Fig. 14.1. The function that does this is called the inverse function of f and is written f^{-1}. We emphasize that

$$f^{-1} \neq (f)^{-1} = \frac{1}{f}.$$

To get the graph of an inverse function in the usual coordinate system with inputs running horizontally, we can lean our heads over to the right so that the y axis looks horizontal. We see the graph shown on the left in Fig. 14.2. Now the problem is that the usual meaning of right and left have

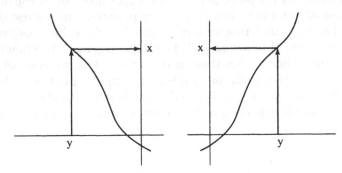

FIGURE 14.2. To get the graph of the inverse function, we first lean our heads over to the right so the y-axis looks horizontal, giving the graph on the left. To get the usual orientation with positive numbers on the right, we have to switch the right and left hands of the graph as shown on the right.

been reversed, with the left hand denoting positive numbers and the right negative numbers. To fix this, we have to switch the right and left hands, as shown on the left in Fig. 14.2. The resulting curve is the graph of the inverse function.

We observe this picture can be obtained by reflecting the graph of the function through the line $y = x$, as shown in Fig. 14.3. Note that we switch

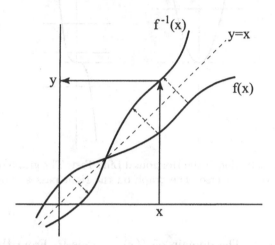

FIGURE 14.3. To get the graph of the inverse function f^{-1}, we reflect the graph of the function f through the line $y = x$.

y and x to fit the axes.

The question is, when do we get a function when we perform this reflection? Recall that a graph represents a function when it passes the **Vertical Line Test**, which says that any vertical line intersects the graph at most at one point. We illustrate in Fig. 14.4. Therefore, the reflected graph must

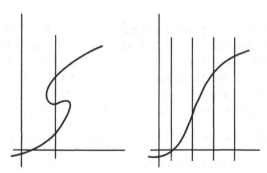

FIGURE 14.4. Illustration of the Vertical Line Test. The graph on the left fails and does not represent a function. The graph on the right passes the test.

pass the Vertical Line Test. Back in the original coordinates, vertical lines correspond to horizontal lines. So we get the **Horizontal Line Test**, which states that a function has an inverse function if any horizontal line intersects its graph at most at one point. We illustrate in Fig. 14.5.

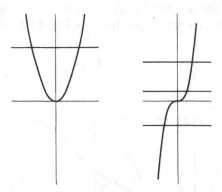

FIGURE 14.5. Illustration of the Horizontal Line Test. The graph on the left fails and does not have an inverse. The graph on the right passes and does have an inverse.

EXAMPLE 14.1. The domain of $f(x) = x^3$ is $\mathbb{R}$. From the graph, it appears that x^3 passes the Horizontal Line Test. We can prove this by arguing that if $x_1 < x_2$, then $x_1^3 = x_1 \times x_1 \times x_1 < x_2^3 = x_2 \times x_2 \times x_2$. Therefore, x^3 has an inverse function, which we denote $f^{-1}(x) = \sqrt[3]{x}$, that is also defined on $\mathbb{R}$. We find

$$f(f^{-1}(x)) = (\sqrt[3]{x})^3 = x$$
$$f^{-1}(f(x)) = \sqrt[3]{x^3} = x$$

for all x.

When does a function pass the Horizontal Line Test? In Fig. 14.5, we show a function that fails on the left and one that passes the test on the right. The difference between these two graphs is that the function on the left is first decreasing and then increasing in value while the function on the right is always increasing as x increases. A function is called **monotone** when it is either always increasing or always decreasing in value.[1]

EXAMPLE 14.2. Any line that is not horizontal is monotone.

EXAMPLE 14.3. The domain of $f(x) = x^2$ is all real numbers, but this function is decreasing for $x < 0$ and increasing for $x > 0$ and therefore fails the Horizontal Line Test and does not have an inverse.

---

[1]Some authors call a function that is either always increasing or decreasing in value **strictly monotone** while a monotone function is merely either nonincreasing or nondecreasing in value.

Geometrically, we have proved

Theorem 14.1 Inverse Function Theorem *A function has an inverse if and only if it is monotone. When a function has an inverse, the graph of the inverse function is obtained by reflecting the graph of the function through the line $y = x$.*

A function that has an inverse is said to be **invertible**.

Note that we can often take a "piece" of a graph to get a function that is invertible. We call the resulting function a **restriction** of the original function.

EXAMPLE 14.4. By taking part of the graph of the function plotted on the left in Fig. 14.5, we get an invertible function (see Fig. 14.6). Note that we can also take the left-hand part of the function and get

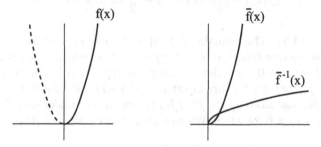

FIGURE 14.6. We take the right half of the graph of the function plotted on the left in Fig. 14.5 and obtain an invertible function $\bar{f}$. We plot $\bar{f}$ and $\bar{f}^{-1}$ on the right.

an invertible function.

EXAMPLE 14.5. The function $f(x) = x^2$ for $x \geq 0$ is monotone on $x \geq 0$ and hence has an inverse, which is $f^{-1}(x) = \sqrt{x}$.

14.2   An Analytic Investigation

Armed with a geometric understanding, we now discuss the subject of inverse functions using analysis. One reason to take a different approach is that when we study functions of several variables, it is difficult to get a good geometric picture.

We begin with the definition. The functions f and g are said to be **inverse functions** if:

1. For every x in the domain of g, $g(x)$ is in the domain of f and $f(g(x)) = x$.

2. For every x in the domain of f, $f(x)$ is in the domain of g and $g(f(x)) = x$.

In this case, we write $g = f^{-1}$ and $f = g^{-1}$.

We can see the idea that g "undoes" the action of f, and vice versa, in this definition. But we have to be careful. In order to evaluate $f(g(x))$, we must assume that $g(x)$ is a value in the domain of f, for example.

EXAMPLE 14.6. The domain of $f(x) = 2x - 1$ is $\mathbb{R}$. The inverse function is $f^{-1}(x) = \frac{1}{2}(x+1)$, which is also defined on $\mathbb{R}$. Therefore, there is no problem to compute $f(f^{-1}(x))$ or $f^{-1}(f(x))$ for all x. We find

$$f(f^{-1}(x)) = 2f^{-1}(x) - 1 = 2 \times \frac{1}{2}(x+1) - 1 = x$$

$$f^{-1}(f(x)) = \frac{1}{2}(f(x) + 1) = \frac{1}{2}(2x + 1 - 1) = x.$$

EXAMPLE 14.7. The domain of $f(x) = 1/(x-1)$ is all real numbers $x \neq 1$. The inverse function is $f^{-1}(x) = 1 + 1/x$, which is defined on all real numbers $x \neq 0$. In order to compute $f(f^{-1}(x))$, we have to make sure that $f^{-1}(x) \neq 1$ for any input x. But $1 + 1/x \neq 1$ for any x, so that is okay. Likewise to compute $f^{-1}(f(x))$, we have to make sure $f(x) \neq 0$. But $1/(x-1) \neq 0$ for all x. We can therefore compute without fear

$$f(f^{-1}(x)) = \frac{1}{1 + 1/x - 1} = x, \quad x \neq 0$$

$$f^{-1}(f(x)) = 1 + \frac{1}{1/(x-1)} = x, \quad x \neq 1.$$

How do we compute an inverse function? The analog of reflecting the graph of a function through the line $y = x$ is to switch the variables y and x in the equation $y = f(x)$ to get $x = f(y)$ and then trying to solve for y.

EXAMPLE 14.8. Given $f(x) = 2x - 1$, we write $y = 2x - 1$. Switching y and x gives $x = 2y - 1$, which finally leads to $f^{-1}(x) = y = \frac{1}{2}(x+1)$. These computations are valid for all x in $\mathbb{R}$.

EXAMPLE 14.9. In fact the inverse function of $f(x) = mx + b$, $m \neq 0$, is $f^{-1}(x) = \frac{1}{m}(x - b)$.

EXAMPLE 14.10. Given $f(x) = 1/(x-1)$, we write $y = 1/(x-1)$. Switching y and x gives $x = 1/(y-1)$, which finally leads to $f^{-1}(x) = y = 1 + 1/x$. These computations are valid for all real $x \neq 0$ and $x \neq 1$.

Note this procedure is not guaranteed to work.

EXAMPLE 14.11. Given $f(x) = x^2$, we write $y = x^2$. Switching y and x gives $x = y^2$. Now when we try to solve, we get $y = \pm\sqrt{x}$; in other words there are two possible values of y for each valid input x. This reflects the fact that $f(x) = x^2$ does not have an inverse function since its graph does not pass the Horizontal Line Test. In fact, this is the function plotted on the left in Fig. 14.5.

When does a function have an inverse? Suppose that f is defined and is Lipschitz continuous on the interval $[a, b]$ and

$$\alpha = f(a) \text{ and } \beta = f(b).$$

We illustrate in Fig. 14.7

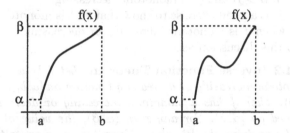

FIGURE 14.7. Two examples of functions on $[a, b]$.

By the Intermediate Value Theorem, f takes on every value between α and β at least once in (a, b). Now f may take on a particular value, including α and β, at more than one point, as shown on the right in Fig. 14.7. In this case, f does not have an inverse function.

We say that a function is **one-to-one**, or 1-1, on an interval $[a, b]$ if each value of f reached on $[a, b]$ is reached at exactly one point. Equivalently, f is 1-1 if for any two points $x_1 \neq x_2$ in $[a, b]$ we have $f(x_1) \neq f(x_2)$. Clearly, a function is 1-1 on an interval if and only if it passes the Horizontal Line Test on that interval. The function on the left in Fig. 14.7 is 1-1 while the function on the right is not. Therefore, *a function has an inverse when it is 1-1.*

When is a function 1-1? Suppose that $\alpha < \beta$ and f is strictly increasing on $[a, b]$, which means that $x_1 < x_2$ implies that $f(x_1) < f(x_2)$. This implies that f is 1-1 on $[a, b]$ and takes on each value between α and β at exactly one point. Therefore, f has an inverse function. The same is true if $\alpha > \beta$ and f is strictly decreasing, so as before we conclude that f has an inverse on an interval if and only if it is monotone on the interval. If y is any number between α and β, then there is exactly one number x with $f(x) = y$. The definition of f^{-1} is simply the function that assigns $f^{-1}(y) = x$.

But this definition is dissatisfying in that it does not say how to find the value of x associated with y. In other words, knowing that a function has an inverse function on an interval is not the same as having a formula for the inverse function. In fact, it is usually the case that we cannot find an explicit formula for an inverse function.

So how do we compute the inverse function of a given function when there is no explicit formula for the inverse? We use the Bisection Algorithm. Suppose that $\alpha = f(a) < \beta = f(b)$ and f is monotone increasing. For each value y in (α, β), the function $f(x) - y$ is Lipschitz continuous and $f(a) - y = \alpha - y < 0$ while $f(b) - y = \beta - y > 0$. Therefore, the Bisection Algorithm applied to $f - y$ starting with the interval $[a, b]$ converges to a root x of $f(x) - y = 0$. The root is unique because f is increasing. In this way, we can compute the value of $f^{-1}(y)$ at any point y in (α, β). The same method works if $\alpha > \beta$ and f is monotone decreasing.[2]

By the way, it is a good exercise to show that f^{-1} is monotone increasing or decreasing when f is monotone increasing or decreasing.

We sum up this discussion as:

Theorem 14.2 Inverse Function Theorem *Let f be a Lipschitz continuous, monotone increasing or decreasing function on $[a, b]$ with $\alpha = f(a)$ and $\beta = f(b)$. Then f has a monotone increasing or decreasing inverse function defined on $[\alpha, \beta]$. For any x in (α, β), the value of $f^{-1}(x)$ can be computed by applying the Bisection Algorithm to compute the root y of $f(y) - x = 0$ starting on the interval $[a, b]$.*

EXAMPLE 14.12. We can use the Inverse Function Theorem to define the **root** function $x^{1/n}$ for a natural number n.

If n is odd, then $f(x) = x^n$ is a monotone increasing function and therefore f is invertible on any interval. We define the function $x^{1/n}$ to be the inverse of f,

$$x^{1/n} = f^{-1}(x) \text{ where } f(x) = x^n.$$

The graph of f indicates that the domain and range of $x^{1/n}$ is all real numbers and moreover $x^{1/n}$ is a monotone increasing function.

If n is even, then $f(x) = x^n$ is not monotone. However, we can restrict $f(x) = x^n$ to have domain $x \geq 0$ and then obtain an increasing function. The range of f is all nonnegative real numbers as well. Once again we define the function $x^{1/n}$ to be the inverse of f,

$$x^{1/n} = f^{-1}(x) \text{ where } f(x) = x^n \text{ for } x \geq 0.$$

The graph of f indicates that the domain and range of $x^{1/n}$ are all nonnegative real numbers and that $x^{1/n}$ is a monotone increasing function.

---

[2]Note that we usually write f^{-1} as a function of x not y. If we want to do this, then we let x denote any value in (α, β) and then compute the root y of $f(y) - x = 0$.

With this definition, we can define the **power** function $f(x) = x^{p/q}$ for any integer p and natural number q as the composition

$$f(x) = \left(x^{1/q}\right)^{p}.$$

We can verify that all the expected properties of exponents hold by using the properties that hold for integer exponents.

Chapter 14 Problems

14.1. Using the internet version of the yellow pages, one can enter the name of a company and obtain the company's phone number. This defines a function from the set of company names to the set of telephone numbers. Describe the corresponding inverse function.

14.2. A survey poll is a function that maps the set of poll takers to their responses in a given set of possible responses. Is this function invertible in general?

14.3. For each of the following functions either make a rough sketch of the inverse of the function or explain why it does not have an inverse.

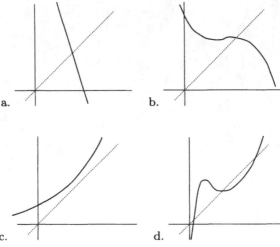

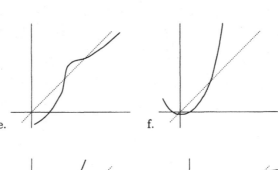

14.4. By restricting the domain of the function shown in Fig. 14.8, obtain three different invertible functions and plot the corresponding inverse functions.

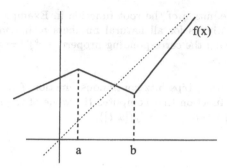

FIGURE 14.8. The function for Problem 14.4.

14.5. Verify that the following functions have the indicated inverses. Be sure to specify the domains of the functions and the inverses!

(a) $f(x) = 3x - 2$ and $f^{-1}(x) = \dfrac{x+2}{3}$.

(b) $f(x) = \dfrac{x}{x-1}$ and $f^{-1}(x) = \dfrac{x}{x-1}$.

(c) $f(x) = (x-4)^5$ and $f^{-1}(x) = x^{1/5} + 4$.

14.6. Compute the inverses of the following functions on the indicated domains:

(a) $f(x) = 3x - 1$, all x (b) $f(x) = x^{1/5}$, all x

(c) $f(x) = -x^4$, $x \geq 0$ (d) $f(x) = (x+1)/(x-1)$, $x > 1$

(e) $f(x) = (1-x^3)/x^3$, $x > 0$ (f) $f(x) = (2+\sqrt{x})^3$, $x > 0$

(g) $f(x) = (1-x^3)^{-1}$, $x > 1$ (h) $f(x) = (x-2)(x-3)$, $2 \leq x \leq 3$.

14.7. Decide if the function $f(x) = x^8$ is 1-1 on the indicated domains:

(a) all x (b) $x \geq 0$ (c) $x < -4$.

14.8. Decide if the function $f(x) = x^2 + x - 1$ is 1-1 on the indicated domains:

(a) all x (b) $x \geq -1/2$ (c) $x \leq -1/2$.

14.9. For each function below, determine an interval on which the function is 1-1 and then find the inverse function on the interval:

(a) $f(x) = (x+1)^2$ (b) $(x-4)(x+5)$ (c) $x^3 + x$.

14.10. Prove geometrically and analytically that if f is monotone increasing or decreasing, then its inverse is also monotone increasing or decreasing.

14.11. Using the definition of the root function in Example 14.12, prove that $(x^{1/n})^{1/m} = x^{1/(nm)}$ holds for all natural numbers n and m and nonnegative real numbers x by using the corresponding property $(x^b)^m = x^{nm}$ that holds for natural number powers.

14.12. Given monotone, Lipschitz continuous function f on the interval $[a, b]$, write a *MATLAB*© function that computes the value of the inverse function at any point y in (α, β) where $[\alpha, \beta] = f([a, b])$.

15

Fixed Points and Contraction Maps

In practice, we are concerned not only with computing a root of a function, but also with how quickly we can compute the root. If we are using an approximation algorithm that computes the root as a limit, then "quickness" is determined mainly by how many iterations the algorithm needs to compute a root to a given accuracy and how much time each iteration takes to compute. In particular, the Bisection Algorithm is a dependable way to approximate a root, but it is slow in practice, requiring many iterations to achieve high accuracy.[1] So we still need to find better ways to solve root problems.

In this chapter, we investigate new ways to solve root problems, motivated in part by a search for faster methods. To look for new ways, we reformulate the root problem into a new form called the fixed point problem. Given a function g, the **fixed point problem** for g is to find $\bar{x}$ such that

$$g(\bar{x}) = \bar{x}. \tag{15.1}$$

Graphically, a fixed point $\bar{x}$ of g is the point of intersection of the graphs of the line $y = x$ and the curve $y = g(x)$ (see Fig. 15.1). It turns out that modeling often results in a fixed point problem rather than a root problem, so it natural to consider the subject of solving fixed point problems. As with root problems, we are rarely able to compute the solution of a fixed point

---

[1]In contrast, the error of the Decasection Algorithm decreases more quickly per iteration than the error of the Bisection Algorithm, but takes so much more time per iteration that it is not faster.

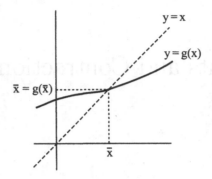

FIGURE 15.1. Illustration of a fixed point $g(\bar{x}) = \bar{x}$.

problem exactly so we devise an algorithm that allows the solution to be computed to any desired accuracy.

To explain how fixed point problems appear in modeling, we consider two models.

15.1   The Greeting Card Sales Model

We model the financial situation of a door-to-door greeting cards salesman that has the following price arrangement with a greeting card company.[2] For each shipment of cards, she pays a flat delivery fee of $25 dollars, and on top of this, for sales of x, where x is measured in units of $100, she pays an additional fee of 25% to the company. In mathematical terms, for sales of x hundreds of dollars, she pays

$$g(x) = \frac{1}{4} + \frac{1}{4}x \qquad (15.2)$$

where g is also given in units of $100. The problem of the *Greeting Card Sales model* is to find the "break-even point:" i.e., the amount of sales $\bar{x}$ where the money that she takes in exactly balances the money she has to pay out. Of course, she expects to clear a profit with each additional sale after this point.

We can visualize this problem by plotting two lines, shown in Fig. 15.2. The first line, $y = x$, represents the amount of money collected for sales of x. In this problem, we measure sales in units of dollars, so we just get $y = x$ for this curve. The second line, $y = \frac{1}{4}x + \frac{1}{4}$, represents the amount of money that has to be paid to the greeting card company. Because of the initial flat fee of $25, the salesman starts with a loss. Then as sales increase,

---

[2]Fixed point problems are essential to mathematical modeling in economics.

she reaches the break-even point $\bar{x}$ and finally begins to see a profit. The picture shows that the break-even point is a fixed point of g.

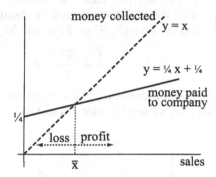

FIGURE 15.2. Illustration of the problem of determining the break-even point for selling greeting cards door-to-door. Sales above the break-even point $\bar{x}$ give a profit to the salesman, but sales below this point mean a loss.

In this example, it is easy to solve for the fixed point $\bar{x}$. We find the intersection of the two lines drawn in Fig. 15.2 by equating their formulas,

$$\bar{x} = g(\bar{x}) = \frac{1}{4}\bar{x} + \frac{1}{4},$$

which gives $\bar{x} = 1/3$.

15.2   The Free Time Model

The second example is more sophisticated and leads to a more complicated fixed point problem.

In an effort to put his life in order, your roommate tries to figure out the optimal balance between working and having fun by constructing a model for his free time.[3] Some activities, like studying and working part-time in the cafeteria, cannot be avoided, and the time spent on these activities has to be set aside. What can be adjusted is the time your roommate spends on discretionary activities like sleep, eating, going to clubs, and so on. The problem is to determine how much free time t he needs to be happy.

In counting up the time spent on discretionary activities, your roommate estimates that he needs 6 hours a day minimum for eating and sleeping. He decides that he should spend half of his free time, $t/2$, on purely fun activities. Finally, he observes that as the amount of free time shrinks, the time

---

[3]Of course, anyone that resorts to mathematics to figure out how to have fun is hopeless anyway, but we ignore that. We are mathematicians after all.

required for doing anything increases dramatically because he is cranky and tired. He models this by assuming the amount of time wasted because of being tired is $.25/t$ of his t free time. To reach a balanced state, the amount of time spent on discretionary activities should equal the amount of free time, i.e., he has to solve the *Free Time model*,

$$D(t) = 6 + \frac{t}{2} + \frac{.25}{t} = t, \tag{15.3}$$

which is just the fixed point problem for the function $D(t)$ giving the time spent on discretionary activities. We plot g and the fixed point in Fig. 15.3.

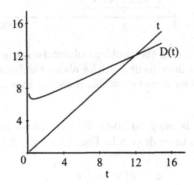

FIGURE 15.3. Plot of the fixed point problem for optimizing your roommate's free time.

Solving (15.3) for the fixed point $\bar{t}$ is not straightforward because the solution is irrational. We need to determine an algorithm for approximating the solution. In any case, we can see the solution is around $t = 12$; so to be truly happy, your roommate should spend around 12 hours on chores, 8 hours sleeping and eating, and 4 hours having fun.

15.3   Fixed Point Problems and Root Problems

As we said, fixed point problems and root problems are closely related. In particular, a given fixed point problem can be rewritten as a root problem and vice versa.

EXAMPLE 15.1.   If we define

$$f(x) = g(x) - x,$$

then

$$f(x) = 0 \quad \text{if and only if} \quad g(x) = x.$$

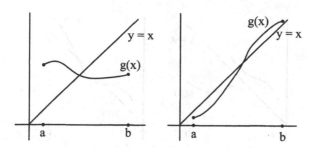

FIGURE 15.4. Two conditions that guarantee that a Lipschitz continuous function g has a fixed point in the interval $[a, b]$. First, $g(a) > a$ and $g(b) < b$; and second, $g(a) < a$ and $g(b) > b$.

With these choices, if we find a root of f, so $f(\bar{x}) = 0$, then we have also found a fixed point of g, $g(\bar{x}) = \bar{x}$.

Notice that in general we can rewrite a fixed point problem as a root problem in many different ways.

EXAMPLE 15.2. The fixed point problem

$$g(x) = x^3 - 4x^2 + 2 = x$$

can be written as the root problems

$$x^3 - 4x^2 - x + 2 = 0$$
$$5(x^3 - 4x^2 - x + 2) = 0$$
$$x^2 - 4x + \frac{2}{x} - 1 = 0.$$

The same is true for writing root problems as fixed point problems.

EXAMPLE 15.3. The root problem

$$f(x) = x^4 - 2x^3 + x - 1 = 0$$

is equivalent to the fixed point problems

$$x = -x^4 + 2x^3 + 1$$
$$x = \frac{2x^3 - x + 1}{x^3}$$
$$x = x^5 - 2x^4 + x^2.$$

This discussion suggests one way to solve a fixed point problem $g(x) = x$. Namely, we rewrite it as a root problem $f(x) = 0$ and then apply the Bisection Algorithm. We know this works provided we find an interval $[a, b]$ for which $f(a)$ and $f(b)$ have opposite signs and f is Lipschitz continuous on $[a, b]$. Whether these properties are true depends on how we change the fixed point problem into a root problem.

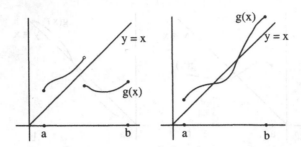

FIGURE 15.5. In the figure on the left, g is not continuous on $[a, b]$ and consequently does not have a fixed point on [a,b] even though $g(a) > a$ and $g(b) < b$. In the figure on the right, g has two fixed points on $[a, b]$ even though $g(a) > a$ and $g(b) > b$ and g is Lipschitz continuous on $[a, b]$.

EXAMPLE 15.4. If g is Lipschitz continuous on an interval $[a, b]$ and we choose $f(x) = g(x) - x$, then f is Lipschitz continuous on $[a, b]$. The condition that $f(a) < 0$ means that $g(a) < a$, and $f(a) > 0$ means that $g(a) > a$. Therefore, g is guaranteed to have a fixed point in the interval $[a, b]$ provided either that $g(a) > a$ and $g(b) < b$ or $g(a) < a$ and $g(b) > b$. We illustrate the two possibilities in Fig. 15.4. Note that if g is not continuous on the interval $[a, b]$, then these conditions do not guarantee that g has a fixed point in $[a, b]$ (see Fig. 15.5). Also, g may have a fixed point on an interval $[a, b]$ even if these conditions do not hold, as shown in Fig. 15.5.

It is certainly possible to convert a fixed point problem into a "bad" root problem.

EXAMPLE 15.5. We can rewrite the fixed point problem $g(x) = x$ as the root problem $f(x) = 0$ by defining $f(x) = (g(x) - x)^2$. f is certainly Lipschitz continuous when g is Lipschitz continuous but there is no way we can find points a and b where f has the opposite signs, since f is always non-negative.

So to use the Bisection Algorithm to solve a fixed point problem for g, we find a way to transform the fixed point problem into a root problem $f(x) = 0$ where f is Lipschitz continuous on an interval $[a, b]$ and $f(a)$ and $f(b)$ have the opposite signs.

EXAMPLE 15.6. For the fixed point problem $\frac{1}{4}x + \frac{1}{4} = x$ in the Greeting Card model, we set $f(x) = \frac{1}{4}x + \frac{1}{4} - x = -\frac{3}{4}x + \frac{1}{4}$. Then $f(0) = 1/4$, $f(1) = -1/2$ and f is Lipschitz continuous on $[0, 1]$. The Bisection Algorithm starting with $x_0 = 0$ and $X_0 = 1$ converges to the root $1/3$.

EXAMPLE 15.7. For the fixed point problem $6 + t/2 + .5/t = t$ in the Free Time model, we set

$$f(t) = \frac{6 + t/2 + .5/t}{t} - 1 = \frac{6}{t} + \frac{.5}{t^2} - \frac{1}{2}.$$

Then $f(1) = 6$, $f(15) = -0.0977 \cdots$ and f is Lipschitz continuous on $[1, 15]$. The Bisection Algorithm starting with $x_0 = 1$ and $X_0 = 15$ converges to the root $12.0415229868 \cdots$.

15.4   Solving the Greeting Card Sales Model

Actually, we already solved for the solution $\bar{x} = 1/3$ of the fixed point problem for the Greeting Card model. But we consider this problem once again in order to figure out a new method for solving fixed point problems. Knowing the fixed point beforehand makes it a lot easier to explain how the new procedure works.

In Fig. 15.6, we plot the function $g(x) = \frac{1}{4}x + \frac{1}{4}$ used in the Card Sale model along with $y = x$ and the fixed point $\bar{x}$. We also plot the value of

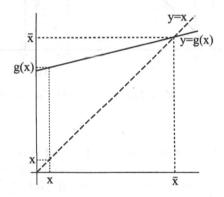

FIGURE 15.6. The value of $g(x)$ is closer to $\bar{x}$ than x.

$g(x)$ for some point x. We choose $x < \bar{x}$ because in the model the sales start at zero and then increase. The plot shows that $g(x)$ is closer to $\bar{x}$ than x, i.e.,

$$|g(x) - \bar{x}| < |x - \bar{x}|.$$

In fact, we can compute the difference exactly using $\bar{x} = 1/3$,

$$|g(x) - \bar{x}| = \left| \frac{1}{4}x + \frac{1}{4} - \frac{1}{3} \right| = \left| \frac{1}{4} \left(x - \frac{1}{3} \right) \right| = \frac{1}{4}|x - \bar{x}|.$$

So the distance from $g(x)$ to $\bar{x}$ is exactly $1/4$ the distance from $\bar{x}$ than x.

But the same argument shows that if we apply g to $g(x)$, i.e. compute $g(y)$ where $y = g(x)$, then the distance from that value to $\bar{x}$ is $1/4$ the distance from $g(x)$ to $\bar{x}$ and $1/16$ the distance from x to $\bar{x}$. In other words,

$$|g(g(x)) - \bar{x}| = \frac{1}{4}|g(x) - \bar{x}| = \frac{1}{16}|x - \bar{x}|$$

where we write $g(y)$ where $y = g(x)$ as $g(g(x))$. We illustrate this in Fig. 15.7. Following the trend, if we apply g to $g(g(x))$ to get $g(g(g(x)))$,

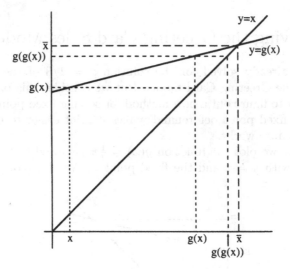

FIGURE 15.7. The distance of $g(g(x))$ to $\bar{x}$ is $1/4$ the distance from $g(x)$ to $\bar{x}$ and $1/16$ the distance from x to $\bar{x}$.

this value is closer to $\bar{x}$ than x by a factor of $1/4 \times 1/16 = 1/64$. This suggests that we can approximate the fixed point simply by choosing an initial point $x \geq 0$ and then continually reapplying g. This is called the **Fixed Point Iteration** for $g(x)$.

We show 7 steps of the Fixed Point Iteration in Fig. 15.8. We also show the values of X_i from the Bisection Algorithm applied to the equivalent root problem $f(x) = -\frac{3}{4}x + \frac{1}{4}$ starting on $[0, 1]$. The numbers suggest that the error of the Fixed Point Iteration decreases by a factor of $1/4$ for each iteration, as opposed to the error of the Bisection Algorithm, which decreases by a factor of $1/2$. Moreover, since both methods require one function evaluation and one storage per iteration but the Bisection Algorithm requires an additional sign check, the Fixed Point Iteration costs less per iteration. For this problem, the Fixed Point Iteration is apparently "faster" than the Bisection Algorithm.

The data in Fig. 15.8 suggests that the sequence generated by the Fixed Point Iteration converges to the fixed point. We can prove this is true in

i	Bisection Algorithm X_i	Fixed Point Iteration x_i
0	1.00000000000000	1.00000000000000
1	0.50000000000000	0.50000000000000
2	0.50000000000000	0.37500000000000
3	0.37500000000000	0.34375000000000
4	0.37500000000000	0.33593750000000
5	0.34375000000000	0.33398437500000
6	0.34375000000000	0.33349609375000
7	0.33593750000000	0.33337402343750
8	0.33593750000000	
⋮	⋮	
13	0.33337402343750	

FIGURE 15.8. Results of the Bisection Algorithm and the Fixed Point Iteration for the Greeting Card model.

this example by computing an explicit formula for the elements. We begin with the first element,

$$x_1 = \frac{1}{4}x_0 + \frac{1}{4},$$

then continuing

$$x_2 = \frac{1}{4}x_1 + \frac{1}{4} = \frac{1}{4}\left(\frac{1}{4}x_0 + \frac{1}{4}\right) + \frac{1}{4} = \frac{1}{4^2}x_0 + \frac{1}{4^2} + \frac{1}{4}.$$

Likewise, we find

$$x_3 = \frac{1}{4^3}x_0 + \frac{1}{4^3} + \frac{1}{4^2} + \frac{1}{4}$$

and after n steps

$$x_n = \frac{1}{4^n}x_0 + \sum_{i=1}^{n}\frac{1}{4^i}. \tag{15.4}$$

The first term on the right-hand side of (15.4), $x_0/4^n$ converges to 0 as n increases to infinity. The second term is equal to

$$\sum_{i=1}^{n}\frac{1}{4^i} = \frac{1}{4}\times\sum_{i=0}^{n-1}\frac{1}{4^i} = \frac{1}{4}\times\frac{1-\frac{1}{4^n}}{1-\frac{1}{4}} = \frac{1-\frac{1}{4^n}}{3},$$

using the formula for the geometric sum. The second term therefore converges to the fixed point $1/3$ as n increases to infinity.

15.5   The Fixed Point Iteration

The Fixed Point Iteration for a general fixed point problem

$$g(x) = x,$$

is simply:

Algorithm 15.1 Fixed Point Iteration Choose x_0 and set

$$x_i = g(x_{i-1}) \text{ for } i = 1, 2, 3, \cdots \qquad (15.5)$$

Showing that the algorithm converges to a fixed point of g and estimating the error at each iteration are more difficult topics. Before tackling these, we present a few examples.

EXAMPLE 15.8. We apply the Fixed Point Iteration to solve the Free Time model with $g(t) = 6 + t/2 + .5/t$ starting with $t = 1$ and show the results in Fig. 15.9. The iteration apparently converges to the fixed

i	x_i
0	1
1	6.75
2	9.41203703703704
3	10.7325802499499
4	11.3895836847879
5	11.7167417228215
⋮	⋮
10	12.0315491941695
⋮	⋮
15	12.0412166444154
⋮	⋮
20	12.0415135775222

FIGURE 15.9. Results of the Fixed Point Iteration applied to the Free Time model.

point in this example.

EXAMPLE 15.9. In computing the solubility of $Ba(IO_3)_2$ in Section 4.5, we solved the root problem (13.3)

$$x(20 + 2x)^2 - 1.57 = 0$$

using the Bisection Algorithm and displayed the results in Fig. 13.3. In this example, we use the Fixed Point Iteration to solve the equivalent fixed point problem

$$g(x) = \frac{1.57}{(20 + 2x)^2} = x. \tag{15.6}$$

We know that g is Lipschitz continuous on any interval that avoids $x = 10$ (and we also know that the fixed point/root is close to 0). We start off the iteration with $x_0 = 1$ and show the results in Fig. 15.10 The iteration appears to converge very quickly in this case.

i	x_i
0	1.00000000000000
1	0.00484567901235
2	0.00392880662465
3	0.00392808593169
4	0.00392808536527
5	0.00392808536483

FIGURE 15.10. Results of the Fixed Point Iteration applied to (15.6).

But the Fixed Point Iteration often fails to converge as well.

EXAMPLE 15.10. The fixed point of

$$g(x) = x^2 + x = x \tag{15.7}$$

is easily computed to be $\bar{x} = 0$. It turns out, however, that the Fixed Point Iteration diverges for any initial value $x_0 \neq 0$. We display the results of the Fixed Point Iteration starting with $x_0 = .1$ in Fig. 15.11.

15.6 Convergence of the Fixed Point Iteration

In this section, we investigate the convergence of the Fixed Point Iteration. The investigation starts with the observation that the Fixed Point Iteration for the Greeting Card model converges because the slope of $g(x) = \frac{1}{4}x + \frac{1}{4}$ is $1/4 < 1$. This produces a factor of $1/4$ in the error after each iteration, forcing the right-hand side of (15.4) to have a limit as n tends to infinity. Recalling that the slope of a linear function is the same thing as its Lipschitz constant, we can say this example worked because the Lipschitz constant of g is $L = 1/4 < 1$.

In contrast if the Lipschitz constant, or slope, of g is larger than 1, then the analog of (15.4) does not converge.

i	x_i
0	.1
1	.11
2	.1221
3	.13700841
4	.155779714410728
5	.180047033832616
6	.212463968224539
7	.257604906018257
8	.323965193622933
9	.428918640302077
10	.612889840300659
11	.988523796644427
12	1.96570309317674
13	5.82969174370134
14	39.8149975702809
15	1625.04902909176
16	2642409.39598115

FIGURE 15.11. Results of the Fixed Point Iteration applied to $g(x) = x^2 + x$.

EXAMPLE 15.11. We demonstrate this graphically in Fig. 15.12 using the function $g(x) = 2x + \frac{1}{4}$. The difference between successive iterates increases with each iteration and the Fixed Point Iteration does not converge. It is clear from the plot that there is no positive fixed point. On the other hand, the Fixed Point Iteration does converge when applied to any linear function with Lipschitz constant $L < 1$. We illustrate the convergence for $g(x) = \frac{3}{4}x + \frac{1}{4}$ in Fig. 15.12. Thinking about (15.4), the reason is simply that the geometric sum for L converges when $L < 1$.

Returning to the general case, we look for conditions on g that guarantee that the Fixed Point Iteration converges to a fixed point of g. Based on the previous examples, it is natural to assume that g is Lipschitz continuous with constant $L < 1$. But we have to be careful now because linear functions are Lipschitz continuous on the entire set of real numbers $\mathbb{R}$, but most functions are not. For example, polynomials of degree larger than 1 are Lipschitz continuous on bounded sets of numbers but not on $\mathbb{R}$. So we assume that there is an interval $[a, b]$ such that g is Lipschitz continuous on $[a, b]$ with Lipschitz constant $L < 1$.

This assumption introduces a complication. If we think about the analysis of the Greeting Card model, we might guess that we have to use the Lipschitz condition on g evaluated at the iterates $\{x_i\}$ produced by the Fixed Point Iteration. To do this, we need all of the x_i to be in the interval $[a, b]$ on which g is Lipschitz continuous. Unfortunately, it is not so easy to check this condition for all i. Since each x_i is given by evaluating $g(x_{i-1})$,

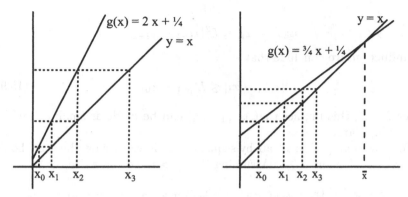

FIGURE 15.12. On the left, we plot the first three fixed point iterates for $g(x) = 2x + \frac{1}{4}$. The iterates increase without bound as the iteration proceeds. On the right, we plot the first three fixed point iterates for $g(x) = \frac{3}{4}x + \frac{1}{4}$. The iteration converges to the fixed point in this case.

a way around this difficulty is to assume that if x is in $[a, b]$, then $g(x)$ is in $[a, b]$. In other words, the image of $[a, b]$ under the transformation g is contained in $[a, b]$. This implies that as long as x_0 is in $[a, b]$ then $x_i = g(x_{i-1})$ is also in $[a, b]$ for every i by induction.

To summarize, we say that g is a **contraction map** on the interval $[a, b]$ if x in $[a, b]$ implies $g(x)$ is in $[a, b]$ and if g is Lipschitz continuous on $[a, b]$ with Lipschitz constant $L < 1$. It turns out that the Fixed Point Iteration for a contraction map always converges to a unique fixed point of g in $[a, b]$.

The first step is to show that the sequence $\{x_i\}$ generated by the Fixed Point Iteration is a Cauchy sequence and therefore converges to a real number $\bar{x}$. We have to show that the difference $x_i - x_j$ can be made arbitrarily small by taking $j \geq i$ both large. We start by showing that the difference $x_i - x_{i+1}$ can be made arbitrarily small. We subtract the equation $x_i = g(x_{i-1})$ from $x_{i+1} = g(x_i)$ to get

$$x_{i+1} - x_i = g(x_i) - g(x_{i-1}).$$

Because x_{i-1} and x_i are both in $[a, b]$ by assumption, we can use the Lipschitz continuity of g to conclude that

$$|x_{i+1} - x_i| \leq L|x_i - x_{i-1}|. \tag{15.8}$$

This says that the difference between x_i and x_{i+1} cannot be larger than a factor of L times the previous difference between x_{i-1} and x_i. This is how we get a decrease in successive iterates. We can use the same argument to show that

$$|x_i - x_{i-1}| \leq L|x_{i-1} - x_{i-2}|,$$

so
$$|x_{i+1} - x_i| \le L^2 |x_{i-1} - x_{i-2}|.$$

By induction, we conclude that

$$|x_{i+1} - x_i| \le L^i |x_1 - x_0|. \tag{15.9}$$

Since $L < 1$, this implies that $|x_{i+1} - x_i|$ can be made as small as we like by taking i large.

To show that $\{x_i\}$ is a Cauchy sequence, we have to show the same holds for $|x_i - x_j|$ for any $j \ge i$. Assuming that $j > i$, we can write

$$|x_i - x_j| = |x_i - x_{i+1} + x_{i+1} - x_{i+2} + x_{i+2} - \cdots + x_{j-1} - x_j|.$$

Using the triangle inequality,

$$|x_i - x_j| \le |x_i - x_{i+1}| + |x_{i+1} - x_{i+2}| + \cdots + |x_{j-1} - x_j|$$
$$= \sum_{k=i}^{j-1} |x_k - x_{k+1}|.$$

Now we use (15.9) on each term in the sum and get

$$|x_i - x_j| \le \sum_{k=i}^{j-1} L^k |x_1 - x_0| = |x_1 - x_0| \sum_{k=i}^{j-1} L^k.$$

Now

$$\sum_{k=i}^{j-1} L^k = L^i(1 + L + L^2 + \cdots + L^{j-i-1}) = L^i \frac{1 - L^{j-i}}{1 - L}$$

by the formula for the geometric sum for L. Since $L < 1$, $1 - L^{j-i} \le 1$ and therefore

$$|x_i - x_j| \le \frac{L^i}{1 - L} |x_1 - x_0|.$$

Since $L < 1$, L^i approaches 0 as i increases, we can make the difference $|x_i - x_j|$ with $j \ge i$ as small as we like by taking i large. In other words, $\{x_i\}$ is a Cauchy sequence and therefore converges to a real number $\bar{x}$.

The second step is to show that the limit $\bar{x}$ is a fixed point of g. Recall that by definition

$$g(\bar{x}) = \lim_{i \to \infty} g(x_i).$$

Now by the definition of the Fixed Point Iteration ,

$$\lim_{i \to \infty} g(x_i) = \lim_{i \to \infty} x_{i+1} = \lim_{i \to \infty} x_i = \bar{x},$$

and $g(\bar{x}) = \bar{x}$ as desired.

In the last step, we show that g can have at most one fixed point in $[a, b]$, so there is no question about which fixed point the Fixed Point Iteration approximates. Suppose that $\bar{x}$ and $\tilde{x}$ are fixed points of g in $[a, b]$, i.e.,

$$g(\bar{x}) = \bar{x} \text{ and } g(\tilde{x}) = \tilde{x}.$$

Subtracting and using the Lipschitz assumption on g, we find

$$|\bar{x} - \tilde{x}| = |g(\bar{x}) - g(\tilde{x})| \le L|\bar{x} - \tilde{x}|.$$

Since $L < 1$, this is only possible if $\bar{x} - \tilde{x} = 0$.

We summarize this discussion as a theorem.

Theorem 15.1 Banach Contraction Mapping Principle   *If g is a contraction map on an interval $[a, b]$, then the Fixed Point Iteration starting with any point x_0 in $[a, b]$ converges to the unique fixed point $\bar{x}$ of g in $[a, b]$.*

The theorem is named after Banach,[4] who first proved this result generalizing the method of successive approximations for differential equations. We present a more general version of the theorem and discuss the relation to the method of successive approximation in Chapter 40.

Most of the time, the hard part in using this theorem is finding a good interval. We finish with some examples that show different possibilities and return to this subject in Chapter 31.

EXAMPLE 15.12.   In the Greeting Card model, $g(x) = \frac{1}{4}x + \frac{1}{4}$ is Lipschitz continuous on $\mathbb{R}$, so we can take any interval $[a, b]$ and apply the theorem. This means that there is only one solution in $\mathbb{R}$.

EXAMPLE 15.13. In the Free Time model, we can show (Problem 15.17) that on the interval $[a, b]$ with $a > 0$, $D(t) = 6 + t/2 + .25/t$ has Lipschitz constant

$$L = \frac{1}{2} + \frac{.25}{a^2}.$$

As long as $a > 1/\sqrt{2} \approx .7072$, then we can use the theorem on $[a, b]$ to guarantee convergence. In practice, the Fixed Point Iteration converges on any interval $[a, b]$ with $a > 0$.

EXAMPLE 15.14. To compute the solubility of Ba(IO$_3$)$_2$, we solve the fixed point problem (15.6)

$$g(x) = \frac{1.57}{(20 + 2x)^2} = x.$$

---

[4]The Polish mathematician Stefan Banach (1892–1945) developed the first systematic theory of functional analysis, as well as making contributions to integration, measure theory, orthogonal series, the theory of sets, and the topological vector spaces. Banach's name is found on several fundamental theorems in normed linear spaces as well as Banach spaces. Banach was a popular, lively, and charming person who often worked in cafes, bars, and restaurants.

It is possible to show (Problem 15.18) g is Lipschitz continuous on $[a, b]$ with $a \geq 0$ and $b \leq 9.07$ with Lipschitz constant $L < 1$ and the theorem applies. In practice, the Fixed Point Iteration converges for any $x_0 \neq 10$.

EXAMPLE 15.15. In Example 15.10, we tried the Fixed Point Iteration on $g(x) = x^2 + x = x$ with $x_0 = .1$ and it failed to converge. On any interval $[a, b]$ with $a > 0$, the Lipschitz constant of g is $L = 1 + 2a > 1$, so we cannot use the theorem.

EXAMPLE 15.16. We can show (Problem 15.19) that $g(x) = x^4/(10 - x)^2$ is Lipschitz continuous on $[-1, 1]$ with $L = .053$ and the theorem implies the Fixed Point Iteration converges to $\bar{x} = 0$ for any x_0 in $[-1, 1]$. However, the Lipschitz constant of g on $[-9.9, 9.9]$ is about 20×10^6 and the Fixed Point Iteration diverges rapidly if $x_0 = 9.9$.

15.7   Rates of Convergence

Recall that one motivation for the Fixed Point Iteration was to find a faster way to solve root problems. This is a good time to discuss what is meant by "how quickly an iteration converges", and to provide a way to compare the speeds at which different iterations converge.

We know that the error of the Bisection Algorithm decreases by at least a factor of $1/2$ with each iteration. For comparison, we need an estimate on the amount of decrease of the error $|x_i - \bar{x}|$ of the Fixed Point Iteration with each iteration. Since $\bar{x} = g(\bar{x})$,

$$|x_i - \bar{x}| = |g(x_{i-1}) - g(\bar{x})| \leq L|x_{i-1} - \bar{x}|. \qquad (15.10)$$

This means that the error decreases by *at least* a factor of $L < 1$ during each iteration. In particular if $L < 1/2$, then the Fixed Point Iteration converges more quickly than the Bisection Algorithm in the sense that the error generally decreases by a larger fraction with each iteration.

Moreover, we can turn (15.10) into an estimate on the error after n steps by using induction to conclude that

$$|x_n - \bar{x}| \leq L^n|x_0 - \bar{x}| \leq L^n|b - a|. \qquad (15.11)$$

Using (15.11), we can decide how many iterations are needed to guarantee a given accuracy in x_n.

There is uncertainty in this discussion because it depends on *estimates* of the errors of the iterations. It is possible that the error of the Fixed Point Iteration actually decreases exactly by a factor of L. This is true for the Greeting Card model in which the error decreases by a factor of $L = 1/4$ for each iteration for example. When the error of an iterative method decreases exactly by a constant factor of L during each iteration,

i	x_i for $\frac{1}{9}x + \frac{3}{4}$	x_i for $\frac{1}{5}x + 2$
0	1.00000000000000	1.00000000000000
1	0.86111111111111	2.20000000000000
2	0.84567901234568	2.44000000000000
3	0.84396433470508	2.48800000000000
4	0.84377381496723	2.49760000000000
5	0.84375264610747	2.49952000000000
$\vdots$	$\vdots$	$\vdots$
10	0.84375000004481	2.49999984640000
$\vdots$	$\vdots$	$\vdots$
15	0.84375000000000	2.49999999995085
$\vdots$	$\vdots$	$\vdots$
20	0.84375000000000	2.49999999999998

FIGURE 15.13. Results of the Fixed Point Iterations for $\frac{1}{9}x + \frac{3}{4}$ and $\frac{1}{5}x + 2$.

we call this **linear convergence** and say the iteration converges at a **linear rate** with **convergence factor** L. The Fixed Point Iteration applied to any linear function $g(x)$ with Lipschitz constant $L < 1$ converges at a linear rate with convergence factor L. When two iterations converge at a linear rate, we can compare the rate at which the iterations converge by comparing the size of the convergence factors.

EXAMPLE 15.17. The Fixed Point Iteration converges more quickly for the function $g(x) = \frac{1}{9}x + \frac{3}{4}$ than for $g(x) = \frac{1}{5}x + 2$. We show the results in Fig. 15.13. The iteration for $\frac{1}{9}x + \frac{3}{4}$ reaches 15 places of accuracy within 15 iterations, while the iteration for $\frac{1}{5}x + 2$ has only 14 places of accuracy after 20 iterations.

It is also possible for an iteration to converge more quickly than at a linear rate. We explain using an example.

EXAMPLE 15.18. The functions $\frac{1}{2}x$ and $\frac{1}{2}x^2$ are both Lipschitz continuous on $[-1/2, 1/2]$ with Lipschitz constant $L = 1/2$. The estimate (15.10) suggests the Fixed Point Iteration for both should converge to $\bar{x} = 0$ at the same rate. We show the results of the Fixed Point Iteration applied to both in Fig. 15.14.

It is clear that the Fixed Point Iteration for $\frac{1}{2}x^2$ converges much more quickly, reaching 15 places of accuracy after 7 iterations.

To explain how this can happen, we look into the argument (15.10) for the particular function $g(x) = \frac{1}{2}x^2$. Computing in the same way for the fixed point $\bar{x} = 0$,

$$x_i - 0 = g(x_{i-1}) - g(0) = \frac{1}{2}x_{i-1}^2 - \frac{1}{2}0^2 = \frac{1}{2}(x_{i-1} + 0)(x_{i-1} - 0)$$

i	x_i for $\frac{1}{2}x$	x_i for $\frac{1}{2}x^2$
0	0.50000000000000	0.50000000000000
1	0.25000000000000	0.25000000000000
2	0.12500000000000	0.06250000000000
3	0.06250000000000	0.00390625000000
4	0.03125000000000	0.00001525878906
5	0.01562500000000	0.00000000023283
6	0.00781250000000	0.00000000000000

FIGURE 15.14. Results of the Fixed Point Iterations for $\frac{1}{2}x$ and $\frac{1}{2}x^2$.

so

$$|x_i - 0| = \frac{1}{2}|x_{i-1}|\,|x_{i-1} - 0|.$$

This says that the error of the Fixed Point Iteration for $\frac{1}{2}x^2$ decreases *exactly* by a factor of $\frac{1}{2}|x_{i-1}|$ during the ith iteration. In other words,

for $i = 1$ the factor is $\frac{1}{2}|x_0|$

for $i = 2$ the factor is $\frac{1}{2}|x_1|$

for $i = 3$ the factor is $\frac{1}{2}|x_2|$,

and so on. This is called a **quadratic convergence rate**. Unlike the case of linear convergence, where the error decreases by a fixed factor each time, in quadratic convergence the factor by which the errors decrease depends on the values of the iteration.

Now consider what happens as the iteration proceeds and the iterates x_{i-1} become closer to zero. The factor by which the error in each step decreases becomes smaller as i increases! In other words, the closer the iterates get to zero, the faster they get close to zero. The estimate in (15.10) significantly *overestimates* the error of the Fixed Point Iteration for $\frac{1}{2}x^2$ because it treats the error as if it decreases by a fixed factor each time. Thus it cannot be used to accurately predict the quadratic convergence for this function. For a function g, the first part of (15.10) tells the same story:

$$|x_i - \bar{x}| = |g(x_{i-1}) - g(\bar{x})|.$$

The error of x_i is determined by the change in g in going from $\bar{x}$ to the previous iterate x_{i-1}. This change can depend on x_{i-1} and when it does, the Fixed Point Iteration does not converge at a linear rate.

A natural question is whether or not it is always possible to write a fixed point problem in such a way that we get quadratic convergence. It turns out that it is often possible, and we discuss this question again in Chapter 31. For now, we give another example that demonstrates quadratic convergence.

EXAMPLE 15.19. The Bisection Algorithm for computing the root of $f(x) = x^2 - 2$ converges at a linear rate with convergence factor $1/2$. We can write this problem as the fixed point problem

$$g(x) = \frac{1}{x} + \frac{x}{2} = x. \qquad (15.12)$$

It is easy to verify that $\bar{x}$ is a fixed point for g if and only if it is a root of f. We claim that the Fixed Point Iteration for g converges at a quadratic rate. We show the result in Fig. 15.15. It takes only 5

i	x_i
0	1.00000000000000
1	1.50000000000000
2	1.41666666666667
3	1.41421568627451
4	1.41421356237469
5	1.41421356237310
6	1.41421356237310

FIGURE 15.15. The Fixed Point Iteration for (15.12).

iterations to reach 15 places of accuracy.

To show that the convergence is indeed quadratic, we compute as in (15.10).

$$|x_i - \sqrt{2}| = |g(x_{i-1}) - g(\sqrt{2})|$$

$$= \left| \frac{x_{i-1}}{2} + \frac{1}{x_{i-1}} - \left(\frac{\sqrt{2}}{2} + \frac{1}{\sqrt{2}} \right) \right|$$

$$= \left| \frac{x_{i-1}^2 + 2}{2x_{i-1}} - \sqrt{2} \right|.$$

Now we find a common denominator for the fractions on the right and then use the fact that

$$(x_{i-1} - \sqrt{2})^2 = x_{i-1}^2 - 2\sqrt{2}x_{i-1} + 2$$

to get

$$|x_i - \sqrt{2}| = \frac{(x_{i-1} - \sqrt{2})^2}{2x_{i-1}}. \qquad (15.13)$$

This says that as long as x_{i-1} is not close to zero, and since it converges to $\sqrt{2}$ this is true for large i, then the error of x_i is the square of the error of x_{i-1}. When the error of x_{i-1} is less than one, then the error of x_i is much smaller. This is quadratic convergence.

Chapter 15 Problems

15.1. A salesman selling vacuum cleaners door-to-door has a franchise with the following payment scheme. For each delivery of vacuum cleaners, the salesman pays a fee of $100 and then a percentage of the sales, measured in units of hundreds of dollars, that increases as the sales increases. For sales of x, the percentage is $20x\%$. Show that this model gives a fixed point problem and make a plot of the fixed point problem that shows the location of the fixed point.

15.2. Rewrite the following fixed point problems as root problems three different ways each:

$$\text{(a)} \ \frac{x^3 - 1}{x + 2} = x \qquad \text{(b)} \ x^5 - x^3 + 4 = x \ .$$

15.3. Rewrite the following root problems as fixed point problems three different ways each:

$$\text{(a)} \ 7x^5 - 4x^3 + 2 = 0 \qquad \text{(b)} \ x^3 - \frac{2}{x} = 0 \ .$$

15.4. If possible, find intervals suitable for application of the Bisection Algorithm to each of the three root problems found in Example 15.2. A suitable interval is one on which the function is Lipschitz continuous and on which the function changes sign.

15.5. (a) If possible, find intervals suitable for application of the Bisection Algorithm to each of the three root problems found in Problem 15.2(a). (b) Do the same for Problem 15.2(b). A suitable interval is one on which the function is Lipschitz continuous and on which the function changes sign.

15.6. (a) Draw a Lipschitz continuous function g on the interval $[0, 1]$ that has three fixed points such that $g(0) > 0$ and $g(1) < 1$. (b) Draw a Lipschitz continuous function g on the interval $[0, 1]$ that has three fixed points such that $g(0) > 0$ and $g(1) > 1$.

15.7. Verify that (15.4) is true.

Do Problems 15.8–15.10 by finding an explicit formula analogous to (15.4).

15.8. (a) Find an explicit formula for the nth fixed point iterate x_n for the function $g(x) = 2x + \frac{1}{4}$. (b) Prove that x_n diverges to ∞ as n increases to ∞.

15.9. (a) Find an explicit formula for the nth fixed point iterate x_n for the function $g(x) = \frac{3}{4}x + \frac{1}{4}$. (b) Prove that x_n converges as n increases to ∞ and compute the limit.

15.10. (a) Find an explicit formula for the nth fixed point iterate x_n for the function $g(x) = mx + b$. (b) Prove that x_n converges as n increases to ∞ provided that $L = |m| < 1$ and compute the limit.

15.11. Write a program that implements Algorithm 15.1. The program should employ three methods for stopping the iteration: (1) when the number of iterations is larger than a user-input number and (2) when the difference between successive iterates $|x_i - x_{i-1}|$ is smaller than a user-input tolerance and (3) (15.11). Test the program by reproducing the results in Fig. 15.13.

Problems 15.12–15.15 involve solving a fixed point problem on a computer. The program from Problem 15.11 would be useful for these.

15.12. In Section 4.5, suppose that K_{sp} for $Ba(IO_3)_2$ is 1.8×10^{-5}. Find the solubility S to 10 decimal places using the Fixed Point Iteration after writing the problem as a suitable fixed point problem. *Hint:* $1.8 \times 10^{-5} = 18 \times 10^{-6}$ and $10^{-6} = 10^{-2} \times 10^{-4}$.

15.13. In Section 4.5, determine the solubility of $Ba(IO_3)_2$ in a .037 mol/L solution of KIO_3 to 10 decimal places using the Fixed Point Iteration after writing the problem as a suitable fixed point problem.

15.14. The power P delivered into a load R of a simple class A amplifier of output resistance Q and output voltage E is

$$P = \frac{E^2 R}{(Q + R)^2}.$$

Find all possible solutions R for $P = 1$, $Q = 3$, and $E = 4$ to 10 decimal places using the Fixed Point Iteration after writing the problem as a fixed point problem.

15.15. Van der Waal's model for one mole of an ideal gas including the effects of the size of the molecules and the mutual attractive forces is

$$\left(P + \frac{a}{V^2}\right)(V - b) = RT,$$

where P is the pressure, V is the volume of the gas, T is the temperature, R is the ideal gas constant, a is a constant depending on the size of the molecules and the attractive forces, and b is a constant depending on the volume of all the molecules in one mole. Find all possible volumes V of the gas corresponding to $P = 2$, $T = 15$, $R = 3$, $a = 50$, and $b = .011$ to 10 decimal places using the Fixed Point Iteration after writing the problem as a fixed point problem.

Problems 15.16–15.21 are concerned with the convergence of the Fixed Point Iteration.

15.16. Draw a Lipschitz continuous function g that does *not* have the property that x in $[0, 1]$ means that $g(x)$ is in $[0, 1]$.

15.17. Verify the details of Example 15.13.

15.18. Verify the details of Example 15.14.

15.19. Verify the details of Example 15.16.

15.20. (a) If possible, find intervals suitable for application of the Fixed Point Iteration to each of the three fixed point problems found in Example 15.3. (b) If possible, find intervals suitable for application of the Fixed Point Iteration to each of the three fixed point problems found in Problem 15.3(a). (c) If possible, find intervals suitable for application of the Fixed Point Iteration to each of the three fixed point problems found in Problem 15.3(b). In each case, a suitable interval is one on which the function is a contraction map.

15.21. Apply Theorem 15.1 to the function $g(x) = 1/(1 + x^2)$ to show that the Fixed Point Iteration converges on any interval $[a, b]$.

Problems 15.22–15.26 are concerned with the rate of convergence of the Fixed Point Iteration.

15.22. Given the following results of the Fixed Point Iteration applied to a function $g(x)$,

i	x_i
0	14.00000000000000
1	14.25000000000000
2	14.46875000000000
3	14.66015625000000
4	14.82763671875000
5	14.97418212890625 ,

compute the Lipschitz constant L for g. *Hint:* consider (15.9).

15.23. Given the following results of the Fixed Point Iteration applied to a function $g(x)$,

i	x_i
0	0.50000000000000
1	0.70710678118655
2	0.84089641525371
3	0.91700404320467
4	0.95760328069857
5	0.97857206208770 ,

decide if the convergence rate is linear or not.

15.24. (a) Show that $g(x) = \frac{2}{3}x^3$ is Lipschitz continuous on $[-1/2, 1/2]$ with Lipschitz constant $L = 1/2$. (b) Use the program from Problem 15.11 to compute 6 Fixed Point Iterations starting with $x_0 = .5$ and compare to the results in Fig. 15.14. (c) Show that the error of x_i is approximately the cube of the error of x_{i-1} for any i.

15.25. Verify that (15.13) is true.

15.26. (a) Show the root problem $f(x) = x^2 + x - 6$ can be written as the fixed point problem $g(x) = x$ with $g(x) = \dfrac{6}{x+1}$. Show that the error of x_i decreases at a linear rate to the fixed point $\bar{x} = 2$ when the Fixed Point Iteration converges to 2 and estimate the convergence factor for x_i close to 2. (b) Show the root problem $f(x) = x^2 + x - 6$ can be written as the fixed point problem $g(x) = x$ with $g(x) = \dfrac{x^2 + 6}{2x + 1}$. Show that the error of x_i decreases at a quadratic rate to the fixed point $\bar{x} = 2$ when the Fixed Point Iteration converges to 2.

15.27. The **Regula Falsi Method** is a variation of the Bisection Algorithm for computing a root of $f(x) = 0$. For $i \geq 1$, assuming $f(x_{i-1})$ and $f(x_i)$ have the opposite signs, define x_{i+1} as the point where the straight line through $(x_{i-1}, f(x_{i-1}))$ and $(x_i, f(x_i))$ intersects the x-axis. Write this method as Fixed Point Iteration by giving an appropriate $g(x)$ and estimate the corresponding convergence factor.

Part II

Differential and Integral Calculus

16

The Linearization of a Function at a Point

Up to this point, we have been concerned with mathematical models whose solutions are *numbers*. The next stage is to consider more sophisticated models[1] whose solutions are *functions*. To do this, we need to investigate functions in greater detail. In this chapter, we begin with the differential calculus.

One of the fundamental tools for investigating the behavior of a nonlinear function is to approximate the function by a linear function. The motivation is that linear functions are well understood, whereas nonlinear functions are not. In this chapter, we explain how to compute an accurate linear approximation of a smooth function.

16.1 The Imprecision of Lipschitz Continuity

The linear approximation of a function is based on a refinement of the idea of Lipschitz continuity. Recall that according to the definition of Lipschitz continuity, the output of a linear function changes when the input changes. If $f(x) = mx + b$ for some constants m and b, then for any two points x and $\bar{x}$,

$$|f(\bar{x}) - f(x)| = |m| \, |\bar{x} - x|$$

and the Lipschitz constant of f is $|m|$. Note that this holds for *any* $\bar{x}$ and x (see Fig. 16.1). The idea behind Lipschitz continuity is to apply this

---

[1]Such as differential equations.

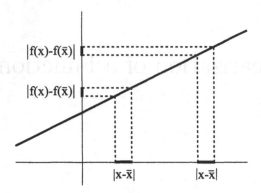

FIGURE 16.1. The change in output of a linear function for a given change in
input $|\bar{x} - x|$ is the same regardless of x and $\bar{x}$.

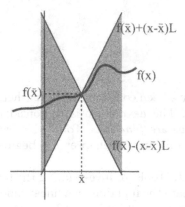

FIGURE 16.2. The Lipschitz condition means that the value of $f(x)$ is "trapped"
between the lines $y = f(\bar{x}) \pm L(x - \bar{x})$.

condition to general nonlinear functions in order to measure how much the
output changes with a small change in input.

The definition of Lipschitz continuity of f on an interval I: namely, there
is a constant L such that

$$|f(x) - f(\bar{x})| \le L|x - \bar{x}|$$

for all x and $\bar{x}$ in I: means that the value of $f(x)$ lies in the sector formed
by the two straight lines

$$y = f(\bar{x}) \pm L(x - \bar{x})$$

through the point $(\bar{x}, f(\bar{x}))$ (see Fig. 16.2). This figure shows that Lipschitz continuity can be a rather imprecise way to describe how a function

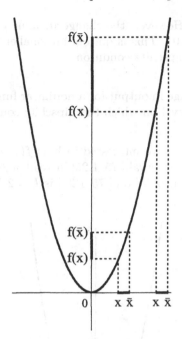

FIGURE 16.3. The change in output of $f(x) = x^2$ corresponding to a given change in input $|\bar{x} - x|$ is larger when $\bar{x}$ and x are larger.

changes. The function can "wiggle" around quite a bit in the region determined by the Lipschitz condition.

In fact for most nonlinear functions, the change in output for a given change in input does depend on the values of the input.

EXAMPLE 16.1. Fig. 16.3 clearly shows that the change in output of $f(x) = x^2$ depends on the values of the input. The reason is simply that $f(x) = x^2$ varies in "steepness" as x increases. We can compute the Lipschitz constant from the equality

$$|\bar{x}^2 - x^2| = |x + \bar{x}|\,|\bar{x} - x|.$$

The factor $|\bar{x} + x|$ is largest when $\bar{x}$ and x are largest. On the interval $[0, 2]$, the Lipschitz constant of $f(x) = x^2$ is therefore $L = 4$.

So showing that a function is Lipschitz continuous does imply that small changes in input cause small changes in output. But the change is measured imprecisely since the Lipschitz constant is determined by the largest change possible in a given interval. If we look at changes for input at points away from where the biggest change occurs, the corresponding change in the output is smaller than predicted by the Lipschitz condition.

EXAMPLE 16.2. Computing the change in x^2 from $x = 1.9$ to $x = 2$, we get $2^2 - 1.9^2 = .39$, while the Lipschitz condition gives $4(2 - 1.9) = .4$, which is not far off. However, the change in value going from $x = 0$ to $x = .1$ is $.1^2 - 0^2 = .01$. This is quite a bit smaller than the change of .4 predicted by the Lipschitz condition.

To measure the change in output for a nonlinear function more precisely, we can shrink the interval on which the Lipschitz condition is applied.

EXAMPLE 16.3. If we are interested in how $f(x) = x^2$ changes for x near 1, we can use the interval $[.75, 1.25]$ instead of $[0, 2]$, (see Fig. 16.4). The Lipschitz constant of x^2 on $[.75, 1.25]$ is $L = 2.5$. The graph shows

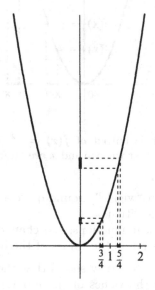

FIGURE 16.4. The Lipschitz condition for $f(x) = x^2$ is more precise if we make the interval smaller.

there is less variation in the change $|f(\bar{x}) - f(x)|$ for a given change $|\bar{x} - x|$ when x and $\bar{x}$ are restricted to the interval $[.75, 1.25]$ as compared to $[0, 2]$.

In other words, $f(x) = x^2$ looks "more" like a linear function on smaller intervals. Figure 16.5 shows that the curve in the graph of x^2 is less noticeable on smaller intervals. Since the idea of Lipschitz continuity is based on how linear functions change, the more a function looks like a linear function, the more precisely the Lipschitz condition determines the change in the function values.

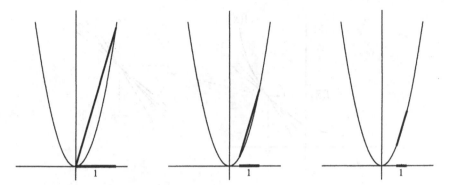

FIGURE 16.5. Comparing $f(x) = x^2$ to linear functions on decreasing intervals from left to right. If we make the interval for the comparison smaller, x^2 looks more like a line.

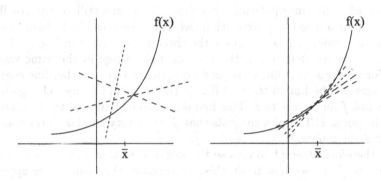

FIGURE 16.6. Some bad linear "approximations" to the function f near $\bar{x}$ are shown on the left and some good linear approximations are shown on the right.

16.2   Linearization at a Point

We construct a linear approximation to a nonlinear function using the idea that a smooth function "looks" more like a line on a small interval. In particular, we construct a linear function that is a good approximation of a nonlinear Lipschitz continuous function f near a particular point $\bar{x}$. In general, there are many lines that are close to a given function f near a point, as illustrated in Fig. 16.6. The question is whether one of the many possible approximate lines is a particularly good choice or not.

Assuming the value of $f(\bar{x})$ is known, it is natural to consider lines that pass through the point $(\bar{x}, f(\bar{x}))$. Such lines are said to **interpolate** f at $\bar{x}$. All such lines have the equation

$$y = f(\bar{x}) + m(x - \bar{x}) \tag{16.1}$$

for some slope m. Several examples are shown in Fig. 16.7.

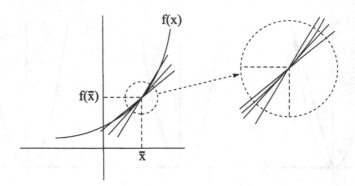

FIGURE 16.7. Linear approximations to a function that pass through the point $(\bar{x}, f(\bar{x}))$. We focus on the region near $(\bar{x}, f(\bar{x}))$ on the right.

Even with the interpolation condition, there are still many possibilities. To find a good approximation, we consider how $f(x)$ changes as x moves away from $\bar{x}$. If we examine the three lines plotted in Fig. 16.7 near $(\bar{x}, f(\bar{x}))$, we see that two of the lines do not change in the same way as $f(x)$ for x near $\bar{x}$. One line changes more quickly and the other line changes more slowly. The line in the middle, on the other hand, does change more or less like f for x close to $\bar{x}$. This line is said to be **tangent** to the graph of f at the point $\bar{x}$. The plot suggests that f looks very similar to the tangent line at $\bar{x}$ for x near $\bar{x}$.

We therefore attempt to choose the slope m so that $f(\bar{x}) + m(x - \bar{x})$ is tangent to f. To see how to do this, we consider the error of the approximation,

$$error = f(x) - \big(f(\bar{x}) + m(x - \bar{x})\big). \qquad (16.2)$$

Naturally, we try to choose m so that the *error* is relatively small. Both the function f and its approximation $f(\bar{x}) + m(x - \bar{x})$ have the same value $f(\bar{x})$ at $\bar{x}$. For x close to $\bar{x}$, we think of $m(x - \bar{x})$ as a small correction to the value of $f(\bar{x})$. Likewise if we rewrite (16.2) as

$$f(x) = f(\bar{x}) + m(x - \bar{x}) + error,$$

then we can think of the *error* as being a correction to the value of $f(\bar{x}) + m(x - \bar{x})$. The linear approximation $f(\bar{x}) + m(x - \bar{x})$ is a good approximation if the correction given by the *error* is small compared to the correction $m(x - \bar{x})$.

To make this more precise, recall that if $|x - \bar{x}| < 1$ and $n \geq 2$, then

$$|x - \bar{x}|^n < |x - \bar{x}|.$$

In fact, $|x - \bar{x}|^n$ is much smaller than $|x - \bar{x}|$ when $|x - \bar{x}|$ is small. For example, $.1^2 = .01$ is quite small compared to $.1$. Therefore, if the slope

m is chosen so that the *error* is roughly proportional to $|x - \bar{x}|^n$ for some $n \geq 2$, then the *error* is relatively small for x close to $\bar{x}$.[2]

Now we are in position to define the linear approximation of a function. The function f is said to be **strongly differentiable** at $\bar{x}$ if there is an open interval $I_{\bar{x}}$ containing $\bar{x}$, a number $f'(\bar{x})$, and a constant $\mathcal{K}_{\bar{x}}$ such that

$$|f(x) - (f(\bar{x}) + f'(\bar{x})(x - \bar{x}))| \leq (x - \bar{x})^2 \mathcal{K}_{\bar{x}} \text{ for all } x \text{ in } I_{\bar{x}}. \qquad (16.3)$$

The approximation $f(\bar{x}) + f'(\bar{x})(x - \bar{x})$ is called the **linearization** of f at the point $\bar{x}$ while the slope of the linearization $f'(\bar{x})$ is called the **derivative** of f at $\bar{x}$. The linearization $f(\bar{x}) + f'(\bar{x})(x - \bar{x})$ is also called the **tangent line** to f at $\bar{x}$. Note that $f'(\bar{x}) = m$ in (16.1).[3] Note that the error term depends on $\bar{x}$ as well as x.

EXAMPLE 16.4. To compute the linearization of x^2 at $\bar{x} = 1$, we compute numbers m and $\mathcal{K}_1$ such that

$$|x^2 - (1 + m(x - 1))| \leq (x - 1)^2 \mathcal{K}_1 \qquad (16.4)$$

for x near 1. Now $x^2 - (1 + m(x - 1)) = x^2 - 1 - m(x - 1)$. Since $x^2 - 1 = (x - 1)(x + 1)$, we get

$$x^2 - (1 + m(x - 1)) = (x - 1)(x + 1) - m(x - 1) = (x - 1)(x + 1 - m).$$

To achieve (16.4), which now reads

$$|x - 1| \, |x + 1 - m| \leq (x - 1)^2 \mathcal{K}_1,$$

for *all* x near 1, we must choose m and $\mathcal{K}_1$ so that

$$|x + 1 - m| \leq |x - 1| \mathcal{K}_1.$$

Now the right-hand side tends to zero as $x \to 1$: hence, the left-hand side must also tend to zero. But this means that $m = 2$. Since $|x + 1 - 2| = |x - 1|$, we conclude that (16.4) holds with $m = 2$ and $\mathcal{K}_1 = 1$. The linearization of $f(x) = x^2$ at $\bar{x} = 1$ is $1 + 2(x - 1)$. The derivative of f at $\bar{x} = 1$ is $f'(1) = 2$. We compare some values of x^2 to $1 + 2(x - 1)$ and plot the functions in Fig. 16.8.

It is illustrative to consider an example of a function that does not have a linearization at one point.

---

[2]Actually, we only require $n > 1$ for the error to be relatively small for x sufficiently close to $\bar{x}$. However, dealing with a fractional power $1 < n < 2$ is more difficult and $n \geq 2$ serves most of the time.

[3]We use $f'(\bar{x})$ to denote the slope of the linearization, even though it is a more complicated symbol than m, because this is useful later on.

x	$f(x)$	$f(\bar{x}) + f'(\bar{x})(x - \bar{x})$	error
.7	.49	.4	.09
.8	.64	.6	.04
.9	.81	.8	.01
1.0	1.0	1.0	0.0
1.1	1.21	1.2	.01
1.2	1.44	1.4	.04
1.3	1.69	1.6	.09

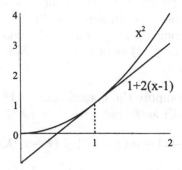

FIGURE 16.8. Some values of the linearization $1 + 2(x - 1)$ of x^2 at $\bar{x} = 1$ along with plots of the functions.

EXAMPLE 16.5. The Lipschitz continuous function $f(x) = |x|$ does not have a linearization at $\bar{x} = 0$. This is intuitively obvious from the graph because the "sharp corner" in the graph of $|x|$ at 0 means that there is no way to draw a single good linear approximation (see Fig. 16.9). If we try to use the definition of the linearization, we immediately run into trouble. For all $x > 0$, $|x| = 0 + x + 0 = |\bar{x}| + x + error$. But for $x < 0$, $|x| = 0 - x + 0$. Hence, there is no *single* number $f'(0)$ for which $|x| = 0 + f'(0)x + error$ for *all* x near 0.

16.3   A Systematic Approach

Computing the linearization and estimating its error can be difficult. However if a function is strongly differentiable, there is a systematic way to determine first the derivative and then estimate the error in principle, if not in practice. Before computing more examples, we describe this method in abstract terms. Consider a Lipschitz continuous function $E_{\bar{x}}$ of x in some open interval I containing $\bar{x}$ and suppose that

$$E_{\bar{x}}(x) \le (x - \bar{x})^2 \mathcal{K}_{\bar{x}} \tag{16.5}$$

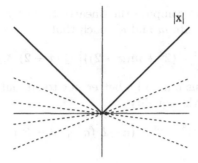

FIGURE 16.9. There is no single good linear approximation of $|x|$ at $\bar{x} = 0$.

for all x in I, where $\mathcal{K}_{\bar{x}}$ is some constant. In the present situation, we can think of $E_{\bar{x}}$ as being the error of the linear approximation, i.e.,

$$E_{\bar{x}}(x) = |f(x) - (f(\bar{x}) + f'(\bar{x})(x - \bar{x}))|;$$

but these comments hold in general. It follows from (16.5) that

$$\lim_{x \to \bar{x}} E_{\bar{x}}(x) = E_{\bar{x}}(\bar{x}) = 0. \tag{16.6}$$

In addition, it follows that

$$\lim_{x \to \bar{x}} \frac{E_{\bar{x}}(x)}{|x - \bar{x}|} = 0 \tag{16.7}$$

since

$$\left| \frac{E_{\bar{x}}(x)}{|x - \bar{x}|} \right| \leq |x - \bar{x}||\mathcal{K}_{\bar{x}}|$$

for all $x \neq \bar{x}$ in I.

Therefore, the statement (16.5) that $E_{\bar{x}}$ is quadratic in $x - \bar{x}$ for x close to $\bar{x}$ means that both (16.6) and (16.7) hold. This gives precise criteria for checking such a statement.[4]

It is important to realize that while (16.6) can be verified by substituting $x = \bar{x}$ into $E_{\bar{x}}$, we cannot check (16.7) by simple substitution. We have divided by $x - \bar{x}$ after all, which is undefined at $x = \bar{x}$. *Condition* (16.7) *must be verified by computing a limit.* It is also important to note that while (16.6) and (16.7) are necessary for (16.5) to hold, they are *not* sufficient. We discuss this point in more detail later.

Note that we actually use these ideas in computing the linearization of x^2 at 1. We now consider a few more examples.

---

[4]Prior experience with the standard definition of the derivative will make the work needed to compute the linearization of a function seem familiar. Showing a function is strongly differentiable does take more work, however, since we are estimating the error of the linearization in addition to computing the linearization.

EXAMPLE 16.6. We compute the linearization of $f(x) = x^3$ at $\bar{x} = 2$. To do this, we compute m and $\mathcal{K}_2$ such that

$$\left| x^3 - \left(2^3 + m(x-2) \right) \right| \leq (x-2)^2 \mathcal{K}_2$$

for x near 2. It turns out to be better not to simplify $2^3 = 8$ just yet. We compute using the equality

$$a^3 - b^3 = (a-b)(a^2 + ab + b^2)$$

to get

$$
\begin{aligned}
\left| x^3 - \left(2^3 + m(x-2) \right) \right| &= \left| x^3 - 2^3 - m(x-2) \right| \\
&= \left| (x-2)(x^2 + 2x + 4) - m(x-2) \right| \\
&= |x-2|\,|x^2 + 2x + 4 - m|.
\end{aligned}
$$

We conclude that

$$\lim_{x \to 2} \left| x^3 - \left(2^3 + m(x-2) \right) \right| = 0$$

for any m as required. We also require

$$\lim_{x \to 2} \frac{\left| x^3 - \left(2^3 + m(x-2) \right) \right|}{|x-2|} = \lim_{x \to 2} |x^2 + 2x + 4 - m| = 0,$$

which forces $m = 12$. The linearization is $8 + 12(x-2)$.

To estimate the error, we compute

$$
\begin{aligned}
\left| x^3 - \left(2^3 + 12(x-2) \right) \right| &= \left| x^3 - 2^3 - 12(x-2) \right| \\
&= \left| (x-2)(x^2 + 2x + 4) - 12(x-2) \right| \\
&= |x-2|\,|x^2 + 2x - 8| = |x-2|\,|x-2|\,|x+4| \\
&= |x-2|^2 |x+4|.
\end{aligned}
$$

On any finite interval I_2 containing 2 of size $|I_2|$, $|x+4| \leq |I_2| + 6$. So for any such interval,

$$\left| x^3 - \left(2^3 + m(x-2) \right) \right| \leq (x-2)^2 (|I_2| + 6)$$

for x in I_2. We conclude that x^3 is strongly differentiable at 2.[5]

We compare some values of x^3 to $8 + 12(x-2)$ and show the linearization in Fig. 16.10 .

---

[5]Note that much of the work in estimating the error duplicates the work used to compute the derivative.

x	$f(x)$	$f(2) + f'(2)(x-2)$	error
1.7	4.913	4.4	.513
1.8	5.832	5.6	.232
1.9	6.859	6.8	.059
2.0	8.0	8.0	0.0
2.1	9.261	9.2	.061
2.2	10.648	10.4	.248
2.3	12.167	11.6	.567

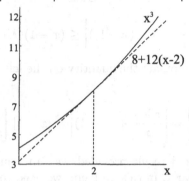

FIGURE 16.10. Some values of the linearization $8 + 12(x-2)$ of x^3 at $\bar{x} = 2$ and plots of the functions.

EXAMPLE 16.7. To compute the linearization of $f(x) = 1/x$ at $\bar{x} = 1$, we compute numbers m and $\mathcal{K}_1$ such that

$$\left| \frac{1}{x} - (1 + m(x-1)) \right| \le (x-1)^2 \mathcal{K}_1 \qquad (16.8)$$

for x near 1. We first compute m and estimate $\mathcal{K}_1$. The strategy in the analysis is to find factors of $|x-1|$ in the expression on the left-hand side of (16.8). Now

$$\left| \frac{1}{x} - (1 + m(x-1)) \right| = \left| \frac{1}{x} - 1 - m(x-1) \right|$$

$$= \left| \frac{1-x}{x} - m(x-1) \right|$$

$$= |1-x| \left| \frac{1}{x} + m \right|.$$

We conclude that

$$\lim_{x \to 1} \left| \frac{1}{x} - (1 + m(x-1)) \right| = 0$$

for any m, as (16.6) requires. For (16.7) to hold, we want

$$\lim_{x \to 1} \frac{\left|\frac{1}{x} - (1 + m(x-1))\right|}{|x-1|} = \lim_{x \to 1} \frac{|1-x|\left|\frac{1}{x} + m\right|}{|x-1|} = \lim_{x \to 1} \left|\frac{1}{x} + m\right| = 0.$$

We conclude that $m = -1$, i.e., $f'(1) = -1$, and the linearization is $1 - (x-1)$. To show that f is strongly differentiable, we need to show that

$$\left|\frac{1}{x} - (1 - (x-1))\right| \le (x-1)^2 \mathcal{K}_1$$

for x near 1. We manipulate the quantity on the left in order to get two factors of $x - 1$,

$$\left|\frac{1}{x} - (1 - (x-1))\right| = \left|\frac{x-1}{x} + (x-1)\right| = |x-1|\left|\frac{1}{x} - 1\right| = \frac{|x-1|^2}{|x|}.$$

This gives the desired result provided we can bound the size of the factor $1/|x|$ for x near 1. In this problem, we need to choose an interval I_1 carefully in order to get the desired bound because the factor $1/|x|$ in the error becomes arbitrarily large as x approaches zero. Since we want to show the linearization is accurate for x near 1, we just choose any interval I_1 that contains 1 but is bounded away from 0. The interval $I_1 = (.5, 2)$ is a suitable choice, for example. Then $1/|x| \le 2$ for x in I_1 and we conclude that

$$\left|\frac{1}{x} - (1 - (x-1))\right| \le (x-1)^2 \, 2$$

for x in I_1. So $1/x$ is strongly differentiable at 1.

We compare some values of $1/x$ to $1-(x-1)$ and show the linearization in Fig. 16.11.

16.4   Strong Differentiability and Smoothness

Example 16.5 shows that some degree of smoothness beyond Lipschitz continuity is required for a function to be strongly differentiable. We investigate this issue in great detail, beginning with this section.

Here, we show the intuitively obvious fact that the graph of a function f that is strongly differentiable at a point $\bar{x}$ cannot have a jump or discontinuity at $\bar{x}$. From the definition, there is an open interval $I_{\bar{x}}$ and constants $f'(\bar{x})$ and $\mathcal{K}_{\bar{x}}$ such that for all x in $I_{\bar{x}}$,

$$|f(x) - (f(\bar{x}) + f'(\bar{x})(x - \bar{x}))| \le |x - \bar{x}|^2 \mathcal{K}_{\bar{x}}.$$

x	$f(x)$	$f(\bar{x}) + f'(\bar{x})(x - \bar{x})$	error
.7	1.428571$\cdots$	1.3	$\approx$ .1286
.8	1.25	1.2	.05
.9	1.111111$\cdots$	1.1	$\approx$ .01111
1.0	1.0	1.0	0.0
1.1	.909090$\cdots$	.9	$\approx$ .00909
1.2	.833333$\cdots$	.8	$\approx$ .03333
1.3	.769230$\cdots$	.7	$\approx$ .06923

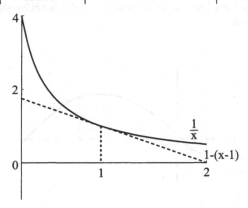

FIGURE 16.11. Some values of the linearization $1 - (x - 1)$ of $1/x$ at $\bar{x} = 1$ along with plots of the functions.

But this means that

$$|f(x) - f(\bar{x})| \le |f'(\bar{x})||x - \bar{x}| + \mathcal{K}_{\bar{x}}|x - \bar{x}|^2$$
$$\le (|f'(\bar{x})| + \mathcal{K}_{\bar{x}}|I_{\bar{x}}|)|x - \bar{x}|$$
$$= L|x - \bar{x}|$$

where we have assumed that $I_{\bar{x}}$ is finite. This essentially says that f is Lipschitz continuous "at a point" where it is strongly differentiable.[6] We summarize as a theorem:

Theorem 16.1 *If f is strongly differentiable at $\bar{x}$, then there is an open interval I containing $\bar{x}$ and a constant L such that*

$$|f(x) - f(\bar{x})| \le L|x - \bar{x}| \text{ for all } x \text{ in } I.$$

---

[6]The quotes are there because we do not define Lipschitz continuity at a point!

Chapter 16   Problems

16.1. Compare the Lipschitz constants for $f(x) = 1/x$ on the intervals $[.01, .1]$ and $[1, 2]$. Explain the reason for the difference using a plot.

16.2. Compare the Lipschitz constants for $f(x) = x^3$ on $[0, 2]$ and $[.9, 1.1]$. Explain the difference using a plot.

16.3. Use a ruler to draw linear approximations to the function shown in Fig. 16.12 at the indicated points.

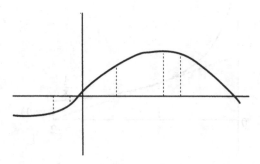

FIGURE 16.12. Figure for Problem 16.3.

16.4. Compute the linearizations for the following functions at the indicated points.

(a) $f(x) = 4x$ at $\bar{x} = 1$ (b) $f(x) = x^2$ at $\bar{x} = 0$

(c) $f(x) = x^2$ at $\bar{x} = 2$ (d) $f(x) = 1/x$ at $\bar{x} = 2$

(e) $f(x) = x^3$ at $\bar{x} = 1$ (f) $f(x) = 1/x^2$ at $\bar{x} = 1$

(g) $f(x) = x + x^2$ at $\bar{x} = 1$ (h) $f(x) = x^4$ at $\bar{x} = 1$.

16.5. (a) Make a table showing the values of the functions, the values of the linearizations, and the values of the error functions at the points 1.7, 1.8, 1.9, 2.0, 2.1, 2.2, 2.3 for the functions $f_1(x) = x^2$ and $f_2(x) = x^3$, where the linearizations are computed at $\bar{x} = 2$. Use a graph to explain why the error of the linearization of x^3 is larger than the error of the linearization for x^2 for most x near $\bar{x}$. (b) Would you expect this is true if $\bar{x} = 0$? Make a table that confirms or refutes your answer.

16.6. Which would you expect to be a worse approximation on the interval $[.1, .3]$, the linearization of $1/x$ at $\bar{x} = .2$ or the linearization of $1/x^2$ at $\bar{x} = .2$? Why?

17

Analyzing the Behavior of a Population Model

To illustrate the power of linearization as a tool for analysis, we use linearization to analyze the behavior of a complicated population model for a certain species of insect. This model encompasses all of the models for insect populations considered so far.

17.1   A General Population Model

We assume that there is a single breeding season during the summer while the adults that breed in one summer die before the next summer. We let P_n denote the population of adults at the start of the nth breeding season and assume that each adult produces, on average, R offspring that survive to breed in the next year. This implies that

$$P_n = RP_{n-1}.$$

In the simplest case, R is constant and induction shows that

$$P_n = R^n P_0,$$

where P_0 is the initial population given in some starting year. In this case, we can determine the behavior of P_n easily. If $0 < R < 1$, then P_n decreases steadily toward 0 as n increases and the population of insects dies out. If $R > 1$, then P_n increases steadily as n increases. If $R = 1$, then $P_n = P_{n-1}$ and P_n remains constant as the population just replenishes itself.

But assuming R is a constant is too simplistic in most cases. Rather, R usually varies with the population. If the population is large, then there is a lot of competition for resources and R tends to be small. If the population is small, then R is also small because, for example, the females may have a difficult time finding mates. On the other hand, for the population to survive, there must be a range of population P with $R(P) > 1$. To make a more realistic model, we have to use a function R that is less than 1, for P small and large, and greater than 1, when P is neither small or large. There are many possible choices of such functions and we plot one choice in Fig. 17.1. It is important to note that the model describes the *qualitative*

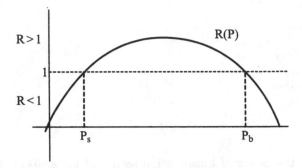

FIGURE 17.1. A possible growth rate function R. $R(P_s) = R(P_b) = 1$, $R(P) < 1$ for $P < P_s$ and $P > P_b$, and $R(P) > 1$ for $P_s < P < P_b$.

features of the coefficient R but does *not* give an exact formula. There are many choices of R that yield the same qualitative features.

Note that the solution of (17.1) is *not* $P_n = R^n P_0$ in general. It is considerably more complicated, and depending on R, we may not even be able to write down the solution as a formula. So we have to be clever when analyzing how the population behaves.

17.2   Equilibrium Points and Stability

We use the linearization to analyze the behavior of the population assuming the relationship

$$P_n = R(P_{n-1})P_{n-1}, \qquad (17.1)$$

where R is a function like that plotted in Fig. 17.1.

The first observation is that the populations P_s and P_b that satisfy $R(P_s) = R(P_b) = 1$ are special. If, for example, $P_{n-1} = P_b$ for some $n - 1$, then $P_n = R(P_{n-1})P_{n-1} = 1 \times P_b = P_b$. Also $P_{n+1} = P_b$, $P_{n+2} = P_b$, and so on. In words, if the population reaches the value of P_b, then it remains at that value for subsequent generations. Likewise, if $P_{n-1} = P_s$, then $P_i = P_s$ for all $i \geq n$. We call these populations **equilibrium points** of the iteration (17.1).

But the behavior of the population model is different around the two equilibrium points P_s and P_b. If $P_{n-1} > P_b$, then $R(P_{n-1}) < 1$ so $P_n < P_{n-1}$. In words, if the population is larger than P_b, then it decreases in the next generation. On the other hand, if $P_{n-1} < P_b$, then $R(P_{n-1}) > 1$ so $P_n > P_{n-1}$. In words, if the population is smaller than P_b, then it increases in the next generation. In short, the population tends toward the value of P_b. We say that P_b is a **stable** equilibrium point of the iteration (17.1). Arguing the same way, we can show that the population tends to move *away* from the value P_s and we call P_s an **unstable** equilibrium point of the iteration (17.1). In particular, if the population becomes less than P_s, the species tends to extinction.

Given that the population tends toward P_b, we would like to get more information about how it behaves as it does this. For example, does the population oscillate in value around P_b or does it steadily increase or decrease to P_b? We assume that R is strongly differentiable at P_b, so there is a constant $\mathcal{K}_{P_b}$ such that

$$R(P) = R(P_b) + R'(P_b)(P - P_b) + (P - P_b)^2 \mathcal{K}_{P_b}(P)$$
$$= 1 + R'(P_b)(P - P_b) + (P - P_b)^2 \mathcal{K}_{P_b}(P).$$

Incidentally, Fig. 17.1 indicates that $R'(P_b) < 0$. For P close to P_b, we use the approximation[1]

$$R(P) \approx 1 + R'(P_b)(P - P_b).$$

In other words, we replace the function R by the linearization of R at P_b for P close to P_b. Substituting into (17.1), we get

$$P_n \approx (1 + R'(P_b)(P_{n-1} - P_b))P_{n-1}.$$

We have made some progress using the linearization since we replaced the possibly complicated factor $R(P_{n-1})$ by the linear factor $(1 + R'(P_b)(P_{n-1} - P_b))$.

We are interested in how P changes relative to P_b, so we rearrange the equations so that the differences $P_n - P_b$ and $P_{n-1} - P_b$ appear:

$$P_n - P_b \approx (1 + R'(P_b)(P_{n-1} - P_b))P_{n-1} - P_b$$
$$\approx P_{n-1} - P_b + R'(P_b)(P_{n-1} - P_b)(P_{n-1} - P_b + P_b)$$
$$\approx P_{n-1} - P_b + R'(P_b)P_b(P_{n-1} - P_b) + R'(P_b)(P_{n-1} - P_b)^2.$$

---

[1] The analysis in this chapter can be made precise using the definition of strong differentiability and carrying along the errors. Doing so makes the analysis a lot harder to read, however, so we give this as Problem 17.6.

In the original approximation, we dropped the term $(P_{n-1} - P_b)^2 \, \mathcal{K}_{P_b}$, which is at least quadratic in $P_{n-1} - P_b$. Therefore, we also drop the quadratic term on the right above to obtain

$$P_n - P_b \approx (1 + R'(P_b)P_b)(P_{n-1} - P_b).$$

Now this is real progress, because the factor $(1 + R'(P_b)P_b)$ is constant! To simplify things, we recall that $R'(P_b) < 0$ and set

$$C = -R'(P_b)P_b > 0$$

to obtain

$$P_n - P_b \approx (1 - C)(P_{n-1} - P_b). \tag{17.2}$$

Based on the previous discussion, there are three cases.

- If $0 < C < 1$, then $|P_n - P_b| < |P_{n-1} - P_b|$, while $P_n - P_b$ has the same sign as $P_{n-1} - P_b$. This means that P_n decreases or increases steadily toward the value P_b, depending on whether the population starts above or below P_b. $0 < C < 1$ means that

$$0 < -R'(P_b) < \frac{1}{P_b}.$$

- If $1 < C < 2$, then $|P_n - P_b| \leq |P_{n-1} - P_b|$, but $P_n - P_b$ has the opposite sign of $P_{n-1} - P_b$. This means that the population tends toward P_b but in an oscillating fashion: if the population of one generation is above P_b, then the population of the next generation is below P_b. $1 < C < 2$ means that

$$\frac{1}{P_b} < -R'(P_b) < \frac{2}{P_b}.$$

- If $C > 2$ or $C < 0$, then the population moves away from P_b.

It is important to note that these conclusions are based on the assumption that the population P_n is close to P_b. If the population is too far away from P_b, then the approximation (17.2) is not valid and these conclusions do not hold.

EXAMPLE 17.1.   To show that these predictions describe what happens, we compute some generations corresponding to the population rate function

$$R(P) = 1 - c(P - 100)(P - 1000) \tag{17.3}$$

which has the shape drawn in Fig. 17.1 for $c > 0$. For this function, $P_s = 100$ and $P_b = 1000$. It is straightforward to verify that we expect the population to decrease steadily to P_b when

$$0 < c < \frac{1}{900000} \tag{17.4}$$

and to approach P_b in an oscillatory fashion when

$$\frac{1}{900000} < c < \frac{1}{450000}. \tag{17.5}$$

We plot both cases in Fig. 17.2 starting with two initial populations in each case.

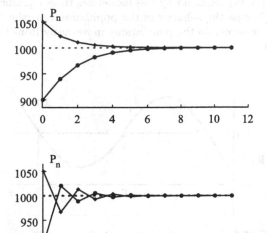

FIGURE 17.2. Eleven generations of the population model (17.1) using the population rate function (17.3). In the upper plot, we use $c = 1/18000000$ and in the lower plot we use $c = 1/600000$.

Chapter 17   Problems

17.1. Verify that (17.4) and (17.5) are correct for the rate function (17.3).

17.2. Repeat the computations shown in Fig. 17.2.

17.3. Consider the population model (17.1) where R is a function like that plotted in Fig. 17.3. (a) Explain why this model has three equilibrium points: P_s, P_c, and P_b. (b) Discuss the behavior of the population near the three points P_s, P_c, and P_b. For example, do the populations move away from these points and in what direction?

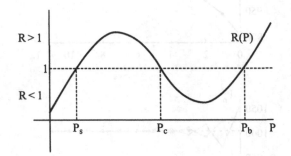

FIGURE 17.3. A population rate with three equilibrium points.

17.4. Consider the population model (17.1) with population rate

$$R(P) = \frac{5}{1 + P/P_c}$$

where $P_c > 0$ is a constant.

(a) Show that P_c is a stable equilibrium point.

(b) Use the derivative of R to discuss the behavior of P for populations near P_c. Find regions of populations when the population decreases or increases steadily toward P_c and regions where it oscillates as it approaches P_c.

(c) Using $P_c = 100$, compute some iterations that illustrate your conclusions.

17.5. Consider the population model (17.1) with population rate

$$R(P) = \frac{1}{1 + P^2}.$$

Discuss the behavior of the population as the iteration proceeds.

17.6. Write out the analysis in this chapter using the definition of strong differentiability and carrying along the error terms. *Hint:* For $p \geq 2$, $(x - \bar{x})^p \leq (x - \bar{x})^2$ for all x sufficiently close to $\bar{x}$. This can be used to simplify things as you go along.

18
Interpretations of the Derivative

Before continuing the investigation of linearization, we present two important interpretations of the derivative. The first is geometric and is related to the idea of the tangent line to a curve. The second involves the rates at which quantities change. Both are important in modeling.

In this chapter, the emphasis is changed a bit. Observe that determining the linearization of a function at a point $\bar{x}$ only requires knowing the derivative of the function at the point. The only purpose of the secondary computation that determines the error is to show that the error is small for x near $\bar{x}$. Therefore, we concentrate on computing the derivative of a function. In Chapter 20, we show that the derivative has several nice properties, which in particular imply that the error of linearization of many functions is small once we have shown it is small for some relatively simple functions.

18.1 A Geometric Picture

Recall that in the process of computing the derivative of a function f at $\bar{x}$, we "match" terms that are the same order in $x - \bar{x}$ by subtracting $f(\bar{x})$ from both sides of (16.3) and dividing by $x - \bar{x}$ to get

$$\left| \frac{f(x) - f(\bar{x})}{x - \bar{x}} - f'(\bar{x}) \right| \leq |x - \bar{x}| \mathcal{K}_{\bar{x}}.$$

At this point, we do some algebra on the expression on the left to "cancel out" the pesky factor of $x - \bar{x}$ in the denominator. After that, we take the limit as $x \to \bar{x}$, the error term drops out, and we find that if f is strongly differentiable at $\bar{x}$, then

$$f'(\bar{x}) = \lim_{x \to \bar{x}} \frac{f(x) - f(\bar{x})}{x - \bar{x}}. \tag{18.1}$$

Of course, we have to prove that this limit exists to compute the derivative.

For x close to $\bar{x}$, we conclude that

$$f'(\bar{x}) \approx \frac{f(x) - f(\bar{x})}{x - \bar{x}}.$$

The quantity on the right can be interpreted as the slope of the **secant line** that connects the points $(\bar{x}, f(\bar{x}))$ and $(x, f(x))$ on the plot of f (see Fig. 18.1).

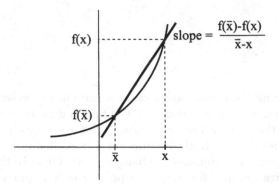

FIGURE 18.1. The secant line joining $(\bar{x}, f(\bar{x}))$ and $(x, f(x))$.

Recalling that $f'(\bar{x})$ is the slope of the tangent line to f at $\bar{x}$, or the linearization of f at $\bar{x}$, (18.1) says that the slope of the secant line between $(\bar{x}, f(\bar{x}))$ and any point $(x, f(x))$ approaches the slope of the tangent line as x approaches $\bar{x}$. We illustrate in Fig. 18.2

Turning this around, we say that a function f is **differentiable** at a point $\bar{x}$ if f is defined on some open interval $I_{\bar{x}}$ containing $\bar{x}$ and if (18.1) holds, i.e.,

$$f'(\bar{x}) = \lim_{x \to \bar{x}} \frac{f(x) - f(\bar{x})}{x - \bar{x}}$$

exists.[1]

EXAMPLE 18.1. We compute the derivative of $f(x) = x^2$ at $\bar{x}$ by taking the limit of the slopes of secant lines. We already know this can be

---

[1]This definition was first given by Bolzano and Cauchy also used this definition.

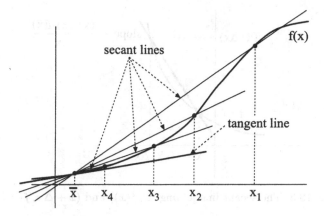

FIGURE 18.2. A sequence of secant lines approaching the tangent line at $\bar{x}$.

done because $f(x) = x^2$ is strongly differentiable at every point. Now we compute

$$f'(\bar{x}) = \lim_{x \to \bar{x}} \frac{x^2 - \bar{x}^2}{x - \bar{x}} = \lim_{x \to \bar{x}} \frac{(x - \bar{x})(x + \bar{x})}{x - \bar{x}} = \lim_{x \to \bar{x}} (x + \bar{x}) = 2\bar{x}.$$

When $\bar{x} = 1$, we get $f'(1) = 2$.

Note that this example uses exactly the same computations used to compute the linearization of f. We have just repackaged the computation for a geometric point of view.[2]

The limit in (18.1) is often written in another way. We replace x by $\bar{x} + \Delta x$, where Δx represents a small change from $\bar{x}$ (see Fig. 18.3). Since $\Delta x = (\bar{x} + \Delta x) - \bar{x}$, the previous limit as $x \to \bar{x}$ is replaced by a limit as $\Delta x \to 0$ and we obtain

$$f'(\bar{x}) = \lim_{\Delta x \to 0} \frac{f(\bar{x} + \Delta x) - f(\bar{x})}{\Delta x}. \tag{18.2}$$

EXAMPLE 18.2. We compute the derivative of $f(x) = x^2$ at x by taking the limit of the slopes of secant lines using the new notation. Now we compute

$$f'(\bar{x}) = \lim_{\Delta x \to 0} \frac{(\bar{x} + \Delta x)^2 - \bar{x}^2}{\Delta x}$$

$$= \lim_{\Delta x \to 0} \frac{\bar{x}^2 + 2\bar{x}\Delta x + \Delta x^2 - \bar{x}^2}{\Delta x} = \lim_{\Delta x \to 0} \frac{2\bar{x}\Delta x + \Delta x^2}{\Delta x}$$

$$= \lim_{\Delta x \to 0} (2\bar{x} + \Delta x) = 2x.$$

---

[2]It is a useful exercise to go back to Chapter 16 and recompute all of the linearizations using the new formulation.

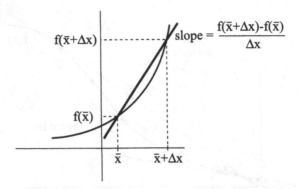

FIGURE 18.3. The secant line joining $(\bar{x}, f(\bar{x}))$ and $(\bar{x} + \Delta x, f(\bar{x} + \Delta x))$.

We introduce this new formulation because it motivates the most common, and indeed the most powerful, notation for the derivative. Since Δx represents the "change" in the variable x, it is natural to define the corresponding change in the function f by

$$\Delta f = f(\bar{x} + \Delta x) - f(\bar{x}).$$

The ratio

$$\frac{\Delta f}{\Delta x},$$

which gives the slope of the secant line through $(x + \Delta x, f(x + \Delta x))$ and $(x, f(x))$ can be interpreted as the change in f over the change in x and the derivative is

$$f'(\bar{x}) = \lim_{\Delta x \to 0} \frac{\Delta f}{\Delta x}.$$

The notation for the derivative reflects the limit process. We write

$$f'(\bar{x}) = Df(\bar{x}) = \frac{df}{dx}.$$

It is common (especially in physics) to interpret df to be the **infinitesimal** change in f corresponding to an infinitesimal change dx in the value x. But these words only mean the derivative is computed by taking the limit in (18.2). If we write $y = f(x)$, then we also write y', by which we mean

$$y' = \frac{dy}{dx} = f'(x).$$

18.2 Rates of Change

The mathematical models of many physical situations involve the rates at which a quantity such as mass or distance changes as some other quantity such as time or position changes.

EXAMPLE 18.3. A model of the gasoline consumption of an automobile involves the rate of change of the position of the car with time because as the car moves faster, it encounters increasingly more air resistance.

A rate of change is determined by a ratio. For example, if we want an idea of how quickly an object is moving, we compute the ratio of elapsed distance to elapsed time. More precisely, assume the object travels in a straight line and is at position s_1 at time t_1 and later at time t_2 is at position s_2. Then its average velocity or average rate of change is

$$\frac{s_2 - s_1}{t_2 - t_1}.$$

EXAMPLE 18.4. If a car is driven 45 miles in 3/4 of an hour, then its average velocity was 60 miles/hour. But clearly the average velocity is rather a crude description of how the car moves. To travel 45 miles in 3/4 of an hour, it might stand still for 15 minutes and then travel at 90 miles/hour for 30 minutes rather than travel at a steady speed of 60 miles/hour for 45 minutes.

If the position of the object is given by a function $s(t)$ at time t, then the **average rate of change** of the object from t to some nearby time $t + \Delta t$ is

$$\frac{s(t + \Delta t) - s(t)}{\Delta t}.$$

From Section 18.1, this is simply the slope of the secant line to $s(t)$ joining $(t, s(t))$ and $(t + \Delta t, s(t + \Delta t))$.

The average rate of change provides only a crude description of how quickly s changes as t changes. We can get a more precise idea by letting $(t, t + \Delta t)$ shrink in size. If s is differentiable at t, we define the **instantaneous rate of change** or **velocity** of the object at t as

$$v(t) = s'(t) = \frac{ds}{dt} = \lim_{\Delta t \to 0} \frac{s(t + \Delta t) - s(t)}{\Delta t}. \tag{18.3}$$

EXAMPLE 18.5. If the position of an object is given by $s(t) = t^2$, then its velocity is $v(t) = s'(t) = 2t$.

EXAMPLE 18.6. Suppose someone is driving a car in a straight line and their position at time t measured from the starting point at $t = 0$ is $s(t) = 3 \times (2t - t^2)$ miles, where t is measured in hours and the positive direction for s is to the right. Later, we show that their speed is $s'(t) = 6 - 6t = 6(1 - t)$ miles/hour. Since the derivative is positive for $0 \le t < 1$, which means that the tangent lines to $s(t)$ have positive slope for $0 \le t < 1$, the car moves to the right up to $t = 1$. At exactly $t - 1$, the car is stopped. After $t > 1$, the car begins to move left again, because the slopes of the tangents are negative. We illustrate in Fig. 18.4.

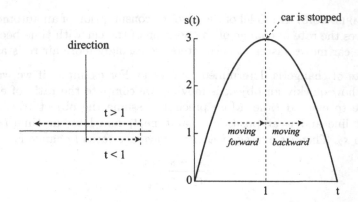

FIGURE 18.4. An illustration of a car moving back and forth on a straight line. On the left, we show how the position of the car and how it moves. On the right, we plot the position versus time. The derivative of position is positive for $0 \le t < 1$, so the car moves to the right during this period. It is stopped at $t = 1$ and then goes left for $t > 1$. At $t = 1$, the speed is zero and therefore the tangent line is horizontal.

EXAMPLE 18.7. Someone throws a ball straight up at 3 feet/second. If they release the ball at a height of 6.2 feet, estimate the height of the ball .4 seconds later.

$$\text{height at } t = .4 \approx \text{ height at } t = 0 + \text{ speed at } t = 0 \times (.4 - 0)$$
$$\approx 6.2 + 3 \times .4 = 7.4$$

18.3 Differentiability and Strong Differentiability

It is important to realize that differentiability and strong differentiability are not equivalent. A function that is strongly differentiable is automatically differentiable. In fact, we compute the limit (18.1) that determines the derivative in the process of computing the linearization of a function. But differentiability does not imply strong differentiability. Strong differentiability requires that the limit in (18.1) converges *at least* at a linear rate, since

$$\left| \frac{f(x) - f(\bar{x})}{x - \bar{x}} - f'(\bar{x}) \right| \le |x - \bar{x}| \mathcal{K}_{\bar{x}}$$

for some constant $\mathcal{K}_{\bar{x}}$. Differentiability only implies that the quantity on the left goes to zero at $x \to \bar{x}$, but does not specify a rate.

Is the difference important? Well, yes and no. It is not an issue for any functions we have met so far and in fact almost all functions that we write down are strongly differentiable anywhere they are differentiable. But there are exceptions and it is important to understand these when they arise. So we return to this discussion in Chapter 32.

Chapter 18 Problems

18.1. Compute the formulas and plot the secant lines to the function $f(x) = x^2$ between $\bar{x} = 1$ and $x = 4$, $x = 2$, $x = 1.5$. What is the average rate of change corresponding to each secant line?

18.2. Compute the formulas and plot the secant lines to the function $f(x) = 1/x$ between $\bar{x} = 1$ and $x = .25$, $x = .5$, $x = .75$. What is the average rate of change corresponding to each secant line?

18.3. Compute the derivatives in Problem 16.4 by taking the limit of slopes of secant lines. Do not compute the error functions again.

18.4. Describe three different situations in your life in which a rate of change is involved.

18.5. A car travels at 35 miles/hour for 30 minutes, 65 miles/hour for 60 minutes, then 35 miles/hour for 30 minutes. What is its average rate of change over the intervals $[0, 30]$, $[0, 60]$, $[0, 90]$, and $[0, 120]$ in minutes?

18.6. A bicyclist rides 15 miles in 1 hour. Find three different riding patterns that lead to this average rate of change. *Hint:* One is of course that the bicyclist rides at a steady 15 miles/hour.

18.7. Verify the computations in Example 18.6.

18.8. Someone rides a unicycle in a straight line so that their distance from an observer in meters at time t is $s(t) = t^5$. What is their velocity at $t = 1$ and $t = 2$?.

18.9. A police officer uses a radar gun to find the speed of a car that is .1 miles away is 80 miles/hour. The radar gun takes .25 seconds to compute the speed. What is the approximate location of the car by that time? What factors could affect the accuracy of your answer?

19
Differentiability on Intervals

So far we have defined the linearization of function at one point. In general however, we are interested in the behavior of a function over an interval. In that case, we may need the linearization of a function at any point in the interval. We say that a function is strongly differentiable on an interval if the function is strongly differentiable at each point in the interval. In this chapter, we investigate the consequences of this definition.

19.1 Strong Differentiability on Intervals

More precisely, a function f is **strongly differentiable** on an open interval I if f is strongly differentiable at each $\bar{x}$ in I. In other words, for each $\bar{x}$ in I, there is an open interval $I_{\bar{x}}$, a number $f'(\bar{x})$, and a constant $\mathcal{K}_{\bar{x}}$ such that

$$\left| f(x) - \left(f(\bar{x}) + f'(\bar{x})(x - \bar{x}) \right) \right| \leq (x - \bar{x})^2 \mathcal{K}_{\bar{x}} \qquad (19.1)$$

for x in $I_{\bar{x}}$.

If a function f is strongly differentiable on open interval I, then each point $\bar{x}$ in I is associated to the slope $f'(\bar{x})$ of the linearization of f at $\bar{x}$. Therefore to a function f that is strongly differentiable on an open interval I, we associate a new function $f'(\bar{x})$ for $\bar{x}$ in I that gives the slope of the linearization at $\bar{x}$. Since $\bar{x}$ is allowed to vary in I, we rename it as simply x, and we define the **derivative** f' of f to be the function that gives the slope $f'(x)$ of the linearization of f at each point x in I. This notation is due to Lagrange.[1] We also use $D(f) = Df = f'$ to denote the derivative, a notation that originates with Johann Bernoulli,[2] who, along with his older brother Jacob Bernoulli,[3] were the first mathematicians to understand and use Leibniz's results in calculus.

EXAMPLE 19.1. For a constant function $f(x) = c$, where c is a real number, we get

$$f'(x) = 0$$

for all real numbers x. This follows from the fact that

$$f(x) - f(\bar{x}) = c - c = 0$$

for any x and $\bar{x}$ and so (19.1) is satisfied with $f'(\bar{x}) = 0$ and $\mathcal{K}_{\bar{x}} \equiv 0$.

EXAMPLE 19.2. If $f(x) = ax + b$ where a and b are real numbers, then

$$f'(x) = a$$

for all x since

$$f(x) - f(\bar{x}) = a(x - \bar{x})$$

for all x and $\bar{x}$ and so (19.1) holds with $f'(\bar{x}) = a$ and $\mathcal{K}_{\bar{x}} \equiv 0$.

In other words, the derivative of a linear function is constant and a linear function has the *same* linearization at each point.

---

[1] Joseph-Louis Lagrange (1736–1813) is claimed as one of their own by both the French and Italian schools of mathematics. Lagrange contributed substantially to the foundations of the calculus of variations and the theory of dynamics. He also made important investigations in astronomy, differential equations, fluid dynamics, number theory, probability, the stability of the solar system, and the theory of sounds, regularly winning prizes for his work. Lagrange was concerned with the foundations of calculus and wrote two textbooks in which he attempted to avoid the use of limits by employing infinite series. Interestingly, Lagrange's approach to the derivative is closely related to the definition of the linearization in this book. Lagrange first stated a general Mean Value Theorem and discovered the formula for the remainder of a Taylor polynomial that carries his name.

[2] The Swiss mathematician Johann Bernoulli (1667–1748) worked in France and Holland before finally returning to Switzerland. Johann Bernoulli made fundamental advances in basic calculus formulas and mechanics, building on Leibniz's results. His legacy is also known indirectly through the lectures he gave to the French mathematician L'Hôpital, who subsequently published the first calculus textbook. See Chapter 35 for more on this. Johann Bernoulli introduced the use of δ to denote a small quantity.

[3] The Swiss mathematician Jacob Bernoulli (1654–1705) traveled and studied in Holland and England before returning to teach in Switzerland. Jacob Bernoulli made particularly important contributions to the foundation of probability, and also contributed to algebra, the calculus of variations, infinite series, and mechanics. He also made fundamental advances in calculus, building on the work of Leibniz. He first used integrals in reference to integration and developed the method of separation of variables.

EXAMPLE 19.3. We next compute the derivative of $f(x) = x^2$. We look for $f'(\bar{x})$ and $\mathcal{K}_{\bar{x}}$ such that

$$\left|x^2 - \left(\bar{x}^2 + f'(\bar{x})\,(x - \bar{x})\right)\right| \leq (x - \bar{x})^2 \mathcal{K}_{\bar{x}} \tag{19.2}$$

for x close to $\bar{x}$. We compute

$$\begin{aligned}
\left|x^2 - \left(\bar{x}^2 + f'(\bar{x})\,(x - \bar{x})\right)\right| &= \left|x^2 - \bar{x}^2 - f'(\bar{x})\,(x - \bar{x})\right| \\
&= \left|(x - \bar{x})(x + \bar{x}) - f'(\bar{x})\,(x - \bar{x})\right| \\
&= |x - \bar{x}||x + \bar{x} - f'(\bar{x})|.
\end{aligned}$$

We conclude that $\lim_{x \to \bar{x}} \left|x^2 - \left(\bar{x}^2 + f'(\bar{x})\,(x - \bar{x})\right)\right| = 0$. We also want

$$\lim_{x \to \bar{x}} |x + \bar{x} - f'(\bar{x})| = 0,$$

which forces $f'(\bar{x}) = 2\bar{x}$. The linearization is $\bar{x}^2 + 2\bar{x}(x - \bar{x})$. Estimating the error

$$\left|x^2 - \left(\bar{x}^2 + 2\bar{x}(x - \bar{x})\right)\right| = |x - \bar{x}||x + \bar{x} - 2\bar{x}| = |x - \bar{x}|^2,$$

so (19.2) holds with $\mathcal{K}_{\bar{x}} = 1$ for any x.

We conclude that

$$f'(x) = 2x$$

for any x. We illustrate in Fig. 19.1.

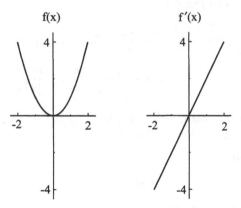

FIGURE 19.1. $f(x) = x^2$ and $f'(x) = 2x$.

EXAMPLE 19.4. We next compute the derivative of $f(x) = 1/x$ for any $x \neq 0$. We look for $f'(\bar{x})$ and $\mathcal{K}_{\bar{x}}$ such that

$$\left|\frac{1}{x} - \left(\frac{1}{\bar{x}} + f'(\bar{x})\,(x - \bar{x})\right)\right| \leq (x - \bar{x})^2 \mathcal{K}_{\bar{x}}$$

for x near $\bar{x}$. Clearly we require that $\bar{x}, x \neq 0$. Since $\bar{x} \neq 0$, we can always find a small open interval $I_{\bar{x}}$ such that 0 is not in $I_{\bar{x}}$. For example, we could use $I_{\bar{x}} = (\bar{x}/2, 2\bar{x})$ when $\bar{x} > 0$. We then restrict x to $I_{\bar{x}}$. Note that the maximum size of $I_{\bar{x}}$ depends on $\bar{x}$!

Computing,

$$\left| \frac{1}{x} - \frac{1}{\bar{x}} - f'(\bar{x})(x - \bar{x}) \right| = \left| \frac{\bar{x} - x}{\bar{x}x} - f'(\bar{x})(x - \bar{x}) \right|$$

$$= |x - \bar{x}| \left| \frac{1}{\bar{x}x} + f'(\bar{x}) \right|.$$

Therefore, $\lim_{x \to \bar{x}} \left| \frac{1}{x} - \left(\frac{1}{\bar{x}} + f'(\bar{x})(x - \bar{x}) \right) \right| = 0$. We also want

$$\lim_{x \to \bar{x}} \left| \frac{1}{\bar{x}x} + f'(\bar{x}) \right| = 0,$$

which forces

$$f'(\bar{x}) = \frac{-1}{\bar{x}^2}.$$

The linearization at $\bar{x}$ is

$$\frac{1}{\bar{x}} + \frac{-1}{\bar{x}^2}(x - \bar{x}).$$

We estimate the error next,

$$\left| \frac{1}{x} - \frac{1}{\bar{x}} - \frac{-1}{\bar{x}^2}(x - \bar{x}) \right| = \left| \frac{\bar{x} - x}{\bar{x}x} + \frac{x - \bar{x}}{\bar{x}^2} \right| = |x - \bar{x}| \left| \frac{1}{\bar{x}x} - \frac{1}{\bar{x}^2} \right|$$

$$= |x - \bar{x}| \left| \frac{\bar{x} - x}{\bar{x}^2 x} \right| = \frac{|\bar{x} - x|^2}{|\bar{x}^2 x|}.$$

On $I_{\bar{x}}$, there is a constant $\mathcal{K}_{\bar{x}}$ such that $1/|x\bar{x}^2| \leq \mathcal{K}_{\bar{x}}$. We conclude that $1/x$ is strongly differentiable at $\bar{x} \neq 0$.

The derivative of $1/x$ is

$$f'(x) = \frac{-1}{x^2}$$

for any $x \neq 0$. We illustrate in Fig. 19.2.

EXAMPLE 19.5. We compute the derivative of the monomial $f(x) = x^n$, $n \geq 1$, using the factorization

$$x^n - \bar{x}^n = (x - \bar{x})\left(x^{n-1} + x^{n-2}\bar{x} + x^{n-3}\bar{x}^2 + \cdots + x\bar{x}^{n-2} + \bar{x}^{n-1} \right)$$

$$= (x - \bar{x}) \sum_{i=0}^{n-1} x^{n-1-i}\, \bar{x}^i.$$

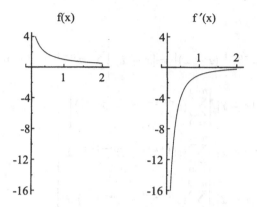

FIGURE 19.2. $f(x) = 1/x$ and $f'(x) = -1/x^2$.

This means that

$$\frac{x^n - \bar{x}^n}{x - \bar{x}} = \sum_{i=0}^{n-1} x^{n-1-i}\, \bar{x}^i. \tag{19.3}$$

We look for $f'(\bar{x})$ and $\mathcal{K}_{\bar{x}}$ such that

$$|x^n - (\bar{x}^n + f'(\bar{x})\,(x - \bar{x}))| \le (x - \bar{x})^2 \mathcal{K}_{\bar{x}}$$

for x near $\bar{x}$. Subtracting, we get

$$|x^n - \bar{x}^n - f'(\bar{x})\,(x - \bar{x})| = \left|(x - \bar{x}) \sum_{i=0}^{n-1} x^{n-1-i}\, \bar{x}^i - f'(\bar{x})\,(x - \bar{x})\right|$$

$$= |x - \bar{x}| \left|\sum_{i=0}^{n-1} x^{n-1-i}\, \bar{x}^i - f'(\bar{x})\right|.$$

so $\lim_{x \to \bar{x}} |x^n - (\bar{x}^n + f'(\bar{x})\,(x - \bar{x}))| = 0$. We also want

$$\lim_{x \to \bar{x}} \left|\sum_{i=0}^{n-1} x^{n-1-i}\, \bar{x}^i - f'(\bar{x})\right| = 0,$$

which forces $f'(\bar{x}) = n\bar{x}^{n-1}$ since there are n terms in the sum.

Estimating the error is a real mess.[4] But it is a good exercise in the summation notation and we ask you to verify each step in Problem 19.2.

---

[4]But we carry out the computation in the spirit of sharing pain.

We compute

$$\left| x^n - \bar{x}^n - n\bar{x}^{n-1}(x - \bar{x}) \right| = |x - \bar{x}| \left| \sum_{i=0}^{n-1} x^{n-1-i}\, \bar{x}^i - n\bar{x}^{n-1} \right|$$

$$= |x - \bar{x}| \left| \sum_{i=0}^{n-1} (x^{n-1-i}\, \bar{x}^i - \bar{x}^{n-1}) \right|$$

$$= |x - \bar{x}| \left| \sum_{i=0}^{n-1} (x^{n-1-i} - \bar{x}^{n-1-i})\, \bar{x}^i \right|$$

$$= |x - \bar{x}| \left| \sum_{i=0}^{n-2} (x^{n-1-i} - \bar{x}^{n-1-i})\, \bar{x}^i \right|$$

$$= |x - \bar{x}| \left| \sum_{i=0}^{n-2} \left(\left(\sum_{j=0}^{n-1-i-1} x^{n-1-i-j}\, \bar{x}^j \right)(x - \bar{x}) \right) \bar{x}^i \right|$$

$$= |x - \bar{x}|^2 \left| \sum_{i=0}^{n-2} \sum_{j=0}^{n-1-i-1} x^{n-1-i-j}\, \bar{x}^{j+i} \right|.$$

$$(19.4)$$

Choosing any finite open interval $I_{\bar{x}}$ containing $\bar{x}$, we can bound

$$\left| \sum_{i=0}^{n-2} \sum_{j=0}^{n-1-i-1} x^{n-1-i-j}\, \bar{x}^{j+i} \right| \leq \left| \sum_{i=0}^{n-2} \sum_{j=0}^{n-1-i-1} |I_{\bar{x}}|^{n-1-i-j}\, |I_{\bar{x}}|^{j+i} \right| = \mathcal{K}_{\bar{x}}$$

and conclude that

$$\left| x^n - \bar{x}^n - n\bar{x}^{n-1}(x - \bar{x}) \right| \leq |x - \bar{x}|^2 \mathcal{K}_{\bar{x}}$$

for x in $I_{\bar{x}}$. Therefore, x^n is strongly differentiable for all x and

$$f'(x) = nx^{n-1}$$

for any x.

19.2   Uniform Strong Differentiability

Note that the interval $I_{\bar{x}}$ and constant $K_{\bar{x}}$ in the definition (19.1) of the derivative of a function f on an open interval I depend on $\bar{x}$ in general.

EXAMPLE 19.6.   This observation is important when differentiating $1/x$ (see Example 19.4). We have to restrict the size of $I_{\bar{x}}$ depending on the distance between $\bar{x}$ and 0 and moreover the factor multiplying $(x - \bar{x})^2$ in the estimate of the error

$$(x - \bar{x})^2 \, \frac{1}{\bar{x}^2 \, x}$$

becomes larger as $\bar{x}$ and x approach zero.

In many situations however, it is possible to choose the *same* interval $I_{\bar{x}}$ for all $\bar{x}$ in I, namely, $I_{\bar{x}} = I$, and moreover the same constant $\mathcal{K}_{\bar{x}} = \mathcal{K}$ for all $\bar{x}$ in the interval I. In this situation, we say that f is uniformly strongly differentiable on the interval I. More precisely, f is **uniformly strongly differentiable** on an open interval I if there is a constant $\mathcal{K}$ such that for any $\bar{x}$ in I

$$\left| f(x) - \big(f(\bar{x}) + f'(\bar{x})(x - \bar{x})\big) \right| \leq (x - \bar{x})^2 \mathcal{K}$$

for x in I.

EXAMPLE 19.7. The function $ax+b$ is uniformly strongly differentiable on any interval I since $\mathcal{K}_{\bar{x}} = 0$ for any $\bar{x}$ and x.

EXAMPLE 19.8. The function x^2 is uniformly strongly differentiable on any interval I, including $(-\infty, \infty)$, since $\mathcal{K}_{\bar{x}} = 1$ for any $\bar{x}$ and x. Hence, uniform strong differentiability on an interval does not imply Lipschitz continuity on the interval.

EXAMPLE 19.9. The monomial x^n, $n \geq 3$, is uniformly strongly differentiable on any bounded interval. The monomial x^n, $n \geq 3$, is not uniformly strongly differentiable on $(0, \infty)$ or any other infinite interval, however.

EXAMPLE 19.10. The function x^{-1} is uniformly strongly differentiable on any interval bounded away from 0 but is not uniformly strongly differentiable on any interval that contains 0 as an endpoint.

19.3 Uniform Strong Differentiability and Smoothness

Uniform strong differentiability on an interval conveys a lot of smoothness properties on a function.

We first show that the derivative $f'(x)$ of a function $f(x)$ that is uniformly strongly differentiable on an interval I with size $|I| > 0$ is Lipschitz continuous on I. We choose two points x, y in I and by assumption

$$|f(y) - (f(x) + f'(x)(y - x))| \leq |y - x|^2 \mathcal{K}$$
$$|f(x) - (f(y) + f'(y)(x - y))| \leq |x - y|^2 \mathcal{K}$$

for some constant $\mathcal{K}$. We first estimate

$$
\begin{aligned}
|(f'(y) - f'(x))(x - y)| &= |f'(x)(y - x) + f'(y)(x - y)| \\
&= |(f(y) + f(x)) - (f(x) + f(y)) + f'(x)(y - x) + f'(y)(x - y)| \\
&= |f(y) - (f(x) + f'(x)(y - x)) + f(x) - (f(y) + f'(y)(x - y))| \\
&\leq |y - x|^2(\mathcal{K} + \mathcal{K}).
\end{aligned}
$$

For all $x \neq y$, we conclude that

$$
|f'(y) - f'(x)| \leq 2\mathcal{K}|y - x|. \tag{19.5}
$$

Since (19.5) is also true when $x = y$ because both sides are zero, we conclude the claim is true.

Using this result, we can also show that f is Lipschitz continuous on a bounded interval I if it is uniformly strongly differentiable on I. This follows from the fact that f' is Lipschitz continuous on I and hence is bounded by some constant, e.g., $|f'(x)| \leq M$ for x in I. Now if x and y are in I, then the definition of uniform strong differentiability means that there is a constant $\mathcal{K}$ such that

$$
|f(y) - (f(x) + f'(x)(y - x))| \leq |y - x|^2 \mathcal{K}.
$$

This means that

$$
\begin{aligned}
|f(y) - f(x)| &\leq |f'(x)||y - x| + \mathcal{K}|y - x|^2 \\
&\leq (M + \mathcal{K}|I|)|y - x| \\
&= L|y - x|
\end{aligned}
$$

for all x, y in I with Lipschitz constant $L = M + \mathcal{K}|I|$.

We can improve this result by showing that if f is uniformly strongly differentiable and $|f'|$ is bounded by M, then in fact f is Lipschitz continuous with constant M. We didn't quite obtain this result above because we ended up with a larger Lipschitz constant $L = M + \mathcal{K}|I|$.

Given $|I| > \delta > 0$, if x and y are restricted to a subinterval I_δ of I of length δ, then the discussion above implies that

$$
|f(x) - f(y)| \leq (L + K\delta)|x - y|.
$$

By choosing δ small, we can make $L + K\delta$ as close to L as desired. The only problem is that x and y may not be in a I_δ with a small δ.

Suppose that x and y in I are given. We choose points $\{x_0, x_1, \cdots, x_N\}$ such that $x = x_0 < x_1 < \cdots < x_N = y$, and $x_i - x_{i-1} \leq \delta$. By the triangle inequality,

$$|f(x) - f(y)| = |\sum_{i=1}^{N}(f(x_i) - f(x_{i-1}))| \le \sum_{i=1}^{N}|f(x_i) - f(x_{i-1})|$$

$$\le (L + K\delta)\sum_{i=1}^{N}|x_i - x_{i-1}|$$

$$= (L + K\delta)|x - y|.$$

Since this inequality holds for any $\delta > 0$, we conclude that indeed

$$|f(x) - f(y)| \le L|x - y|, \quad \text{for } x, y \text{ in } I.$$

We summarize as a useful theorem.

Theorem 19.1 *If f is uniformly strongly differentiable on an interval I, then f and f' are Lipschitz continuous on I. If in addition $|f'(x)| \le M$ for all x in I, then f is Lipschitz continuous with constant M.*[5]

19.4   Closed Intervals and One-Sided Linearization

We say that a function is strongly differentiable on an open interval if it is strongly differentiable at each point in the interval. To extend this definition to a closed interval, a new idea is needed.[6] The difficulty can be understood by considering three examples. The function x^2 is differentiable on the interval $(0, 1)$ and moreover we can also define the derivative at $x = 0$ by noting that it is also differentiable on $(-1, 1)$ with 0 being in the middle of the interval. The step function (6.8) is another matter. This is clearly differentiable on any of the intervals $(-\infty, 0)$, $(0, 1)$, and $(1, \infty)$, with the linearizations being the constant functions 0, 1, and 0, respectively; but we also have great difficulty defining a meaningful linearization at either $x = 0$ or $x = 1$ because of the discontinuity. Finally, x^{-1} is differentiable on the intervals $(-\infty, 0)$ and $(0, \infty)$, but we cannot define a linearization at $x = 0$ since the function is not even defined there.

We deal with these difficulties using the idea of a one-sided linearization, which is closely related to the idea of a one-sided limit. A one sided limit is like a usual limit, except we restrict to one side of the point in question. So the one-sided limit of f from the right at x, which is denoted as $f(x^+)$, is defined by

$$f(x^+) = \lim_{z \downarrow x} f(z) = \lim_{z \to x, \, z > x} f(z).$$

---

[5]This result does *not* hold in general for functions that are merely strongly differentiable on an interval. This is discussed further in Chapter 32.

[6]This section is rather technical and can be skipped on first reading if the claim that differentiability can be extended to closed intervals is palatable.

Likewise the one-sided limit $f(x^-)$ of f from the left at x is defined by

$$f(x^-) = \lim_{z \uparrow x} f(z) = \lim_{z \to x,\ z < x} f(z).$$

EXAMPLE 19.11. Consider the step function $I(x)$ defined in (6.8) and $J(x) = xI(x)$ (see Fig. 19.3). We see that $I(0^-) = \lim_{z \uparrow 0} I(z) =$

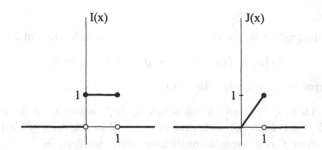

FIGURE 19.3. $I(x)$ and $J(x) = xI(x)$.

$\lim_{z \uparrow 0} 0 = 0$ while $I(0^+) = \lim_{z \downarrow 0} I(z) = \lim_{z \downarrow 0} 1 = 1$. Note that we define $I(0) = 1$, which in this case is equal to $I(0^+)$ but not $I(0^-)$. Likewise, $J(0^-) = \lim_{z \uparrow 0} J(z) = \lim_{z \uparrow 0} 0 = 0$, while $J(0^+) = \lim_{z \downarrow 0} J(z) = \lim_{z \downarrow 0} z = 0$ as well.

A one-sided linearization of a function $f(x)$ at a point $\bar{x}$ is a linearization that is a good approximation of f for x on one side of $\bar{x}$, i.e., either for $x > \bar{x}$ or $x < \bar{x}$. In order to define this condition precisely, we have to take into account the possibility that $f(x)$ might be discontinuous at $\bar{x}$. Consider the step function $I(x)$ at 0, for example. A linear approximation on the left of 0 should have the value 0 at 0, while a linear approximation on the right should have the value 1 at 0. But a discontinuous function can only have one value at each point, and in this case $I(0) = 1$. Hence, we can define a linear approximation on the right of 0 but not on the left.

We say that $f(x)$ is **strongly differentiable on the right** at $\bar{x}$ if there is an interval $[\bar{x}, b)$, a number $f'(\bar{x}^+)$, and a constant $\mathcal{K}_{\bar{x}}$ such that

$$\left| f(x) - \left(f(\bar{x}) + f'(\bar{x}^+)(x - \bar{x}) \right) \right| \le (x - \bar{x})^2 \mathcal{K}_{\bar{x}} \quad \text{for all } \bar{x} \le x < b. \ (19.6)$$

Note that $f(\bar{x})$ has to be equal to $\lim_{x \downarrow \bar{x}} f(x)$. We call $f(\bar{x}) + f'(\bar{x}^+)(x - \bar{x})$ the **right linearization** of f at $\bar{x}$ and $f'(\bar{x}^+)$ the **derivative from the right** or the **right derivative** of f at $\bar{x}$. The function $f(x)$ is **strongly differentiable on the left** at $\bar{x}$ if there is an interval $(a, \bar{x}]$, a number $f'(\bar{x}^-)$, and a constant $\mathcal{K}_{\bar{x}}$ such that

$$\left| f(x) - \left(f(\bar{x}) + f'(\bar{x}^-)(x - \bar{x}) \right) \right| \le (x - \bar{x})^2 \mathcal{K}_{\bar{x}}(x) \quad \text{for all } a < x \le \bar{x}. \ (19.7)$$

Note again that $f(\bar{x})$ has to be equal to $\lim_{x \uparrow \bar{x}} f(x)$. We call $f(\bar{x}) + f'(\bar{x}^-)(x - \bar{x})$ the **left linearization** of f at $\bar{x}$ and $f'(\bar{x}^-)$ the **derivative from the left** or the **left derivative** of f at $\bar{x}$.

By Theorem 16.1, if $f(x)$ is strongly differentiable on the right at $\bar{x}$, then there is an interval I and a constant L such that $|f(x) - f(\bar{x})| \le L|x - \bar{x}|$ for all $x \ge \bar{x}$ in I. Likewise, if $f(x)$ is strongly differentiable on the left at $\bar{x}$, then there is an interval I and a constant L such that $|f(x) - f(\bar{x})| \le L|x - \bar{x}|$ for all $x \le \bar{x}$ in I.

EXAMPLE 19.12. The step function (6.8) has a right linearization at 0, namely, 1, but not a left linearization.

EXAMPLE 19.13. The right linearization of x^2 at $x = 0$ is $0 + 0(x - 0) = 0$ and the left linearization of x^2 at $x = 0$ is $0 + 0(x - 0) = 0$. The right and left derivatives are both 0. In this case, the right and left linearizations at 0 are equal if the functions are extended in the obvious way past 0.

EXAMPLE 19.14. The function $|x|$ is strongly differentiable at any point $x \ne 0$. This is easy to see since $|x| = x$ when $x > 0$ and $|x| = -x$ when $x < 0$ and we can differentiate x and $-x$. There is a problem at $x = 0$, of course, because of the sharp "point" in the graph of $|x|$. However, the right linearization of $|x|$ at 0 is $0 + 1(x - 0) = x$ and the left linearization is $0 - 1(x - 0) = -x$.

This example shows that being Lipschitz continuous does not imply a function is strongly differentiable.

EXAMPLE 19.15. The function $1/x$ has neither a left or right linearization at 0.

Note that it follows from these definitions that a function is strongly differentiable at a point $\bar{x}$ if and only if it is strongly right differentiable and strongly left differentiable at $\bar{x}$, and moreover the right and left linearizations of f at $\bar{x}$ are equal if these functions are extended past $\bar{x}$. The extensions of the right and left linearizations are equal when the right and left derivatives are equal. We conclude:

Theorem 19.2 *A function is strongly differentiable at a point $\bar{x}$ if and only if it is strongly right and left differentiable at $\bar{x}$ and the right and left derivatives are equal.*

Now we can define differentiability on closed intervals. A function f is strongly differentiable on an interval $[a, b)$ if it is strongly differentiable on (a, b) and it is strongly differentiable on the right at a. A function f is strongly differentiable on an interval $(a, b]$ if it is strongly differentiable on (a, b) and it is strongly differentiable on the left at b. A function f is strongly differentiable on an interval $[a, b]$ with length $b - a > 0$ if it is strongly differentiable on (a, b), it is strongly differentiable on the right at a, and it is strongly differentiable on the left at b.

EXAMPLE 19.16. The function x^2 is strongly differentiable on any open or closed interval.

EXAMPLE 19.17. The function $|x|$ is strongly differentiable on $[0, \infty)$ and $(-\infty, 0]$ but is not differentiable on any open interval that contains 0.

EXAMPLE 19.18. The step function (6.8) is strongly differentiable on the intervals $(-\infty, 0)$, $[0, 1]$, and $(1, \infty)$.

19.5   Differentiability on Intervals

We can carry over all the extensions of strong differentiability to differentiability. We say that a function is **differentiable** on an open interval I if it is differentiable at each point in I. In this case, we define the **derivative** $f'(x)$ of $f(x)$ as the function that gives the derivative of $f(x)$ at each point x in I.

We define right and left differentiability of a function f at a point $\bar{x}$ by using one-sided limits again. We say f is right differentiable at $\bar{x}$ if it is defined on a small interval $[\bar{x}, b)$ and if

$$f'(\bar{x}^+) = \lim_{\Delta x \downarrow 0} \frac{f(\bar{x} + \Delta x) - f(\bar{x})}{\Delta x}$$

is defined. By $\Delta x \downarrow 0$ we mean take the limit as $\Delta x \to 0$ for $\Delta x > 0$. Likewise, we say f is left differentiable at $\bar{x}$ if it is defined on a small interval $(a, \bar{x}]$ and if

$$f'(\bar{x}^-) = \lim_{\Delta x \uparrow 0} \frac{f(\bar{x} + \Delta x) - f(\bar{x})}{\Delta x}$$

is defined, where $\Delta x \uparrow 0$ means take the limit as $\Delta x \to 0$ for $\Delta x < 0$. We call the resulting limits the right and left derivatives of f at $\bar{x}$. The following theorem is immediate.

Theorem 19.3 *A function f is differentiable at point x if and only if it is right and left differentiable at x and the right and left derivatives are equal.*

Given these definitions, we can define differentiability on different kinds of intervals. For example, f is differentiable on $[a, b]$ if it is differentiable on (a, b), is right differentiable at a, and left differentiable at b.

EXAMPLE 19.19.   Consider the function $J(x) = xI(x)$ graphed in Fig. 19.3 $J(x)$ is clearly differentiable on $(\infty, 0)$, $(0, 1)$, and $(1, \infty)$, since in those regions it is simply 0, x, and 0 respectively. Since J is not continuous at $x = 1$, it is clearly not differentiable there. It does have a left derivative at 1, however, since

$$\lim_{\Delta x \uparrow 0} \frac{J(1 + \Delta x) - J(1)}{\Delta x} = \lim_{\Delta x \uparrow 0} \frac{1 + \Delta x - 1}{\Delta x} = 1.$$

Chapter 19 Problems

19.1. Verify the details in Example 19.1 and Example 19.2.

19.2. Verify the details in (19.4).

19.3. Compute the linearizations of the following functions at a point $\bar{x}$ in the indicated sets.

(a) $f(x) = 7x$ on $\mathbb{R}$ 　　　　　　　　(b) $f(x) = 1/x^2$ on $(0, \infty)$

(c) $f(x) = 2x^3$ on $\mathbb{R}$ 　　　　　　　　(d) $f(x) = 2x^2 - 5x$ on $\mathbb{R}$

(e) $f(x) = 1/(1+x)$ on $(-1, \infty)$ 　　　(f) $f(x) = (x+2)^2$ on $\mathbb{R}$.

Problems 19.4–19.6 have to do uniform strong differentiability.

19.4. Verify the argument in Example 19.9.

19.5. Verify the claim in Example 19.10.

19.6. Prove that if f and g are uniformly strongly differentiable on an interval I, then so is $f + g$, fg, and $f(g(x))$.

Problems 19.7–19.10 have to do with one-sided differentiability.

19.7. Verify the claim in Example 19.12.

19.8. Verify the claim in Example 19.13.

19.9. Verify the claim in Example 19.14.

19.10. Discuss the strong differentiability properties of the function $J(x) = xI(x)$ graphed in Fig. 19.3. This means find intervals on which the function is strongly differentiable. If it is not strongly differentiable at a point, indicate whether it is strongly differentiable from the right or left or both. Do the same for $K(x) = x^2I(x)$.

Problems 19.11–19.13 have to do with differentiability on intervals.

19.11. Define

$$f(x) = \begin{cases} x^2, & x \le 1, \\ 2 - x^2, & x > 1. \end{cases}$$

(a) Is $f(x)$ differentiable at $x < 1$, $x > 1$, $x = 1$? (b) What is the derivative of $f(x)$ at $x < 1$? At $x > 1$? (b) Compute the derivatives from the right and the left at $x = 1$.

19.12. Draw a function that is piecewise differentiable on $[0, 4]$, but not differentiable at 1, 2, and 3.

19.13. Discuss the differentiability of the function $f(x) = 1/(1-x)$ using a plot.

20
Useful Properties of the Derivative

As we have seen, computing the derivative and the linearization of a function using the definitions is tedious. Luckily, the derivative has some properties that can help to make the computations easier and we explore these in this chapter. Unfortunately, verifying these properties leads to some of the ugliest estimates in the book. Everything worthwhile has a price. Nonetheless, it is useful to go through the arguments and try to understand each step. Over and over, the analysis uses the same idea: *rewrite a given quantity in terms of differences of quantities that can be estimated.*[1]

20.1 Linear Combinations of Functions

We first consider differentiation of a function that is a linear combination of two functions with known derivatives. Suppose that $h = f + g$, where f and g are differentiable at $\bar{x}$. This means that f and g are defined on some interval $I_{\bar{x}}$ (strictly speaking there are different intervals associated to f and g, but we take $I_{\bar{x}}$ to be the intersection of these two intervals). Now h is also defined on $I_{\bar{x}}$ and hence is differentiable if

$$h'(\bar{x}) = \lim_{x \to \bar{x}} \frac{h(x) - h(\bar{x})}{x - \bar{x}}$$

---

[1]It may be useful to go back to Chapter 8 and review the techniques used to prove facts about Lipschitz continuous functions.

exists. But,

$$\lim_{x \to \bar{x}} \frac{h(x) - h(\bar{x})}{x - \bar{x}} = \lim_{x \to \bar{x}} \frac{f(x) + g(x) - (f(\bar{x}) + g(\bar{x}))}{x - \bar{x}}$$

$$= \lim_{x \to \bar{x}} \frac{f(x) - f(\bar{x})}{x - \bar{x}} + \lim_{x \to \bar{x}} \frac{g(x) - g(\bar{x})}{x - \bar{x}}$$

$$= f'(\bar{x}) + g'(\bar{x})$$

by assumption. Thus,

$$h'(\bar{x}) = f'(\bar{x}) + g'(\bar{x}). \tag{20.1}$$

To show that h is strongly differentiable at $\bar{x}$, we estimate

$$|h(x) - (h(\bar{x}) + h'(\bar{x})(x - \bar{x}))|$$
$$= |f(x) + g(x) - (f(\bar{x}) + g(\bar{x}) + (f'(\bar{x}) + g'(\bar{x}))(x - \bar{x})|$$
$$\leq |f(x) - (f(\bar{x}) + f'(\bar{x})(x - \bar{x}))| + |g(x) - (g(\bar{x}) + g'(\bar{x})(x - \bar{x}))|$$
$$\leq (\mathcal{K}_f + \mathcal{K}_g)|x - \bar{x}|^2$$

for some constants $\mathcal{K}_f$ and $\mathcal{K}_g$.

We ask you to treat the case cf for a constant c in the same way in Problem 20.1. We summarize:

Theorem 20.1 Linearity of Differentiation *Suppose f and g are differentiable at $\bar{x}$ and c is a constant. Then $f + g$ and cf are differentiable at $\bar{x}$ and*

$$D(f(\bar{x}) + g(\bar{x})) = Df(\bar{x}) + Dg(\bar{x}) \tag{20.2}$$

and

$$D(cf(\bar{x})) = cDf(\bar{x}). \tag{20.3}$$

If f and g are strongly differentiable at $\bar{x}$, then so is $f + g$ and cf.

EXAMPLE 20.1. For all $x \neq 0$,

$$\frac{d}{dx}\left(2x^3 + 4x^5 + \frac{7}{x}\right) = 6x^2 + 20x^4 - \frac{7}{x^2}.$$

EXAMPLE 20.2. Using this theorem and the derivative of the monomial, $Dx^i = ix^{i-1}$, we find that the derivative of

$$f(x) = a_0 + a_1 x + a_2 x^2 + \cdots + a_n x^n = \sum_{i=0}^{n} a_i x^i$$

is

$$f'(x) = a_1 + 2a_2 x^2 + \cdots + na_n x^{n-1} = \sum_{i=1}^{n} ia_i x^{i-1}$$

for all x.

20.2   Products of Functions

Next, we look at the product of two differentiable functions, $h = fg$. If f and g are defined on an interval $I_{\bar{x}}$, then so is h. To discover the formula for the derivative of h, we first estimate

$$|f(x)g(x) - (f(\bar{x})g(\bar{x}) + h'(\bar{x})(x - \bar{x}))|$$
$$= |f(x)g(x) - f(\bar{x})g(\bar{x}) - h'(\bar{x})(x - \bar{x})|.$$

We insert the linear approximations of f and g and multiply out to get

$$|f(x)g(x) - (f(\bar{x})g(\bar{x}) + h'(\bar{x})(x - \bar{x}))|$$
$$\approx |(f(\bar{x}) + f'(\bar{x})(x - \bar{x}))(g(\bar{x}) + g'(\bar{x})(x - \bar{x}))$$
$$- f(\bar{x})g(\bar{x}) - h'(\bar{x})(x - \bar{x})|$$
$$= |g(\bar{x})f'(\bar{x})(x - \bar{x}) + f(\bar{x})g'(\bar{x})(x - \bar{x})$$
$$+ f'(\bar{x})g'(\bar{x})(x - \bar{x})^2 - h'(\bar{x})(x - \bar{x})|.$$

Finally, we drop the term that is quadratic in $(x - \bar{x})$ because it is smaller than the other terms when x is close to $\bar{x}$. We get

$$|f(x)g(x) - (f(\bar{x})g(\bar{x}) + h'(\bar{x})(x - \bar{x}))|$$
$$\approx |g(\bar{x})f'(\bar{x})(x - \bar{x}) + f(\bar{x})g'(\bar{x})(x - \bar{x}) - h'(\bar{x})(x - \bar{x})|.$$

Based on (16.7), we conclude that $h'(\bar{x}) = f(\bar{x})g'(\bar{x}) + f'(\bar{x})g(\bar{x})$.

Armed with the formula for h', we now verify that

$$h'(\bar{x}) = \lim_{x \to \bar{x}} \frac{f(x)g(x) - f(\bar{x})g(\bar{x})}{x - \bar{x}}$$

is defined. Based on the formula for the derivative, we compute

$$\lim_{x \to \bar{x}} \frac{f(x)g(x) - f(\bar{x})g(\bar{x})}{x - \bar{x}}$$
$$= \lim_{x \to \bar{x}} \frac{f(x)g(x) - f(\bar{x})g(x) + f(\bar{x})g(x) - f(\bar{x})g(\bar{x})}{x - \bar{x}}.$$

We add and subtract $f(\bar{x})g(x)$ in the numerator because the properties of limits then imply that

$$\lim_{x \to \bar{x}} \frac{f(x)g(x) - f(\bar{x})g(\bar{x})}{x - \bar{x}}$$
$$= \lim_{x \to \bar{x}} g(x)\frac{f(x) - f(\bar{x})}{x - \bar{x}} + \lim_{x \to \bar{x}} f(\bar{x})\frac{g(x) - g(\bar{x})}{x - \bar{x}}$$
$$= f'(\bar{x})g(\bar{x}) + f(\bar{x})g'(\bar{x})$$

as expected.

To show that $h = fg$ is strongly differentiable, we estimate the error of the linearization

$$|f(x)g(x) - (f(\bar{x})g(\bar{x}) + (f'(\bar{x})g(\bar{x}) + f(\bar{x})g'(\bar{x}))(x - \bar{x})|$$
$$= |f(x)g(x) - f(\bar{x})g(\bar{x}) - f'(\bar{x})g(\bar{x})(x - \bar{x}) - f(\bar{x})g'(\bar{x})(x - \bar{x})|$$

using the error estimates for the linearizations of f and g. We add and subtract terms to get these errors

$$|f(x)g(x) - (f(\bar{x})g(\bar{x}) + (f'(\bar{x})g(\bar{x}) + f(\bar{x})g'(\bar{x}))(x - \bar{x})|$$
$$= |f(x)g(\bar{x}) - f(\bar{x})g(\bar{x}) - f'(\bar{x})g(\bar{x})(x - \bar{x})$$
$$+ f(x)g(x) - f(x)g(\bar{x}) - f(x)g'(\bar{x})(x - \bar{x})$$
$$+ f(x)g'(\bar{x})(x - \bar{x}) - f(\bar{x})g'(\bar{x})(x - \bar{x})|.$$

Now we start estimating,

$$|f(x)g(x) - (f(\bar{x})g(\bar{x}) + (f'(\bar{x})g(\bar{x}) + f(\bar{x})g'(\bar{x}))(x - \bar{x})|$$
$$\leq |f(x) - f(\bar{x}) - f'(\bar{x})(x - \bar{x})| \, |g(\bar{x})|$$
$$+ |f(x)| \, |g(x) - g(\bar{x}) - g'(\bar{x})(x - \bar{x})|$$
$$+ |f(x) - f(\bar{x})| \, |g'(\bar{x})(x - \bar{x})|$$
$$\leq (x - \bar{x})^2 |g(\bar{x})| \mathcal{K}_f + (x - \bar{x})^2 M_f \mathcal{K}_g + (x - \bar{x})^2 |g'(\bar{x})| L_f$$
$$= (x - \bar{x})^2 (|g(\bar{x})| \mathcal{K}_f + M_f \mathcal{K}_g + |g'(\bar{x})| L_f),$$

where L_f is the Lipschitz constant of f and M_f is a bound on $|f|$ for x in $I_{\bar{x}}$ and $\mathcal{K}_f$ and $\mathcal{K}_g$ are constants given by the assumption that f and g are strongly differentiable at $\bar{x}$. This proves that fg is strongly differentiable.

We summarize:

Theorem 20.2 The Product Rule *If f and g are differentiable at $\bar{x}$, then fg is differentiable at $\bar{x}$ and*

$$D(f(\bar{x})g(\bar{x})) = f(\bar{x})Dg(\bar{x}) + Df(\bar{x})g(\bar{x}). \tag{20.4}$$

If f and g are strongly differentiable at $\bar{x}$, then so is fg.

EXAMPLE 20.3.

$$D\left((10 + 3x^2 - x^6)(x - 7x^4)\right)$$
$$= (10 + 3x^2 - x^6)(1 - 28x^3) + (6x - 6x^5)(x - 7x^4).$$

We often write the product rule in the form

$$(fg)' = fg' + f'g.$$

20.3   Composition of Functions

Next, we consider the composition $z = h = f \circ g(x)$ of two differentiable functions $y = g(x)$ and $z = f(y)$. We assume that g is differentiable at a point $\bar{x}$ with associated interval $I_{\bar{x}}$ and f is differentiable at $\bar{y} = g(\bar{x})$ with associated interval $I_{\bar{y}}$. We decrease the interval $I_{\bar{x}}$ if necessary to guarantee that $g(x)$ is in $I_{\bar{y}}$ for all x in $I_{\bar{x}}$. We can do this because Theorem 16.1 implies that g varies continuously on $I_{\bar{x}}$ (see Problem 20.2).

To discover the formula for the derivative of h, we first estimate using the approximation

$$f(y) \approx f(\bar{y}) + f'(\bar{y})(y - \bar{y}).$$

Substituting $y = g(x)$ and $\bar{y} = g(\bar{x})$ and using the approximation

$$y - \bar{y} = g(x) - g(\bar{x}) \approx g'(\bar{x})(x - \bar{x}),$$

we get

$$f(g(x)) \approx f(g(\bar{x})) + f'(g(\bar{x}))g'(\bar{x})(x - \bar{x}).$$

We conclude that $h'(\bar{x}) = f'(g(\bar{x}))g'(\bar{x})$.

We use this formula to prove

$$h'(\bar{x}) = \lim_{x \to \bar{x}} \frac{f(g(x)) - f(g(\bar{x}))}{x - \bar{x}}$$

is defined. It suggests that we rewrite

$$\lim_{x \to \bar{x}} \frac{f(g(x)) - f(g(\bar{x}))}{x - \bar{x}} = \lim_{x \to \bar{x}} \frac{f(g(x)) - f(g(\bar{x}))}{g(x) - g(\bar{x})} \frac{g(x) - g(\bar{x})}{x - \bar{x}}$$

so that Theorem 16.1 apparently implies

$$h'(\bar{x}) = \lim_{g(x) \to g(\bar{x})} \frac{f(g(x)) - f(g(\bar{x}))}{g(x) - g(\bar{x})} \lim_{x \to \bar{x}} \frac{g(x) - g(\bar{x})}{x - \bar{x}} = f'(g(\bar{x}))g'(\bar{x}).$$

The only difficulty with this argument is that $g(x) - g(\bar{x})$ may be zero at some point, in which case we cannot compute like this. We have to play a little game to get around this sticking point.

Recalling the notation $\Delta y = y - \bar{y}$, we define

$$\epsilon_f(\Delta y) = \begin{cases} \frac{f(y) - f(\bar{y})}{\Delta y} - f'(\bar{y}), & \Delta y \neq 0, \\ 0, & \Delta y = 0. \end{cases}$$

Notice that

$$\lim_{\Delta y \to 0} \epsilon_f(\Delta y) = \epsilon_f(0) = 0$$

by the assumption that f is differentiable at $\bar{y}$. Now we have

$$f(y) - f(\bar{y}) = (f'(y) + \epsilon_f(\Delta y))\Delta y$$

for *all* $\Delta y \geq 0$. In the same way, we define

$$\epsilon_g(\Delta x) = \begin{cases} \frac{g(x)-g(\bar{x})}{\Delta x} - g'(\bar{x}), & \Delta x \neq 0, \\ 0, & \Delta x = 0, \end{cases}$$

so

$$g(x) - g(\bar{x}) = (g'(x) + \epsilon_g(\Delta x))\Delta x$$

for all $\Delta x \geq 0$. Now with $y = g(x)$, we compute

$$\frac{f(g(x)) - f(g(\bar{x}))}{\Delta x} = (f'(g(x)) + \epsilon_f(g(x) - g(\bar{x})))\frac{g(x) - g(\bar{x})}{\Delta x}$$
$$= (f'(g(x)) + \epsilon_f(g(x) - g(\bar{x})))(g'(x) + \epsilon_g(\Delta x)).$$

We conclude that

$$\lim_{\Delta x \to 0} \frac{f(g(x)) - f(g(\bar{x}))}{\Delta x} = f'(g(\bar{x}))g'(\bar{x}).$$

To verify strong differentiability, we estimate the error of the linearization of $f \circ g$ at $\bar{x}$ by first adding and subtracting a term that yields an expression giving the error of the linearization of $f(g(x))$ around $g(\bar{x})$,

$$|f(g(x)) - (f(g(\bar{x})) + f'(g(\bar{x}))g'(\bar{x})(x - \bar{x})|$$
$$= |f(g(x)) - f(g(\bar{x})) - f'(g(\bar{x}))(g(x) - g(\bar{x}))$$
$$+ f'(g(\bar{x}))(g(x) - g(\bar{x})) - f'(g(\bar{x}))g'(\bar{x})(x - \bar{x})|.$$

Now estimating,

$$|f(g(x)) - (f(g(\bar{x})) + f'(g(\bar{x}))g'(\bar{x})(x - \bar{x})|$$
$$\leq |f(g(x)) - f(g(\bar{x})) - f'(g(\bar{x}))(g(x) - g(\bar{x}))|$$
$$+ |f'(g(\bar{x}))(g(x) - g(\bar{x}) - g'(\bar{x})(x - \bar{x}))|$$
$$\leq |g(x) - g(\bar{x})|^2 \mathcal{K}_f + |x - \bar{x}|^2 \mathcal{K}_g |f'(g(\bar{x}))|$$
$$\leq (x - \bar{x})^2 (L_g^2 \mathcal{K}_f + \mathcal{K}_g |f'(g(\bar{x}))|)$$

for some constants $\mathcal{K}_f$ and $\mathcal{K}_g$ given by the strong differentiability of f and g and with L_g denoting the constant given by Theorem 16.1.

Theorem 20.3 The Chain Rule *Assume that g is differentiable at $\bar{x}$ and f is differentiable at $\bar{y} = g(\bar{x})$. Then the composite function $f \circ g$ is differentiable at $\bar{x}$ and*

$$D(f(g(\bar{x}))) = Df(g(\bar{x}))Dg(\bar{x}).$$

If g is strongly differentiable at $\bar{x}$ and f is strongly differentiable at $\bar{y} = g(\bar{x})$, then $f \circ g$ is strongly differentiable at $\bar{x}$.[2]

---

[2]It is useful to state the Chain Rule in words. We compute the derivative of a composite function $f(g(x))$ by taking the derivative of the outside function f leaving the inside $g(x)$ unchanged and then multiplying by the derivative of the inside function.

EXAMPLE 20.4. Let $f(x) = x^5$ and $g(x) = 9 - 8x$. Then $Df(x) = 5x^4$ and $Dg(x) = -8$ so

$$D(f(g(x))) = Df(g(x))Dg(x) = 5(g(x))^4 Dg(x) = 5(9 - 8x)^4 \times -8.$$

We often write the Chain Rule in the form

$$(f(g(x)))' = f'(g(x))g'(x);$$

and with $y = g(x)$ and $z = f(y)$,

$$\frac{dz}{dx} = \frac{dz}{dy}\frac{dy}{dx}.$$

The last equation is very suggestive as it appears to indicate that we "cancel" the infinitesimal changes dy from the denominator and numerator in the two factors on the right. Of course, such cancellation is completely meaningless! Nonetheless it does serve make the Chain Rule easier to remember.

EXAMPLE 20.5. We compute the following derivative

$$\frac{d}{dx}\left(7x^3 + 4x + 6\right)^{18} = 18\left(7x^3 + 4x + 6\right)^{17}\frac{d}{dx}\left(7x^3 + 4x + 6\right)$$
$$= 18\left(7x^3 + 4x + 6\right)^{17}\left(21x^2 + 4\right)$$

by identifying $g(x) = 7x^3 + 4x + 6$ and $f(y) = y^18$.

EXAMPLE 20.6. The Chain Rule can also be used recursively:

$$\frac{d}{dx}((((1 - x)^2 + 1)^3 + 2)^4 + 3)^5$$
$$= 5((((1 - x)^2 + 1)^3 + 2)^4 + 3)^4 \times 4(((1 - x)^2 + 1)^3 + 2)^3$$
$$\times 3((1 - x)^2 + 1)^2 \times 2(1 - x) \times -1.$$

20.4 Quotients of Functions

We can use the Chain Rule to deal with quotients of functions. Consider first the reciprocal of a function.

EXAMPLE 20.7. From Example 19.4, it follows that

$$D\frac{1}{f} = D((f)^{-1}) = \frac{-1}{(f)^2}Df = \frac{-Df}{f^2}$$

as long as f is differentiable and $f \neq 0$.

EXAMPLE 20.8. Using Example 20.7 and the Chain Rule, we get the formula for $n \geq 1$:

$$Dx^{-n} = D\left(\frac{1}{x^n}\right) = \frac{-1}{(x^n)^2}Dx^n$$

$$= \frac{-1}{x^{2n}} \times nx^{n-1}$$

$$= -nx^{-n-1}.$$

The Chain Rule can also be used to prove the following theorem:

Theorem 20.4 The Quotient Rule *Assume that f and g are differentiable at $\bar{x}$ and $g(\bar{x}) \neq 0$. Then*

$$D\left(\frac{f(\bar{x})}{g(\bar{x})}\right) = \frac{Df(\bar{x})g(\bar{x}) - f(\bar{x})Dg(\bar{x})}{g(\bar{x})^2}.$$

If f and g are strongly differentiable at $\bar{x}$ and $g(\bar{x}) \neq 0$, then f/g is strongly differentiable at $\bar{x}$.

We often write this as

$$\left(\frac{f}{g}\right)' = \frac{f'g - fg'}{g^2}.$$

EXAMPLE 20.9.

$$\frac{d}{dx}\left(\frac{3x+4}{x^2-1}\right) = \frac{3 \times (x^2-1) - (3x+4) \times 2x}{(x^2-1)^2}.$$

EXAMPLE 20.10.

$$\frac{d}{dx}\left(\frac{x^3+x}{(8-x)^6}\right)^9$$

$$= 9\left(\frac{x^3+x}{(8-x)^6}\right)^8 \frac{d}{dx}\left(\frac{x^3+x}{(8-x)^6}\right)$$

$$= 9\left(\frac{x^3+x}{(8-x)^6}\right)^8 \frac{(8-x)^6\frac{d}{dx}(x^3+x) - (x^3+x)\frac{d}{dx}(8-x)^6}{\left((8-x)^6\right)^2}$$

$$= 9\left(\frac{x^3+x}{(8-x)^6}\right)^8 \frac{(8-x)^6(3x^2+1) - (x^3+x)6(8-x)^5 \times -1}{(8-x)^{12}}.$$

20.5 Derivatives of Derivatives: Descent into Despair

Continuing the discussion about the smoothness properties of derivatives, it is natural to consider situations in which the derivative itself is differentiable. The derivative of the derivative describes how quickly the rate of change of a function is changing. The derivative of the derivative of a function f is called the **second derivative** of f and is denoted by

$$f'' = D^2 f = \frac{d^2 f}{dx^2}.$$

EXAMPLE 20.11. For $f(x) = x^2$, $f'(x) = 2x$ and $f''(x) = 2$.

EXAMPLE 20.12. For $f(x) = 1/x$, $f'(x) = -1/x^2$ and $f''(x) = 2/x^3$.

Second order derivatives are particularly important in modeling. The derivative of velocity is called **acceleration**. Velocity indicates how quickly the position of an object is changing with time and acceleration indicates how quickly the object is speeding up or slowing down with respect to time.

When the second derivative of a function is differentiable, then we can compute a third derivative. In general, we define the nth derivative of f recursively. The nth derivative is found by taking the derivative of the $n - 1$st derivative, which is found by taking the derivative of the $n - 2$ derivative and so on. We denote this by

$$f^{(n)} = D^n f = \frac{d^n f}{dx^n}.$$

Note that it is too awkward to use $'$ to denote higher order derivatives, so instead we use (n) with parentheses. Note that $f^{(n)}(x) \neq (f(x))^n$.

EXAMPLE 20.13. If $f(x) = x^4$, then $Df(x) = 4x^3$, $D^2 f(x) = 12x^2$, $D^3 f(x) = 24x$, $D^4 f(x) = 24$ and $D^5 f(x) \equiv 0$.

EXAMPLE 20.14. The $n + 1$st derivative of a polynomial of degree n is zero.

EXAMPLE 20.15. If $f(x) = 1/x$, then

$$f(x) = x^{-1}$$
$$Df(x) = -1 \times x^{-2}$$
$$D^2 f(x) = 2 \times x^{-3}$$
$$D^3 f(x) = -6 \times x^{-4}$$
$$\vdots$$
$$D^n f(x) = (-1)^n \times 1 \times 2 \times 3 \times \cdots \times n x^{-n-1}$$
$$= (-1)^n n! x^{-n-1}.$$

Chapter 20   Problems

20.1. Show that $D(cf) = cDf$ for any differentiable function f and constant c. Discuss strong differentiability as well.

20.2. Suppose that g is Lipschitz continuous on an open interval $I_{\bar{x}}$ containing a point $\bar{x}$. Let $I_{\bar{y}}$ denote any open interval containing $\bar{y} = g(\bar{x})$. Prove that it is possible to find a new interval I contained in $I_{\bar{x}}$ such that g is in $I_{\bar{y}}$ for all x in I.

20.3. Prove the Quotient Rule Theorem 20.4 first using the Chain Rule and then directly with an argument similar to that used for the Product Rule.

20.4. *Without* using the definition, compute the derivatives of the following functions. In each case, indicate any restrictions needed on the domain.

(a) $7x^{13} + \dfrac{14}{x^8}$

(b) $(1 + 9x)^5 + (2x^5 + 1)^4$

(c) $(-1 + x^{-4})^{-6}$

(d) $\dfrac{4}{1+x}$

(e) $\dfrac{x}{x^2 + 3}$

(f) $\dfrac{2x - 2}{1 + 6x}$

(g) $\left(2x - (4 - 7x)^3\right)^{10}$

(h) $\left(((2 - x^2)^3 + 2)^{-1} - x\right)^9$

(i) $\dfrac{1}{1 + \dfrac{1}{2 + \dfrac{1}{3 + \dfrac{1}{4 + 2x}}}}$

(j) $\dfrac{(1 + 7x - 81x^2)^5}{\left(1 + \dfrac{4}{x}\right)^7}$

(k) $1 + x^2 + x^4 + \cdots + x^{100}$

(l) $\displaystyle\sum_{i=0}^{n} ix^i$.

20.5. Compute the second order derivatives of

(a) $9x^3 - \dfrac{4}{x}$

(b) $(1 + 4x^5)^3$

(c) $\dfrac{1}{1 - x}$

(d) $\displaystyle\sum_{i=0}^{n} ix^i$.

20.6. Compute the third derivatives of (a) $x^5 - 78x^2$ and (b) $1/x$.

20.7. Show that the nth derivative of x^n is zero and use this to explain Example 20.14.

20.8. Compute the nth derivative of (a) $1/x^2$, (b) $1/(1 + 2x)$, and (c) $x/(1 - x)$.

21
The Mean Value Theorem

We previously found that the tangent line to a function at a point can be interpreted as the limit of secant lines to the function. In this chapter, we explore the converse relationship: that is, how a given secant line is related to tangent lines of a function. In doing so, we touch on the basic issue of the relation between the *global* behavior of a function over an interval, for example, as given by a secant line over the interval, and the *local* behavior of the function, as in the tangent line at each point. The result we derive is the celebrated Mean Value Theorem. While the Mean Value Theorem is rarely used for practical purposes, it is *very* useful for proving all kinds of interesting and useful facts about functions. We conclude this chapter with a simple application and we present many others in the following chapters.

The Mean Value Theorem is a simple observation. Consider the secant line of a function f between two points $(a, f(a))$ and $(b, f(b))$, for example, as illustrated in Fig. 21.1. When the graph of f is smooth, there is always at least one point c in $[a, b]$ where the tangent line of f is parallel to the secant line. Intuitively, for the graph of f to bend around to connect the points $(a, f(a))$ and $(b, f(b))$, it has to become "parallel" to the secant line at least at one point. Stated precisely:

Theorem 21.1 Mean Value Theorem   *Suppose that f is uniformly strongly differentiable on an interval $[a, b]$. There is at least one point c in $[a, b]$ such that*

$$\frac{f(b) - f(a)}{b - a} = f'(c). \qquad (21.1)$$

Note, as Fig. 21.1 shows, there may be more than one such point c.

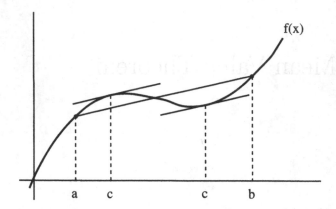

FIGURE 21.1. Illustration of the Mean Value Theorem. The tangent lines to the function at the points marked c are both parallel to the secant line through $(a, f(a))$ and $(b, f(b))$.

Another way to write (21.1) is that if f is uniformly strongly differentiable on $[a, b]$, then there is a point c in $[a, b]$ such that

$$f(b) = f(a) + f'(c)(b - a).$$

This means that the Mean Value Theorem has direct application to the approximation of functions. We investigate this further in Chapters 37 and 38.

Sometimes we can find the point c directly.

EXAMPLE 21.1. For $f(x) = x^2$ on $[1, 4]$, we find

$$\frac{f(4) - f(1)}{4 - 1} = \frac{15}{3} = 5.$$

Since $f'(x) = 2x$, we solve for c in $2c = 5$ to get $c = 2.5$.

But most of the time, c can only be approximated using an iterative algorithm. Therefore, the first proof of the Mean Value Theorem involves constructing an algorithm for approximating the point c that is based on the Bisection Algorithm. Though the individual steps are simple enough, the proof is rather long.[1]

In fact, the way in which the Mean Value Theorem is most often used, it is only the existence of the point c that is important, not its value. In Chapter 32, we present a non-constructive proof.

---

[1] The best way to learn the proof is to implement it on the computer.

21.1   A Constructive Proof

We begin by simplifying things a bit. The equation of the secant line through the points $(a, f(a))$ and $(b, f(b))$ is

$$\frac{f(b) - f(a)}{b - a}(x - a) + f(a),$$

and the distance between the secant line and the value of $f(x)$ is

$$g(x) = \frac{f(b) - f(a)}{b - a}(x - a) + f(a) - f(x).$$

We illustrate in Fig. 21.2.

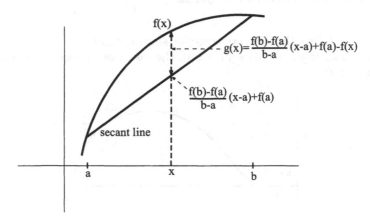

FIGURE 21.2. The formulas for the secant line to f and the distance between the secant line and the graph of f.

Of course, $g(a) = g(b) = 0$. Now suppose we find a c in $[a, b]$ with $g'(c) = 0$. This means

$$g'(c) = \frac{f(b) - f(a)}{b - a} - f'(c) = 0$$

or

$$f'(c) = \frac{f(b) - f(a)}{b - a}.$$

In other words, the Mean Value Theorem 21.1 follows from:

Theorem 21.2 Rolle's Theorem   *If g is uniformly strongly differentiable on an interval $[a, b]$ and $g(a) = g(b) = 0$, then there is a point c in $[a, b]$ such that $g'(c) = 0$.*

This theorem, named after the mathematician Rolle,[2] says that there must be at least one point in $[a, b]$ where g has a horizontal tangent when $g(a) = g(b) = 0$. We illustrate in Fig. 21.3.

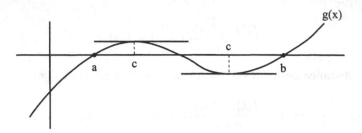

FIGURE 21.3. Illustration of Rolle's theorem. The tangent lines to the function at the points marked c are both horizontal.

To prove Rolle's theorem, we consider several cases.

First case: If $g'(a) = 0$ or $g'(b) = 0$, then we just choose $c = a$ or $c = b$ (see Fig. 21.4).

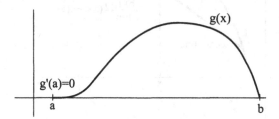

FIGURE 21.4. The case when $g'(a) = 0$.

So we only have to consider the cases when $g'(a) \neq 0$ and $g'(b) \neq 0$.

Second case: We next assume that $g'(a)$ and $g'(b)$ have opposite signs (see Fig. 21.5). By Theorem 19.1, g' is Lipschitz continuous on $[a, b]$ and therefore the Intermediate Value Theorem 13.2 implies there is a c in $[a, b]$ such that $g'(c) = 0$. Moreover, we can compute c by using the Bisection Algorithm applied to g' starting with the interval $[a, b]$. We ask you to explain how in Problem 21.3.

Third case: In the final case, we assume that $g'(a)$ and $g'(b)$ have the same sign. This case is more complicated. We can assume that $g'(a) > 0$ and $g'(b) > 0$, as for example shown in Fig. 21.3. Otherwise, we replace g by $-g$.

---

[2]Michel Rolle (1652–1719) was a French mathematician who worked on algebra, geometry, and number theory. He stated a purely algebraic form of his namesake theorem but did not prove it.

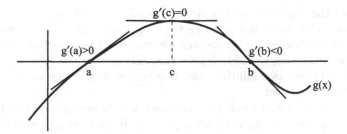

FIGURE 21.5. The case when $g'(a) > 0$ and $g'(b) < 0$.

We set $a_0 = a$ and $b_0 = b$. In the first step, we show that there is a point c_0 in (a_0, b_0) such that $g(c_0) = 0$ using the Intermediate Value Theorem (see Fig. 21.6). To do this, we first show there are two points $\tilde{a}_0 < \tilde{b}_0$ in $[a_0, b_0]$ such that g has different signs at these points. In fact, $g(\tilde{a}_0) > 0$ for all $\tilde{a}_0$ close to a_0 with $\tilde{a}_0 > a_0$ and likewise $g(\tilde{b}_0) < 0$ for all $\tilde{b}_0$ close to b_0 with $\tilde{b}_0 < b_0$. We can see this is true in Fig. 21.6:

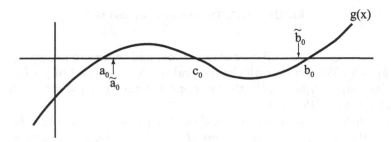

FIGURE 21.6. The choice of $\tilde{a}_0$ and $\tilde{b}_0$.

We can prove this using the definition of the derivative. Because g is uniformly strongly differentiable, there is a constant $\mathcal{K}$ such that for $x > a_0$ close to a_0

$$|g(x) - (g(a_0) + g'(a_0)(x - a_0))| = |g(x) - g'(a_0)(x - a_0)| \leq (x - a_0)^2 \mathcal{K}.$$

This implies that

$$g(x) \geq g'(a_0)(x - a_0) - (x - a_0)^2 \mathcal{K} = (g'(a_0) - (x - a_0)\mathcal{K})(x - a_0).$$

Now,

$$g'(a_0) - (x - a_0)\mathcal{K} \geq \frac{1}{2}g'(a_0) \tag{21.2}$$

for all $x \geq a_0$ with $|x - a_0| \leq \frac{1}{2\mathcal{K}}|g'(a_0)|$. We choose $x = \tilde{a}_0 > a_0$ sufficiently close to a_0 so

$$g(\tilde{a}_0) \geq (g'(a_0) - (\tilde{a}_0 - a_0)\mathcal{K})(\tilde{a}_0 - a_0) \geq \frac{1}{2}g'(a_0)(\tilde{a}_0 - a_0) > 0.$$

The proof that $\tilde{b}_0$ exists is very similar (see Problem 21.5).

It follows that there is a point c_0 with $a_0 < \tilde{a}_0 < c_0 < \tilde{b}_0 < b_0$ such that $g(c_0) = 0$ as claimed and moreover we can use the Bisection Algorithm applied to g starting with $[\tilde{a}_0, \tilde{b}_0]$ to compute c_0.

There are three possibilities now. If $g'(c_0) = 0$, then we set $c = c_0$ and the theorem is proved.

If $g'(c_0) < 0$, then we can use the Bisection Algorithm applied to g' to compute a point c in $[a_0, c_0]$ with $g'(c) = 0$. In fact, we can also compute another such point c in $[c_0, b_0]$! This is illustrated in Fig. 21.6.

In the third case, which is more complicated, $g'(c_0) > 0$. We illustrate in Fig. 21.7. We define a new interval $[a_1, b_1]$ by setting $a_1 = a_0$ and $b_1 = c_0$

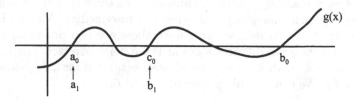

FIGURE 21.7. The choice of a_1 and b_1.

if $c_0 - a_0 \leq b_0 - c_0$, i.e., if c_0 is closer to a_0, and otherwise setting $a_1 = c_0$ and $b_1 = b_0$. We end up with the interval $[a_1, b_1]$ contained in $[a_0, b_0]$ such that $|b_1 - a_1| \leq \frac{1}{2}|b_0 - a_0|$ with the properties that $g(a_1) = g(b_1) = 0$ and $g'(a_1) > 0$ and $g'(b_1) > 0$.

Now repeating the argument, we choose points $\tilde{a}_1 > a_1$ and $\tilde{b}_1 < b_1$ with $g(\tilde{a}_1) > 0$ and $g(\tilde{b}_1) < 0$ and then use the Bisection Algorithm applied to g to compute c_1 in $[\tilde{a}_1, \tilde{b}_1]$ with $g(c_1) = 0$. Again if $g'(c_1) = 0$, then we stop with $c = c_1$ or if $g'(c_1) < 0$, we can compute the point c by applying the Bisection Algorithm to g' on the interval $[a_1, c_1]$. Otherwise, we define a new interval $[a_2, b_2]$ contained in $[a_0, b_0]$ with $|b_2 - a_2| \leq 2^{-2}|b_0 - a_0|$ and $g(a_2) = g(b_2) = 0$ and $g'(a_2) > 0$ and $g'(b_2) > 0$.

Continuing in this way, we compute a sequence of intervals $[a_i, b_i]$ with $[a_i, b_i] \subset [a_{i-1}, b_{i-1}] \subset [a_0, b_0]$ for $i \geq 1$ and $[a_0, b_0]$ with $|b_i - a_i| \leq 2^{-i}|b_0 - a_0|$, $g(a_i) = g(b_i) = 0$, and $g'(a_i) > 0$ and $g'(b_i) > 0$ together with a sequence of points c_i in (a_i, b_i) such that $g(c_i) = 0$. We stop the iteration if $g'(c_i) = 0$, so we set $c = c_i$, or if $g'(c_i) < 0$, in which case we compute c in $[a_i, c_i]$ using the Bisection Algorithm applied to g'. If these two conditions never occur, then we obtain, as with the Bisection Algorithm, two Cauchy sequences $\{a_i\}$ and $\{b_i\}$ that converge to the same limit in $[a, b]$. We claim this limit is c. In other words, if

$$c = \lim_{i \to \infty} a_i = \lim_{i \to \infty} b_i,$$

then $g'(c) = 0$.

We prove this by showing $g'(c) \neq 0$ contradicts the construction of the sequences $\{a_i\}$ and $\{b_i\}$. We begin by observing that because g is Lipschitz continuous, say, with constant L, we get for any i

$$|g(c)| = |g(c) - g(a_i)| \leq L|c - a_i|,$$

which implies that $g(c) = 0$ since $|c - a_i| \to 0$.

Suppose first that $g'(c) < 0$. Since g' is Lipschitz continuous, there is a $\delta > 0$ such that $g'(x) < 0$ for all x with $|x - c| < \delta$. Recall that we use a similar result (see Fig. 13.4), to show that the Bisection Algorithm converges. But for i sufficiently large, $|a_i - c| < \delta$ while $g'(a_i) > 0$ by construction. Thus $g'(c) < 0$ is impossible.

Suppose next that $g'(c) > 0$. This turns out to be more complicated, so we describe the idea of the proof before giving the details. Since $g'(c) > 0$, the graph of $g(x)$ must increase as x moves from left to right near the value of c (see Fig. 21.8). But this means that $g(x) > 0$ for all $x > c$ sufficiently

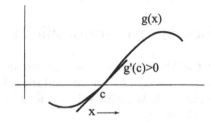

FIGURE 21.8. If $g'(c) > 0$, then $g(x)$ increases as x moves from left to right near c.

close to c. Now, we get a contradiction since $g(x) < 0$ for all $x < b_i$ close to b_i and we can make b_i arbitrarily close to c by taking i large.

To make this argument precise, we use the fact that for x close to c,

$$|g(x) - (g(c) + g'(c)(x - c))| = |g(x) - g'(c)(x - c)| \leq (x - c)^2 \mathcal{K},$$

which implies

$$g(x) \geq g'(c)(x - c) - (x - c)^2 \mathcal{K} = (g'(c) - (x - c)\mathcal{K})(x - c).$$

As before, because $g'(c) > 0$ there is a δ such that for all x with $|x - c| \leq \delta$,

$$g'(c) - (x - c)\mathcal{K} \geq \frac{1}{2}g'(c). \tag{21.3}$$

Now we choose i so that $[a_i, b_i]$ is contained in $[c - \delta, c + \delta]$ and set

$$\tilde{\delta} = \text{ the smaller of } |c - a_i|/2 \text{ and } |c - b_i|/2$$

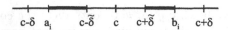

FIGURE 21.9. The definitions of $\tilde{\delta}$ and J. J is marked by the thick line segments.

and define the set

$$J = \{x \text{ in } [a_i, b_i] \text{ but not in } [c - \tilde{\delta}, c + \tilde{\delta}]\}.$$

We illustrate these definitions in Fig. 21.9. By these choices and (21.3), for any x in J we have

$$g(x) > (g'(c) + (x - c)\mathcal{K})(x - c) \geq \frac{1}{2}g'(c) \times \tilde{\delta} > 0.$$

But this gives a contradiction because $g(x) < 0$ for all $x < b_i$ sufficiently close to b_i. Thus $g'(c) > 0$ is also impossible and therefore $g'(c) = 0$.

21.2   An Application to Monotonicity

As an application of the Mean Value Theorem, we show that a function whose derivative has only one sign on an interval must be either monotone increasing or decreasing. This is intuitively obvious and it follows easily from the Mean Value Theorem.

Suppose that f is uniformly strongly differentiable on an interval I and $f'(x) > 0$ for all x in I. We want to show that $x_1 < x_2$ implies $f(x_1) < f(x_2)$ for any x_1 and x_2 in I. Consider the difference $f(x_2) - f(x_1)$. The Mean Value Theorem says there is a c between x_1 and x_2 such that

$$f(x_2) - f(x_1) = f'(c)(x_2 - x_1) > 0.$$

The case when $f'(x) < 0$ for all x in I is similar (see Problem 21.7). We have proved:

Theorem 21.3 *Suppose that f is uniformly strongly differentiable on an interval I. If $f'(x) > 0$ for all x in I, then f is monotone increasing on I. If $f'(x) < 0$ for all x in I, then f is monotone decreasing on I.*

Chapter 21   Problems

Do Problems 21.1 and 21.2 by forming an equation and solving it directly.

21.1. Find the point c given by the Mean Value Theorem for the function $f(x) = x^2 - 2x$ on the interval $[1, 3]$ and make a plot illustrating the theorem on this example.

21.2. Find the point c given by the Mean Value Theorem for the function $f(x) = 6/(1 + x)$ on the interval $[0, 1]$ and make a plot illustrating the theorem on this example.

21.3. Explain how to use the Bisection Algorithm to compute the point c in the second case of the proof of Rolle's theorem.

21.4. Explain why (21.2) and (21.3) are true.

21.5. Prove that the point $\tilde{b}_0$ used in the proof of Rolle's theorem exists as claimed.

21.6. (a) Write down an algorithm that implements the proof of Rolle's theorem for determining the point c. (b) Implement this algorithm in a program that computes the point c given by the Mean Value Theorem. (c) Find the points c for $f(x) = x^3 - 4x^2 + 3x$ on $[0, 3]$ and $f(x) = x/(1 + x)$ on $[0, 1]$.

21.7. Prove the claim about monotone decreasing functions in Theorem 21.3.

21.8. Use the Mean Value Theorem to give a simple proof of the second part of Theorem 19.1.

22

Derivatives of Inverse Functions

As an application of the Mean Value Theorem, we investigate the continuity and differentiability of the inverse function to a given function that is Lipschitz continuous or differentiable. It is a good idea to review Chapter 14.

22.1 The Lipschitz Continuity of an Inverse Function

We begin by investigating whether a given Lipschitz continuous function f that has an inverse always has a *Lipschitz continuous* inverse function f^{-1}. The short answer is no!

EXAMPLE 22.1. Consider $f(x) = x^3$ with $f^{-1}(x) = \sqrt[3]{x} = x^{1/3}$. x^3 is Lipschitz continuous on any interval containing the origin, say $[-1, 1]$. But $x^{1/3}$ is not Lipschitz continuous on any interval containing the origin. For suppose that there is a constant L such that

$$|x^{1/3} - y^{1/3}| \leq L|x - y|$$

for all x and y in $[0, 1]$. Now the identity

$$x^3 - y^3 = (x - y)(x^2 + xy + y^2)$$

means that

$$x - y = (x^{1/3} - y^{1/3})(x^{2/3} + x^{1/3}y^{1/3} + y^{2/3}).$$

So L must satisfy

$$L \geq \frac{|x^{1/3} - y^{1/3}|}{|x - y|} = \frac{1}{|x^{2/3} + x^{1/3}y^{1/3} + y^{2/3}|}.$$

But we can make the right-hand side of this inequality as large as desired by taking x and y close to zero. So L cannot exist.

The problem with x^3 is that this function is flat near $x = 0$ (see the graph on the right in Fig. 14.5). This means that the graph of $\sqrt[3]{x}$ is steep near $x = 0$. Too steep in fact for $\sqrt[3]{x}$ to be Lipschitz continuous! Lipschitz continuity controls how much a function can change on an interval, but not how little a function can change.

To get a handle on how little a function changes, we can use the Mean Value Theorem. We assume that f is uniformly strongly differentiable on the interval $[a, b]$ and moreover that f is monotone increasing on $[a, b]$ so that f has an inverse on $[a, b]$ defined on $[\alpha, \beta]$ with $\alpha = f(a) < \beta = f(b)$. Given two points x_1 and x_2 in (α, β), we want to estimate $|f^{-1}(x_1) - f^{-1}(x_2)|$ in terms of $|x_1 - x_2|$. Let $y_1 = f^{-1}(x_1)$ and $y_2 = f^{-1}(x_2)$ so that

$$|x_1 - x_2| = |f(y_1) - f(y_2)|.$$

By the Mean Value Theorem, there is a c between y_1 and y_2 such that

$$|f(y_1) - f(y_2)| = |f'(c)||y_1 - y_2|.$$

This means that

$$|x_1 - x_2| = |f'(c)||f^{-1}(x_1) - f^{-1}(x_2)|.$$

Turning this around, we get

$$|f^{-1}(x_1) - f^{-1}(x_2)| = \frac{1}{|f'(c)|}|x_1 - x_2|$$

provided $f'(c) \neq 0$. Clearly, the Lipschitz constant for f^{-1} depends on the size of $1/|f'(c)|$.

Since we want the Lipschitz condition on f^{-1} to hold for all x_1 and x_2 in the interval (α, β), c could possibly take on any value in $[a, b]$. Therefore, we assume that there is a constant d with

$$|f'(x)| \geq d > 0 \text{ for all } x \text{ in } [a, b].$$

Under this assumption, we conclude that

$$|f^{-1}(x_1) - f^{-1}(x_2)| \leq L|x_1 - x_2|$$

for all x_1 and x_2 in (α, β) where $L = 1/d$. Of course the same argument also works if f is monotone decreasing.

Note that the condition $|f'(x)| \geq d > 0$ for all x in $[a, b]$ means that either $f'(x) \geq d > 0$ for all x or $f'(x) < -d < 0$ for all x. This means that f is either strictly increasing or strictly decreasing on the interval $[a, b]$. We summarize as a theorem:

Theorem 22.1 *If f is uniformly strongly differentiable on $[a, b]$, $[\alpha, \beta] = f([a, b])$ and if there is a constant d such that $|f'(x)| \geq d > 0$ for all x in $[a, b]$, then f has a Lipschitz continuous inverse function defined in $[\alpha, \beta]$ with Lipschitz constant $1/d$.*

EXAMPLE 22.2. $f(x) = x^3$ is strictly increasing on any interval that contains the origin, yet $f'(0) = 0$. So there is no $d > 0$ with $|f'(x)| \geq d$ for all x in any interval containing the origin. The theorem does not apply, luckily!

22.2   The Differentiability of an Inverse Function

There are two natural questions that follow from the assumptions of this last theorem. Since we have assumed that f is uniformly strongly differentiable, does it follow that f^{-1} is differentiable? If f^{-1} is differentiable, what is its derivative?

First, we conduct a geometric investigation. If f is smooth enough to have a linearization at each point, then the reflection of the graph of f through the line $y = x$ is also sufficiently smooth to have a linearization at each point. The only possible problem is, if the linearization of f is horizontal at some point, f' is zero there. Then the reflection of the graph of f has a linearization that is vertical at the corresponding reflected point and so the derivative of the inverse is undefined at that point. The assumption that $|f'(x)| \geq d > 0$ for all x in $[a, b]$ prevents this. We conclude that f^{-1} is also differentiable.

We can also compute the value of Df^{-1} at any point once we realize that the linearization of f at any point and the linearization of f^{-1} at the corresponding reflected points are themselves reflections of each other through the line $y = x$. We illustrate in Fig. 22.1. The means that the two linearizations are inverse functions, and by Example 14.9, the slopes of the two lines are reciprocal. In other words, if $y_0 = f(x_0)$, then

$$\frac{df}{dx}(x_0) = \frac{1}{\dfrac{df^{-1}}{dx}(y_0)} \quad \text{or} \quad Df(x_0) = \frac{1}{Df^{-1}(y_0)}. \qquad (22.1)$$

EXAMPLE 22.3.   If $f(x) = x^3$, then $f(2) = 8$. Therefore,

$$Df^{-1}(8) = \frac{1}{Df(2)} = \frac{1}{3 \times 2^2} = \frac{1}{12}.$$

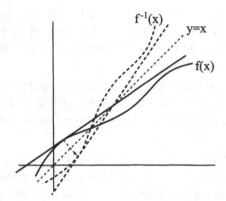

FIGURE 22.1. The linearization of f at a point and the linearization of f^{-1} at the corresponding reflected point are reflections of each other through the line $y = x$.

We can also derive (22.1) using an analytic argument. We suppose that f is uniformly strongly differentiable on $[a, b]$ and that $|f'(x)| \geq d > 0$ for all x in $[a, b]$. By the definition, there is a constant $\mathcal{K}$ such that

$$|f(x) - (f(\bar{x}) + f'(\bar{x})(x - \bar{x}))| \leq (x - \bar{x})^2 \mathcal{K} \qquad (22.2)$$

for all $x, \bar{x}$ in $[a, b]$. Now let

$$y = f(x) \text{ or } x = f^{-1}(y)$$
$$\bar{y} = f(\bar{x}) \text{ or } \bar{x} = f^{-1}(\bar{y}).$$

Substituting these into (22.2) yields

$$|y - (\bar{y} + f'(\bar{x})(f^{-1}(y) - f^{-1}(\bar{y})))| \leq (f^{-1}(y) - f^{-1}(\bar{y}))^2 \mathcal{K}.$$

Dividing both sides by $f'(\bar{x})$, we get

$$|\frac{1}{f'(\bar{x})}(y - \bar{y}) - (f^{-1}(y) - f^{-1}(\bar{y}))| \leq (f^{-1}(y) - f^{-1}(\bar{y}))^2 \frac{\mathcal{K}}{|f'(\bar{x})|}.$$

Rearranging and using the assumptions and Theorem 22.1, we get

$$|f^{-1}(y) - (f^{-1}(\bar{y})) + \frac{1}{f'(\bar{x})}(y - \bar{y})|$$
$$\leq (f^{-1}(y) - f^{-1}(\bar{y}))^2 \frac{\mathcal{K}}{|f'(\bar{x})|} \leq (y - \bar{y})^2 \frac{\mathcal{K}}{d^3}.$$

Since this holds for every y and $\bar{y}$ in $[\alpha, \beta]$, we conclude:

Theorem 22.2 *If f is uniformly strongly differentiable on $[a,b]$, $[\alpha, \beta] = f([a,b])$, and if there is a constant d such that $|f'(x)| \geq d > 0$ for all x in $[a,b]$, then f has a uniformly strongly differentiable inverse function defined in $[\alpha, \beta]$. If $\bar{y} = f(\bar{x})$ for $\bar{x}$ in $[a,b]$, then $Df^{-1}(\bar{y}) = 1/Df(\bar{x})$.*

EXAMPLE 22.4.   We can use Theorem 22.2 to compute the derivative of $x^{1/n}$ for a natural number n. Let $f(x) = x^n$ so $f^{-1}(x) = x^{1/n}$ by Example 14.12. If $y_0 = x_0^n$ or $x_0 = y_0^{1/n}$, then

$$Df^{-1}(y_0) = \frac{1}{Df(x_0)} = \frac{1}{nx_0^{n-1}} = \frac{x_0}{nx_0^n} = \frac{y_0^{1/n}}{ny_0} = \frac{1}{n}y_0^{1/n-1}.$$

This is nothing more than the usual power rule for derivatives!

Using the Chain Rule, we can extend the formula

$$Dx^r = rx^{r-1}$$

to any rational number r.

EXAMPLE 22.5.

$$D(2x - 1)^{5/7} = \frac{5}{7}(2x - 1)^{-2/7} \times 2.$$

Chapter 22   Problems

22.1. Prove that $y = x^{1/2}$ is not Lipschitz continuous on $[0, 1]$.

22.2. Assume that f is Lipschitz continuous and monotone on $[a, b]$ and further-more that there is a constant $l > 0$ such that

$$|f(x) - f(y)| \geq l|x - y|$$

for all x and y in $[a, b]$. (Note the direction of the inequality!) If $[\alpha, \beta] = f([a, b])$, prove that f has a Lipschitz continuous inverse on $[\alpha, \beta]$ and compute a Lipschitz constant.

22.3. Prove that the fact that f is strictly increasing on an interval $[a, b]$ does not imply there is a constant d such that $|f'(x)| \geq d > 0$ for all x in $[a, b]$. *Hint:* Consider $f(x) = x^3$ on $[-1, 1]$.

22.4. If f and g are strictly monotone functions with $f(2) = 7$ and $f'(2) = -1$, compute $g'(7)$.

22.5. (a) Prove that $f(x) = x^3 + 2x$ is strictly increasing, so it has an inverse function defined for all x. (b) Compute $Df^{-1}(12)$ given that $f(2) = 12$.

22.6. (a) Prove that $f(x) = 1 - 3x^3 - x^5$ is strictly decreasing, so it has an inverse function defined for all x. (b) Compute $Df^{-1}(-3)$ given that $f(1) = -3$.

22.7. (a) Prove that $f(x) = x^2 - x + 1$ is strictly increasing for $x > 1/2$, so it has an inverse function defined for all $x > 1/2$. (b) Compute $Df^{-1}(3)$.

22.8. (a) Prove that $f(x) = x^3 - 9x$ is strictly decreasing for $-\sqrt{3} < x < \sqrt{3}$, so it has an inverse function defined for $-\sqrt{3} < x < \sqrt{3}$. (b) Compute $Df^{-1}(0)$.

22.9. Compute the derivatives of the following functions:

(a) $f(x) = (x + 1)^{1/2}$ (b) $f(x) = (x^{1/3} - 3)^4$ (c) $f(x) = x^{-3/4}$.

22.10. Prove that if f is uniformly strongly differentiable in an interval con-taining a point x_0 and if $f'(x_0) \neq 0$, then there is some interval containing x_0 on which f is 1-1.

23
Modeling with Differential Equations

The discussion of derivatives so far has emphasized the approximation of a function through linearization. This is one of the two main uses for the derivative. The other main use is for mathematical modeling in the form of differential equations. A **differential equation** describes a function, called the solution, by specifying a relationship between the derivatives of the solution and the physical world. Differential equations arise in all areas of science and engineering and an enormous amount of time and energy is spent trying to analyze and compute solutions of differential equations. In this chapter, we introduce the subject of posing, analyzing, and solving differential equations.

We begin by presenting a couple of models to make the discussion concrete. The first model is Newton's Law of Motion, which describes an object moving in a straight line under the influence of a force. This is perhaps the quintessential model in physics and one of Newton's main motivations for inventing the calculus. We apply Newton's law to model the motion of a mass connected to a spring. After that we describe Einstein's Law of Motion, which is a modern version of Newton's law. We then introduce some general language for describing differential equations and make some observations about the existence and uniqueness of solutions. We conclude by encapsulating these ideas in the solution of Galileo's model for a falling object.

We continue the discussion about modeling with differential equations in Chapter 39 after developing integration as an important tool for solving differential equations in the following chapters.

23.1   Newton's Law of Motion

Newton's Law of Motion is one of the cornerstones of the Newtonian physics that accurately describes much of the world in which we live. It is a relationship between the mass and acceleration of an object and the forces acting on the object.

> The acceleration a of an object of mass m moving in a straight line while being acted on by a force F satisfies

$$ma = F. \qquad (23.1)$$

If $s(t)$ gives the position of the object at time t relative to some initial starting position $s_0 = s(0)$, then (23.1) is

$$m\frac{d^2s}{dt^2} = F \text{ or } ms'' = F. \qquad (23.2)$$

This is an example of a differential equation. Solving the differential equation means finding a function s that satisfies (23.2) for all time in some specified interval.

EXAMPLE 23.1.   By differentiating twice, we can show that $s(t) = 1/(1+t)$ is a solution of the differential equation

$$ms'' = 2ms^3,$$

which is (23.2) with $F = 2ms^3$, for all $t > -1$.

Differential equations can often be written in different ways. For example, if $v = s'$ denotes the velocity of the object, then (23.2) can be written as a differential equation for v,

$$\frac{dv}{dt} = F \text{ or } v' = F. \qquad (23.3)$$

A model of the motion of an object in a specific physical setting involves combining Newton's law (23.2) with a description of the forces involved in the system.

EXAMPLE 23.2. Galileo's[1] law says

> A free-falling object, that is, an object that is acted upon by gravity and no other force, always has the same acceleration regardless of its mass, position, or the time.

---

[1]Galileo Galilei (1564–1642) was an Italian mathematician and scientist. Galileo is perhaps best known for his work in astronomy and mechanics. But, he also used a concept akin to functions and dealt with infinite sums.

In other words, the acceleration of a free-falling object is constant.

We assume that the object moves vertically and let $s(t)$ denote the height of the object at time t starting from some initial height $s(0) = s_0$ where the positive direction corresponds to moving upward and the negative direction corresponds to moving downward (see Fig. 23.1). In

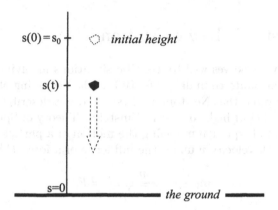

FIGURE 23.1. The coordinate system describing the position of a free-falling object with the initial height $s(0) = s_0$.

this coordinate system, a positive velocity $v = s' > 0$ corresponds to the object moving upwards, while a negative velocity means the object is falling. The **weight** W of a object is the absolute value of the force acting on the object that tends to make it fall. Since the force of gravity tends to make the object move downward, in this coordinate system, Newton's law takes the form

$$ms'' = -W.$$

This is usually rewritten by assuming[2] there is a constant g such that

$$W = mg,$$

which gives

$$s'' = -g. \tag{23.4}$$

The constant g is called the **acceleration of gravity** and has the value ≈ 9.8 m/sec^2.

In general the force acting on the object can depend on the time t, the position s, the velocity s', and on physical characteristics such as the mass

---

[2] A valid assumption near the Earth's surface.

m and size of the object. For example, the force might be a function of the form $F = F(t, s, s', m)$ so that (23.2) becomes

$$ms''(t) = F(t, s(t), s'(t), m), \qquad (23.5)$$

which is an equation that is supposed to hold for *all* time t in some interval.

23.2   Einstein's Law of Motion

Newtonian physics serves well to describe situations involving low speeds, such as those encountered in daily life. But at the beginning of the century, Einstein[3] discovered that Newtonian physics did not describe the behavior of particles moving at high speeds. In Einstein's Theory of Special Relativity, the differential equation modeling the motion of a particle moving in a straight line with velocity v under the influence of a force F is

$$m_0 \frac{d}{dt} \frac{v}{\sqrt{1 - v^2/c^2}} = F, \qquad (23.6)$$

where m_0 is the mass of the particle at rest and $c \approx 3 \times 10^8$ m/s is the speed of light in a vacuum. This differential equation is considerably more complicated than the Newton's law (23.3).[4] In general, (23.6) is sufficiently complicated that we almost never can write down a solution in the form of an explicit function.

23.3   Describing Differential Equations

As we learn more about differential equations, we begin to discover patterns among different kinds of problems and their solutions. This kind of information is important when deciding on ways to compute solutions and analyze their properties. There is a lot of notation and language associated with differential equations that help to classify different problems according to their form and the behavior of their solutions.

One way that differential equations are classified is by their **order**, which refers to the highest order derivative that appears in the equation.

EXAMPLE 23.3. The order of

$$y^{(4)} - 45(y^{(3)})^{10} = \frac{1}{y + 1}$$

---

[3] The famous Albert Einstein (1879–1955) was a physicist of course. But he made use of the most current mathematics in his research and corresponded closely with leading mathematicians during his most active period of research.

[4] Among other things, it involves the $\sqrt{}$ which we have not differentiated yet.

is 4, while the order of

$$\left(y^{(2)}\right)^5 = y$$

is 2. The order of Newton's law is 2 for the position, though we changed it into a differential equation of order 1 for the velocity.

In general, we expect complications to increase with order.

Differential equations are also classified according to how the unknown variable appears in the equation. A nth order differential equation is called **linear** if it can be written in the form

$$a_n(x)\frac{d^n y}{dx^n} + a_{n-1}(x)\frac{d^{n-1} y}{dx^{n-1}} + \cdots + a_1(x)\frac{dy}{dx} + a_0(x)y = f(x), \quad (23.7)$$

i.e., if it is a linear function of the unknown variable and its derivatives; otherwise it is called **nonlinear**.

EXAMPLE 23.4. The differential equation

$$x^7 u'' + 2u' + 4u = \frac{1}{x}$$

is linear and second order, while

$$y\frac{dy}{dx} = x^3$$

is nonlinear and first order. The differential equations in Example 23.3 are also nonlinear. Newton's law may or may not be linear depending on the force F. Hooke's model of a spring is linear. Einstein's law is nonlinear.

In general, we expect nonlinear differential equations to pose additional difficulties.

By the way, when we write a differential equation in the form (23.7), we implicitly assume that $a_n(x) \neq 0$ on the interval where we want to solve the differential equation. If $a_n(x) = 0$ at some points in the interval of solution, the differential equation is said to be **degenerate**. In general, a differential equation is called degenerate if the order of the problem changes at some point(s).

Degenerate problems are particularly difficult.

With a couple of exceptions, we focus on first order differential equations in this book. The most general first order differential equation can be written

$$G(y', y, x) = 0$$

for some function G. First order equations are classified by whether they can be rewritten in the form

$$h(y(x))y'(x) = g(x)$$

for some functions h and g, in which case they are called **separable**, or not, in which case they are called **nonseparable**.

EXAMPLE 23.5. The following problems are separable:

$$y' - y = 0 \rightarrow \frac{y'}{y} = 1$$

$$y' = 4x^2y^3 \rightarrow \frac{y'}{y^3} = 4x^2$$

$$(y')^3 = 2x^3/(1+y) \rightarrow (1+y)^{1/3}y' = 2^{1/3}x,$$

while

$$(y')^2 + y - x = 0$$
$$y' + (y')^3 = 1$$
$$y' - y = x$$

are all nonseparable.

There is a general technique for solving separable equations in principle described in Chapter 39.

> *In general, nonseparable differential equations are more difficult to solve.*

23.4   Solutions of Differential Equations

The difference between solving differential equations and algebraic equations is worth pointing out again. A solution of an algebraic equation like the Dinner Soup or Muddy Yard model is a single number $\bar{x}$. A **solution** of a differential equation is a function that satisfies the differential equation at *all* points in a given interval.

EXAMPLE 23.6. By differentiation, we can verify that $s(t) = t^3 - 2t + 4$ satisfies $s'(t) = 3t^2 - 2$ for all t.

EXAMPLE 23.7. By differentiation, we can verify that $f(x) = 6x^2$ satisfies $f'(x) = 2f(x)/x$ for all $x > 0$ and all $x < 0$ since computing we get $f'(x) = 12x = 2 \times 6x^2/x$.

EXAMPLE 23.8.   By differentiation, we can verify that $y(x) = x^3/3$ satisfies $\left(y''(x)\right)^2 - 9y(x) = 5x^3$ for all x. This follows because $\left(y''(x)\right)^2 - 9y(x) = \left(2x\right)^3 - 3x^3 = 5x^3$.

Note that a function is *not* necessarily a solution of a differential equation just because it satisfies the differential equation at a single point.

EXAMPLE 23.9.   The function $y = 2x^2 - 4$ does not satisfy the differential equation $y' = 2(x - 1)^2$ since $y' = 4x - 4$ is not equal to $2(x - 1)^2$ at all points in any interval. Note that $y'(x)$ does agree with the differential equation at isolated points like $x = 1$.

Putting this in perspective, there are many distinct functions that have the same derivative at a single point, as illustrated in Fig. 23.2. Therefore, *a*

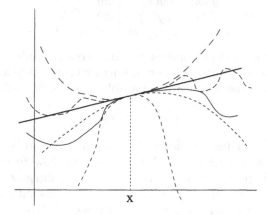

FIGURE 23.2. All of these functions have the same derivative at the point x but cannot all solve the differential equation (23.8) on any interval containing x.

differential equation should always be accompanied by an domain on which the solution is to be computed. If a domain is not specified, then it is understood that the solution formula holds on the entire real line.

We consider two approaches to studying differential equations in this book. The first approach is to simply guess the solution. For some specific kinds of differential equations, it is possible to make the technique of guessing solutions fairly sophisticated and systematic. In fact, a more old-fashioned calculus text concentrates on developing the art of guessing to a high level. We do not do this. For one thing, symbolic manipulation programs like *MAPLE*© have made it unnecessary. For another thing, the fact is that most differential equations that arise in engineering and science cannot be solved by guesswork.

Therefore, we concentrate on developing algorithms that produce an approximation of a given solution to any desired accuracy and developing techniques of analysis that give information about a solution without knowing a formula for the solution. This approach is entirely analogous to the approach used for studying root and fixed point problems.

However, the study of differential equations is a much more involved and complicated subject than the study of equations whose solutions are numbers. The fact that the solutions are functions makes all the difference. Therefore, we can only begin the study of differential equations in this book. We do this by considering a set of specific problems that illustrate fundamental ideas. A roadmap of the problems we consider looks like:

Type of problem	Chapter(s)
$y'' = $ constant	23
$y'(x) = f(x)$	24–27, 34
$y'(x) = c/x$	28
$y'(x) = cy(x)$	29
$y''(x) = cy(x)$	30
$y'(x) = f(y(x), x)$	39–41

The problems are arranged more or less in increasing order of difficulty.

Over the next few chapters, we concentrate on the simplest differential equation, which is the first order, separable, linear differential equation,

$$y'(x) = f(x). \tag{23.8}$$

Equation (23.8) specifies the slope of the tangent line to the solution $y(x)$ at every point x in an interval. After discussing how to guess solutions for some specific problems, we turn to developing integration, which is a general constructive method for solving (23.8). It turns out that integration is a fundamental tool for solving and analyzing all differential equations, and the subsequent material uses integration heavily.

23.5   Uniqueness of Solutions

So far the discussion has focused on the existence of solutions, but there is another topic of practical importance, namely, the **uniqueness** of solutions. By uniqueness, we mean that there should be only one solution of a given equation. This is desirable because if there is more than one solution, then we have to decide which solution is the correct description for the situation being modeled. The possibility of multiple solutions can easily lead to predicting incorrect and nonphysical behavior of a model.[5] In the case of multiple solutions, we have to use additional information about the physical situation to pick out the "physically meaningful" solution. In

---

[5]Fluid mechanics, which is the study of the physics of fluids, is notorious for this problem.

addition, when constructing approximations to a solution, we have to make sure that the correct solution is approximated.

EXAMPLE 23.10. The Dinner Soup model has a unique solution in contrast to the Muddy Yard model, $x^2 = 2$, which has solutions $x = \pm\sqrt{2}$. In other words, mathematically both $x = \sqrt{2}$ and $x = -\sqrt{2}$ are valid solutions of the model equation. As a general rule, nonlinear problems have multiple solutions. In the Muddy Yard model, we simply eliminate $x = -\sqrt{2}$ because it is meaningless in physical terms since a yard cannot have a negative diagonal. If we use the Bisection Algorithm to compute the solution, we have to start with an interval that contains the positive root as opposed to the negative root. The Bisection Algorithm converges to $-\sqrt{2}$ as readily as to $\sqrt{2}$ if given the chance.

The discussion of uniqueness of solutions of differential equations is complicated by the fact that differentiation destroys information about functions. This is simply the observation that if the graph of one function can be obtained by translating the graph of a second function vertically, then the two functions have the same derivative at every point. We illustrate in Fig. 23.3. Therefore when we try to recover a function that satisfies a

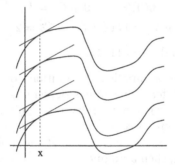

FIGURE 23.3. All of these functions have the same derivative at every point x.

differential equation, we often have to specify some additional information in order to pick out a particular solution.

For example, suppose $y = F(x)$ is a solution of (23.8), $y' = f(x)$; i.e., suppose that $F(x)$ satisfies $F'(x) = f(x)$ in some interval. Then any function of the form

$$y = F(x) + C$$

for a constant C also satisfies the differential equation. If we specify a value of y at one point, then we can determine a specific value of the constant. For example, if we specify that the solution of (23.8) has the value y_0 at the point x_0, then we can determine C by solving

$$y_0 = F(x_0) + C$$

for C.

EXAMPLE 23.11. $y = x^2$ satisfies $y' = 2x$, as does $y = x^2 + 1$, $y = x^2 - 6$, and so on. The only solution of the form $y = x^2 + C$ of the differential equation that satisfies $y(0) = 1$ is $y = x^2 + 1$.

Referring to Fig. 23.3, if we give a value of the solution at one point, we can pick out which of the translated graphs satisfies the differential equation.

EXAMPLE 23.12. Any function of the form $y = x^4/12 + C_1 x + C_2$ for constants C_1 and C_2 satisfies the differential equation

$$y'' = x^2.$$

To determine a unique solution of this form, we can specify the value of y and/or some of its derivatives at one or more points. For example, if we specify that $y(0) = 1$ and $y'(0) = 2$, we get the equations

$$y(0) = 0 + 0 + C_2 = 1$$
$$, y'(0) = 0 + C_1 = 2.$$

or $C_2 = 1$ and $C_1 = 2$. If we specify that $y(0) = 1$ and $y(1) = 2$, we find

$$y(0) = 0 + 0 + C_2 = 1$$
$$y(1) = 1/12 + C_1 + C_2 = 2,$$

which gives $C_2 = 1$ and $C_1 = 11/12$.

The term uniqueness as applied to solutions of differential equations therefore has slightly different meanings depending on the context. If just the differential equation is given, then having a unique solution means that the solution is unique up to some constants that can be determined by specifying information about the solution. If the differential equation and additional data is given, then a unique solution means there can be at most one function that satisfies both the differential equation and the additional data.

Establishing uniqueness can often be difficult. In fact, as with algebraic models, nonlinear differential equations may not have unique solutions at all.

EXAMPLE 23.13. The functions $y(t) = 0$ for all $t \geq 0$ and

$$y(t) = \begin{cases} 0, & 0 \leq t \leq a, \\ \dfrac{(t-a)^2}{4}, & t \geq a, \end{cases}$$

for any $a \geq 0$ all satisfy the differential equation and data

$$\begin{cases} y' = \sqrt{y}, & 0 \leq t, \\ y(0) = 0. \end{cases}$$

In this situation, we need to use additional information from the model to determine the meaningful solution.

However, it is relatively straightforward to establish uniqueness for the simplest differential equation (23.8). We can interpret uniqueness of a solution of (23.8) in two equivalent ways:

- If $y(x)$ satisfies (23.8) for all x in an interval, then any other solution has the form $y(x) + C$ for some constant C.

- There can be at most one function $y(x)$ that satisfies (23.8) and an additional condition $y(x_0) = y_0$ for some point x_0 in the interval of solution and some value y_0.

The theorem we prove is:

Theorem 23.1 *If f is Lipschitz continuous on an interval I, then there is at most one uniformly strongly differentiable solution of (23.8) on I.*[6]

We use the Mean Value Theorem 21.1 to prove Theorem 23.1. Suppose that $y = F(x)$ and $y = G(x)$ are two uniformly strongly differentiable solutions of $y' = f(x)$ in the interval I. This means that

$$F'(x) = G'(x) = f(x) \text{ for all } x \text{ in } I,$$

and we want to prove that there is a constant C such that $F(x) = G(x) + C$ for all x in I.

If we define the function $E(x) = F(x) - G(x)$, then $y = E(x)$ satisfies the differential equation $y' = 0$ for all x in I and moreover is uniformly strongly differentiable on I. We want to show that $E(x)$ is constant in I. Choose two points x_1 and x_2 in $[a, b]$. By the Mean Value Theorem there is a point c between x_1 and x_2 such that

$$E(x_2) - E(x_1) = E'(c)(x_2 - x_1).$$

But $E'(c) = 0$ for any such c, so $E(x_2) = E(x_1)$ for any x_1 and x_2. In other words, E is constant.[7]

EXAMPLE 23.14. The only solution of the differential equation $y' = 2x$ in Example 23.11 is $y = x^2 + C$ for a constant C.

---

[6]By the way, this theorem does *not* say that there is a solution, just that there can be at most one solution.

[7]Basically, this proof says that (23.8) has a unique solution because the only function with a derivative that is zero everywhere is the constant function.

23.6   Solving Galileo's Model of a Free-Falling Object

We encapsulate the ideas in this chapter by solving Galileo's model of a free-falling body (23.4). We begin by rewriting (23.4) as a first order differential equation for the velocity,

$$v' = -g. \tag{23.9}$$

This is simpler because it involves one derivative as opposed to two derivatives.

It turns out that one solution of (23.9) can be found fairly easily. Since linear functions have constant derivatives, we might guess that a solution of (23.9) is a linear function,

$$v = ct + b, \tag{23.10}$$

where c and b are constants. We immediately find by substitution that $v' = c = -g$ and $v(t) = -gt + b$. We cannot determine the value of b using the differential equation (23.9) since differentiating a constant gives zero. To pick out a particular solution, we suppose that the initial speed at time $t = 0$, $v(0) = v_0$, is given. Substituting $t = 0$ and $a = -g$ into (23.10) gives $b = v_0$ and

$$v = -gt + v_0, \tag{23.11}$$

where g and v_0 are constants, is a solution of the differential equation (23.9). Recall that Theorem 23.1 implies that (23.11) is the only solution of the differential equation (23.9) that satisfies the **initial condition** $v(0) = v_0$. Of course, we have to know this to predict how the object falls.

Returning to the model (23.4), it remains to solve the differential equation

$$s' = -gt + v_0, \tag{23.12}$$

for constants g and v_0. Recalling that $(t^2)' = 2t$ and that $(cf(t))' = cf'(t)$, it is natural to guess that one solution of (23.12) is a quadratic function

$$s(t) = dt^2 + et + f$$

for some constants d, e, f. Differentiating and substituting into (23.12) gives

$$2dt + e = -gt + v_0$$

for all t, which means that $d = -g/2$ and $e = v_0$. As in the previous problem, we cannot determine the value of f from the differential equation. However, if we specify an initial height $s(0) = s_0$, we find that $f = s_0$ and

$$s(t) = -\frac{g}{2}t^2 + v_0 t + s_0 \tag{23.13}$$

is a solution of (23.4) with an initial velocity at $t = 0$ of v_0 and an initial height of s_0. Moreover, Theorem 23.1 again implies that this is the unique solution.

We can now determine the position and velocity of the falling object as desired.

EXAMPLE 23.15. If the initial height of the object is 15 m and it is dropped from rest, what is the height at $t = .5$ s? We have

$$s(3) = -\frac{9.8}{2}(.5)^2 + 0 \times .5 + 15 = 13.775\,m.$$

If initially it is thrown upward at 2 m/s, the height at $t = .5$ is

$$s(3) = -\frac{9.8}{2}(.5)^2 + 2 \times .5 + 15 = 14.775\,m.$$

If initially it is thrown downward at 2 m/s, the height at $t = .5$ is

$$s(3) = -\frac{9.8}{2}(.5)^2 - 2 \times .5 + 15 = 12.775\,m.$$

EXAMPLE 23.16. An object starting from rest is dropped and hits the ground at $t = 5$ s. What was its initial height? We have

$$s(5) = 0 = -\frac{9.8}{2}5^2 + 0 \times 5 + s_0,$$

so $s_0 = 122.5$ m.

Chapter 23 Problems

Problems 23.1–23.2 have to do with modeling using Newton's Law of Motion.

23.1. A bicyclist experiences force due to wind resistance that is proportional to the square of the velocity and force due to friction from the wheels that is proportional to the weight of the bicycle and rider. Write down a differential equation modeling the motion of a bicyclist coasting from an initial velocity of v_0.

23.2. Left alone, a cube of wood 1 centimeter on a side floats in a basin of water with one side parallel to the water surface and with 2/3 of the cube submerged. If the cube experiences an upward force equal to the amount of water that it displaces, write down a differential equation modeling the motion of the cube when it is disturbed in a vertical direction.

Problems 23.3–23.5 have to do with classifying differential equations.

23.3. Indicate the order of the following differential equations:

(a) $(y^{(5)} - 2yy^{(2)})^3 = y' + x^2$ (b) $y'' + 45(y')^4 = x/(1+x)$

(c) $yy'y'' = 2$ (d) $(y^{(3)})^5 + (y^{(5)})^3 = y$.

23.4. Indicate whether the following differential equations are linear or nonlinear:

(a) $y^{(5)} - 2xy^{(2)} = y' + x^2$ (b) $y'' + 45(y')^4 = x$

(c) $xy'' + x^2y' + x^3y = 2$ (d) $y' = x(1+y)$

(e) $y^{(4)} + (1+y)y' = x$ (f) $y' = x + x^2 - 2x^3y$.

23.5. Determine whether the following differential equations are separable or nonseparable:

(a) $y' + xy = 4x^2$ (b) $y' = x^2y^3$

(c) $y' + xy = y$ (d) $(1+x)yy' = (2+y)(1-x)$

(e) $(y' + y^{1/3})^3 = xy$ (f) $y' - 1 = y^2$.

Problems 23.6–23.11 have to do with existence and uniqueness of solutions of differential equations.

23.6. Determine whether the indicated functions satisfy the indicated differential equations on some interval.

(a) $y = x^2 - x$ and $y'' = 3$

(b) $y = 1/(x+1)$ and $y''' = -6y^4$

(c) $y = x^2 + 1/x$ and $(yx - 1)y' = 2x^4 - x$

(d) $y = x^4/4 + 4x^2$ and $y'' + y' = 2x^2 - x$

(e) $y = \dfrac{1}{12}x^4 - x^2$ and $\left(\dfrac{d^3y}{dx^3}\right)^2 + 4y = \dfrac{1}{3}x^4$

(f) $y = 6x^3 + 4x$ and $y'' - y' + y = 4x$

(g) $y = \dfrac{1}{x^2+1}$ and $y' = -2xy^2$.

23.7. (a) Verify that any function of the form $y = 2x^2 + C$ with C constant satisfies $y' = 4x$ for all x. (b) Determine the solution that satisfies $y(0) = 1$. (c) Determine the solution that satisfies $y(2) = 3$.

23.8. (a) Verify that any function of the form $y = x^3/3 + C_1 x + C_0$ with C_1 and C_0 constant satisfies $y^{(3)} + y^{(2)} = 2 + 2x$ for all x. (b) Determine the solution that satisfies $y(0) = 1$ and $y'(0) = 2$. (c) Determine the solution that satisfies $y(0) = 3$ and $y(1) = 1$.

23.9. Verify that any function of the form

$$y = \frac{x^5}{5!} + C_4 x^4 + C_3 x^3 + C_2 x^2 + C_1 x + C_0$$

with $C_0, \cdots, C_5$ constant satisfies $y^{(5)} = 1$ for all x.

23.10. Verify the claims in Example 23.13.

23.11. Verify that the functions $y = x^2$ and $y = -x^2$ both satisfy the differential equation $(y')^2 = 4x^2$ and the data $y(0) = 0$.

Problems 23.12–23.15 have to do with Galileo's model of a free-falling object.

23.12. A car drives with constant acceleration of 30 miles/hour^2. How fast is it moving after 2 hours starting from a dead stop?

23.13. An object is thrown up in the air at 2.5 meters/second from a height of 120 meters. (a) How high is the object after 1 second? (b) When does the object hit the ground?

23.14. An object is thrown downward from a height of 95 meters and hits the ground in 4 seconds. How fast was it thrown down?

23.15. A ball is thrown upward from the ground at 20 meters/second. What is its maximum height?

Problem 23.16 is a relatively difficult modeling problem which is realistic in the sense that it begins with data measured in a laboratory experiment.

23.16. The goal of this problem is to devise a differential equation that describes how much of a certain drug present in the bloodstream has been absorbed into the body as time passes. A quantity of the drug is injected into the bloodstream of a rabbit and then the concentration of the drug measured in micrograms per milliliter ($\mu g/mL$) left after that is determined from blood samples taken periodically. This gives the following results:

Time (sec)	Concentration of Drug ($\mu g/mL$)
3.00	1.639
3.05	1.613
3.10	1.587
3.15	1.563
3.20	1.538
3.25	1.515
3.30	1.493
3.35	1.471
3.40	1.449
3.45	1.429
3.50	1.408

If the drug is infused intravenously at a steady rate of r $\mu g/mL$/sec, write down a differential equation modeling the amount of drug in the bloodstream. *Hint:* Use the data above to figure out a modeling assumption about how quickly the drug is absorbed into the body. To do this, compute average rates of change at the various times. Assuming the rate of change due to absorption is proportional to some power of the amount in the blood, use logarithms and a least squares line fit to determine the power and proportionality constant from the average rates of change.

24

Antidifferentiation

In this chapter, we consider the solution of some first order, linear, separable equations

$$y'(x) = f(x) \tag{24.1}$$

for which we can find solutions explicitly, i.e., as a formula involving known functions. We call the process of guessing an explicit solution of (24.1) **antidifferentiation** and a solution y is called an **antiderivative** of f.

The solution of differential equations shares many characteristics with the solution of algebraic models. Recall that there are two kinds of algebraic models. The first kind, like the Dinner Soup model (1.1), have rational solutions computable using a finite amount of arithmetic. In contrast, the second kind, like the Muddy Yard model, have irrational solutions that can only be approximated using an iterative algorithm. The fact that most algebraic models fall into the second category motivated much of the subsequent material on sequences, convergence, real numbers, fixed point iterations, et cetera, as we strove to develop systematic methods for approximating solutions.

So it goes with differential equations. There are a few problems for which we can determine solutions explicitly by the simple technique of guessing the correct answer. We can actually make guessing into a fairly refined and systematic tool and in this chapter we develop some ideas that help out. Unfortunately, there are relatively few differential equations for which we can guess at explicit solutions. So for the vast majority of differential equations arising in engineering and science, we have to resort to some constructive algorithm for approximating solutions. Starting with Chapter 25, we concentrate on developing constructive techniques for approximating solutions.

24.1 Antidifferentiation

We now develop the general method of guessing the solution of (24.1). The idea is to compute antiderivatives for some basic simple problems and then to develop ways to use this "library" of solutions to solve more complicated problems. We obtain the basic antiderivatives simply by taking some function F and differentiating to get $f = F'$. Because the antiderivative of a given function is unique, we conclude any antiderivative of f can be written $F + C$ for some constant C.

EXAMPLE 24.1. First,
$$\frac{d}{dx}(x^2) = 2x$$
means that any antiderivative of $y' = 2x$ has the form $y = x^2 + C$ for some constant C. Second,
$$\frac{d}{dx}(x^{-1}) = -x^{-2}$$
means that any antiderivative of $y' = -x^{-2}$ has the form $y = x^{-1} + C$ for some constant C.

In fact, this argument leads immediately to the general rule. Since
$$\frac{d}{dx}\frac{x^{m+1}}{m+1} = x^m, \qquad m \neq -1,$$
any antiderivative of $y' = x^m$ is $y = x^{m+1}/(m+1) + C$ for $m \neq -1$.

24.2 The Indefinite Integral

At this point, we need to get a good notation for the antiderivative of a given function. We use the notation invented by Leibniz. The reason for the notation becomes clearer when we study constructive methods for computing antiderivatives.

Given a function f, we use
$$\int f(x)\, dx$$
to denote *all* antiderivatives of f. We also call $\int f(x)\, dx$ the **indefinite integral**, or **integral**, of f. The process of computing an antiderivative of a function f is called integrating f, or simply **integration**. If f has an antiderivative, we say that f is **integrable**.

EXAMPLE 24.2. By Example 24.1, we conclude that

$$\int 2x\, dx = x^2 + C$$

$$\int -x^{-2}\, dx = x^{-1} + C$$

for some constants C.

Extrapolating from these examples, we also get the general rule

$$\int x^m\, dx = \frac{x^{m+1}}{m+1} + C, \quad m \neq -1, \tag{24.2}$$

for some constant C. This **power rule** is the first entry into the library of integration formulas that we carry around in our heads.

Since the antiderivative of y' is $y + C$ for any function C, we get the nice formula

$$\int y'(x)\, dx = \int \frac{dy(x)}{dx}\, dx = y(x) + C. \tag{24.3}$$

This is the motivation for the name antiderivative.

In the integral of f, $\int$ is called the **integral sign** and x is called the **integration variable**. The **integrand** is f. Note that the integration variable is a dummy variable in the sense that using a different name just corresponds to renaming the independent variable.

EXAMPLE 24.3.

$$\int 2z\, dz = z^2 + C$$

$$\int -r^{-2}\, dr = r^{-1} + C$$

for some constants C.

It does not change the antiderivative.

24.3 Sophisticated Guesswork

In the rest of this chapter, we show how to leverage a few known integration formulas like (24.2) into a technique for computing integrals of more complicated functions. To do this, we derive properties of the integral based on properties of the derivative.

To begin with, if f and g are differentiable and c is a constant, then $D(cf) = cDf$ and $D(f + g) = Df + Dg$. We conclude the following:

Theorem 24.1 Linearity of Integration *If f and g are integrable functions on a common interval and c_1 and c_2 are constant, then*

$$\int (c_1 f + c_2 g)(x)\, dx = \int (c_1 f(x) + c_2 g(x))\, dx$$

$$= c_1 \int f(x)\, dx + c_2 \int g(x)\, dx. \quad (24.4)$$

This is very useful for computing some complicated integrals.

EXAMPLE 24.4.

$$\int 5s^{10}\, ds = 5 \int s^{10}\, ds = \frac{5}{11} s^{11} + C$$

$$\int (x^2 - 8x)\, dx = \int x^2\, dx - 8 \int x\, dx = \frac{x^3}{3} - 4x^2 + C$$

$$\int \left(t - \frac{1}{t^2}\right) dt = \int t\, dt - \int \frac{1}{t^{-2}}\, dt = \frac{t^2}{2} + \frac{1}{t} + C.$$

We can always check the answer of course:

$$\frac{d}{dx}\left(\frac{x^3}{3} - 4x^2 + C\right) = x^2 - 8x.$$

Note there is some ambiguity about the constant C. It is natural to think we need two or more constants in some of these examples. For example, when we compute the integrals

$$\int (x^2 - 8x)\, dx = \int x^2\, dx - 8 \int x\, dx,$$

we apparently get

$$\frac{x^3}{3} + C_1 - 4x^2 + C_2$$

for some constants C_1 and C_2. But we just sum these constants to get $C = C_1 + C_2$. Any time there is a sum of constants arising from different integrals, we simply rename the sum as a new constant C.

EXAMPLE 24.5. We can even do an abstract example. If $a_0, \cdots, a_n$ are constants, then

$$\int (a_0 + a_1 x + a_2 x^2 + \cdots a_n x^n)\, dx$$

$$= a_0 x + \frac{a_1}{2} x^2 + \frac{a_{n+1}}{n+1} x^n + C,$$

or using the Σ notation,

$$\int \left(\sum_{i=0}^{n} a_i x^i\right) dx = \sum_{i=0}^{n} \frac{a_i}{i+1} x^{i+1} + C.$$

In other words, combining the power rule and the linearity property (24.4) allows us to integrate any polynomial.

24.4 The Method of Substitution

We saw how the linearity properties of the derivative allow more complicated integrals to be computed. In this section, we show how to use the Chain Rule for the derivative to compute integrals. Recall that the Chain rule says that if g and u are differentiable functions,

$$\frac{d}{dx}g(u(x)) = g'(u(x))u'(x).$$

We conclude immediately from (24.3) that

$$\int g'(u(x))u'(x)\,dx = g(u(x)) + C \tag{24.5}$$

for a constant C. This is called the **method of substitution**.

EXAMPLE 24.6. Consider

$$\int (x^2 + 1)^{10}\, 2x\,dx.$$

We set $g'(u) = u^{10}$ and $u(x) = x^2 + 1$; so $u'(x) = 2x$ and the integral has precisely the form

$$\int (x^2 + 1)^{10}\, 2x\,dx = \int g'(u(x))u'(x)\,dx.$$

We know that

$$g'(u) = u^{10} \rightarrow g(u) = \frac{u^{11}}{11} + C;$$

so from (24.5) we conclude that

$$\int (x^2 + 1)^{10}\,dx = g(u) + C = \frac{u^{11}}{11} + C = \frac{(x^2 + 1)^{11}}{11} + C.$$

We can check,

$$\frac{d}{dx}\left(\frac{(x^2 + 1)^{11}}{11} + C\right) = (x^2 + 1)^{10} \times 2x.$$

Note that typically we can compute an integral in several different ways. In this case, we could multiply out $(x^2+1)^{10}$ and use the tricks from the previous section. That would take a lot more time and effort, however.

The method of substitution, or substitution for short, is the most powerful tool for computing integrals. Using it effectively, however, requires a lot of practice so that patterns can be recognized easily. In general, the idea is to choose a function g' that we know how to integrate.

EXAMPLE 24.7. Consider

$$\int \frac{3s^2 + 1}{(s^3 + s)^2}\, ds.$$

We can choose

$$g'(u) = \frac{1}{u^2} \rightarrow g(u) = \int g'(u)\, du = \frac{-1}{u} + C$$
$$u(s) = s^3 + s \rightarrow u'(s) = 3s^2 + 1$$

so the integral has the form required for (24.5), i.e.,

$$\int \frac{3s^2 + 1}{(s^3 + s)^2}\, ds = \int \frac{1}{(s^3 + s)^2}\, (3s^2 + 1)\, ds = \int g'(u(s))u'(s)\, ds.$$

We conclude that

$$\int \frac{3s^2 + 1}{(s^3 + s)^2}\, ds = g(u) + C = \frac{-1}{u} + C = \frac{-1}{x^3 + x} + C.$$

Note that we can combine substitution with the linearity properties of the integral.

EXAMPLE 24.8. Consider

$$\int \frac{\left(x^{-2} - 4\right)^4}{x^3}\, dx.$$

To use substitution, it is tempting to choose

$$g'(u) = u^4 \rightarrow g(u) = \int g'(u)\, du = \frac{u^5}{5} + C$$
$$u(x) = x^{-2} - 4 \rightarrow u'(x) = -2x^{-3}.$$

The problem is that we don't quite have u' in the integral

$$\int \frac{\left(x^{-2} - 4\right)^4}{x^3}\, dx = \int \left(x^{-2} - 4\right)^4 x^{-3}\, dx$$

because we are missing a factor of -2. However, $\int cf(x)\, dx = c\int f(x)\, dx$ for any *constant* c. So we can write

$$\int \left(x^{-2} - 4\right)^4 x^{-3}\, dx = \frac{-1}{2} \times -2 \times \int \left(x^{-2} - 4\right)^4 x^{-3}\, dx$$
$$= \frac{-1}{2} \int \left(x^{-2} - 4\right)^4 \times -2x^{-3}\, dx$$
$$= \frac{-1}{2} \int g'(u(x))u'(x)\, dx$$

and we conclude that

$$\int \frac{\left(x^{-2}-4\right)^4}{x^3}\,dx = \frac{-1}{2} \times \frac{\left(x^{-2}-4\right)^5}{5} + C = \frac{-1}{10}\left(x^{-2}-4\right)^5 + C.$$

Remember, we can factor constants "through" an integral sign but we cannot factor functions through an integral sign.

24.5   The Language of Differentials

Using substitution as above is a little awkward because the notation is cumbersome. We can improve the notation to make it easier using the language of differentials, which is due to Leibniz. If u is a differentiable function, we define the **differential** du of u to be

$$du = u'\,dx.$$

The differential of the function x is nothing more than dx of course.

EXAMPLE 24.9. If $u(x) = (x^4 - x^3 + 3)^9$, then

$$du = 9(x^4 - x^3 + 3)^8\left(4x^3 - 3x^2\right)dx.$$

Also

$$d(4 - x^3)^2 = 2(4 - x^3) \times -3x^2\,dx.$$

It is tempting to think that we get this notation by multiplying both sides of

$$\frac{du}{dx} = u'$$

by dx and then "canceling" the dx on the left. But we cannot actually do this of course; the dx in the denominator of the derivative is not part of a fraction. It only indicates the variable by which we differentiate. Nonetheless, it is useful to think of differentials as quantities that can be manipulated using simple arithmetic.

We define the arithmetic operations involving differentials to be consistent with the properties of the derivatives. For example, if u and v are differentiable, then $(u + v)' = u' + v'$. Hence, we define

$$d(u + v) = du + dv.$$

Likewise, if c is a constant, then

$$d(cu) = c\,du.$$

With these properties, the Product Rule implies that

$$d(uv) = (uv)'dx = (uv' + vu')\,dx = u\,dv + v\,du.$$

In the language of differentials, the Chain Rule applied to $g \circ u$, where g and u are differentiable, reads

$$dg \circ u = g'(u)u' \, dx = g'(u) \, du.$$

The last equation is connected to substitution.

EXAMPLE 24.10. Consider again

$$\int \frac{3s^2 + 1}{(s^3 + s)^2} \, ds.$$

Recall that we chose

$$g'(u) = \frac{1}{u^2} \text{ and } u(s) = s^3 + s \to du = (3s^2 + 1) \, ds.$$

Using differentials, we get

$$\int \frac{3s^2 + 1}{(s^3 + s)^2} \, ds = \int \frac{1}{u^2} \, du = \frac{-1}{u} + C = \frac{-1}{x^3 + x} + C.$$

EXAMPLE 24.11. Consider again

$$\int \frac{\left(x^{-2} - 4\right)^4}{x^3} \, dx.$$

We choose

$$g'(u) = u^4 \text{ and } u(x) = x^{-2} - 4.$$

Since $du = -2x^{-3} \, dx$, we write

$$\int \left(x^{-2} - 4\right)^4 x^{-3} \, dx = \frac{-1}{2} \int \left(x^{-2} - 4\right)^4 \times -2x^{-3} \, dx$$

$$= \frac{-1}{2} \int u^4 \, du = \frac{-1}{2} \times \frac{u^5}{5} + C$$

$$= \frac{-1}{10} (x^{-2} - 4)^5 + C.$$

Note that the differential notation makes it clear that we can change the name of the integration variable at will without affecting the results. In other words,

$$\int g(u) \, du = \int g(s) \, ds = \int g(x) \, dx,$$

and so on. For this reason, we call the integration variable a dummy variable.

We can do quite complicated integrals using the method of substitution.

EXAMPLE 24.12. Consider

$$\int (2x^7 - 4x^3)((x^4 - 2)^2 + 3)^7 \, dx.$$

We set

$$g'(u) = u^7 \text{ and } u = (x^4 - 2)^2 + 3$$

and get

$$\int (2x^7 - 4x^3)((x^4 - 2)^2 + 3)^7 \, dx$$

$$= \frac{1}{4} \int ((x^4 - 2)^2 + 3)^7 \times 2(x^4 - 2) \times 4x^3 \, dx$$

$$= \frac{1}{4} \int u^7 \, du.$$

Therefore,

$$\int (2x^7 - 4x^3)((x^4 - 2)^2 + 3)^7 \, dx = \frac{((x^4 - 2)^2 + 3)^8}{32} + C.$$

24.6   The Method of Integration by Parts

The last method that we study is based on the Product Rule. If u and v are differentiable functions, then

$$(uv)' = uv' + u'v,$$

which immediately gives

$$\int u(x)v'(x) \, dx + \int u'(x)v(x) \, dx = \int (u(x)v(x))' \, dx = u(x)v(x).$$

This is usually rewritten as the **integration by parts** formula

$$\int u(x)v'(x) \, dx = u(x)v(x) - \int u'(x)v(x) \, dx. \qquad (24.6)$$

Using differentials, we get

$$\int u \, dv = uv - \int v \, du. \qquad (24.7)$$

EXAMPLE 24.13. We compute the integral

$$\int (x^2 + 4)^7 x^3 \, dx.$$

Trying substitution directly leads nowhere. If we set $u = x^2 + 4$, for example, then we have $du = 2x\,dx$. But there is an x^3, not x, factor in the integrand and we cannot factor functions through the integral sign.

However, this abortive attempt does show the way to using integration by parts. Recognizing that we can integrate $(x^2 + 4)^7 x$, we write the integral as

$$\int x^2 \left(x^2 + 4\right)^7 x\,dx.$$

Now we choose

$$u = x^2 \qquad dv = (x^2 + 4)^7 x\,dx$$

$$du = 2x\,dx \qquad v = \frac{1}{2}\frac{(x^2 + 4)^8}{8}$$

Note that we do not include a constant in the antiderivative of dv, since the constant is added after the last integral in the integration by parts formula. Now (24.7) implies that

$$\int (x^2 + 4)^7 x^3\,dx = uv - \int v\,du$$

$$= x^2 \times \frac{1}{2}\frac{(x^2 + 4)^8}{8} - \int \frac{1}{2}\frac{(x^2 + 4)^8}{8} 2x\,dx$$

$$= \frac{1}{16}x^2(x^2 + 4)^8 - \frac{1}{8}\int (x^2 + 4)^8 x\,dx.$$

We do the last integral using substitution with $u = x^2 + 8$ and get

$$\int (x^2 + 4)^7 x^3\,dx = \frac{1}{16}x^2(x^2 + 4)^8 - \frac{1}{144}(x^2 + 4)^9 + C,$$

where $144 = 8 \times 2 \times 9$.

24.7 Definite Integrals

Since the integral of f represents all of f's antiderivatives, we also call it the **general solution** of the differential equation (24.1). But sometimes we want to compute the antiderivative of a given function f that not only satisfies (24.1) but in addition takes on a specific value at a specified point. A solution of the differential equation (24.1) that takes on a specified value at a specified point is called a **particular solution** of the differential equation.

Recall that we solve

$$\begin{cases} y'(x) = f(x), \\ y(a) = y_a, \end{cases}$$

where y_a is the value we specify for y at the point a, by first finding any antiderivative of f, say, F so $F' = f$, and then using the fact that any other antiderivative of f, including the one we want, can be written as $y = F + C$ for some constant C. Plugging in, we solve

$$y(a) = y_a = F(a) + C$$

for C and get $y = F + (y_a - F(a))$.

EXAMPLE 24.14. We solve

$$\begin{cases} y' = 2x, \\ y(0) = 4, \end{cases}$$

by first noting that $y = \int 2x\, dx = x^2 + C$ is the antiderivative and then solving $4 = 0^2 + C$ for C to get $y = x^2 + 4$.

Note that solving for the value of C is particularly easy when the antiderivative F that we compute has the property $F(a) = 0$. Then $C + F(a) = C = y_a$ and

$$y = F + y_a.$$

We modify the integral notation to indicate this particular antiderivative. The **definite integral**

$$\int_a^x f(s)\, ds$$

denotes the antiderivative of f that has the value zero at $x = a$. That is,

$$F(x) = \int_a^x f(s)\, ds \text{ if and only if } F'(x) = f(x) \text{ and } F(a) = 0.$$

This notation is due to Fourier[1].

EXAMPLE 24.15.

$$\int_0^x 2s\, ds = x^2$$

$$\int_1^x 2s\, ds = x^2 - 1$$

$$\int_2^x 2s\, ds = x^2 - 4$$

---

[1]The French mathematician Jean Baptiste Joseph Fourier (1768–1830) had an interesting and varied career. Not only was he considered a leading mathematician of his generation, but he was also a political administrator who was highly valued by Napoleon. Fourier is best known for his theory of heat, which used the trigonometric series now called the Fourier series. However, Fourier's analysis was sometimes not completely rigorous and so was controversial. Yet, his work was an important step in the eventual rigorous treatment of functions and infinite series.

If F is any antiderivative of f, then $F(x) - F(a)$ is the antiderivative of f that is zero at a. Hence, we conclude

$$F(x) - F(a) = \int_a^x f(s)\, ds$$

when F is an antiderivative of f. In particular,

$$y(x) - y(a) = \int_a^x y'(x)\, dx. \tag{24.8}$$

EXAMPLE 24.16. We compute the height of a particle that falls from an initial height of 37 m with initial velocity 0 m/s at time t s. In these variables, (24.8) becomes

$$s(t) - s(0) = \int_0^t s'(r)\, dr,$$

so by (23.12)

$$s(t) - s(0) = \int_0^t (-gr)\, dr = -g\frac{t^2}{2},$$

or

$$s(t) = 37 - 4.9t^2.$$

The subscript on the integral sign, here a, is called the **lower limit** of the integral and denotes the point where the antiderivative is zero. The superscript on the integral sign, here x, is called the **upper limit** of the integral and denotes the independent variable used for the antiderivative. Sometimes we actually need the value of the antiderivative only at one point, and will substitute that point in for x.

EXAMPLE 24.17.

$$\int_{-1}^3 x^3\, dx = \frac{3^4}{4} - \frac{(-1)^4}{4} = 20.$$

When we use substitution to compute a definite integral, we have to change the limits as well.

EXAMPLE 24.18. To compute

$$\int_1^x (2s^3 + 6s)^4 (s^2 + 1)\, ds$$

we use the substitution

$$u = 2s^3 + 6s \rightarrow du = (6s^2 + 6)\, ds = 6(s^2 + 1)\, ds.$$

So we can compute the integral, but the limits for the integration variable s are not the same as the integration limits for the variable u. So we have to change the limits as well. But this is straightforward. Since $u = 2s^3 + 6s$, when $s = 1$, $u = 2 + 6 = 8$ and when $s = x$, $u = 2x^3 + 6x$. So

$$\int_1^x (2s^3 + 6s)^4 \, (s^2 + 1) \, ds = \frac{1}{6} \int_8^{2x^3 + 6x} u^4 \, du$$

$$= \frac{1}{6} \left(\frac{(2x^3 + 6x)^5}{5} - \frac{8^5}{5} \right).$$

Chapter 24 Problems

24.1. Compute the following antiderivatives:

(a) $\int x^3 \, dx$

(b) $\int \frac{1}{s^3} \, ds$

(c) $\int t^{1256} \, dt$

(d) $\int 7u^5 \, du$

(e) $\int (2r^{-99} - 4r^9) \, dr$

(f) $\int \left(\frac{74}{5x^{23}} + \frac{3}{x^6} + x^{61} \right) dx$

(g) $\int (x - 5)(x^3 - 2x^2) \, dx$

(f) $\int \left(\frac{4}{x^2} - x \right) \frac{1}{x^3} \, dx$.

24.2. Given $u(x) = 2x^4 - x^2$ and $v(x) = x^{-3}$, compute

(a) du (b) dv (c) $d(uv)$ (d) $d(u(v(x)))$.

24.3. Compute the following integrals:

(a) $\int (x^3 - 2)^8 \, 3x^2 \, dx$

(b) $\int \frac{4s^3 + 2}{(s^4 + 2s)^3} \, ds$

(c) $\int (6r^2 + 1)^8 \, r \, dr$

(d) $\int \frac{1}{u^9} \left(\frac{1}{u^8} + 4 \right)^{13} \, du$

(e) $\int (t - t^{-3})(t^2 + t^{-2})^{11} \, dt$

(f) $\int \frac{7x^6 + x^2}{(3x^7 + x^3)^{14}} \, dx$

(g) $\int \left((4x - 3)^8 - 6 \right)^9 (4x - 3)^7 \, dx$

(f) $\int \frac{1}{x^2} \frac{\left(\frac{1}{x} + 4 \right)^2}{\left(\left(\frac{1}{x} + 4 \right)^3 + 92 \right)^5} \, dx$.

24.4. Compute the following integrals:

(a) $\int (3 - 2x)^{23} \, x \, dx$ (b) $\int (x^3 + 4)^{19} \, x^5 \, dx$ (c) $\int \frac{x^3}{(3x^2 + 1)^3} \, dx$.

24.5. Compute the following definite integrals:

(a) $\int_0^x s^3 \, ds$ (b) $\int_2^x s^3 \, ds$ (c) $\int_{-1}^x s^3 \, ds$.

24.6. Compute the following definite integrals:

(a) $\int_3^x \frac{u^2}{(u^3 + 1)^8} \, du$ (b) $\int_4^7 (3t^2 - 1)(3t^3 - 3t)^4 \, dt$

(c) $\int_0^t (2 + x)^{81} \, dx$ (d) $\int_x^1 s^3 \, ds$.

25
Integration

The technique of guessing the solution of a first order, linear, separable equation (24.1),

$$y'(x) = f(x),$$

discussed in Chapter 24, works only for a limited number of examples. Most of the differential equations that arise in engineering and science have solutions that are so complicated, they defy any attempt to express them as combinations of known functions.

EXAMPLE 25.1. A simple example is

$$\int \frac{dm}{m},$$

which arises in modeling the motion of a rocket. Recalling the differentiation formulas from Chapter 20, we do not yet know a function whose derivative is x^{-1}. It turns out that this requires a new function called the logarithm. This particular problem is discussed in Chapter 28.

When we encountered algebraic models that could not be solved with simple arithmetic, we devised approximation methods like the Bisection Algorithm and the Fixed Point Iteration to approximate the solution using the computer. For the same reason, we devise a method for computing approximations of the solution of (24.1). This method is in fact very general and is used to solve a nonlinear differential equation in Chapter 41.

The approximation method yields values of a particular solution at particular points. To use the method, we therefore have to specify a particular solution to approximate. For arbitrary reasons, we choose to approximate the solution of

$$\begin{cases} y'(x) = f(x), & x_0 < x, \\ y(x_0) = 0. \end{cases} \tag{25.1}$$

From any approximate value of the solution of (25.1) at a point x, the corresponding approximate value of the solution of

$$\begin{cases} y'(x) = f(x), & x_0 < x, \\ y(x_0) = y_0, \end{cases} \tag{25.2}$$

at x can be obtained by simply adding y_0 to the approximate value of the solution of (25.1). Problem (25.2) is called an **initial value problem** for y.

EXAMPLE 25.2. To get the solution of

$$\begin{cases} y' = 12x^3 - 4x, \\ y(1) = 0, \end{cases}$$

we first compute the antiderivative

$$y = \int (12x^3 - 4x)\, dx = 3x^4 - 2x^2 + C,$$

and then solve

$$y(1) = 0 = 3 - 2 + C \to C = -1$$

to get $y(x) = 3x^4 - 2x^2 - 1$.

The solution of

$$\begin{cases} y' = 12x^3 - 4x, \\ y(1) = 4, \end{cases}$$

is simply

$$y = 3x^4 - 2x^2 - 1 + 4 = 3x^4 - 2x^2 + 3.$$

Note that we do not know yet if the solution of (25.1) exists, only that it is unique, if it does exist. So along with approximating values of the solution, we also have to show that it does indeed exist. We faced the same issue when solving root and fixed point problems.

25.1   A Simple Case

Before writing down the approximation procedure for (25.1), we first consider a case where we know there is a solution. Namely, suppose that f is constant. The solution of (25.1) is then

$$y = f \times (x - x_0).$$

It is interesting to interpret the solution with a graph, as in Fig. 25.1. The

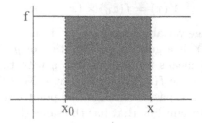

FIGURE 25.1. The solution y of (25.1) when f is constant gives the area underneath f from x_0 to x.

figure indicates that $y(x)$ gives the area underneath the graph of f from x_0 to x.[1]

25.2   A First Attempt at Approximation

The idea is to replace the general f, for which we do not know if there is an antiderivative, by an approximation of f for which an antiderivative can be computed. For example, we can try to replace f by a constant that is close to f. We might choose the value $f(x_0)$ of f at x_0, for instance (see Fig. 25.2). This is called the constant **interpolant** of f that interpolates

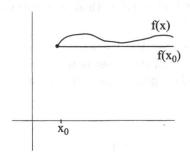

FIGURE 25.2. The constant approximation of f that **interpolates** f at x_0.

f at x_0.[2] We can certainly solve the problem

$$\begin{cases} Y' = f(x_0), \\ Y(x_0) = 0, \end{cases}$$

---

[1]This observation is the basis for several interesting applications of integration discussed in Chapter 27.

[2]Interpolation is discussed in great detail in Chapter 38.

since the solution is simply

$$Y(x) = f(x_0) \times (x - x_0).$$

Note that we change variables from y to Y to mark that Y is not y. The question is whether Y is a good approximation of y. We understand that solving $y'(x) = f(x)$ means finding a curve y with the property that the linearization of y has slope $f(x)$ at every point x. Now Y is a linear function that has slope $f(x_0)$ and passes through the point $(x_0, y(x_0)) = (x_0, 0)$. In other words, Y is a straight line that has the same slope as the linearization of y (if it exists) at x_0 and agrees with y at x_0. But this simply means that Y *is* the linearization of y at x_0 if it exists (see Fig. 25.3).

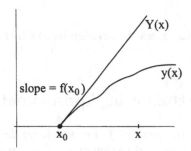

FIGURE 25.3. Y is the linearization of y at x_0, if it exists.

EXAMPLE 25.3. For $f(x) = 3x^2$ with $x_0 = 1$, we compute $y(x) = x^3 - 1$ while $Y(x) = 3(x - 1)$.

We therefore expect that $Y(x)$ is a good approximation to $y(x)$, if it exists, for x close to x_0. On the other hand, as illustrated in Fig. 25.3, we do not expect $Y(x)$ to be a good approximation of $y(x)$ for x far away from x_0.

25.3   Approximating the Solution on a Large Interval

We have constructed a reasonable way to approximate the unknown function y close to x_0. But we do not expect that the approximation $Y(x)$ is accurate for x in a relatively large interval. To overcome this difficulty, we divide up a large interval $[a, b]$ into a number of small pieces and then use the approximation properties of the linearization on those small pieces.

Suppose that we want to solve (25.1) for x in an interval $[a, b]$. We create a **mesh** of equally spaced points $\{x_{N,i}\}$ in $[a, b]$ by setting

$$\Delta x_N = (b - a)/2^N \text{ for } N \text{ in } \mathbb{N}$$

and

$$x_{N,i} = a + i \times \Delta x_N, \quad i = 0, 1, \cdots, 2^N$$

(see Fig. 25.4). Note in particular that $x_{N,0} = a$, $x_{N,2^N} = b$, and given the mesh size Δx_N, we can determine N. The reason we use 2^N to define the

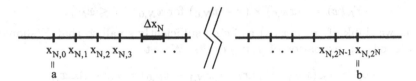

FIGURE 25.4. A mesh for $[a, b]$.

number of points in the mesh is the fact that if $M > N$ are two natural numbers, then each node in the mesh corresponding to 2^N is automatically a node in the mesh corresponding to 2^M (see Fig. 25.5). Such meshes

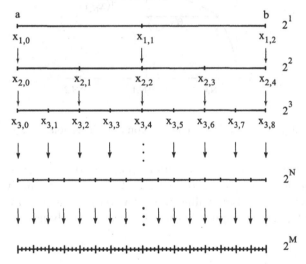

FIGURE 25.5. By construction, the mesh for 2^N is nested in the mesh for 2^M when $M \geq N$.

are called **nested**. Using nested meshes makes it much easier to compare approximations corresponding to different meshes, which we have to do below.[3]

We construct the approximation Y_N on the mesh with $2^N + 1$ points "interval-by-interval." First on $[x_{N,0}, x_{N,1}]$, we replace $f(x)$ by the constant

---

[3]General meshes are considered in Chapter 34.

interpolant $f(x_{N,0})$ (see Fig. 25.6) and solve

$$\begin{cases} Y_N' = f(x_{N,0}), & x_{N,0} \le x \le x_{N,1}, \\ Y_N(x_{N,0}) = 0, \end{cases}$$

to get

$$Y_N(x) = f(x_{N,0}) \times (x - x_{N,0}) \text{ for } x_{N,0} \le x \le x_{N,1}.$$

Here, we think of $[x_{N,0}, x_{N,1}]$ as being sufficiently small so that Y should be a good approximation of y, if y exists. We set

$$Y_{N,1} = Y_N(x_{N,1}) = f(x_{N,0})(x_{N,1} - x_{N,0}) = f(x_{N,0})\Delta x_N$$

to be the "nodal" value of Y_N at the node $x_{N,1}$.

On the next interval $[x_{N,1}, x_{N,2}]$, we approximate f by the constant interpolant $f(x_{N,1})$ (see Fig. 25.6). Ideally, we would solve $Y' = f(x_{N,1})$

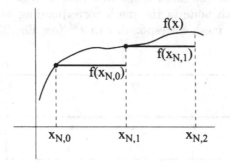

FIGURE 25.6. The piecewise constant interpolant of f.

with initial value $y(x_{N,1})$. This would make Y_N the linearization of y at $x_{N,1}$, if we knew that y existed (see Fig. 25.7). Unfortunately, we would have to know y to get that value. Since the only value we have at $x_{N,1}$ is $Y_N(x_{N,1}) = Y_{N,1}$, we compute Y on $[x_{N,1}, x_{N,2}]$ by solving

$$\begin{cases} Y_N' = f(x_{N,1}), & x_{N,1} \le x \le x_{N,2}, \\ Y_N(x_{N,1}) = Y_{N,1}, \end{cases}$$

to get

$$Y_N(x) = Y_{N,1} + f(x_{N,1}) \times (x - x_{N,1}) \text{ for } x_{N,1} \le x \le x_{N,2}.$$

We show Y_N in Fig. 25.7. The plot indicates that Y_N is a continuous function that is linear on each of the intervals $[x_{N,0}, x_{N,1}]$ and $[x_{N,1}, x_{N,2}]$; i.e., Y_N is piecewise linear. We define the next nodal value to be

$$Y_{N,2} = Y_N(x_{N,2}) = Y_{N,1} + f(x_{N,1})(x_{N,2} - x_{N,1})$$
$$= f(x_{N,0})\Delta x_N + f(x_{N,1})\Delta x_N.$$

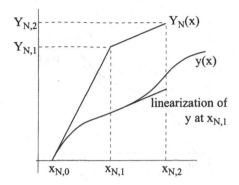

FIGURE 25.7. The computation of Y_N on the intervals $[x_{N,0}, x_{N,1}]$ and $[x_{N,1}, x_{N,2}]$. Note that Y_N is parallel to the linearization of y at x_1 on $[x_{N,1}, x_{N,2}]$, if y exists.

Note that the difference, or error, between Y and y on the second interval is due not only to the fact that Y is linear but also to the fact that we started with the "wrong" initial value $Y_{N,1}$.

Now we proceed interval by interval. Given the nodal value $Y_{N,n-1}$, we solve

$$\begin{cases} Y_N' = f(x_{N,n-1}), & x_{N,n-1} \leq x \leq x_{N,n}, \\ Y_N(x_{N,n-1}) = Y_{N,n-1}, \end{cases} \tag{25.3}$$

to get

$$Y_N(x) = Y_{N,n-1} + f(x_{N,n-1}) \times (x - x_{N,n-1})$$

for $x_{N,n-1} \leq x \leq x_{N,n}$. In computing Y_N this way, we have replaced the function f by a **piecewise constant** interpolant, as in Fig. 25.8 This

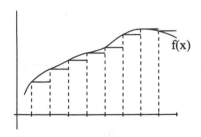

FIGURE 25.8. The piecewise constant interpolant of f used to compute Y_N.

defines a continuous, piecewise linear function $Y_N(x)$ like that shown in Fig. 25.9. Working backward, the value of $Y_N(x)$ for x in $[x_{n-1}, x_n]$ is

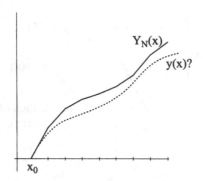

FIGURE 25.9. The continuous, piecewise linear function Y_N.

$$Y_N(x) = Y_{N,n-1} + f(x_{N,n-1})(x - x_{N,n-1})$$
$$= Y_{N,n-2} + f(x_{N,n-2})\Delta x_N + f(x_{N,n-1})(x - x_{N,n-1})$$
$$= Y_{N,n-3} + f(x_{N,n-3})\Delta x_N + f(x_{N,n-2})\Delta x_N$$
$$+ f(x_{N,n-1})(x - x_{N,n-1}),$$

Using induction, we conclude that for $x_{N,n-1} \le x < x_{N,n}$,

$$Y_N(x) = \sum_{i=1}^{n-1} f(x_{N,i-1})\Delta x_N + f(x_{N,n-1})(x - x_{N,n-1}), \qquad (25.4)$$

Likewise, the nodal value of Y_N at $x_{N,n}$ is

$$Y_{N,n} = Y_N(x_{N,n}) = \sum_{i=1}^{n} f(x_{N,i-1})\Delta x_N. \qquad (25.5)$$

EXAMPLE 25.4. For $f(x) = x$ on $[0,1]$, we have $\Delta x_N = 1/2^N$, $x_{N,i} = i/2^N$, and on $[x_{N,i-1}, x_{N,i})$, $f(x)$ is replaced by

$$f(x_{N,i-1}) = \frac{i-1}{2^N}.$$

This means that

$$Y_N(x_{N,i}) = Y_N(x_{N,i-1}) + \frac{i-1}{2^N} \times \frac{1}{2^N}$$

and by induction

$$Y_N(x_{N,n}) = \sum_{i=1}^{n} \frac{i-1}{2^{2N}} = \frac{1}{2^{2N}} \frac{n(n-1)}{2}.$$

In particular,

$$Y_N(1) = Y_N(x_{N,2^N}) = \frac{1}{2^{2N}} \frac{2^N(2^N-1)}{2} = \frac{1}{2} - \frac{1}{2^{N+1}}.$$

Now that we have constructed the function Y_N, the next step is to determine if it approximates the unknown solution y, if it exists. It is clear that it takes more work to compute Y_N when N increases, so we presumably hope that Y_N is a better approximation to y when N increases. In algebraic root problems, we show that the Bisection Algorithm and the Fixed Point Iteration produce approximations of a root by showing the sequences they produce converge to the root. Now we have constructed a sequence of *functions* $\{Y_N(x)\}_{N=1}^{\infty}$ and we need to show that this sequence of functions converges to the solution $y(x)$.

The difficulty with talking about Y_N converging to y is that we do not know if y even exists. Recall that the same problem crops up when solving algebraic root problems. In that setting, we introduce the idea of a Cauchy sequence to avoid having to use the limit of a sequence when checking if a sequence converges or not. We do the same thing here.

25.4   Uniform Cauchy Sequences of Functions

Before showing $\{Y_N(x)\}$ converges to the solution y, we develop some basic facts about sequences of functions. A sequence of functions $\{f_n(x)\}_{n=1}^{\infty}$ is a set of functions that depend on the index n in some fashion. Some examples are

$$\{x^n\}_{n=1}^{\infty} = \{x, x^2, x^3, \cdots\}$$
$$\{nx^3\}_{n=1}^{\infty} = \{x^3, 2x^3, 3x^3, \cdots\}$$
$$\left\{\left(1 + \frac{1}{n}\right)x^3 + 5x - 3\right\}_{n=1}^{\infty} = \left\{2x^3 + 5x - 3, \frac{3}{2}x^3 + 5x - 3, \frac{5}{2}x^3 + 5x - 3\right\}$$
$$\{2 + x\}_{n=1}^{\infty} = \{2 + x, 2 + x, 2 + x, \cdots\}$$
$$\{5 - 1/n^2\}_{n=1}^{\infty} = \{4, 4.75, 4.888\cdots, \cdots\}.$$

Luckily, $\{Y_N(x)\}$ has special properties that make convergence relatively easy to show, and we concentrate on sequences that share these properties in this chapter.[4] A sequence of functions $\{f_n(x)\}_{n=1}^{\infty}$ **converges uniformly** to a function $f(x)$ on an interval I if for every $\epsilon > 0$ there is an N such that for all x in I,

$$|f_n(x) - f(x)| < \epsilon \text{ for } n \geq N. \tag{25.6}$$

In this case, we write

$$\lim_{n \to \infty} f_n(x) = f(x).$$

---

[4]More general sequences are discussed in Chapter 33.

EXAMPLE 25.5. $\left\{(1+\frac{1}{n})x^3 + 5x - 3\right\}_{n=1}^{\infty}$ converges uniformly to $x^3 + 5x - 3$ on any interval $[a, b]$ since

$$\left|(1 + \frac{1}{n})x^3 + 5x - 3 - (x^3 + 5x - 3)\right| = \frac{1}{n}|x|^3 \leq \frac{1}{n}M^3,$$

where $M = \max\{|a|, |b|\}$. Therefore for any $\epsilon > 0$,

$$\left|(1 + \frac{1}{n})x^3 + 5x - 3 - (x^3 + 5x - 3)\right| \leq \epsilon$$

for all $n \geq N = M^3/\epsilon$.

EXAMPLE 25.6. $\{x^n\}_{n=1}^{\infty}$ converges uniformly to the zero function $f(x) = 0$ on any interval $I = [-a, a]$ with $0 < a < 1$ since for any $\epsilon > 0$

$$|x^n - 0| \leq a^n \leq \epsilon$$

for all n sufficiently large.

The qualifier "uniformly" refers to the fact that the values of $\{f_n(x)\}$ converge to the corresponding value $f(x)$ at the same rate for all x in the interval I. It is possible to have non-uniform convergence and of course not all sequences converge. A sequence that does not converge is said to **diverge**.

EXAMPLE 25.7. $\{nx^3\}_{n=1}^{\infty}$ converges to 0 for $x = 0$ but diverges for any $x \neq 0$.

EXAMPLE 25.8. $\{x^n\}_{n=1}^{\infty}$ converges to 0 for each x in $I = (0, 1)$, but the convergence is not uniform since for any n, we can find values of x for which x^n is arbitrarily close to 1.

If I includes points larger in magnitude than 1, the sequence diverges.

With these definitions, it is straightforward to prove that uniformly convergent sequences share some of the important properties of convergent sequences. We ask you to prove the following theorem in Problem 25.10.

Theorem 25.1 *Suppose that $\{f_n(x)\}$ and $\{g_n(x)\}$ are uniformly convergent sequences on $[a, b]$ that converge to f and g, respectively, and c is a constant. Then*

- *$\{f_n + g_n\}$ converges uniformly to $f + g$ on $[a, b]$.*

- *$\{cf_n\}$ converges uniformly to cf on $[a, b]$.*

In addition if the sequences $\{f_n\}$ and $\{g_n\}$ are uniformly bounded, i.e., there is a constant M such that for all n and x in $[a, b]$, $|f_n(x)| \leq M$ and $|g_n(x)| \leq M$, then

- $\{f_n g_n\}$ converges uniformly to fg on $[a, b]$.

In addition if the sequences $\{f_n\}$ and $\{g_n\}$ are uniformly bounded and there is a constant C such that $|g_n(x)| \geq C > 0$ for all x in $[a, b]$ and n,[5]

- $\{f_n/g_n\}$ converges uniformly to f/g on $[a, b]$.

EXAMPLE 25.9. The sequence $\{f_n(x)\} = \{x + 1/n\}$ on $(-\infty, \infty)$ shows that we need the assumption about uniform boundedness when dealing with products and quotients of sequences. In Problem 25.11, we ask you to show that f_n converges uniformly to x for x in $(-\infty, \infty)$ but f_n^2 does not converge uniformly to x^2 for x in $(-\infty, \infty)$.

As mentioned, Cauchy sequences are useful when the limit is unknown. A sequence of functions $\{f_n(x)\}_{n=1}^{\infty}$ is a **uniform Cauchy sequence** on an interval I if for any $\epsilon > 0$ there is an N such that for all x in I,

$$|f_n(x) - f_m(x)| \leq \epsilon \text{ for } m \geq n \geq N. \tag{25.7}$$

EXAMPLE 25.10. $\left\{\frac{1}{n}x^2 + 3x - 1\right\}_{n=1}^{\infty}$ is a uniform Cauchy sequence on any bounded interval since for $m \geq n$,

$$\left|(\frac{1}{n}x^2 + 3x - 1) - (\frac{1}{m}x^2 + 3x - 1)\right| = (\frac{1}{n} - \frac{1}{m})|x|^2$$
$$= \frac{m-n}{mn}|x|^2$$
$$\leq \frac{1}{n}|x|^2,$$

so the difference can be made arbitrarily small by taking n large provided $|x|$ is bounded by some constant.

EXAMPLE 25.11. $\{x^n\}_{n=1}^{\infty}$ is a uniform Cauchy sequence on any interval $I = [-a, a]$ with $0 < a < 1$ since

$$|x^m - x^n| = |x|^n |x^{m-n} - 1| \leq 2a^n \text{ for } m \geq n,$$

and the difference can be made arbitrarily small by taking n large.

EXAMPLE 25.12. $\{nx^3\}_{n=1}^{\infty}$ is not a uniform Cauchy sequence on any interval except the point 0 and $\{x^n\}_{n=1}^{\infty}$ is not a uniform Cauchy sequence on any interval that contains points with magnitude greater than or equal to 1.

---

[5] In words, $\{g_n\}$ is bounded away from zero uniformly on I.

With these definitions, a uniform Cauchy sequence of functions on an interval I converges to a function on I. This follows because for each x in I, the sequence of *numbers* $\{f_n(x)\}$ is a Cauchy sequence and therefore has a limit by Theorem 11.6. We define the unique limit function $f(x)$ on I by setting

$$f(x) = \lim_{n \to \infty} f_n(x) \text{ for each } x \text{ in } I.$$

This convergence is defined pointwise for each x, but the definitions above imply that $\{f_n\}$ converges to f uniformly as functions as well. We summarize as a theorem that we ask you to prove in Problem 25.13.

Theorem 25.2 Uniform Cauchy Criterion for Sequences of Functions *A uniform Cauchy sequence of functions on an interval I converges uniformly to a unique limit function on I. Conversely, a uniformly convergent sequence of functions is a uniform Cauchy sequence.*

An important issue is determining which properties of a uniform Cauchy sequence of functions $\{f_n(x)\}$ are inherited by the limit $f(x)$. To show that integration works in particular, we want to find conditions that guarantee the limit is Lipschitz continuous if the functions in the sequence are Lipschitz continuous.[6]

A sequence of functions $\{f_n(x)\}$ is **uniformly Lipschitz continuous** on an interval I if there is a constant L such that

$$|f_n(x) - f_n(y)| \le L|x - y| \text{ for all } n \text{ and } x, y \text{ in } I.$$

EXAMPLE 25.13. We can use Theorem 19.1 and the Mean Value Theorem to show that $\{f_n(x)\}_{n=1}^{\infty} = \left\{\left(1 + \frac{1}{n}\right)x^3 + 5x - 3\right\}_{n=1}^{\infty}$ is uniformly Lipschitz continuous on any bounded interval. First the functions are uniformly strongly differentiable on any bounded interval I so by the Mean Value Theorem, for any n and x, y in I, there is a c between x and y with

$$|f_n(x) - f_n(y)| = |f_n'(c)||x - y|.$$

But,

$$|f_n'(x)| = \left|3\left(1 + \frac{1}{n}\right)x^2 + 5\right| \le 6|x|^2 + 5,$$

independent of n. Hence, there is a constant L depending only on I such that $|f_n(x) - f_n(y)| \le L|x - y|$ for all n and x, y in I.

Now suppose that the uniform Cauchy sequence $\{f_n(x)\}$ with limit $f(x)$ is uniformly Lipschitz continuous on an interval I. We want to show that

---

[6]We discuss the inheritance of other properties in Chapter 33.

$f(x)$ is Lipschitz continuous as well, so we choose two points x and y in I and compute using the old tricks

$$
\begin{aligned}
|f(x) - f(y)| &= |f(x) - f_n(x) + f_n(x) - f_n(y) + f_n(y) - f(y)| \\
&\leq |f(x) - f_n(x)| + |f_n(x) - f_n(y)| + |f_n(y) - f(y)|.
\end{aligned} \tag{25.8}
$$

The point of this argument is to get the term $|f_n(x) - f_n(y)|$ in the middle because the sequence is Lipschitz continuous. Hence there is a constant L such that

$$
|f_n(x) - f_n(y)| \leq L|x - y| \text{ for all } x, y \text{ in } I \text{ and } n.
$$

As far as the remaining terms on the right of (25.8), for any $\epsilon > 0$ there is an N such that

$$
|f(x) - f_n(x)| \leq \epsilon \text{ and } |f(y) - f_n(y)| \leq \epsilon \text{ for all } x, y \text{ in } I \text{ and } n \geq N.
$$

Using this in (25.8), we get

$$
|f(x) - f(y)| \leq 2\epsilon + L|x - y| \text{ for all } x, y \text{ in } I.
$$

But $\epsilon > 0$ can be made arbitrarily small, so this means that

$$
|f(x) - f(y)| \leq L|x - y| \text{ for all } x, y \text{ in } I.
$$

We summarize:

Theorem 25.3 *A uniform Cauchy sequence of uniformly Lipschitz continuous functions on an interval I converges to a Lipschitz continuous function on I with the same Lipschitz constant.*

25.5   Convergence of the Integration Approximation

To show that the sequence of functions $\{Y_N\}$ converges, we show that $\{Y_N\}$ is a uniform Cauchy sequence when the function f is Lipschitz continuous.[7] Moreover, we show that $\{Y_N\}$ is uniformly Lipschitz continuous, so the limit is also a Lipschitz continuous function.

We assume that f is Lipschitz continuous on $[a, b]$ with Lipschitz constant L. We show that given any $\epsilon > 0$ there is a $\tilde{N}$ such that for all x in $[a, b]$,

$$
|Y_N(x) - Y_M(x)| \leq \epsilon \text{ for } M \geq N \geq \tilde{N}.
$$

The main difficulty is working through the notation needed to compare functions on different meshes. We choose $M \geq N$ and show the two meshes

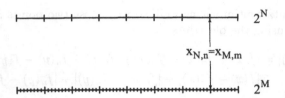

FIGURE 25.10. The meshes corresponding to $M \geq N$.

in Fig. 25.10. We first estimate $|Y_N(x) - Y_M(x)|$ with $x = x_{N,n}$ for some n by using the formula (25.5) to get

$$Y_N(x_{N,n}) = \sum_{i=1}^{n} f(x_{N,i-1})\Delta x_N.$$

Because the meshes are nested, $x_{N,n} = x_{M,m}$ for some m, hence

$$Y_M(x_{N,n}) = Y_M(x_{M,m}) = \sum_{j=1}^{m} f(x_{M,j-1})\Delta x_M.$$

To compare these two values, we rewrite the sums for each to be the same. We can do this by using the fact that adding up the right number of Δx_M gives exactly Δx_N. In fact, the number is

$$2^{M-N} = \frac{\frac{b-a}{2^N}}{\frac{b-a}{2^M}}.$$

To make use of this, we define $\mu(i)$ to be the set of indices j such that $[x_{M,j-1}, x_{M,j}]$ is contained in $[x_{N,i-1}, x_{N,i}]$ (see Fig. 25.11). We can then

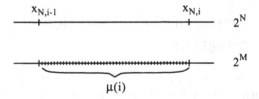

FIGURE 25.11. The definition of $\mu(i)$.

write

$$Y_M(x_{N,n}) = \sum_{i=1}^{n} \sum_{j \text{ in } \mu(i)} f(x_{M,j-1})\Delta x_M.$$

---

[7]This means we can consider $Y_N(x)$ to be an approximation of the limit.

There are 2^{M-N} indices in $\mu(i)$ for each i, so

$$\sum_{j \text{ in } \mu(i)} \Delta x_M = \Delta x_N,$$

and we can also write

$$Y_N(x_{N,n}) = \sum_{i=1}^{n} f(x_{N,i-1})\Delta x_N = \sum_{i=1}^{n} \sum_{j \text{ in } \mu(i)} f(x_{N,i-1})\Delta x_M.$$

We now estimate

$$|Y_M(x_{N,n}) - Y_N(x_{N,n})| = \left| \sum_{i=1}^{n} \sum_{j \text{ in } \mu(i)} (f(x_{M,j-1}) - f(x_{N,i-1}))\Delta x_M \right|$$

$$\leq \sum_{i=1}^{n} \sum_{j \text{ in } \mu(i)} |f(x_{M,j-1}) - f(x_{N,i-1})|\Delta x_M.$$

Since $|x_{M,j-1} - x_{N,i-1}| \leq \Delta x_N$ for j in $\mu(i)$ and f is Lipschitz continuous, we find

$$|f(x_{M,j-1}) - f(x_{N,i-1})| \leq L|x_{M,j-1} - x_{N,i-1}| \leq L\Delta x_N.$$

We conclude that

$$|Y_M(x_{N,n}) - Y_N(x_{N,n})| \leq \sum_{i=1}^{n} \sum_{j \text{ in } \mu(i)} L\Delta x_N \Delta x_M$$

$$= \sum_{i=1}^{n} L(\Delta x_N)^2 \qquad (25.9)$$

$$= (x_{N,n} - x_{N,0})L\Delta x_N.$$

This estimate certainly implies the differences between $Y_N(x)$ and $Y_M(x)$ can be made arbitrarily small at the nodes $x_{N,n}$.

We have to show that a similar result holds for any x in $[a, b]$. We choose x in $[a, b]$ and choose n and m so that $x_{N,n-1} \leq x \leq x_{N,n}$ and $x_{M,m-1} \leq x \leq x_{M,m}$ and also choose $\tilde{m}$ so that $x_{M,\tilde{m}-1} = x_{N,n-1}$. These definitions are illustrated in Fig. 25.12. By (25.4),

$$Y_N(x) = Y_N(x_{N,n-1}) + (x - x_{N,n-1})f(x_{N,n-1}),$$
$$Y_M(x) = Y_M(x_{M,m-1}) + (x - x_{M,m-1})f(x_{M,m-1}).$$

Using induction as before, we can write the second equation as

$$Y_M(x) = Y_M(x_{M,\tilde{m}-1}) + \sum_{i=\tilde{m}}^{m-1} f(x_{M,i-1})\Delta x_M$$

$$+ (x - x_{M,m-1})f(x_{M,m-1}). \qquad (25.10)$$

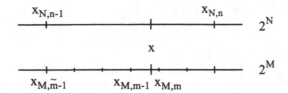

FIGURE 25.12. The choice of nodes near x.

We can also write

$$x - x_{N,n-1} = \sum_{i=\tilde{m}}^{m-1} \Delta x_M + (x - x_{M,m-1}),$$

and therefore

$$Y_N(x) = Y_N(x_{N,n-1}) + \sum_{i=\tilde{m}}^{m-1} f(x_{N,n-1})\Delta x_M$$

$$+ (x - x_{M,m-1})f(x_{N,n-1}). \quad (25.11)$$

Now subtracting (25.11) from (25.10) and estimating gives

$$|Y_M(x) - Y_N(x)| \leq |Y_M(x_{M,\tilde{m}-1}) - Y_N(x_{N,n-1})|$$

$$+ \sum_{i=\tilde{m}}^{m-1} |f(x_{M,i-1}) - f(x_{N,n-1})|\Delta x_M$$

$$+ (x - x_{M,m-1})|f(x_{M,m-1}) - f(x_{N,n-1})|.$$

Using the Lipschitz continuity of f and (25.9), we get

$$|Y_M(x) - Y_N(x)| \leq L(x_{N,n-1} - x_{N,0})\Delta x_N$$

$$+ \sum_{i=\tilde{m}}^{m-1} L\Delta x_N \Delta x_M + (x - x_{M,m-1})L\Delta x_N$$

$$= (x - x_{N,0})L\Delta x_N.$$

In other words, for any x in $[a, b]$ we can bound

$$|Y_M(x) - Y_N(x)| \leq (b - a)L\Delta x_N. \quad (25.12)$$

This implies that $\{Y_N\}$ is a uniform Cauchy sequence since for any $\epsilon > 0$, there is a $\tilde{N}$ such that for all x,

$$|Y_M(x) - Y_N(x)| \leq \epsilon \text{ for } M \geq N \geq \tilde{N}$$

Namely, we choose $\tilde{N}$ so that

$$(b - a)L\Delta x_N = (b - a)^2 L/2^{\tilde{N}} \le \epsilon.$$

We conclude that there is a function, that we call $y(x)$, such that

$$\lim_{N \to \infty} Y_N(x) = y(x) \tag{25.13}$$

uniformly for x in $[a, b]$. In Problem 25.18, we ask you to show that $\{Y_N(x)\}$ is a uniform Lipschitz continuous sequence using the same kind of arguments used to show that $\{Y_N(x)\}$ converges. So Theorem 25.3 implies that $y(x)$ is Lipschitz continuous.

EXAMPLE 25.14. In Example 25.4, we computed

$$Y_N(1) = \frac{1}{2} - \frac{1}{2^{N+1}};$$

so clearly

$$\lim_{N \to \infty} Y_N(1) = \frac{1}{2}.$$

The solution of $y' = x$, $y(0) = 0$, is $y = x^2/2$ and therefore $y(1) = 1/2$.

25.6   The Limit Solves the Differential Equation

We know that $\{Y_N\}$ converges uniformly to a Lipschitz continuous function $y(x)$, but we still have to show that this limit in fact solves the differential equation (25.1). Actually, we also have to show that y has a derivative. This is not obvious because the function $Y_N(x)$ clearly does not have a derivative at every point in $[a, b]$ because of the "corners" at the nodes $\{x_{N,j}\}$. However, intuition suggests that as we increase the number of points in the mesh, the angle of change in these corners could become smaller and the limit could be smooth.

To show that y is strongly differentiable at each $\bar{x}$ in $[a, b]$ with derivative $f(\bar{x})$, we show that there is a constant $\mathcal{K}$ such that for any $\bar{x}$ in $[a, b]$ [8]

$$|y(x) - (y(\bar{x}) + f(\bar{x})(x - \bar{x}))| \le \mathcal{K}|x - \bar{x}|^2 \text{ for all } x \text{ in } I. \tag{25.14}$$

This implies that y is differentiable at each $\bar{x}$ and $y'(\bar{x}) = f(\bar{x})$.

We begin by estimating $Y_N(x) - Y_N(\bar{x})$ for x and $\bar{x}$ in $[a, b]$. We first assume that $x > \bar{x}$ and for each N, choose m_N so that $x_{N,m_N-1} < \bar{x} \le x_{N,m_N}$ and n_N so that $x_{N,n_N-1} < x \le x_{N,n_N}$ (see Fig. 25.13). By this

---

[8]With the obvious one-sided interpretation when $\bar{x} = a$ or b.

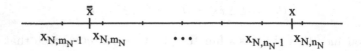

FIGURE 25.13. The choice of m_N and n_N.

choice, it follows that

$$x - \bar{x} = (x - x_{N,n_N-1}) + \sum_{j=m_N}^{n_N-1} \Delta x_N - (\bar{x} - x_{N,m_N-1}), \qquad (25.15)$$

and

$$\lim_{N\to\infty} x_{N,m_N-1} = \lim_{N\to\infty} x_{N,m_N} = \bar{x} \text{ and } \lim_{N\to\infty} x_{N,n_N-1} = \lim_{N\to\infty} x_{N,n_N} = x.$$
$$(25.16)$$

Moreover, using (25.4),

$$Y_N(\bar{x}) = Y_N(x_{N,m_N-1}) + f(x_{N,m_N-1})(\bar{x} - x_{N,m_N-1}),$$

and

$$Y_N(x) = Y_N(x_{N,m_N-1}) + \sum_{j=m_N}^{n_N-1} f(x_{N,j-1})\Delta x_N$$
$$+ f(x_{N,n_N-1})(x - x_{N,n_N-1}). \quad (25.17)$$

Subtraction gives

$$Y_N(x) - Y_N(\bar{x}) = f(x_{N,n_N-1})(x - x_{N,n_N-1}) + \sum_{j=m_N}^{n_N-1} f(x_{N,j-1})\Delta x_N$$
$$- f(x_{N,m_N-1})(\bar{x} - x_{N,m_N-1}).$$

Using (25.15), we can rewrite this as

$$Y_N(x) - Y_N(\bar{x}) = f(\bar{x})(x - \bar{x})$$
$$+ (f(x_{N,n_N-1}) - f(\bar{x}))(x - x_{N,n_N-1})$$
$$+ \sum_{j=m_N}^{n_N-1} (f(x_{N,j-1}) - f(\bar{x}))\Delta x_N$$
$$- (f(x_{N,m_N-1}) - f(\bar{x}))(\bar{x} - x_{N,m_N-1}). \quad (25.18)$$

Finally we estimate

$$|Y_N(x) - Y_N(\bar{x}) - f(\bar{x})(x - \bar{x})|$$
$$\leq |f(x_{N,n_N-1}) - f(\bar{x})|\,|x - x_{N,n_N-1}|$$
$$+ \sum_{j=m_N}^{n_N-1} |f(x_{N,j-1}) - f(\bar{x})|\Delta x_N$$
$$+ |f(x_{N,m_N-1}) - f(\bar{x})|\,|\bar{x} - x_{N,m_N-1}|. \quad (25.19)$$

This looks like a mess, but if we use the Lipschitz continuity of f and the fact that all the x points in (25.18) are in $[x_{N,m_N-1}, x_{N,n_N}]$, it simplifies to

$$|Y_N(x) - Y_N(\bar{x}) - f(\bar{x})(x - \bar{x})| \leq 3L|x_{N,n_N} - x_{N,m_N-1}|^2. \quad (25.20)$$

For example,

$$|f(x_{N,n_N-1}) - f(\bar{x})|\,|x - x_{N,n_N-1}| \leq L|x_{N,n_N} - x_{N,m_N-1}|^2.$$

Taking the limit of both sides as $N \to \infty$, we conclude that

$$|y(x) - y(\bar{x}) - f(\bar{x})(x - \bar{x})| \leq 3L|x - \bar{x}|^2$$

for $x > \bar{x}$ in $[a, b]$. It is straightforward to treat the cases $\bar{x} > x$ and $\bar{x} = a$ or b in the same way. Thus, (25.14) holds.

25.7   The Fundamental Theorem of Calculus

We summarize these results of the previous analysis into an important theorem, which was first proved by Cauchy.

Theorem 25.4 Fundamental Theorem of Calculus *If f is a Lipschitz continuous function on $[a, b]$ with Lipschitz constant L, then there is a unique solution $y(x)$ of*

$$\begin{cases} y'(x) = f(x), & a \leq x \leq b, \\ y(a) = 0. \end{cases}$$

Moreover, the sequence $\{Y_n\}$ with

$$Y_N(x) = \sum_{i=1}^{n-1} f(x_{N,i-1})\Delta x_N + f(x_{N,n-1})(x - x_{N,n-1}),$$

where $\Delta x_N = (b - a)/2^N$ for a natural number N, $x_{N,i} = a + i \times \Delta x_N$ for $i = 0, 1, \cdots, 2^N$, and $x_{N,n-1} < x \leq x_{N,n}$, converges uniformly to y on $[a, b]$. For any N, the error is bounded by

$$|y(x) - Y_N(x)| \leq (b - a)L\Delta x_N. \quad (25.21)$$

We can rewrite this result using the notation for definite integrals. By definition,

$$y(x) = \int_a^x f(s)\, ds.$$

If we take $x = b$ for simplicity,

$$y(b) = \int_a^b f(x)\, dx.$$

Moreover,

$$Y_N(b) = Y_N(x_{N,2^N}) = \sum_{i=1}^{2^N} f(x_{N,i-1})\Delta x_N.$$

So we have

Theorem 25.5 Fundamental Theorem of Calculus *If f is a Lipschitz continuous function on $[a, b]$ with Lipschitz constant L, then*

$$\int_a^b f(x)\, dx$$

exists and

$$\left| \int_a^b f(x)\, dx - \sum_{i=1}^{2^N} f(x_{N,i-1})\Delta x_N \right| \leq (b-a)L\Delta x_N,$$

where $\Delta x_N = (b-a)/2^N$ for a natural number N and $x_{N,i} = a + i \times \Delta x_N$ for $i = 0, 1, \cdots, 2^N$.

This theorem provides the motivation for the notation we use for the integral. Since $N \to \infty$ is the same as $\Delta x_N \to 0$, we can write *formally*

$$\text{``}\lim_{\Delta x \to 0} \sum f(x_i)\Delta x = \int f(x)\, dx\text{''}$$

so in the limit $\sum \to \int$ and $\Delta x \to dx$.

The theory of integration described in this chapter is a simplification of the general theory first suggested by Cauchy and later systematized and generalized by Riemann.[9] Cauchy's goal was to prove that integration is

---

[9]The German mathematician Georg Friedrich Bernhard Riemann (1826–1866) was one of the most original and creative thinkers in mathematics. Riemann depended heavily on intuitive arguments that sometimes were not completely correct, yet his discoveries have had a profound impact on mathematics and physics. His major achievements were in abelian functions, complex analysis, electromagnetism, geometry, number theory, the theory of integration, and topology. His name is remembered in the Riemann integral, Riemann surfaces, and the Riemann zeta function. He was led to define a rigorous theory of integration in the course of investigating the convergence of Fourier series.

defined for continuous integrands. Riemann reversed the question: given this process for defining integration, find a general class of functions for which it works. A sum of the form

$$\sum_{i=1}^{2^N} f(x_{N,i-1})\Delta x_N$$

is known as a **Riemann sum**.

Chapter 25 Problems

25.1. Suppose $f(x)$ is a continuously differentiable function on an interval containing x_0. Use the Mean Value Theorem to prove the following error estimate on the error of the constant interpolant of f,

$$|f(x) - f(x_0)| \leq \max_{[x_0, x]} |f'| \, |x - x_0|.$$

Problems 25.2–25.6 have to do with forming integration approximations.

25.2. Compute formulas for the value of $x_{N,i}$ for i between 0 and 2^N when (a) $[a, b] = [0, 1]$ and (b) $[a, b] = [3, 7]$.

25.3. Mark the mesh points on the interval $[a, b]$ corresponding to $\Delta x_N = (b - a)/3^N$ for $N = 1, 2, 3, 4$.

25.4. Draw the integration approximation $Y_N(x)$ corresponding to the $y(x)$ shown in Fig. 25.14 on the indicated mesh.

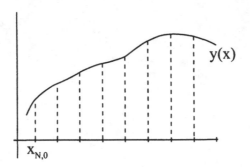

FIGURE 25.14. Plot for Problem 25.4.

25.5. Compute $Y_2(x)$ and then plot $Y_2(x)$ together with $y(x)$ for (a) $f(x) = x$, (b) $f(x) = x^2$, and (c) $f(x) = x^3$ on $[0, 1]$.

25.6. Compute a formula for $Y_N(1)$ for (a) $f(x) = 2x$, (b) $f(x) = x^2$, and (c) $f(x) = x^3$ on $[0, 1]$.

Problems 25.7–25.13 have to do with uniform Cauchy sequences of functions.

25.7. Write down five different sequences of functions.

25.8. Use the definition to determine an interval on which the following sequences converge or explain why they diverge for all x:

(a) $x^3 - \left(2 + \frac{1}{n^2}\right)x^2 - 3$ (b) $(3n+1)x - 2$

(c) $(x-3)^n$ (d) $4^n + x^2$

(e) $x^n/2^n$ (f) $(nx+3)/(n+2)$

(g) $x^n + x^2 + 2$ (h) $\left(\frac{1}{2}\right)^n - 5x$.

25.9. Use the definition to determine an interval on which the following sequences are Cauchy or explain why they are not Cauchy sequences for any x:

(a) $x^2 - \left(1 + \frac{2}{n}\right)x + 4$ (b) $(2n-3)x^2 + 5$

(c) $(x-2)^n$ (d) $2^n + 4x$

(e) $x^n/3^n$ (f) $(nx-1)/(n+2)$

(g) $x^n + x + 1$ (h) $\left(\frac{1}{5}\right)^n + 2x$.

25.10. Prove Theorem 25.1. *Hint:* To deal with products and quotients, it is useful to show that if $\{f_n\}$ converges uniformly on I to f and $\{f_n\}$ is uniformly bounded on I by M, then f is also bounded by M on I.

25.11. Verify the claim in Example 25.9.

25.12. State and prove the analog of Theorem 25.1 for Cauchy sequences.

25.13. Prove Theorem 25.2.

In Problems 25.14–25.25, we ask you to verify details of the proof that the approximate integral converges.

25.14. Find a formula relating the values of the common node $x_{N,n} = x_{M,m}$ in the nested meshes for $M \geq N$.

25.15. Prove (25.9).

25.16. Prove (25.10).

25.17. Prove (25.11).

25.18. Prove that $\{Y_N(x)\}$ is a sequence of uniformly Lipschitz continuous functions with Lipschitz constant equal to the maximum value of f on the $[a, b]$.

25.19. Prove (25.15) is true.

25.20. Prove (25.16) is true.

25.21. Show that (25.17) is valid.

25.22. Show that (25.18) is valid.

25.23. Show that (25.19) implies (25.20).

25.24. Derive the analog of (25.20) when $\bar{x} > x$.

25.25. Prove (25.21).

25.26. Write a program that uses a user-defined function $f(x)$ and computes $Y_N(x)$ at any point x in the user-defined interval $[a, b]$ for a user-defined number of mesh points 2^N and the user-defined interpolation points of either the left- or right-hand endpoints or the midpoints of each interval as well as computes a bound on the error. Test your program on $y = x$ and compare to the results above. Then run the program to compute $\int_1^x t^{-1} dt$ at $x = 2$ and $x = 3$ for $N = 4, 8, 16, 32, 64$. Compare the results to $\ln(2)$ and $\ln(3)$, respectively. Compare the accuracy of using the midpoint interpolant versus either of the endpoint interpolants.

26

Properties of the Integral

In Chapter 24, we derived some properties of the integral by using the properties of the derivative. In this brief chapter, we derive some important properties of the definite integral by using properties of Riemann sums and taking the limit, i.e.,

$$\int_a^b f(x)\, dx = \lim_{N \to \infty} \sum f(x_{N,i}) \Delta x_N.$$

26.1 Linearity

To illustrate the idea, we begin with a property already derived. Recall that if f and g are Lipschitz continuous functions and c is a constant, then

$$\int (f(x) + cg(x))\, dx = \int f(x)\, dx + c \int g(x)\, dx.$$

This property corresponds directly to the linearity property of the derivative. The corresponding property for definite integrals, namely,

$$\int_a^b (f(x) + cg(x))\, dx = \int_a^b f(x)\, dx + c \int_a^b g(x)\, dx \qquad (26.1)$$

follows from the properties of limits of sequences. For example, if $\{a_n\}$ and $\{b_n\}$ are convergent sequences, then $\lim_n (a_n + cb_n) = \lim_n a_n + c \lim_n b_n$.

If $\{x_{N,i}\}$ are the nodes in a mesh, then

$$\sum_{i=1}^{2^N}(f(x_{N,i}) + cg(x_{N,i}))\Delta x_N = \sum_{i=1}^{2^N} f(x_{N,i})\Delta x_N + c\sum_{i=1}^{2^N} g(x_{N,i})\Delta x_N.$$

Taking limits gives (26.1).

EXAMPLE 26.1. We can compute

$$\int_1^2 (3x^2 + 4x)\, dx = (x^3 + 2x^2)\Big|_{x=1}^{x=2} = 16 - 3 = 13$$

directly or by computing

$$\int_1^2 x^2\, dx = \frac{x^3}{3}\Big|_{x=1}^{x=2} = \frac{8}{3} - \frac{1}{3} = \frac{7}{3}$$

and likewise

$$\int_1^2 x\, dx = \frac{x^2}{2}\Big|_{x=1}^{x=2} = \frac{4}{2} - \frac{1}{1} = \frac{3}{2}$$

and then adding

$$3 \times \frac{7}{3} + 4 \times \frac{3}{2} = 13.$$

26.2   Monotonicity

It occasionally happens that we need to compare the sizes of the solutions of the differential equations $y_1'(x) = f(x)$ and $y_2'(x) = g(x)$ when, say, $f(x) \leq g(x)$ for all x in $[a, b]$. Intuitively, this means that the slope of the linearization of y_1 is always smaller than the slope of the linearization of y_2 at every point x; hence, $y_1(x)$ must not increase as quickly as $y_2(x)$ as x increases. If $y_1(a) = y_2(a)$, then $y_1(x)$ must be less than $y_2(x)$ for $x \geq a$. We illustrate in Fig. 26.1

Written in terms of integrals, we prove the following:

Theorem 26.1 Monotonicity of Integration *If f and g are Lipschitz continuous functions on $[a, b]$ and $f(x) \leq g(x)$ for all $a \leq x \leq b$, then*

$$\int_a^b f(x)\, dx \leq \int_a^b g(x)\, dx. \tag{26.2}$$

This follows from properties of limits of sequences. If $\{x_{N,i}\}$ are the nodes in a mesh, then

$$f(x_{N,i}) \leq g(x_{N,i})$$

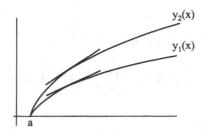

$y_2(x)$

$y_1(x)$

a

FIGURE 26.1. The derivative of y_1 is smaller than the derivative of y_2 at every point x.

for all i and therefore

$$\sum_{i=1}^{2^N} f(x_{N,i})\Delta x_N \le \sum_{i=1}^{2^N} g(x_{N,i})\Delta x_N.$$

Taking limits as $N \to \infty$ does the trick.

EXAMPLE 26.2. Since $x^2 \le x$ for $0 \le x \le 1$,

$$\int_0^1 x^2 \, dx \le \int_0^1 x \, dx.$$

We can easily check that

$$\int_0^1 x^2 \, dx = \frac{1}{3} < \int_0^1 x \, dx = \frac{1}{2}.$$

26.3 Playing with the Limits

So far we have discussed solving the differential equation $y'(x) = f(x)$ by specifying that $y = 0$ at a point a and then computing the solution for $a \le x \le b$. But this is an arbitrary choice. It can just as well happen that we specify that $y(b) = 0$ and then solve for $b \ge x \ge a$, which is naturally written

$$\int_b^a f(x) \, dx.$$

To compute this integral using the approximation, we create a mesh for $[a, b]$ as before, except that now $x_{N,0} = b$, $x_{N,2^N} = a$, and $\Delta x_N = (a - b)/2^N$. But the only difference in the approximation sum is that we have relabeled the node points from right to left, instead of left to right, and the new Δx_N is minus the old Δx_N. Therefore, the new approximation sum is just minus the old approximation sum. But this means that

$$\int_a^b f(x) \, dx = -\int_b^a f(x) \, dx. \tag{26.3}$$

EXAMPLE 26.3. We can compute

$$\int_1^3 3x^2 \, dx = x^3 \Big|_{x=1}^{x=3} = 27 - 1 = 26,$$

while

$$\int_3^1 3x^2 \, dx = x^3 \Big|_{x=3}^{x=1} = 1 - 27 = -26,$$

as predicted.

We approximate the value of

$$\int_a^b f(x) \, dx$$

by defining a mesh on $[a, b]$ and then computing the corresponding integration approximation. If c is a point between a and b, we can also approximate the integral by defining meshes on $[a, c]$ and $[c, b]$, computing integration approximations for each mesh, and then adding the two results. In other words, if $a \le c \le b$, then

$$\int_a^b f(x) \, dx = \int_a^c f(x) \, dx + \int_c^b f(x) \, dx. \qquad (26.4)$$

The formula generalizes to arbitrary a, b, and c.

EXAMPLE 26.4. We can compute

$$\int_1^5 6x \, dx = 3x^2 \Big|_{x=1}^{x=5} = 75 - 3 = 72$$

by computing the two integrals

$$\int_1^3 6x \, dx = 3x^2 \Big|_{x=1}^{x=3} = 27 - 3 = 24$$

and

$$\int_3^5 6x \, dx = 3x^2 \Big|_{x=3}^{x=5} = 75 - 27 = 48$$

and adding the results.

We can interpret (26.4) as saying that the value $y(b)$ of the solution of the differential equation $y'(x) = f(x)$ with $y(a) = 0$ can be computed by first solving to $x = x$ to get $y(c)$ for an intermediate point c and then starting with that value at $x = c$ and continuing on to b.

26.4   More on Definite and Indefinite Integrals

Recall that the general solution of $y'(x) = f(x)$,

$$y(x) = \int f(x)\, dx,$$

can be computed by adding a constant C to any antiderivative of f. This means in particular that the general solution can also be written as

$$y(x) = \int_a^x f(s)\, ds + C,$$

where C is an undetermined constant.

One consequence of this relation between the indefinite and definite integral is yet another way to write the Fundamental Theorem of Calculus:

Theorem 26.2 Fundamental Theorem of Calculus   *If f is a Lipschitz continuous function on $[a, b]$, then*

$$\int_a^x f(s)\, ds$$

defines a differentiable function for $a < x \leq b$ and

$$\frac{d}{dx} \int_a^x f(s)\, ds = f(x). \tag{26.5}$$

EXAMPLE 26.5.   Using Theorem 26.2, we compute

$$\frac{d}{dx} \int_1^x 3s^2\, ds = 3x^2.$$

We can also compute

$$\frac{d}{dx} \int_1^x 3s^2\, ds = \frac{d}{dx} \left(s^3 \big|_{s=1}^{s=x} \right) = \frac{d}{dx} \left(x^3 - 1 \right) = 3x^2.$$

Chapter 26 Problems

26.1. Prove (26.1) by assuming that F is an antiderivative of f and G is an antiderivative of g and rewriting the definite integrals in terms of F and G.

26.2. Compute the following integrals directly and using (26.1):

$$\text{(a)} \int_1^2 (4x^2 - 8x)\, dx \quad \text{(b)} \int_0^1 (4x + 6x^2 - 1)\, dx.$$

26.3. Draw a graph illustrating the fact that it is possible for one function $y_1(x)$ to be less than another function $y_2(x)$ at every x, where $y_1(a) = y_2(a)$, even if the derivative of y_1 is larger than the derivative of y_2.

26.4. Since the function $x^2/(1 + x^2)$ is Lipschitz continuous on any bounded interval,

$$\int_0^1 \frac{x^2}{x^2 + 1}\, dx$$

is defined even if we cannot compute it. Prove that this integral is less than or equal to one.

26.5. Prove that

$$\int_1^b (1 + x^2)^{100}\, dx \le \int_1^b (1 + x^3)^{100}\, dx$$

for all $b \ge 1$ (without computing the integrals!).

26.6. Compute the following integrals directly and using (26.3):

$$\text{(a)} \int_2^1 12x^3\, dx \quad \text{(b)} \int_0^{-1} 2x\, dx.$$

26.7. Evaluate $\int_0^4 2x^3\, dx$ directly and using (26.4) with $c = 1$.

26.8. Evaluate $\int_0^2 2x\, dx$ directly and using (26.4) with $c = -1$.

26.9. Explain how Theorem 26.2 follows from Theorem 25.4.

26.10. Compute the following derivatives using Theorem 26.2:

$$\text{(a)} \frac{d}{dx} \int_1^x (4s + 1)\, ds \quad \text{(b)} \frac{d}{dx} \int_0^{2x} 3s^2\, ds.$$

Note that (b) requires some thought!

27
Applications of the Integral

Recall that the derivative was introduced as part of the linearization of a function, but the derivative itself has a geometric interpretation as the limit of the slopes of secant lines that is important for modeling. The situation with integration is similar. We introduced integration as a method of solving the simplest differential equation. But integration also has a geometric interpretation that is important for a variety of applications. In fact, integration was originally developed based on its geometric interpretation by Greek mathematicians long before the derivative was developed. Even today in modern calculus books, integration is usually introduced by means of its geometric interpretation.

We discuss this interpretation and a couple of applications in this chapter. Geometrically, integration is a method for computing quantities like length, area, and volume as a limit of approximating sums of "small" quantities. In fact, this is exactly how Archimedes[1] used integration to approximate the area of the unit circle, which is π. In his "method of exhaustion," Archimedes approximated the area of the circle by the areas of regular polygonal figures, which are figures made up of straight edge sides, that just barely encompass the unit circle. We illustrate in Fig. 27.1. The polygonal figures in the method of exhaustion are made up of triangular sections and the area of the figure is given by summing the areas of the tri-

---

[1]Archimedes of Sicily (287–212 B.C.) is considered to be one of the greatest mathematicians and engineers of all time. His "method of exhaustion" was very important to the later development of calculus because early versions depended heavily on infinite series.

4 sides   8 sides   16 sides

FIGURE 27.1. Illustration of the method of exhaustion. The areas of the polygonal figures are, respectively, 4,   3.313708498985, and   3.182597878075, while $\pi \approx 3.14159265359$. The black regions represent the error of the approximations to the circle.

angular sections. As the number of sides of the polygonal figure increases, the number of triangles also increases while the areas of the triangles decrease (see Fig. 27.2). Hence, the approximations to the area of the unit

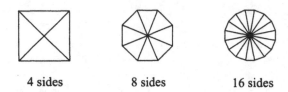

4 sides   8 sides   16 sides

FIGURE 27.2. The polygonal figures in the method of exhaustion are made up of an increasing number of smaller and smaller triangles.

circle can be written as sums of increasing numbers of ever smaller terms.

27.1   Area Under a Curve

We use a technique similar to Archimedes' method of exhaustion to define the area underneath the curve of a Lipschitz continuous function f on a bounded interval $[a, b]$ and to approximate this area to any desired degree of accuracy. It turns out that this is just the integral of f over $[a, b]$.

Geometrically, we have little trouble believing that the idea of the area under the curve makes sense (see Fig. 27.3). But at this point, we do not know that the area is well-defined mathematically. This is similar to the situation with $\sqrt{2}$. Geometrically, we defined $\sqrt{2}$ to be the length of the diagonal of a square with sides of length 1, but we were unhappy until we found a method to compute $\sqrt{2}$ to any desired accuracy and used this method to give an analytic definition of $\sqrt{2}$.

If $f(x) = c$ is constant, then there is no problem defining the area under f from a to b precisely (see Fig. 27.4). The problem in general is that f might be curved. Recall that we dealt with the same problem in defining

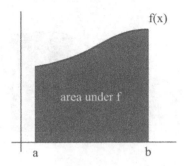

FIGURE 27.3. The area under the curve of f.

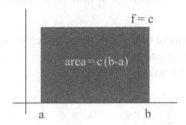

FIGURE 27.4. The area under the curve of a constant function f.

the integral by using the idea that on sufficiently small intervals a curved function looks "flat."

More precisely, we choose a **mesh** of equally spaced points $\{x_{N,i}\}$ in $[a, b]$ by setting $\Delta x_N = (b-a)/2^N$ for a natural number N and $x_{N,i} = a + i \times \Delta x_N$ for $i = 0, 1, \cdots, 2^N$. Note in particular that $x_{N,0} = a$ and $x_{N,2^N} = b$ (see Fig. 27.5). We compute an approximate area underneath f by summing

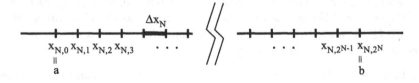

FIGURE 27.5. A mesh for $[a, b]$.

the areas of the rectangles given by the piecewise constant interpolant of f on the mesh just created (see Fig. 27.6).

The area of the rectangle on $[x_{N,i-1}, x_{N,i}]$ is $f(x_{N,i-1})\Delta x_N$ and so the area underneath the piecewise constant interpolant of f is

$$\sum_{i=1}^{2^N} f(x_{N,i-1})\Delta x_N.$$

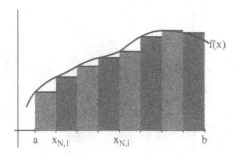

FIGURE 27.6. The area underneath the piecewise constant interpolant of f. We alternate the shading to distinguish contributions from neighboring rectangles.

The Fundamental Theorem of Calculus implies that this sum converges to a unique limit, namely, the integral of f, as the number of intervals increases, i.e., as $N \to \infty$. In other words,

$$\lim_{N \to \infty} \sum_{i=1}^{2^N} f(x_{N,i-1}) \Delta x_N = \int_a^b f(x)\, dx. \qquad (27.1)$$

So we *define* this limit, $\int_a^b f(x)\, dx$, to be the **area underneath the curve** f **from** a **to** b.[2]

EXAMPLE 27.1.   The area underneath x^2 from -1 to 1 is $\int_{-1}^1 x^2\, dx = x^3/3|_{-1}^1 = 2/3$.

This definition agrees with the couple of cases in which we can compute the area using geometric identities, as when f is constant or linear.

EXAMPLE 27.2.   The area underneath x from 0 to 1 is $\int_0^1 x\, dx = x^2/2|_0^1 = 1/2$. This agrees with the area of the triangle with corners at $(0,0)$, $(1,0)$, and $(1,1)$ given by the formula for the area of a triangle as one half the base times the height.

In situations in which f cannot be integrated exactly, we can approximate the area underneath f to any desired accuracy by computing the sum in (27.1) with a sufficiently large N.

Note that this definition of area has one peculiarity that does not fit intuition. Namely, it might very well be negative.

EXAMPLE 27.3.   The area underneath the curve $y = -x^2$ on $[1,2]$ is $\int_1^2 -x^2\, dx = -x^3/3|_1^2 = -7/3$.

---

[2] The approach of defining geometric quantities like area underneath a curve and length of a curve as the integrals of functions originates with Cauchy.

This is because the "height" of the rectangles in the sum in (27.1) can be negative or positive depending on the sign of f. In general, we could expect that f is positive on part of $[a, b]$ and negative on the rest (see Fig. 27.7). So we sometimes call $\int_a^b f(x)\, dx$ the **net area underneath the curve** f

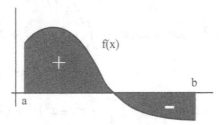

FIGURE 27.7. A function that has regions of both "positive" and "negative" areas under its graph.

from a to b.

If we want the total area between f and the x-axis summed without cancelation, then we can compute $\int_a^b |f(x)|\, dx$. We call this the **absolute area underneath** f **from** a **to** b. The absolute value has the effect of reflecting those parts of f that are negative above the x-axis, resulting in a function that is always nonnegative (see Fig. 27.8).

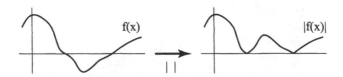

FIGURE 27.8. The action of taking the absolute value of f.

EXAMPLE 27.4. The absolute area underneath the curve $y = -x^2$ on $[1, 2]$ is $\int_1^2 |-x^2|\, dx = \int_1^2 x^2\, dx = x^3/3|_1^2 = 7/3$.

EXAMPLE 27.5. To compute the absolute area underneath $f(x) = x^3 - 3x^2 + 2x = x(x - 1)(x - 2)$ from $x = 0$ to $x = 3$, we first note that $f(x)$ is positive when $0 < x < 1$, negative when $1 < x < 2$, and positive when $2 < x < 3$. Note that these regions are bordered by the zeros of f. Since f is Lipschitz continuous, there must always be a point where

f is zero between regions on which f changes sign. So,

$$\int_0^3 |f(x)|\,dx = \int_0^1 |f(x)|\,dx + \int_1^2 |f(x)|\,dx + \int_2^3 |f(x)|\,dx$$

$$= \int_0^1 f(x)\,dx - \int_1^2 f(x)\,dx + \int_2^3 f(x)\,dx$$

$$= \int_0^1 (x^3 - 3x^2 + 2x)\,dx - \int_1^2 (x^3 - 3x^2 + 2x)\,dx$$

$$+ \int_2^3 (x^3 - 3x^2 + 2x)\,dx$$

$$= (\frac{x^4}{4} - x^3 + x^2)|_0^1 - (\frac{x^4}{4} - x^3 + x^2)|_1^2 + (\frac{x^4}{4} - x^3 + x^2)|_2^3$$

$$= \frac{1}{4} - \left(-\frac{1}{4}\right) + \frac{9}{4} = \frac{11}{4}.$$

Based on this discussion, we define the **area between the curves f and g from a to b** as

$$\int_a^b (f(x) - g(x))\,dx$$

(see Fig. 27.9). We define the **absolute area between f and g from a**

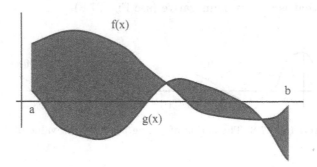

FIGURE 27.9. The area between the curves f and g.

to b as

$$\int_a^b |f(x) - g(x)|\,dx.$$

EXAMPLE 27.6.   We compute the absolute area between the curves $8 - x^2$ and x^2. When the interval is not specified, by default we take the interval defined by the points of intersection of the two curves. Here

$8 - x^2 = x^2$ at $x^2 = 4$ or $x = -2$ and 2. Since $8 - x^2 \geq x^2$ on $[-2, 2]$, we compute

$$\int_{-2}^{2} (8 - x^2 - x^2) \, dx = (8x - \frac{2}{3}x^3)\big|_{-2}^{2} = \frac{64}{3}.$$

27.2   Average Value of a Function

As another application of the idea that integration is a sum of small quantities, we consider the problem of defining the average value of a function.

The **average value** $\bar{f}$ of a set of N numbers $\{f_1, f_2, \cdots, f_N\}$ is defined

$$\bar{f} = \frac{f_1 + f_2 + \cdots + f_N}{N} = f_1 \frac{1}{N} + f_2 \frac{1}{N} + \cdots + f_N \frac{1}{N}.$$

The average is an important quantity in statistics, as any student who has been graded on a curve knows.

We can think of the N numbers $\{f_1, f_2, \cdots f_N\}$ as the N values of a function f with domain $\{1, 2, \cdots, N\}$. So we have defined the average value of a function with a discrete domain. We want to extend this idea to functions $f(x)$ of a real variable x on $[a, b]$. The difficulty is that there are an infinite number of x in $[a, b]$, so the analog of summing the values is not so obvious.

A way around this difficulty is to sample f at a set of N points $\{x_1, \cdots, x_N\}$ and then let N tend to infinity. We want to choose the sample points more-or-less well-distributed across the interval $[a, b]$. So we choose a **mesh** of equally spaced points $\{x_{N,i}\}$ in $[a, b]$ by setting $\Delta x_N = (b - a)/2^N$ for a natural number N and $x_{N,i} = a + i \times \Delta x_N$ for $i = 0, 1, \cdots, 2^N$. Note in particular that $x_{N,0} = a$ and $x_{N,2^N} = b$. This is the usual mesh we use for integration. The average of f's values on the nodes of the mesh is therefore

$$\sum_{i=1}^{2^N} f(x_{N,i-1}) \frac{1}{N}.$$

From the definition of Δx_N,

$$\sum_{i=1}^{2^N} f(x_{N,i-1}) \frac{1}{N} = \frac{1}{b-a} \sum_{i=1}^{2^N} f(x_{N,i-1}) \frac{b-a}{N} = \frac{1}{b-a} \sum_{i=1}^{2^N} f(x_{N,i-1}) \Delta x_N.$$

Since the quantity on the right tends to a unique limit, namely, the integral of f, as $N \to \infty$, we define the **average value** of f on $[a, b]$ to be

$$\bar{f} = \frac{1}{b-a} \int_a^b f(x) \, dx.$$

EXAMPLE 27.7. The average value of $f(x) = x^3$ on $[0,2]$ is

$$\frac{1}{2} \int_0^2 x^3 \, dx = \frac{16}{8} = 2.$$

EXAMPLE 27.8. The average value of $f(x) = x$ on $[-1,1]$ is

$$\frac{1}{2} \int_{-1}^1 x \, dx = \frac{1}{2} \frac{x^2}{2} \Big|_{-1}^1 = 0.$$

Recall that sometimes we want to "weight" some of the numbers more than others when computing an average. We call the numbers $\{\omega_1, \omega_2, \cdots, \omega_N\}$ a set of **weights** if they are nonnegative and $\omega_1 + \omega_2 + \cdots + \omega_N = N$. In this case, we define the **weighted average** of $\{f_1, f_2, \cdots, f_N\}$ as

$$\bar{f} = f_1 \frac{\omega_1}{N} + f_2 \frac{\omega_2}{N} + \cdots + f_N \frac{\omega_N}{N}.$$

To generalize this to functions of real variables, we call a Lipschitz continuous function $\omega(x)$ a **normalized weight function**, or **weight function**, if $\omega(x) \geq 0$ for all x in $[a,b]$ and $\int_a^b \omega(x) \, dx = b-a$. We define the **weighted average** of f on $[a,b]$ with respect to ω as

$$\bar{f} = \frac{1}{b-a} \int_a^b f(x)\omega(x) \, dx.$$

EXAMPLE 27.9. The average values of $f(x) = 2 - x$ on $[0,1]$ with respect to weights 1 and $2x$ are

$$\frac{1}{1} \int_0^1 (2-x) \, dx = \frac{3}{2} \quad \text{and} \quad \frac{1}{1} \int_0^1 (2-x) \, 2x \, dx = \frac{4}{3}.$$

It is sometimes inconvenient to choose ω so that $\int_a^b \omega(x) \, dx = b-a$. If we assume only that $\int_a^b \omega(x) \, dx > 0$, then we define the **weighted average** of f to be

$$\bar{f} = \frac{\int_a^b f(x) \, dx}{\int_a^b \omega(x) \, dx}$$

EXAMPLE 27.10. The average value of $f(x) = 2 - x$ on $[0,1]$ with respect to x is

$$\frac{\int_0^1 (2-x) \, x \, dx}{\int_0^1 x \, dx} = \frac{2/3}{1/2} = \frac{4}{3}.$$

Chapter 27   Problems

27.1. Verify the formula for half the length of a side of a n-sided regular polygonal figure in Fig. 27.10 and use the formula to compute approximations to π using $n = 4, 8, 16, 32$ sides.

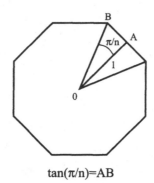

$$\tan(\pi/n) = AB$$

FIGURE 27.10. Archimedes' computation for the area of the polygonal figure with n sides.

27.2. Compute the area and absolute area of the following functions on the indicated intervals:

(a) $f(x) = x - 2$ on $[-1, 4]$

(b) $f(x) = x(x - 2)(x + 3)$ on $[-4, 4]$.

27.3. Compute the area and absolute area between the curves $f(x) = 2x^3$ and $g(x) = x^2 + 4x - 3$.

27.4. Compute the average values of $f(x) = x + x^2$ with on $[0, 2]$ with respect to 1, x, and x^2.

27.5. Prove that $\overline{|f|} \leq |\bar{f}|$ for any Lipschitz continuous function.

27.6. Assume that ω_1 and ω_2 are two weight functions on $[a, b]$. Suppose that the average value of some function f with respect to ω_1 is less than or equal to its average with respect to ω_2. Does it follow that $\omega_1(x) \leq \omega_2(x)$ for all x in $[a, b]$? Does it follow that $\int_a^b \omega_1(x)\, dx \leq \int_a^b \omega_2(x)\, dx$? When answering these questions, either provide a proof or a counterexample. Repeat the questions, but now assume that the average value of f with respect to ω_1 is less than the average value with respect to ω_2 for *all* Lipschitz continuous functions.

27.7. Prove that the average value of f on $[a, b]$ with respect to 1 satisfies

$$\min_{[a,b]} f \leq \bar{f} \leq \max_{[a,b]} f$$

assuming the min and max values are well-defined. Does the result hold for arbitrary weight functions?

28

Rocket Propulsion and the Logarithm

Over the next three chapters, we analyze and solve three important differential equations that appear repeatedly in mathematical modeling. In each case, solving the equation requires defining a "new" function; i.e., the solution is not to be found among the set of rational functions. These new functions are essential to analysis.

The first problem is $y'(x) = 1/x$, and the solution is called the logarithm. The logarithm was actually invented by Napier[1] in the seventeenth century as a method for simplifying complicated computations and it immediately became an essential tool in engineering and science. The slide rule, which was the standard calculating device before the electronic calculator, is based on the logarithm. However, the advent of the calculator has reduced the need to use logarithms and engineering students are saved from having to memorize a table of logarithm values and learning to use a slide rule. Nonetheless, the logarithm is still useful as a computational tool in many situations and remains important in modeling.

28.1 A Model of Rocket Propulsion

The Law of Conservation of Momentum says,

---

[1] John Napier (1550–1617) was a wealthy Scottish landowner who studied mathematics as a hobby.

If the sum of external forces acting on a system of particles is zero, the total momentum of the system remains constant.

We apply this law to obtain a differential equation describing the acceleration of a rocket in terms of its mass, the rate that fuel is being consumed, and the rate the gases coming from the burnt fuel are ejected. We assume that the rocket is in space sufficiently far from any planets that the effect of gravitation can be ignored.[2]

When the rocket's engine is fired, the exhaust gases from the burnt fuel shoot backward at high speed and the rocket moves forward so as to balance the momentum of the gases (see Fig. 28.1). Neglecting external forces, the total momentum of the exhaust+rocket should remain zero.

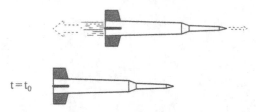

FIGURE 28.1. A model of rocket propulsion.

We let $m(t)$ denote the mass of the rocket and $v(t)$ its velocity at time t and assume that the exhaust gases are ejected at a constant rate u, which is approximately true in practice. We derive a differential equation for v by considering the changes that occur in the rocket-exhaust system from time t, greater than some initial time t_0, to a later time $s \geq t$, from the point of view of an observer watching the rocket from a stationary position. We assume that the mass m and the velocity v remain strongly differentiable functions for $t \geq t_0$.

The change in momentum associated with the expelled gas is equal to the mass of the gas ejected times the velocity of the gas. By the conservation of mass, the mass of the gas must equal the change of mass of the rocket, which is $m(t) - m(s)$.[3] The velocity of the gas is the sum of the velocity of the gas relative to the rocket and the velocity of the rocket relative to the observer, $u + v$. Hence, the change in momentum of the expelled gas is

$$(m(t) - m(s))(u + v(s)).$$

The change in momentum of the rocket is simply

$$m(s)v(s) - m(t)v(t).$$

---

[2] A rocket lifting off from the Earth's surface is affected by the force of gravity and the friction created by the atmosphere, so the equation describing launch is more complicated.

[3] Note that $m(t) > m(s)$.

The conservation of momentum therefore implies that

$$(m(t) - m(s))(u + v(s)) + m(s)v(s) - m(t)v(t) = 0.$$

Dividing by $s - t$, assuming that $s > t$, gives

$$-(u + v(s))\frac{m(s) - m(t)}{s - t} + \frac{m(s)v(s) - m(t)v(t)}{s - t} = 0.$$

By the assumption of strong differentiability, letting $s \downarrow t$, we conclude that v and m satisfy

$$-(u + v(t))m'(t) + \frac{d(m(t)v(t))}{dt} = 0.$$

By the product rule

$$-(u + v(t))m'(t) + \frac{d(m(t)v(t))}{dt}$$
$$= -um'(t) - v(t)m'(t) + m(t)v'(t) + m'(t)v(t)$$
$$= -um'(t) + m(t)v'(t).$$

So assuming that $m(t) > 0$, we obtain

$$\frac{v'(t)}{u} = \frac{m'(t)}{m(t)} \text{ for } t \geq t_0. \tag{28.1}$$

To specify a particular solution, (28.1) is supplemented with an initial condition

$$v(t_0) = v_0, \quad m(t_0) = m_0 > 0. \tag{28.2}$$

Equation (28.1) can be used in various ways to determine information about a rocket flight.

EXAMPLE 28.1. To obtain an acceleration of $5g$ while expending the fuel at 2000 m/s (a typical value), the fuel must be consumed so that

$$\frac{m'(t)}{m(t)} = -.0025.$$

A solution of this equation indicates how long a rocket of a given initial mass can maintain an acceleration of $5g$.

We can use (28.2) to conclude there is a solution as long as $m(t) > 0$. The Fundamental Theorem of Calculus implies

$$\frac{1}{u}(v(t) - v_0) = \int_{t_0}^{t} \frac{m'(s)}{m(s)} ds, \tag{28.3}$$

provided the integral on the right exists. But m'/m is a Lipschitz continuous function as long as there is a constant $c > 0$ such that $m(t) \geq c > 0$. Since $m_0 > 0$, there is a $c > 0$ such that this holds for some time interval starting at t_0, if not for all time. Therefore the integral does exist as long as $m(t) > 0$; and moreover it defines a strongly differentiable function, which is v of course.

We can use substitution to rewrite (28.3) in a more tractable form. We define $x(t) = m(t)$ so that $dx = m' \, dt$ and

$$\frac{1}{u}(v(t) - v_0) = \int_{m_0}^{m(t)} \frac{dx}{x}. \tag{28.4}$$

This motivates the study of the integral

$$\int \frac{dx}{x} = \int x^{-1} \, dx.$$

Note that the integrand does *not* fit into the formulas for differentiating powers of x from Chapter 20. Somehow x^{-1} is a special case. In fact, we have to define a new function, which turns out to be the logarithm. This is an entirely new situation. Up to now, we have dealt with integrals of functions that give new functions that can be written down explicitly. Now, we are faced with an integral that is defined but for which we do not know the resulting function.

28.2   The Definition and Graph of the Logarithm

The **natural logarithm function**, which is called the **logarithm** for short and written as $y = \log(x)$, is defined by

$$\log(x) = \int_1^x \frac{1}{s} \, ds. \tag{28.5}$$

Equivalently, log is the unique solution of the differential equation

$$\begin{cases} y'(x) = 1/x, \\ y(1) = 0. \end{cases} \tag{28.6}$$

We know the integral is defined but we cannot use standard formulas to evaluate it. We do know however that the logarithm is strongly differentiable for $x > 0$.

We can use Riemann sums to approximate values of log at any desired point.[4] However, we would also like to determine general information about

---

[4]The efficient and accurate approximation of functions like log is an interesting and difficult subject that we cannot discuss here in sufficient detail to make it worthwhile. We leave it as a mystery: what exactly does a calculator do when you press the log or any other function key?

the function. It turns out that a surprising amount of information can be determined from the fact that log satisfies the differential equation (28.6).

For example, since $1/s > 1/x$ on the interval $[1, x]$, it follows from the monotonicity property of integration that for $x > 1$,

$$\log(x) = \int_1^x \frac{1}{s}\, ds \geq \int_1^x \frac{1}{x}\, ds = \frac{1}{x}(x - 1) = 1 - \frac{1}{x},$$

which implies that $\log(x) > 0$.[5] For $x < 1$,

$$\log(x) = \int_1^x \frac{1}{s}\, ds = -\int_x^1 \frac{1}{s}\, ds$$

so the same argument shows that $\log(x) < 0$. Summarizing,

$$\begin{cases} \log(x) > 0, & 1 < x, \\ \log(1) = 0, & \\ \log(x) < 0, & 0 < x < 1. \end{cases} \qquad (28.7)$$

We can also find out a lot about the shape of the graph of log. For example,

$$\frac{d}{dx}\log(x) = \frac{1}{x} > 0;$$

hence log is a monotone increasing function, i.e.,

$$x_1 < x_2 \text{ implies } \log(x_1) < \log(x_2). \qquad (28.8)$$

Going a little deeper, the second derivative of log is

$$\frac{d^2}{dx^2}\log(x) = \frac{d}{dx}\frac{1}{x} = \frac{-1}{x^2} < 0.$$

Since the second derivative is negative, this means the first derivative is a monotone decreasing function.[6] This means that while log is an increasing function, it increases at a steadily decreasing rate!

We finish by presenting a plot of log in Fig. 28.2.

28.3   Two Important Properties of the Logarithm

If a is a constant, then the Chain Rule implies

$$\frac{d}{dx}\log(ax) = \frac{1}{ax}a = \frac{1}{x} = \frac{d}{dx}\log(x),$$

---

[5]Note that the integration variable in these integrals is s, and x is treated like any other constant when we compute the integral.
[6]This can also be seen by plotting $1/x$ of course.

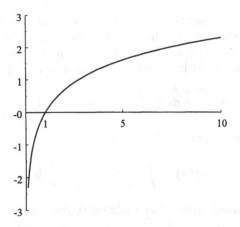

FIGURE 28.2. A plot of the logarithm log.

or in other words,

$$\frac{d}{dx}(\log(ax) - \log(x)) = 0$$

for all x. So $\log(ax) - \log(x)$ is a constant. Setting $x = 1$ gives

$$\log(ax) - \log(x) = \log(a \times 1) - 0 = \log(a).$$

Renaming a to be y, we obtain a **functional relation** for log,

$$\log(xy) = \log(x) + \log(y), \tag{28.9}$$

which holds for all $y, x > 0$ This equation is the reason that the logarithm is so useful for some kinds of computations.

One consequence is another functional equation. Setting $x = x^{n-1}$ and using (28.9) twice gives

$$\begin{aligned}
\log(x^n) &= \log(x) + \log(x^{n-1}) \\
&= \log(x) + \log(x \times x^{n-2}) \\
&= \log(x) + \log(x) + \log(x^{n-2}) \\
&= 2\log(x) + \log(x^{n-2}).
\end{aligned}$$

Induction yields

$$\log(x^n) = n\log(x) \tag{28.10}$$

for all $x > 0$ and positive integers n. If n is a negative integer, we set $m = -n$ so m is a positive integer, and use (28.9) and (28.10) to find

$$\log(x^n) = \log(x^{-m}) = \log\left(\frac{1}{x^m}\right) = 0 - \log(x^m)$$

$$= -m\log(x) = n\log(x).$$

Since $x = (x^{1/n})^n$, (28.10) implies

$$\log(x) = \log\big((x^{1/n})^n\big) = n\log(x^{1/n})$$

or solving

$$\log(x^{1/n}) = \frac{1}{n}\log(x).$$

Putting this all together, if p and q are integers, then

$$\log(x^{p/q}) = \log\big((x^{1/q})^p\big) = p\log(x^{1/q}) = \frac{p}{q}\log(x);$$

or in other words

$$\log(x^r) = r\log(x) \qquad\qquad (28.11)$$

for all $x > 0$ and rational numbers r.

EXAMPLE 28.2. Given that $\log(2) \approx .693$ and $\log(3) \approx 1.10$, we can estimate

$$\log(12) = \log(2^2 \times 3) = 2\log(2) + \log(3) \approx 2.49.$$

EXAMPLE 28.3. Given $\log(a) = 2$ and $\log(b) = -.1$, we can compute

$$\log\left(\frac{a^3}{b^2}\right) = 3\log(a) - 2\log(b) = 6 + .2 = 6.2.$$

28.4   Irrational Exponents

One consequence of the functional equation (28.10) is

$$\lim_{x\to 0^-}\log(x) = -\infty \quad\text{and}\quad \lim_{x\to\infty}\log(x) = \infty.$$

To prove the second limit for example, we use the fact that $\log(2) > 0$ which means that

$$\log(2^n) = n\log(2)$$

can be made as large as desired by taking n large. The first limit follows from the fact that $\log(x^{-1}) = -\log(x)$.

By the Intermediate Value Theorem, this means that log takes on every value between $-\infty$ and ∞. In other words, given a number a there is a number x such that $\log(x) = a$. Moreover, x is unique because log is strictly increasing and cannot have the same value at two different points.

Using the logarithm makes it easy to define b^a for irrational values of a and $b > 0$ or $b \geq 0$. For example, we might want to compute $3^{\sqrt{2}}$. The definition is based on (28.11). For any $b > 0$ and real number a, the number

b^a is *defined* to be the unique number whose natural logarithm is $a \log(b)$.[7] In other words, b^a is the unique number such that

$$\log(b^a) = a \log(b). \tag{28.12}$$

When $b = 0$ and $a > 0$, we define $b^a = 0$. We know (28.12) is defined based on the discussion above. With this definition, it turns out that all the standard properties of exponents hold. For all real a_1 and a_2 and positive reals b_1 and b_2,

$$1^a = 1, \ b_1^{a_1 + a_2} = b_1^{a_1} b_1^{a_2}, \ b_1^{a_1 - a_2} = \frac{b_1^{a_1}}{b_2^{a_2}}, \ (b_1^{a_1})^{a_2} = b_1^{a_1 a_2}, \ (b_1 b_2)^{a_1} = b_1^{a_1} b_2^{a_1}.$$

$$\tag{28.13}$$

In each case, the property is true when both sides of the equality have the same logarithm.

EXAMPLE 28.4. To verify the last property in (28.13), we first note that by the definition $(b_1 b_2)^{a_1}$ is the unique number such that $\log((b_1 b_2)^{a_1}) = a_1 \log(b_1 b_2)$. But (28.9) therefore implies that $\log((b_1 b_2)^{a_1}) = a_1 \log(b_1) + a_1 \log(b_2)$, which by definition means $\log((b_1 b_2)^{a_1}) = \log(b_1^{a_1}) + \log(b_2^{a_1})$. Using (28.9) again gives the desired $\log((b_1 b_2)^{a_1}) = \log(b_1^{a_1} b_2^{a_1})$. We ask you to verify the others in Problem 28.10.

28.5 Power Functions

Given that we have defined b^a uniquely for each $b > 0$ and real a and for $b = 0$ when $a > 0$, we are in a position to define the power function $f(x) = x^a$ for any real a. The value of x^a is determined uniquely by

$$\log(x^a) = a \log(x)$$

when $x > 0$ (and is zero when $x = 0$ and $a > 0$). The domain of x^a is the positive reals for any a and the nonnegative reals for any real $a > 0$.

From this definition, it follows that x^a is strongly differentiable when $x > 0$ and the derivative of x^a can be computed by differentiating both sides of (28.12)

$$\frac{d}{dx} \log(x^a) = \frac{d}{dx} a \log(x),$$

or using the Chain Rule,

$$\frac{1}{x^a} \frac{d}{dx} (x^a) = a \frac{1}{x}.$$

---

[7] It is important to note (28.11) implies this definition coincides with the value for x^n defined earlier for integer n.

Solving gives

$$\frac{d}{dx}(x^a) = ax^{a-1}$$

which is the same power rule that holds for rational powers!

EXAMPLE 28.5.

$$\frac{d}{dx}x^{\sqrt{2}} = \sqrt{2}\,x^{\sqrt{2}-1}.$$

When $a > 0$, determining if x^a is strongly differentiable at $x = 0$ is more difficult. If a is any integer, then x^a is strongly differentiable at 0. So consider the case when $0 < a < 1$ first. Then for $x > 0$,

$$\frac{d}{dx}x^\alpha = \alpha x^{\alpha-1} = \frac{\alpha}{x^{1-\alpha}},$$

where $1 - \alpha > 0$. Hence,

$$\lim_{x\downarrow 0}\frac{d}{dx}x^\alpha$$

is undefined. We conclude that x^a is *not* differentiable at $x = 0$ when $0 < \alpha < 1$. Next, we consider $\alpha > 1$. Now $D(x^\alpha) = \alpha x^{\alpha-1}$ is defined at 0 and $\lim_{x\downarrow 0} \alpha x^\alpha = 0$, so x^α is differentiable at $x = 0$. But when we try to verify the definition of strong differentiability, we get

$$|x^\alpha - (0^\alpha + \alpha 0^{\alpha-1}(x - 0)| = |x|^\alpha$$

for all $x > 0$. Clearly $|x|^\alpha \le |x|^2 \mathcal{K}$ for some constant $\mathcal{K}$ and $x > 0$ only if $\alpha \ge 2$. Hence we conclude that for $\alpha > 0$,

$$x^\alpha \text{ is not differentiable at 0 when } 0 < \alpha < 1$$
$$x^\alpha \text{ is differentiable but not strongly differentiable at 0 when } 1 < \alpha < 2$$
$$x^\alpha \text{ is strongly differentiable at 0 when } \alpha = 1 \text{ and } \alpha \ge 2.$$

We also get the integration formula for $\alpha \ne 1$,

$$\int x^\alpha \, dx = \frac{x^{\alpha+1}}{\alpha + 1} + C.$$

28.6   Change of Base

Let x and b be positive numbers with $b \ne 1$. The classical definition of the logarithm of x to the base b, is the number y such that $x = b^y$. We write this as $y = \log_b(x)$.

EXAMPLE 28.6.

$$\log_3(27) = 3, \ \log_2\left(\frac{1}{64}\right) = -6, \ \log_{10}(100) = 2.$$

Without using calculus, it is difficult to show that this definition is meaningful. However we note that if the classic definition holds, then $\log(x) = y\log(b)$ or $y = \log(x)/\log(b)$. We can go back in the opposite direction of course, and we conclude that $\log_b(x)$ is well defined and, moreover,

$$\log_b(x) = \frac{\log(x)}{\log(b)}. \tag{28.14}$$

We can define a unique number e with the property that $\log(e) = 1$ and then we have $\log(x) = \log_e(x)$.[8] This means in particular that

$$\log(e^x) = x \text{ for all } x \text{ and } e^{\log(x)} = x \text{ for } x > 0. \tag{28.15}$$

In Problem 28.13, we ask you to prove the following properties of $\log_b(x)$:

$$\log_b(1) = 0, \ \log_b(xy) = \log_b(x) + \log_b(y),$$
$$\log_b(a^x) = x\log_b(a), \ \log_b(b^x) = x. \tag{28.16}$$

EXAMPLE 28.7.   To verify that $\log_b(a^x) = x\log_b(a)$, we convert to $\log(x)$:

$$\log_b(a^x) = \frac{\log(a^x)}{\log(b)} = \frac{x\log(a)}{\log(b)} = x\frac{\log(a)}{\log(b)} = x\log_b(a).$$

Finally, (28.14) implies

$$\frac{d}{dx}\log_b(x) = \frac{1}{\log(b)x}. \tag{28.17}$$

28.7   Solving the Model of Rocket Propulsion

From (28.4), we conclude that the velocity of the rocket is given by

$$\frac{1}{u}(v(t) - v_0) = \log(m(t)) - \log(m_0)$$

or

$$v(t) = v_0 + u\log\left(\frac{m(t)}{m_0}\right).$$

Since $m(t) < m_0$ and $u < 0$ and moreover $m(t)$ decreases to zero as fuel is burned, $v(t)$ increases from the initial value v_0 as t increases.

---

[8]It is common to denote the function $\log_e(x) = \log(x)$ by $\ln(x)$ and the function $\log_{10}(x)$ by $\log(x)$.

EXAMPLE 28.8.   Continuing Example 28.1, to obtain an acceleration of $5g$ while expending the burnt fuel at a rate of 2000 m/s, we need to consume fuel so that

$$\log\left(\frac{m(t)}{m_0}\right) = -.0025t.$$

To determine m, we need to compute the inverse function to log, which we do in the next chapter.

28.8   Derivatives and Integrals Involving the Logarithm

We conclude by discussing some derivatives and integrals that involve the logarithm.

By the Chain Rule, if u is a differentiable function, then

$$\frac{d}{dx}\log(u(x)) = \frac{u'(x)}{u(x)}.$$

EXAMPLE 28.9.
$$\frac{d}{dx}\log(x^2 + x) = \frac{2x + 1}{x^2 + x}.$$

In fact, we can differentiate $\log(|u|)$ for any differentiable function u *that is never zero*. If u is never zero, then either $u(x) > 0$ for all x, in which case

$$\frac{d}{dx}\log(|u(x)|) = \frac{d}{dx}\log(u(x)) = \frac{u'(x)}{u(x)},$$

or $u(x) < 0$ for all x, so that $-u(x) > 0$ and

$$\frac{d}{dx}\log(|u(x)|) = \frac{d}{dx}\log(-u(x)) = \frac{-u'(x)}{-u(x)} = \frac{u'(x)}{u(x)}.$$

Putting these results together, we get

$$\frac{d}{dx}\log(|u(x)|) = \frac{u'(x)}{u(x)}. \tag{28.18}$$

This means that if u is never zero, then

$$\int \frac{u'(x)}{u(x)}\, dx = \log(|u(x)|) + C. \tag{28.19}$$

EXAMPLE 28.10.

$$\int \frac{1}{x-1}\, dx = \log|x-1| + C.$$

EXAMPLE 28.11.

$$\int \frac{x}{x^2-1}\, dx = \frac{1}{2}\int \frac{2x}{x^2-1}\, dx = \frac{1}{2}\log|x^2-1| + C.$$

EXAMPLE 28.12. To compute the following integral, we note that $1/x$ is never zero on $[-2, -1]$, so

$$\int_{-2}^{-1} \frac{1}{x}\, dx = \log|x|\,\Big|_{x=-2}^{x=-1} = \log|-1| - \log|-2| = -\log(2).$$

In this example, we need the absolute value signs!

Chapter 28   Problems

28.1. Write down a model of a rocket launching from the Earth's surface that includes the effects of gravity but ignores the effect of wind resistance.

28.2. Show that $\log(x) \leq x$ for all $x \geq 1$. *Hint:* First show that $1/x \leq 1$ for $x \geq 1$.

28.3. Prove that $\log(1+x) < x$ for $x > 0$. *Hint:* First show that $(1+x)^{-1} < 1$ for $x > 0$.

28.4. Show that $\log(2) \geq 1/2$ without using a calculator. *Hint:* Apply the Mean Value Theorem to $\log(x)$ on the interval $[1, 2]$.

28.5. Suppose we define a new function called lug by

$$\text{lug}(x) = \int_2^x \frac{1}{s}\, ds.$$

Find an equation relating $\text{lug}(x)$ to $\log(x)$ for all x.

28.6. (a) For $0 < x_1 < x_2$, show that

$$\log(x_2) - \log(x_1) = \int_{x_1}^{x_2} \frac{1}{s}\, ds.$$

(b) Use this to show that log is monotone increasing. *Hint:* on $[x_1, x_2]$, $1/s \geq 1/x_2$.

28.7. Make rough sketches of the following functions after specifying their domain and range:

(a) $\log(x - 2)$ (b) $\log(1 + x^2)$ (c) $-\log(x) + 1$.

28.8. Given that $\log(2) \approx .693$, $\log(3) \approx 1.10$, and $\log(5) \approx 1.61$, estimate

(a) 250   (b) $6e^2$ (c) 10/3 .

28.9. Given that $\log(a) = -1$, $\log(b) = 2$, and $\log(c) = .4$, estimate

(a) ab^2/c (b) a^b (c) ea/b .

28.10. Verify the properties (28.13).

28.11. Compute the following derivatives:

(a) x^e (b) $(x^{\sqrt{3}} - x)^3$ (c) $(x^2 + x)^{\sqrt{10}}$.

28.12. Simplify the following:

(a) $\log_5(25)$ (b) $\log_{25}(5)$ (c) $\log_{27}\left(\frac{1}{3}\right)$ (d) $\log_4(2^{1/3})$

(e) $\log_{z^3}(z^{12})$ (f) $\log_{v^2+1}(v^4 + 2v^2 + 1)$ (g) $3^{3\log_3(8)}$ (h) $u^{\log_u 2\,(25)}$.

28.13. Verify the properties (28.16).

28.14. Differentiate the following functions:

(a) $y = \log(|x^2 - x|)$ (b) $y = \log(\log(x))$ (c) $y = \log_3(9 - x)$.

28.15. Compute the following:

(a) $\displaystyle\int \frac{1}{x+1}\,dx$ (b) $\displaystyle\int_{-1}^{0} \frac{1}{2-4x}\,dx$ (c) $\displaystyle\int_{-3}^{-2} \frac{x^2}{x^3+1}\,dx$

(d) $\displaystyle\int \frac{2x - x^3}{4x^2 - x^4}\,dx$ (e) $\displaystyle\int \frac{1}{\sqrt{x}(1+\sqrt{x})}\,dx$ (f) $\displaystyle\int \frac{(\log(2x) - 4)^2}{x}\,dx$.

29

Constant Relative Rate of Change and the Exponential

Continuing the investigation of three important differential equations, the second problem we consider is $y'(x) = cy(x)$ for a constant c. The solution leads to the exponential function.

It would be hard to overstate the importance of the exponential function for analysis and mathematical modeling.[1] Exponential functions appear everywhere in the mathematics of science and engineering. Moreover, it is possible to derive both the logarithm and the trigonometric functions using the exponential function.[2]

29.1 Models Involving a Constant Relative Rate of Change

There are many models in which it is more natural to look at the relative rate of change instead of the rate of change. The **relative change** of a quantity that undergoes some change in size is defined as the change divided by the size of the quantity. We can get the percent change by multiplying the relative change by 100.

---

[1]The author recalls a final exam that his teacher Lipman Bers gave in calculus, which consisted solely of the question, "Write down everything you can about the exponential function."

[2]The relation of the exponential function to the logarithm and the trigonometric functions is a standard topic in complex analysis.

If the quantity is given by $P(t)$ at the value of the variable t, then the relative change in P from t to $t + \Delta t$ is

$$\frac{P(t + \Delta t) - P(t)}{P(t)}.$$

The **relative rate of change** is therefore naturally defined to be

$$\frac{\frac{P(t+\Delta t)-P(t)}{P(t)}}{(t + \Delta t) - t} = \frac{\frac{P(t+\Delta t)-P(t)}{P(t)}}{\Delta t} = \frac{1}{P(t)} \frac{P(t + \Delta t) - P(t)}{\Delta t}.$$

There are many situations in which the relative rate of change is constant: i.e., there is a constant k such that

$$\frac{1}{P(t)} \frac{P(t + \Delta t) - P(t)}{\Delta t} = k$$

or

$$\frac{P(t + \Delta t) - P(t)}{\Delta t} = kP(t). \tag{29.1}$$

Note that we can interpret a constant relative rate of change as saying that the rate of change of a quantity is proportional to the quantity.

If P is a differentiable function, then if we let $t + \Delta t$ approach t, i.e., let $\Delta t \to 0$, we obtain the differential equation modeling constant relative rate of change,

$$P'(t) = kP(t). \tag{29.2}$$

Many physical situations are modeled by (29.2). Among these are

- Earnings from compound interest

- Newton's Law of Cooling governing the temperature difference between an object and the surrounding medium

- Radioactive decay

- Population growth

We describe a model of the growth of a biological population.

EXAMPLE 29.1. Models of biological populations usually begin with discrete equations because the population is an integer and therefore increases or decreases by integer changes while we measure the changes at discrete time intervals. For example, we have already modeled the population of insects in Model 3.4 using a variable P_n denoting the population in year n based on an equation $P_n = RP_{n-1}$ relating P_n to the population in the previous year. This kind of model is called a **discrete** model.

Nonetheless, it is often more convenient to model the population using a differential equation. For one thing, the population might not synchronize well so there may be no natural time at which to measure the population. This is true of human populations, for example. For another thing, the population might be very numerous so that the changes in the population are relatively small, e.g., "almost" infinitesimal. For example, a typical Petri dish contains on the order of millions of cells and a change in population by a single cell is practically speaking undetectable. Lastly, and not the least important, we have many calculus tools for dealing with smooth differentiable functions that do not apply to functions that change by discrete amounts!

If the population at some discrete set of times $t_0 < t_1 < t_2 < \cdots$ is given by $P_0, P_1, P_2, \cdots$ respectively, we model the population by a differentiable function P with $P(t_n) \approx P_n$ for all n. We determine P as the solution of a differential equation constructed to model the characteristics of a particular species and set of environmental conditions. We call P a **continuous** model.

The continuous version of the standard Malthus[3] model for reproducing populations in the absence of competition for resources and predators says that the rate at which the population reproduces is proportional to the population. This gives precisely (29.2). The constant k is called the **intrinsic growth rate**. The differential equation (29.2) is usually combined with the initial population $P(0)$ in order to determine a unique function P.

29.2   The Exponential Function

The solution of (29.2) turns out to exist, however, the solution is not given by a formula involving simple functions like polynomials. Instead, we have to figure out its properties indirectly and compute its values, for example, by approximating the solution of (29.2). We begin by defining the exponential function as the inverse function to the logarithm and then using properties of the logarithm to show that the exponential function solves (29.2) and other properties.

Recall the fact that $\log_e(x) = \log(x)$, where e is the unique number such that $\log(e) = 1$ implies (28.15),

$$\log(e^x) = x \text{ for all } x \text{ and } e^{\log(x)} = x \text{ for all } x > 0.$$

---

[3]Thomas Malthus (1766–1834) was an English political economist and clergyman.

These are exactly the relationships that should hold for log and its inverse function. Recall that

$$\frac{d}{dx}\log(x) = \frac{1}{x} > 0 \text{ for all } x > 0$$

and consequently log is a monotone increasing function on the positive real numbers. The Inverse Function Theorem implies $\log(x)$ has a strongly differentiable inverse function for $x > 0$. We define the **exponential function** exp to be the inverse function of log, i.e.,

$$\exp(x) = \log^{-1}(x).$$

Because of (28.15),

$$\exp(x) = e^x,$$

where e is the unique number such that $\log(e) = 1$.

We can derive several properties of exp using the properties of log. We first show the plot of exp in Fig. 29.1. The function exp is clearly monotone

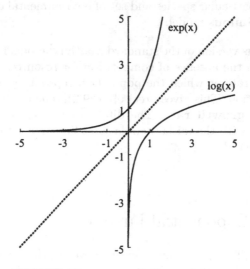

FIGURE 29.1. exp is the inverse of log.

increasing. The domain of exp is $\mathbb{R}$ while the range is all positive real numbers. Also

$$\lim_{x \to \infty} \exp(x) = \infty \text{ and } \lim_{x \to -\infty} \exp(x) = 0.$$

To compute the derivative of exp, we let $x = \log(y)$ so $y = \exp(x)$. The Inverse Function Theorem 22.2 implies that

$$\frac{d}{dx}\exp(x) = \frac{1}{\dfrac{d\log(y)}{dy}} = \frac{1}{\dfrac{1}{y}} = y = \exp(x).$$

In short,

$$\frac{d}{dx}\exp(x) = \exp(x) \text{ or } \frac{d}{dx}e^x = e^x. \qquad (29.3)$$

The exponential function is its own derivative! The Chain Rule immediately implies that

$$\frac{d}{dx}\exp(u(x)) = \exp(u(x))u'(x)$$

$$\frac{d}{dx}e^{u(x)} = e^{u(x)}u'(x),$$

and therefore

$$\int e^{u(x)}u'(x)\,dx = e^{u(x)} + C.$$

In differential notation, we have

$$de^u = e^u\,du \text{ and } \int e^u\,du = e^u + C.$$

EXAMPLE 29.2.

$$\frac{d}{dx}e^{x^3} = e^{x^3}\,3x^2.$$

EXAMPLE 29.3.

$$\frac{d}{dx}(x - 2e^{4x})^9 = 9(x - 2e^{4x})^8\,(1 - 2e^{4x} \times 4).$$

EXAMPLE 29.4.

$$\int e^{3x}\,dx = \frac{1}{3}\int e^{3x}3\,dx = \frac{1}{3}\int e^u\,du = \frac{1}{3}e^u + C = \frac{1}{3}e^{3x} + C.$$

EXAMPLE 29.5. To compute

$$\int \frac{e^x}{1 + e^x}\,dx,$$

we set $u = 1 + e^x$ so $du = e^x\,dx$ and the integral becomes

$$\int \frac{e^x}{1 + e^x}\,dx = \int \frac{du}{u} = \log(u) + C = \log(1 + e^x) + C.$$

Recall that we earlier derived properties of the exponential function such as

$$e^{-x} = 1/e^x \text{ or } \exp(-x) = 1/\exp(x)$$
$$e^{x+y} = e^x e^y \text{ or } \exp(x + y) = \exp(x)\exp(y)$$
$$(e^x)^y = e^{xy} \text{ or } (\exp(x))^y = \exp(xy).$$

29.3   Solution of the Model for Constant Relative Rate of Change

Returning to the model problem (29.2), we show that

$$y = y_0 e^{kx}$$

is the unique solution of the differential equation describing a constant relative rate of change,

$$\begin{cases} y'(x) = ky(x), & 0 \le x, \\ y(0) = y_0. \end{cases} \tag{29.4}$$

Actually, differentiation shows that y satisfies the differential equation immediately. But we need to show that there is no other solution. Note that (29.4) is different from the problem $y'(x) = f(x)$ that we solved previously because now the derivative of y depends on the value of y. In particular, we cannot apply the Fundamental Theorem of Calculus to compute the solution directly. However, there is a trick that allows (29.4) to be rewritten so the Fundamental Theorem of Calculus can be used. This trick works on some differential equations that have a special form, such as (29.4).

We know that the solution y of (29.4) satisfies

$$y'(x) - ky(x) = 0. \tag{29.5}$$

The trick for finding the solution of this equation is based on the observation that the Product Rule implies that

$$\frac{d}{dx}\left(e^{-kx}y(x)\right) = e^{-kx}y'(x) - ke^{-kx}y(x)$$

for any strongly differentiable function y. Therefore, we multiply (29.5) by the **integrating factor** e^{-kx} to get

$$e^{-kx}y'(x) - ke^{-kx}y(x) = 0,$$

which means that

$$\frac{d}{dx}\left(e^{-kx}y(x)\right) = 0. \tag{29.6}$$

The Fundamental Theorem of Calculus says that the only strongly differentiable solution of the differential equation $z' = 0$ is $z = $ a constant. Therefore, (29.6) implies that there is a constant C with

$$e^{-kx}y(x) = C \text{ or } y = Ce^{kx}.$$

Using the condition $y(0) = y_0$ proves the result.

EXAMPLE 29.6. We apply this result to analyze a model of radioactive decay. If a sample of material has P atoms of a radioactive element such as radium, it is found that after a short time Δt, approximately $kP\Delta t$ of the atoms disintegrate. Recalling (29.1), we therefore approximate the amount of atoms $P(t)$ at time t by a continuous function satisfying $P'(t) = -kP(t)$. Note that k is arbitrarily assumed to be a positive constant in this subject and therefore this introduces a minus sign into the differential equation to guarantee that P is a decreasing function.

The standard way to measure how much of the radioactive element has decayed is to measure the **half-life**, which is the length of time required for a given amount to be reduced by half. If P_0 is the initial amount, then the half-life $t_{1/2}$ is determined by the equation

$$P(t_{1/2}) = \frac{1}{P_0} = P_0 e^{-kt_{1/2}}. \tag{29.7}$$

Note that the initial amount P_0 factors out of this equation, so the half-life is independent of the initial amount,

$$\frac{1}{2} = e^{-kt_{1/2}}$$

or solving for k

$$k = \frac{\log(2)}{t_{1/2}}.$$

Using the half-life formula, we can rewrite the solution as

$$P(t) = P_0 e^{-\log(2)t/t_{1/2}} = \frac{P_0}{2^{t/t_{1/2}}}.$$

EXAMPLE 29.7. The half-life of radium is 1656 years. How much radium is left after 2834 years in an object that initial contained 3 grams of radium? The answer is $3/2^{2834/1656} \approx .92$ grams. How long does it take for the same object to contain .1 gram? We solve

$$.1 = 3e^{-\log(2)t/1656}$$

to get $t \approx 8126$ years.

29.4 More on Integrating Factors

We can use the technique of multiplying by an integrating factor to solve many differential equations in the form

$$y'(x) + a(x)y = b(x). \tag{29.8}$$

We cannot apply the Fundamental Theorem of Calculus to (29.8) directly because of the presence of both y' and y. But the Chain Rule and the Fundamental Theorem of Calculus imply

$$\frac{d}{dx}e^{\int a(x)\,dx} = e^{\int a(x)\,dx} \times \frac{d}{dx}\int a(x)\,dx = e^{\int a(x)\,dx}a(x).$$

EXAMPLE 29.8.
$$\frac{d}{dx}e^{\int x^3\,dx} = e^{\int x^3\,dx} \times 3x^2.$$

If we multiply both sides of (29.8) by the integrating factor $e^{\int a(x)\,dx}$, then we get

$$\frac{d}{dx}\left(e^{\int a(x)\,dx}y(x)\right) = e^{\int a(x)\,dx}y'(x) + e^{\int a(x)\,dx}a(x)y(x)$$
$$= e^{\int a(x)\,dx}b(x).$$

Since this holds for any fundamental solution given by $\int a(x)\,dx$, we use the solution that is zero at $x = 0$. Now the Fundamental Theorem of Calculus says that the only solution of $z' = f(x)$ is $z = \int f(x)\,dx$. If we apply this to

$$\frac{d}{dx}\left(e^{\int a(x)\,dx}y(x)\right) = e^{\int a(x)\,dx}b(x),$$

we conclude that

$$e^{\int a(x)\,dx}y(x) = \int e^{\int a(x)\,dx}b(x)\,dx$$

or

$$y(x) = e^{-\int a(x)\,dx}\int e^{\int a(x)\,dx}b(x)\,dx. \qquad (29.9)$$

EXAMPLE 29.9. To solve

$$y' + 3x^2y = x^2$$

we multiply both sides by

$$e^{\int 3x^2\,dx} = e^{x^3}$$

to get

$$\frac{d}{dx}\left(e^{x^3}y\right) = e^{x^3}y' + 3x^3e^{x^3}y = e^{x^3}x^2.$$

This means that

$$e^{x^3}y = \int e^{x^3}x^2\,dx.$$

To do this integral, we set $u = x^3$ so $du = 3x^2\,dx$ and we get

$$e^{x^3}y = \frac{1}{3}e^{x^3} + C$$

for some constant C. Dividing we conclude that

$$y = \frac{1}{3} + Ce^{-x^3}.$$

Note that in all of these examples, we should be careful to specify the domains of the functions involved. We leave that as an exercise (of course).

29.5   General Exponential Functions

In the same way that we define logarithms to different bases based on the logarithm, we define general exponential functions in terms of $\exp(x)$. Since

$$a = e^{\log(a)}$$

for $a > 0$ and

$$\left(e^{\log(a)}\right)^x = e^{x\log(a)}$$

for $a > 0$ and all x, we define the general exponential function

$$a^x = e^{x\log(a)}.$$

Using this definition, we can prove several properties of a^x such as

$$a^0 = 1, \quad a^x a^y = a^{x+y}, \quad (a^x)^y = a^{xy}, \quad a^{-x} = 1/a^x \qquad (29.10)$$

by using the corresponding properties of e^x.

EXAMPLE 29.10. By the definition and properties of e^x,

$$a^x a^y = e^{x\log(a)}e^{y\log(a)} = e^{(x+y)\log(a)} = a^{x+y}.$$

We immediately obtain the derivative formula

$$\frac{d}{dx}a^x = \log(a) \times a^x$$

or

$$da^u = \log(a) \times a^u\,du.$$

This means that

$$\int a^u\,du = \frac{1}{\log(a)}a^u + C.$$

EXAMPLE 29.11.

$$\frac{d}{dx}3^{x^2} = \log(3) \times 3^{x^2} \times 2x.$$

EXAMPLE 29.12.

$$\int 10^x \, dx = \frac{1}{\log(10)}10^x + C.$$

We can define more general exponential functions using the same idea. If $f(x) > 0$ for all x in a domain, then we define

$$f(x)^{g(x)} = e^{g(x)\log(f(x))}.$$

The derivative of such a function is therefore

$$\frac{d}{dx}f(x)^{g(x)} = \frac{d}{dx}e^{g(x)\log(f(x))}$$

$$= e^{g(x)\log(f(x))}\left(g(x)\frac{f'(x)}{f(x)} + g'(x)\log(f(x))\right)$$

$$= f(x)^{g(x)}\left(g(x)\frac{f'(x)}{f(x)} + g'(x)\log(f(x))\right).$$

EXAMPLE 29.13. For $x > 0$,

$$x^x = e^{x\log(x)}$$

and

$$\frac{d}{dx}x^x = \frac{d}{dx}e^{x\log(x)} = e^{x\log(x)}\left(1 + \log(x)\right) = x^x\left(1 + \log(x)\right).$$

We can also compute the derivative of such functions using **logarithmic differentiation**.

EXAMPLE 29.14. To compute the derivative of $y = x^x$, we first take the logarithm of both sides to get

$$\log(y) = \log(x^x) = x\log(x).$$

Now differentiation gives

$$\frac{1}{y}y'(x) = 1 + \log(x),$$

which means that

$$y'(x) = x^x\left(1 + \log(x)\right).$$

29.6   Rates of Growth of the Exponential and Logarithm

The plot of $\exp(x)$ in Fig. 29.1 suggests that the exponential function increases more and more rapidly as x increases. In fact, we show that $\exp(x)$ grows faster than any power of x. To be precise, given any $p > 0$, we prove that

$$e^x > x^p \text{ for all sufficiently large } x. \tag{29.11}$$

The first step to show (29.11) is to show that for natural number n,

$$e^x > 1 + x + \frac{x^2}{2!} + \cdots + \frac{x^n}{n!} + \frac{x^{n+1}}{(n+1)!}, \tag{29.12}$$

where for a natural number $n \geq 1$, we define $n!$, or n **factorial**,[4] by

$$n! = n \times (n-1) \times (n-2) \times \cdots \times 1$$

with $0! = 1$.

EXAMPLE 29.15.

$$1! = 1, \quad 2! = 2, \quad 3! = 6, \quad 4! = 24, \quad 5! = 120.$$

This is just an exercise in induction. For the first case, we use the monotonicity of the integral and the fact that $e^x > 1$ for $x > 0$ to get

$$e^x = 1 + \int_0^x e^s \, ds > 1 + \int_0^x 1 \, ds = 1 + x.$$

Now we repeat the argument, using the newly acquired fact that $e^x > 1 + x$ for $x > 0$,

$$e^x = 1 + \int_0^x e^s \, ds > 1 + \int_0^x (1 + s) \, ds = 1 + x + \frac{x^2}{2}.$$

Continuing in this way, induction gives (29.12). The result (29.11) for an integer power follows easily since

$$1 + x + \frac{x^2}{2!} + \cdots + \frac{x^n}{n!} + \frac{x^{n+1}}{(n+1)!} > \frac{x^{n+1}}{(n+1)!} = x^n \frac{x}{(n+1)!}$$

so

$$e^x > x^n \frac{x}{(n+1)!} > x^n$$

---

[4]The quantity $n!$ can be interpreted as the number of different ways that n objects can be arranged in a sequence from right to left. For we can choose any one of n objects for the first position, then any one of $n-1$ for the second, and so on to get $n \times (n-1) \times (n-2) \times \cdots \times 1$ possible arrangements.

for $x > (n+1)!$. The proof for a general power p follows by choosing an integer $n > p$ for any given p.

It is a good exercise to show that (29.11) implies that for any $p > 0$,

$$x^p > \log(x) \text{ for all sufficiently large } x. \tag{29.13}$$

which means that $\log(x)$, while increasing monotonically, grows slower than any power of x.

29.7   Justification of the Continuous Model

We began this chapter by describing situations in which a quantity that changes by discrete amounts can be approximated by a continuously changing quantity that solves a differential equation. In this section, we justify this approximation in the case of a constant relative rate of change.

We already discussed the modeling of populations. As another example, we consider compound interest.

EXAMPLE 29.16. We model the amount of money in a savings account in which the interest is compounded at a stated periodic interval, which means that the amount of money earned from interest is added back to the current amount in the account at regular time intervals. Suppose we open an account with an initial amount P_0 in a bank that uses an annual interest rate of α percent which is compounded at a regular time interval. The problem is to determine how much we have in the account after one year.

If the interest rate is compounded annually, then we earn

$$\frac{\alpha}{100} P_0$$

at the end of the year, and we end up with

$$(1 + \frac{\alpha}{100}) P_0$$

in the account. To simplify things, we set $\beta = \alpha/100$ so we get

$$(1 + \beta) P_0$$

at the end of a year in an account in which the interest is compounded annually.

If instead the interest rate is compounded bi-annually, then after six months we have

$$\left(1 + \frac{\beta}{2}\right) P_0$$

in the account, which reflects that idea that after six months we only earn half the interest. But this amount stays in the account, so after the second six months, we have

$$\left(1 + \frac{\beta}{2}\right)\left(1 + \frac{\beta}{2}\right)P_0 = \left(1 + \frac{\beta}{2}\right)^2 P_0$$

in the account. This is certainly an improvement over an account in which the interest is compounded yearly since

$$\left(1 + \frac{\beta}{2}\right)^2 = 1 + \beta + \frac{\beta^2}{4} > 1 + \beta.$$

Continuing this idea, if the interest is compounded quarterly, then we end up with

$$\left(1 + \frac{\beta}{4}\right)^4 P_0$$

in the account after a year, and if it is compounded daily, then we end up with

$$\left(1 + \frac{\beta}{365}\right)^{365} P_0.$$

We conclude that if the interest is compounded n times during the year, then we end up with

$$\left(1 + \frac{\beta}{n}\right)^n P_0.$$

In general, the more frequently the interest is compounded the more money we earn after a year.

It is clear that in this situation, the amount we earn changes by a discrete amount at each time that the interest is compounded. Now we try to approximate the amount we earn using a continuously changing function $P(t)$. Presumably this approximation is valid when n is large and the changes are very small and occur very frequently.

Since the time interval is $1/n$, we let $\Delta t = 1/n$ and we compute the change in the amount $P(t)$ in the account at t to the amount $P(t + \Delta t)$ at $t + \Delta t$ after the interest has been compounded exactly once. We get

$$P(t + \Delta t) = P(t)\left(1 + \frac{\beta}{n}\right) = P(t)(1 + \beta \Delta t),$$

which means that

$$\frac{P(t + \Delta t) - P(t)}{\Delta t} = \beta P(t).$$

This is nothing more than (29.1) of course. We know that if $P(t)$ is a differentiable function and we let Δt tend to zero, or equivalently n tend to infinity, then P solves $P' = \beta P$ and therefore

$$P(t) = P_0 e^{\beta t}.$$

After one year, we end up with

$$P(1) = P_0 e^{\beta}.$$

EXAMPLE 29.17. If the annual interest rate is 9% on an account with a balance of \$2500 and the interest is compounded annually we end up with \$2725. If the interest is compounded continuously, we end up with \$2735.44.

The important modeling question about the continuous approximation is whether the continuous approximation $P(t)$ really does approximate the true quantity that changes by discrete amounts. In other words, does

$$P_0 e^{\beta} \approx \left(1 + \frac{\beta}{n}\right)^n P_0$$

for n large? Note that P_0 factors out of both sides and is irrelevant to the discussion.

Mathematically speaking, we want to prove that

$$\lim_{n \to \infty} \left(1 + \frac{\beta}{n}\right)^n = e^{\beta} \tag{29.14}$$

for any β.[5] Note that if we let $h = 1/n$ so $h \to 0^+$ as $n \to \infty$, then this is equivalent to showing that

$$\lim_{h \to 0^+} (1 + \beta h)^{1/h} = e^{\beta}. \tag{29.15}$$

We actually prove the equivalent limit

$$\lim_{h \to 0^+} \log (1 + \beta h)^{1/h} = \beta \tag{29.16}$$

from which (29.15) follows upon exponentiating both sides.

For any β, $\log(1 + \beta x)$ is strongly differentiable for x near 0. This means that there is a constant $\mathcal{K}$ such that for x near 0,

$$\left| \log(1 + \beta x) - \left(\log(1) + \frac{\beta}{1 + \beta x}(x - 0) \right) \right| \leq (x - 0)^2 \mathcal{K}.$$

---

[5] Euler was the first person to prove this.

Setting $x = h$ and simplifying gives

$$\left|\log(1 + \beta h) - \frac{\beta}{1 + \beta h} h\right| \leq h^2 \mathcal{K}.$$

Dividing by h and using the properties of log give

$$\left|\log((1 + \beta h)^{1/h}) - \frac{\beta}{1 + \beta h}\right| \leq h\mathcal{K}.$$

Now we can take the limit of the function on the right as $h \to 0$ to conclude (29.16).

EXAMPLE 29.18. We can compute

$$\lim_{x \to \infty} \left(1 - \frac{2}{x}\right)^{2x} = \left(\lim_{x \to \infty} \left(1 + \frac{-2}{x}\right)^x\right)^2 = (e^{-2})^2 = e^{-4}.$$

Chapter 29   Problems

29.1. Newton's Law of Cooling states,

The rate of change of the temperature difference between an object and its surrounding medium is proportional to the temperature difference.

Write down a differential equation modeling this situation.

29.2. The isotope ^{14}C decays at a rate proportional to its mass, giving up an electron to form a stable nitrogen ^{14}N atom. The basis of "carbon dating" of once living organisms is that the amount of ^{14}C in the organism is replenished while the organism is alive but the replenishing stops once the organism dies. Write down a differential equation modeling the amount of ^{14}C left starting from a fixed initial amount.

29.3. A certain bank pays 4% interest compounded instantaneously. Write down a differential modeling this.

29.4. The "carrying capacity" of a lake with respect to a species of fish is the maximum number of that species of fish the lake can sustain. If the relative rate of change of the population of a species of fish is proportional to the unused capacity, write down a differential equation modeling the population.

29.5. A running motor generates heat at a constant rate and radiates it away at a rate proportional to the temperature. Write down a differential equation modeling the temperature in the engine.

29.6. A certain species grows so that its relative birth rate is a positive constant while the relative death rate is proportional to the population. Write down a differential equation modeling the population.

29.7. Explain why 2^x is the inverse function for $\log_2(x)$.

29.8. Compute derivatives of the following functions:

$$\text{(a) } e^{7x^4 - 27x} \quad \text{(b) } \log(e^x + e^{-x}) \quad \text{(c) } \left(e^{3x} - x\right)^8$$

$$\text{(d) } e^{e^{e^x}} \quad\quad \text{(e) } x^e \quad\quad\quad \text{(f) } 7^x \ .$$

29.9. Compute the following integrals:

$$\text{(a) } \int \frac{e^{4x}}{1 - e^{4x}}\, dx \quad\quad \text{(b) } \int \frac{e^x - e^{-x}}{e^x + e^{-x}}\, dx$$

$$\text{(c) } \int e^x \frac{\log(1 + e^x)}{1 + e^x}\, dx \quad \text{(d) } \int \sqrt{e^x}\, dx$$

$$\text{(e) } \int \frac{e^{\sqrt{x}}}{\sqrt{x}}\, dx \quad\quad \text{(f) } \int x e^x\, dx.$$

29.10. The half-life of polonium is around 140 days. How much polonium is left of a 15 gram sample after 2.5 years?

29.11. It takes 4 years for $1/4$ of a given amount of a radioactive material to decay. What is the half-life?

29.12. Twenty-seven grams of a radioactive material are reduced to 9 grams after 1.5 years. What is the half-life of the material?

29.13. The half-life of a radioactive substance is 21 days. How long does it take for 80 grams of the substance to reduce to 2 grams?

29.14. A certain species of bacteria reproduces at rate proportional to the number of bacteria present and an initial colony of 1000 bacteria increase to 1500 bacteria after 50 minutes. How long does it take for the bacteria population to quadruple?

29.15. A certain species of bacteria reproduces at rate proportional to the number of bacteria present. A colony of bacteria is observed to reach a population of 20,000,000 after 24 hours. What was the original population (at the start of the 24 hour period)?

29.16. Compute the solution of the following differential equations by multiplying the equation by a suitable integration factor and then simplifying the problem:

$$\text{(a) } y' + 5y = x \qquad \text{(b) } y' + 2xy = x$$

$$\text{(c) } y' + \frac{1}{x}y = x^3 \qquad \text{(d) } y' = y + \frac{1}{1+e^{-x}}.$$

29.17. Compute derivatives of the following functions:

$$\text{(a) } x^{x^2} \qquad \text{(b) } (2x-4)^x \qquad \text{(c) } (\log(x))^x$$

$$\text{(d) } x^{\log(x)} \qquad \text{(e) } x^{x^x} \qquad \text{(f) } \frac{1}{x^x} .$$

29.18. Prove the formulas in (29.10).

29.19. Compute the following integrals:

$$\text{(a) } \int 3^x \, dx \qquad \text{(b) } \int \frac{1}{9^{5x}} \, dx$$

$$\text{(c) } \int 11^{x^2} x \, dx \qquad \text{(d) } \int 10^{10^x} 10^x \, dx .$$

29.20. Carry out the induction step to prove (29.12).

29.21. Prove that (29.12) is true for any p by using the result that we prove for integer p.

29.22. Assuming initial balances of $1000, compare the amounts earned after 10 years in two savings accounts with annual interests rates of 10% where the interest is compounded 6 times a year in one account and continuously in the other account.

29.23. A parent starting a college fund for her new baby wants to have $80,000 available on the 18th birthday of the child. Assuming an annual interest rate of 7% compounded continuously, what is the initial investment required to reach this amount?

29.24. Evaluate the following limits:

(a) $\displaystyle\lim_{x\to\infty}\left(1+\frac{3}{x}\right)^x$ (b) $\displaystyle\lim_{x\to\infty}\left(1-\frac{1}{4x}\right)^x$ (c) $\displaystyle\lim_{x\to\infty}\left(1+\frac{1}{x+1}\right)^x$

(d) $\displaystyle\lim_{x\to\infty}\left(1+\frac{1}{x}\right)^{2x+1}$ (e) $\displaystyle\lim_{x\to0^+}\left(1+\frac{x}{5}\right)^{1/x}$ (f) $\displaystyle\lim_{x\to\infty}\left(1+\frac{1}{x}\right)^{(x+1)/x}$.

29.25. Prove (29.13).

30

A Mass-Spring System and the Trigonometric Functions

Concluding the investigation of three important differential equations, the third problem we consider is $y''(x) = cy(x)$ for a constant c. The solution leads to the trigonometric functions.

The trigonometric functions are usually introduced as a way of describing geometric situations involving angles and lengths. Recall that a defining characteristic of the trigonometric functions is that they are periodic. We show the graphs of sin and cos in Fig. 30.1. For this reason, the trigonometric functions often appear in situations in which there is some quantity that varies in a repetitive fashion.

30.1 Hooke's Model of a Mass-Spring System

Hooke's law for a spring says

> The restoring force exerted by a spring stretched or compressed a distance s from the rest position is proportional to s.

Hooke's law, named for the English scientist Robert Hooke,[1] is a linear approximation of what actually happens with a spring and is valid for small s.

---

[1] Robert Hooke (1635–1703) was a true general scientist. While holding a chair in geometry for most of his career, he also made numerous important scientific observations and worked as an architect and surveyor.

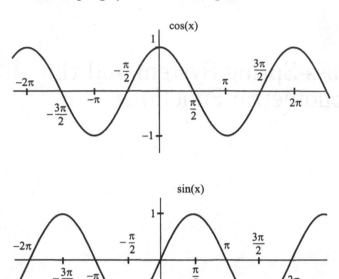

FIGURE 30.1. Plots of cos (top) and sin (bottom).

We model a spring that has one end attached to a wall and the other end to a mass m that is allowed to slide freely back and forth on a table, neglecting friction. We choose coordinates so that when the spring is at its rest position the mass is located at $s = 0$ and $s > 0$ corresponds to stretching the spring to the right (see Fig. 30.2). In this coordinate system,

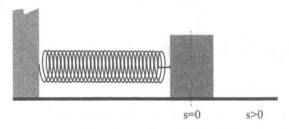

FIGURE 30.2. Illustration of the coordinate system used to describe a spring-mass system. The mass is allowed to slide freely back and forth with no friction.

Hooke's law for the force F reads,

$$F = -ks,$$
(30.1)

where the constant of proportionality $k > 0$ is called the **spring constant**. Combining Hooke's law with Newton's Law of Motion yields the equation

$$m\frac{d^2s}{dt^2} = -ks \qquad (30.2)$$

determining the motion of the mass.

This equation is usually rewritten as

$$s'' + \omega^2 s = 0, \quad \omega = \sqrt{\frac{k}{m}}. \qquad (30.3)$$

To describe specific solution, we also give some initial conditions at time $t = 0$. We can specify the initial position of the object, $s(0) = s_0$, and the initial velocity, $s'(0) = s_1$, for example. The complete problem is therefore

$$\begin{cases} s''(t) + \omega^2 s(t) = 0, & t > 0, \\ s(0) = s_0, \ s'(0) = s_1. \end{cases} \qquad (30.4)$$

This corresponds to pulling or pushing the object to some position, then giving it a shove. The goal is to describe how the object moves after that.

In this chapter, we show that the solution of (30.4) is given in terms of the trigonometric functions.

30.2 The Smoothness of Trigonometric Functions

We first show that $\sin(x)$ and $\cos(x)$ are Lipschitz continuous and then we show they are strongly differentiable for all x. Note that only the smoothness of sin has to be discussed, since

$$\cos(x) = \sin(x + \pi/2) \qquad (30.5)$$

means that cos is simply the composition of sin with the linear function $x + \pi/2$.

To show that sin is Lipschitz continuous for all x, we show there is a constant L such that

$$|\sin(x_2) - \sin(x_1)| \le L|x_2 - x_1|$$

for all x_2 and x_1. We consider the case $x_2 \ge x_1 \ge 0$ and the other cases follow from standard trigonometric identities. We illustrate the proof in Fig. 30.3. We draw a right triangle in the section of the circle between the rays defining x_1 and x_2 sides parallel to the axes and hypotenuse connecting the two points on the unit circle associated to x_1 and x_2. The triangle is drawn in Fig. 30.3. The height of the triangle, which is $|\sin(x_2) - \sin(x_1)|$,

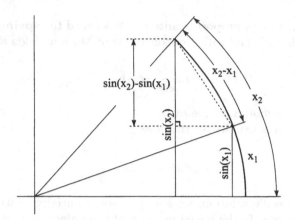

FIGURE 30.3. Illustration of the proof that sin is Lipschitz continuous.

is smaller than the hypotenuse which in turn is smaller than the distance along the part of the circle connecting the endpoints of the hypotenuse, which is $x_2 - x_1$. In other words,

$$|\sin(x_2) - \sin(x_1)| \leq |x_2 - x_1| \qquad (30.6)$$

for all x_1 and x_2.

Since $x + \pi/2$ is Lipschitz continuous for all x with Lipschitz constant 1, (30.5) implies that

$$|\cos(x_2) - \cos(x_1)| \leq |x_2 - x_1| \qquad (30.7)$$

for all x_1 and x_2 as well. This means that tan is Lipschitz continuous on any interval that avoids the points $\cdots, -5\pi/2, -3\pi/2, -\pi/2, \pi/2, 3\pi/2, 5\pi/2, \cdots$, though the Lipschitz constant is not 1.

Choosing $x_1 = 0$ and $x_2 = x$ in (30.6) and (30.7) yields the useful estimates

$$|\sin(x)| \leq |x| \qquad (30.8)$$

and

$$|1 - \cos(x)| \leq |x|, \qquad (30.9)$$

which hold for all sx. The second estimate can be improved by using standard trigonometric identities. Since

$$1 - \cos(x) = \frac{1 + \cos(x)}{1 + \cos(x)} \, (1 - \cos(x)) = \frac{1 - \cos^2(x)}{1 + \cos(x)} = \frac{\sin^2(x)}{1 + \cos(x)},$$

(30.8) implies that

$$|1 - \cos(x)| = \frac{|\sin^2(x)|}{|1 + \cos(x)|} \leq \frac{|x|^2}{|1 + \cos(x)|}.$$

If x is restricted so $|x| \leq \pi/2$, then $\cos(x) \geq 0$ and

$$|1 - \cos(x)| \leq |x|^2 \text{ for } |x| \leq \pi/2. \tag{30.10}$$

It turns out that $\sin$ is strongly differentiable and $D\sin(x) = \cos(x)$ for all x. To show this, we have to produce a constant $\mathcal{K}_{\bar{x}}$ that for all x near $\bar{x}$,

$$|\sin(x) - (\sin(\bar{x}) + \cos(\bar{x})(x - \bar{x}))| \leq (x - \bar{x})^2 \mathcal{K}_{\bar{x}}. \tag{30.11}$$

To do this, we use the addition formula

$$\sin(x_1 + x_2) = \cos(x_1)\sin(x_2) + \sin(x_1)\cos(x_2). \tag{30.12}$$

We let $s = x - \bar{x}$ and compute

$$\sin(x) = \sin(\bar{x} + s) = \sin(\bar{x})\cos(s) + \cos(\bar{x})\sin(s).$$

The first step is to rearrange the right-hand side so it looks like the right-hand side of (30.11). By adding and subtracting, we get

$$\sin(x) = \sin(\bar{x}) + \cos(\bar{x})s + \big(\sin(\bar{x})(\cos(s) - 1) + \cos(\bar{x})(s - \sin(s))\big).$$

Now if we define R by

$$R(s) = \sin(\bar{x})(\cos(s) - 1) + \cos(\bar{x})(s - \sin(s))$$

the result is

$$|\sin(x) - (\sin(\bar{x}) + \cos(\bar{x})s)| = |R(s)|.$$

This implies (30.11) with $x - \bar{x} = s$ provided there is a constant $\mathcal{K}_{\bar{x}}$ such that

$$|R(s)| \leq |s|^2 \mathcal{K}_{\bar{x}}.$$

To estimate $|R|$, we begin by using the triangle inequality,

$$|R(s)| \leq |\sin(\bar{x})|\,|\cos(s) - 1| + |\cos(\bar{x})|\,|s - \sin(s)|.$$

Since $|\sin(\bar{x})| \leq 1$ and $|\cos(\bar{x})| \leq 1$,

$$|R(s)| \leq |\cos(s) - 1| + |s - \sin(s)|.$$

The first term on the right is bounded quadratically for small s by (30.10). We draw another picture to show that $|s - \sin(s)|$ is bounded quadratically. In Fig. 30.4, we draw two "nested" sections of circles. The smaller section is determined from the unit circle by the angle s. We denote this section by $\angle BOA$ where A and B are the two endpoints of the section and O is the origin. To draw the larger circle, indicated by $\angle COD$, we draw a vertical line from the point A up to the point C, where this line intersects the line passing through O and B and then draw the circle that has center through

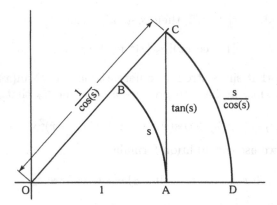

FIGURE 30.4. Illustration of the estimate on $|s - \sin(s)|$.

the origin and passes through the point of intersection C. We also need to refer to the triangle with endpoints C, O, and A, which we denote by $\triangle COA$.

The estimate is based on the observation that the area of the smaller section of the circle $\angle BOA$ is smaller than the area of the triangle $\triangle COA$, which in turn is smaller than the area of the section of the larger circle $\angle COD$. In other words,

$$\text{area of } \angle BOA \quad \leq \quad \text{area of } \triangle COA \quad \leq \quad \text{area of } \angle COD.$$

Classic results in geometry state that the area of a section of a circle is half the radius times the length along the arc of the section and the area of a triangle is half of the base times the height.

The area of $\angle BOA$ is therefore $\frac{1}{2} \times 1 \times s = s/2$. The base of $\triangle COA$ has length 1 and its height is $\tan(s)$. So the area of $\triangle COA$ is $\tan(s)/2$. Finally, the radius of $\angle COD$ is $1/\cos(s)$, which implies by similarity that the length of the arc from D to C is $s/\cos(s)$. So the area of $\angle COD$ is $s/(2\cos^2(s))$. We conclude that

$$\frac{s}{2} \leq \frac{\tan(s)}{2} \leq \frac{s}{2\cos^2(s)}$$

or

$$s \leq \tan(s) \leq \frac{s}{\cos^2(s)}.$$

Multiplying through by $\cos(s)$ gives

$$s\cos(s) \leq \sin(s) \leq \frac{s}{\cos(s)}.$$

Finally subtracting s yields

$$s\cos(s) - s \leq \sin(s) - s \leq \frac{s}{\cos(s)} - s.$$

This pair of inequalities imply that $|\sin(s) - s|$ is smaller than the larger of

$$|s\cos(s) - s| \text{ and } \left|\frac{s}{\cos(s)} - s\right|.$$

Now (30.10) implies that

$$|s\cos(s) - s| \le |s|^3$$

for $|s| \le \pi/2$. To estimate the other term, we write

$$\frac{s}{\cos(s)} - s = s\frac{1 - \cos(s)}{\cos(s)}.$$

If s is restricted so that $|s| \le \pi/6$, then $\cos(s) \ge 1/2$ and

$$\left|\frac{s}{\cos(s)} - s\right| \le 2|s|^3.$$

This finishes the proof that

$$\frac{d}{dx}\sin(x) = \cos(x). \tag{30.13}$$

Using (30.5), we conclude immediately that

$$\frac{d}{dx}\cos(x) = -\sin(x), \tag{30.14}$$

and so

$$\frac{d}{dx}\tan(x) = \frac{1}{\cos^2(x)} = \sec^2(x). \tag{30.15}$$

The Chain Rule then implies that

$$\frac{d}{dx}\sin(u) = \cos(u)u', \quad \frac{d}{dx}\cos(u) = -\sin(u)u',$$

$$\frac{d}{dx}\tan(u) = \sec^2(u)u'. \tag{30.16}$$

EXAMPLE 30.1.

$$\frac{d}{dx}\sin(e^x) = \cos(e^x)e^x$$

EXAMPLE 30.2.

$$\frac{d}{dx}\log(\tan(x)) = \frac{1}{\tan(x)}\frac{1}{\cos^2(x)}$$

The Fundamental Theorem of Calculus implies that

$$\int \sin(u)\, du = -\cos(u) + C, \qquad (30.17)$$

$$\int \cos(u)\, du = \sin(u) + C, \qquad (30.18)$$

and

$$\int \sec^2(u)\, du = \tan(u) + C. \qquad (30.19)$$

EXAMPLE 30.3. To integrate

$$\int \frac{\sin(\log(x))}{x}\, dx$$

we set $u = \log(x)$ so $du = dx/x$ and

$$\int \frac{\sin(\log(x))}{x}\, dx = \int \sin(u)\, du$$

$$= -\cos(u) + C = -\cos(\log(x)) + C.$$

EXAMPLE 30.4. To integrate

$$\int \tan(x)\, dx = \int \frac{\sin(x)}{\cos(x)}\, dx$$

we set $u = \cos(x)$ so $du = -\sin(x)\, dx$ and

$$\int \frac{\sin(x)}{\cos(x)}\, dx = -\int \frac{du}{u} = \log|u| + C = -\log|\cos(x)| + C.$$

30.3 Solving the Model for a Mass-Spring System

It is straightforward to verify that the function

$$s(t) = A\sin(\omega t) + B\cos(\omega t)$$

satisfies the differential equation (30.3) where A and B are constants. Differentiating, we compute

$$s'(t) = A\omega \cos(\omega t) - B\omega \sin(\omega t)$$

and therefore

$$s''(t) = -A\omega^2 \sin(\omega t) - B\omega^2 \cos(\omega t) = \omega^2 s.$$

We can solve for the constants A and B using the initial conditions,

$$s(0) = B = s_0 \text{ and } s'(0) = \omega A = s_1.$$

We conclude that the function

$$s(t) = \frac{s_1}{\omega} \sin(\omega t) + s_0 \cos(\omega t) \qquad (30.20)$$

is a solution of the problem (30.4).

We plot two examples of this solution for specific ω, s_1, and s_0 in Fig. 30.5
In general, ω determines the frequency of the oscillations of the spring, while

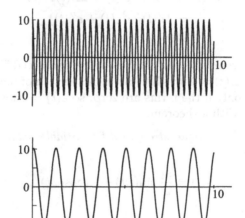

FIGURE 30.5. Two solutions of (30.4) for $0 \le t \le 10$. The solution in the upper plot has $\omega = 2$, $s_0 = 10$ and $s_1 = 1$. The solution in the lower plot has $\omega = .5$, $s_0 = 10$, and $s_1 = 1$.

s_0 and s_1 determine the magnitude of the oscillations. Large ω mean faster oscillations, as can be seen in Fig. 30.5. ω is large when the spring constant k is large relative to the mass.

Now that we have one solution of (30.4), the important question is whether there are other solutions. We have to determine this in order to predict how the spring-mass system behaves. To show that (30.20) is the only solution, we use what is called an **energy argument**.

If we suppose that s and r are two strongly differentiable solutions of (30.4), then the object is to show that $s(t) = r(t)$ for all t. Another way to view this is to define $\varepsilon = s - r$ and then to show that $\varepsilon(t) = 0$ for all t. First we show that $\varepsilon(t)$ satisfies (30.3). This follows because

$$\varepsilon''(t) + \omega^2 \varepsilon(t) = s''(t) - r''(t) + \omega^2(s(t) - r(t))$$
$$= s''(t) + \omega^2 s(t) - (r''(t) + \omega^2 r(t))$$
$$= 0 - 0 = 0.$$

Moreover, $\varepsilon(0) = s(0) - r(0) = 0$ and likewise $\varepsilon'(0) = 0$. In other words, ε satisfies

$$\begin{cases} \varepsilon'' + \omega^2\varepsilon = 0 & t > 0, \\ \varepsilon(0) = 0, \ \varepsilon'(0) = 0. \end{cases}$$

We define a new function, called the "energy,"

$$E(t) = \omega^2\varepsilon^2(t) + (\varepsilon'(t))^2.$$

Differentiating E, we find

$$\begin{aligned} E'(t) &= \omega^2 2\varepsilon(t)\varepsilon'(t) + 2\varepsilon'(t)\varepsilon''(t) \\ &= 2\varepsilon'(t)(\omega^2\varepsilon(t) + \varepsilon''(t)) \\ &= 0. \end{aligned}$$

In other words E remains constant: i.e., it is "conserved." Since $E(0) = 0$, we conclude that $E(t) = 0$ for all t. But E is the sum of nonnegative terms, so it can be zero only if the terms are zero, so $\varepsilon(t) = 0$ for all t.

We summarize with a theorem.

Theorem 30.1 *The unique strongly differentiable solution of the initial value problem (30.4) is*

$$s(t) = \frac{s_1}{\omega}\sin(\omega t) + s_0\cos(\omega t).$$

30.4   Inverse Trigonometric Functions

Now we turn to defining the inverse trigonometric functions and deriving some of their properties. It is not surprising that the need for the inverses pops up wherever there is a need for the trigonometric functions. We provide a particular example in an application to the polar coordinate system.

The polar coordinate system is an alternative to the rectangular coordinate system for describing points in the plane. The idea is to mark a point in the plane by the distance from the origin to the point together with the angle the ray from the origin to the point makes with the positive x-axis. This is a natural way to describe an object that is orbiting the origin for example. We illustrate in Fig. 30.6. If we let $r \geq 0$ denote the distance between the origin and the point P and s the corresponding angle, then we can denote a point P by (r, s). Note that a point P does not have a unique representation in this system, since $(r, s) = (r, s + 2n\pi)$ for any integer n.

A natural problem is to convert between the rectangular coordinates of a point P, (x, y), and its polar coordinates (r, s) (see Fig. 30.6). Given (r, s), it is easy to determine x and y since the properties of similar triangles and the definition of sin and cos mean that

$$x = r\cos(s) \text{ and } y = r\sin(s). \tag{30.21}$$

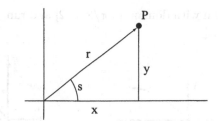

FIGURE 30.6. The polar coordinate system (r, s).

EXAMPLE 30.5.   Given $(r, s) = (3, \pi/4)$, we compute $x = 3/\sqrt{2}$ and $y = 3/\sqrt{2}$ while if $(r, s) = (3, 5\pi/4)$, we compute $x = -3/\sqrt{2}$ and $y = -3/\sqrt{2}$.

It is more difficult to go back. By Pythagorean's theorem, $r = \sqrt{x^2 + y^2}$. But can we recover s from

$$\sin(s) = x/r \ ?$$

To do this we need the inverse to sin.

We work out the inverse to sin first, then present the conclusions for cos and leave the details as an exercise.

Of course the first thing we notice when trying to compute an inverse to sin is that $\sin(x)$ does not pass the Horizontal Line Test, recall Fig. 30.1, and does not have an inverse. As with x^2, we have to restrict the domain of sin in order to get an invertible function.

There are many possibilities for restricting the domain of sin. We plot some examples in Fig. 30.7. We make the arbitrary choice of choosing the

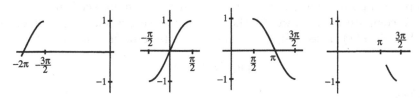

FIGURE 30.7. Four possibilities for restricting the domain of sin in order to get an invertible function.

largest possible domain that is closest to the origin. Therefore, we consider the "new" function sin with domain $[-\pi/2, \pi/2]$ and range $[-1, 1]$ pictured in Fig. 30.8.[2]

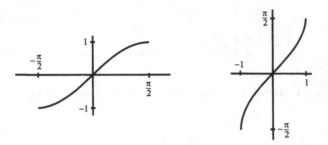

FIGURE 30.8. The restricted function sin defined on $[-\pi/2, \pi/2]$ (left) and its inverse $\sin^{-1}$ (right).

EXAMPLE 30.6.
$$\sin^{-1}(1/2) = \pi/6.$$

We plot the inverse of sin in Fig. 30.8, which is obtained by reflection. The domain is $[-1, 1]$ and the range is $[-\pi/2, \pi/2]$. With this choice,

$$\sin^{-1}(\sin(x)) = x \text{ for } -\frac{\pi}{2} \le x \le \frac{\pi}{2}$$
$$\sin(\sin^{-1}(x)) = x \text{ for } -1 \le x \le 1.$$

EXAMPLE 30.7.   We have to be careful using $\sin^{-1}$ sometimes, for example
$$\sin^{-1}(\sin(5\pi/6)) = \pi/6.$$

Since $D\sin(x) = \cos(x) > 0$ for $-\pi/2 < x < \pi/2$, $\sin^{-1}(x)$ is differentiable for $-1 < x < 1$. We can also see this from the plot of course. It does not have one-sided derivatives at $x = 1$ or $x = -1$ since the linearizations at points approaching 1 and -1 become more and more vertical.

To compute $D\sin^{-1}$, we use the fact that since $y = \sin^{-1}(x)$ is differentiable, $\sin(y(x))$ is as well, so we can differentiate both sides of

$$\sin(y(x)) = x$$

to get

$$\cos(y(x))\, y'(x) = 1$$

---

[2]This function is sometimes written Sin but we do not do this. We assume that whenever there is a $\sin^{-1}$ floating around, we are talking about the restricted sin function shown in Fig. 30.8.

or

$$y'(x) = \frac{1}{\cos(y(x))}.$$

Recall that we used this same trick to differentiate e^x. It is called **implicit differentiation**. Normally we do not like to have a y in the formula for the derivative, so to get rid of $\cos(y)$ we use the identity

$$\sin^2(x) + \cos^2(x) = 1 \text{ for all } x$$

to get

$$\cos(y) = \pm\sqrt{1 - \sin^2(y)} = \pm\sqrt{1 - x^2}.$$

Since the plot of $\cos^{-1}$ shows that its derivative is positive, we conclude that

$$\frac{d}{dx}\sin^{-1}(x) = \frac{1}{\sqrt{1 - x^2}}. \tag{30.22}$$

It follows that

$$\frac{d}{dx}\sin^{-1}(u) = \frac{u'}{\sqrt{1 - u^2}} \tag{30.23}$$

and therefore

$$\int \frac{1}{\sqrt{1 - u^2}}\, du = \sin^{-1}(u) + C. \tag{30.24}$$

EXAMPLE 30.8.

$$\frac{d}{dx}\sin^{-1}(e^x) = \frac{e^x}{\sqrt{1 - e^{2x}}}.$$

EXAMPLE 30.9. To integrate

$$\int \frac{1}{\sqrt{1 - 4t^2}}\, dt,$$

we use $u = 2t$ and $du = 2dt$ so that

$$\int \frac{1}{\sqrt{1 - 4t^2}}\, dt = \int \frac{1}{\sqrt{1 - (2t)^2}}\, dt = \frac{1}{2}\int \frac{1}{\sqrt{1 - u^2}}\, du$$
$$= \sin^{-1}(u) + C = \sin^{-1}(2t) + C.$$

EXAMPLE 30.10. To integrate

$$\int \frac{z^{3/2}}{\sqrt{1 - z^5}}\, dz,$$

we first write this as

$$\int \frac{z^{3/2}}{\sqrt{1 - (z^{5/2})^2}}\, dz.$$

We use $u = z^{5/2}$ and $du = \frac{5}{2}z^{3/2}dz$ so that

$$\frac{z^{3/2}}{\sqrt{1-z^5}}\,dz = \frac{2}{5}\int \frac{1}{\sqrt{1-u^2}}\,du = \frac{2}{5}\sin^{-1}(z^{3/2}) + C.$$

With this definition, we can solve the problem of converting from rectangular coordinates (x,y) to polar coordinates (r,s). If the point P is in the first or fourth quadrants, which means that $x \geq 0$, then we compute

$$r = \sqrt{x^2 + y^2} \text{ for all } x, y, \tag{30.25}$$

and then

$$s = \sin^{-1}(y/r) \text{ for } x \geq 0. \tag{30.26}$$

Note that $|y/r| \leq 1$ for all x. This does not work if the point P is in the second or third quadrants, when $x < 0$, because the range of $\sin^{-1}(x)$ is $[-\pi/2, \pi/2]$. We can still compute r the same way, but when we compute $s = \sin^{-1}(y/r)$, we get the angle between the negative x-axis and the line joining the origin to the point P. To get the angle to the positive x-axis, we have to set

$$s = \pi - \sin^{-1}(y/r) \text{ for } x < 0. \tag{30.27}$$

We can develop $\cos^{-1}$ in the same way. We take $\cos$ on the arbitrary restricted domain $[0, \pi]$, as shown in Fig. 30.9, to get an invertible function. We also plot $\cos^{-1}$. The domain of $\cos^{-1}$ is $[-1, 1]$, while the range is $[0, \pi]$.

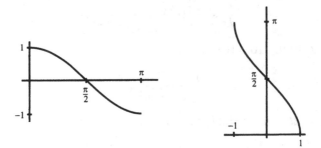

FIGURE 30.9. The restricted function $\cos$ defined on $[-0, \pi]$ (left) and its inverse $\cos^{-1}$ (right).

Arguing as for $\sin^{-1}$ we compute

$$\frac{d}{dx}\cos^{-1}(x) = \frac{-1}{\sqrt{1-x^2}}. \tag{30.28}$$

It follows that

$$\frac{d}{dx}\cos^{-1}(u) = \frac{-u'}{\sqrt{1-u^2}} \tag{30.29}$$

and therefore

$$\int \frac{-1}{\sqrt{1-u^2}}\, du = \cos^{-1}(u) + C. \qquad (30.30)$$

We usually use (30.24) instead of (30.30).

EXAMPLE 30.11. To integrate

$$\int \frac{\cos^{-1}(x)}{\sqrt{1-x^2}}\, dx,$$

we use $u = \cos^{-1}(x)$ so $du = -dx/\sqrt{1-x^2}$ and

$$\int \frac{\cos^{-1}(x)}{\sqrt{1-x^2}}\, dx = -\int u\, du = -\frac{u^2}{2} + C = -\frac{\left(\cos^{-1}(x)\right)}{2} + C.$$

Finally we compute an inverse for tan. We begin by restricting the domain to $(-\pi/2, \pi/2)$ as shown in Fig. 30.10 to get an invertible function. The plot shows that the new function is invertible and we can also compute $D\tan(x) = 1/\cos^2(x) > 0$ for $-\pi/2 < x < \pi/2$. We plot $\tan^{-1}$, obtained by reflection, in Fig. 30.11. The domain of $\tan^{-1}$ is $\mathbb{R}$, while the range is

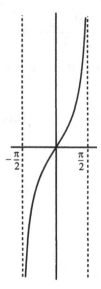

FIGURE 30.10. The restricted function tan defined on $[-\pi/2, \pi/2]$.

$(-\pi/2, \pi/2)$.

Arguing as for $\sin^{-1}$, it is a good exercise to derive

$$\frac{d}{dx}\tan^{-1}(x) = \frac{1}{1+x^2}. \qquad (30.31)$$

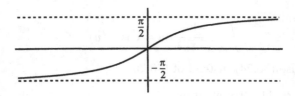

FIGURE 30.11. The plot of $\tan^{-1}$.

It follows that

$$\frac{d}{dx}\tan^{-1}(u) = \frac{u'}{1+u^2} \tag{30.32}$$

and therefore

$$\int \frac{1}{1+u^2}\,du = \tan^{-1}(u) + C. \tag{30.33}$$

EXAMPLE 30.12.

$$\frac{d}{dx}\tan^{-1}(x^2) = \frac{2x}{1+x^4}.$$

EXAMPLE 30.13. To integrate

$$\int_0^1 \frac{s}{1+s^4}\,ds,$$

we use $u = s^2$, so $s = 0 \rightarrow u = 0$ and $s = 1 \rightarrow u = 1$ and $du = 2sds$ so that

$$\int_0^1 \frac{s}{1+s^4}\,ds = \frac{1}{2}\int_0^1 \frac{1}{1+u^2}\,du = \frac{1}{2}(\tan^{-1}(1) - \tan^{-1}(0)) = \frac{\pi}{8}.$$

EXAMPLE 30.14. To integrate

$$\int \frac{1}{9+x^2}\,dx,$$

we first rewrite the integrand so that it has a 1 like the integrand for $\tan^{-1}$, i.e.,

$$\int \frac{1}{9+x^2}\,dx = \frac{1}{9}\int \frac{1}{1+x^2/9}\,dx.$$

Now we use $u = x/3$ and $du = dx/3$ so that

$$\int \frac{1}{9+x^2}\,dx = \frac{1}{3}\int \frac{1}{1+u^2}\,du$$

$$= \frac{1}{3}\tan^{-1}(u) + C = \frac{1}{3}\tan^{-1}(x/3) + C.$$

Chapter 30 Problems

30.1. A more realistic model of a spring and a sliding mass includes the damping effect of friction. The force due to damping is experimentally determined to be proportional to the velocity of the mass on the end of the spring. Write down a differential equation model describing this situation.

30.2. Write down a differential equation modeling the motion of a spring and mass hanging from a ceiling neglecting the weight of the spring.

30.3. Using a geometric argument, prove that cos is Lipschitz continuous with Lipschitz constant 1.

30.4. Determine the smallest possible value of a Lipschitz constant for tan on $[-\pi/4, \pi/4]$.

30.5. Determine the smallest possible value of a Lipschitz constant for sin on $[-\pi/6, \pi/6]$.

30.6. Show that $D\cos = -\sin$ directly, i.e., using the definition and not by using the derivative formula for sin.

30.7. Referring to Fig. 30.4, show that the segment AC has length $\tan(s)$.

30.8. Referring to Fig. 30.4, show that the radius of the angle $\angle COD$ has length $1/\cos(s)$.

30.9. Evaluate the following limits:

$$\text{(a) } \lim_{s \to 0} \frac{1 - \cos(s)}{s} \qquad \text{(b) } \lim_{s \to 0} \frac{\sin(s)}{s} .$$

30.10. Compute derivatives of the following functions:

(a) $\sin(9x^3)$ (b) $\cos(3x - x^7)$ (c) $\sin^2(\tan(t))$ (d) $\tan(x^{1/2})$

(e) $\log(\tan(x))$ (f) $\sin\left(\dfrac{1+t}{1-t}\right)$ (g) $e^{\tan(t)}$ (h) $\dfrac{1 - \sin(u)}{1 + \sin(u)}$.

30.11. Compute the following integrals:

(a) $\displaystyle\int \cos(8x)\, dx$ (b) $\displaystyle\int x^2 \sin(x^3)\, dx$

(c) $\displaystyle\int e^{\sin(x)} \cos(x)\, dx$ (d) $\displaystyle\int \sin(2x) \cos(2x)\, dx$

(e) $\displaystyle\int \tan(x)\, dx$ (f) $\displaystyle\int_{-\pi/4}^{\pi/4} \sin^4(s + \pi/4) \cos(s + \pi/4)\, ds.$

30.12. Develop the details of the derivation of $\cos^{-1}$ by following these steps:

1. Plot the restricted function $\cos(x)$, $0 \le x \le \pi$ and verify it is invertible.
2. Draw the plot of $\cos^{-1}$ using reflection and determine its domain and range.
3. Prove that $\cos^{-1}(x)$ is differentiable for $-1 < x < 1$.
4. Derive the derivative of $\cos^{-1}$.

30.13. Work out the details for the inverse function to the restricted function $\sin(x)$ with $\pi/2 \le x \le 3\pi/2$.

30.14. Work out the details for the inverse function to a suitable restriction of cot.

30.15. Prove (30.27).

30.16. Convert the following rectangular coordinates to polar coordinates:

$$\text{(a) } (4,8) \quad \text{(b) } (-3,7) \quad \text{(c) } (-1,-9).$$

30.17. Convert the following polar coordinates to rectangular coordinates:

$$\text{(a) } (4,\pi/4) \quad \text{(b) } (9,-7\pi/3) \quad \text{(c) } (2,5\pi/4).$$

30.18. Compute the following integrals:

$$\text{(a) } \int \frac{1}{\sqrt{9-x^2}}\,dx \qquad \text{(b) } \int \frac{x}{\sqrt{1-x^4}}\,dx$$

$$\text{(c) } \int \frac{1}{\sqrt{1-(x+5)^2}}\,dx \qquad \text{(d) } \int \left(\frac{\sin^{-1}(x)}{1-x^2}\right)^{1/2}\,dx$$

$$\text{(e) } \int \frac{1}{x^2+2x+2}\,dx \qquad \text{(f) } \int \frac{s^2}{1+s^6}\,ds.$$

30.19. Determine and plot (for $0 \le t \le 10$) the solutions of the spring-mass system (30.4) corresponding to

(a) $s_0 = 10$, $s_1 = 1$, $w = 2$ and $\omega = .2$

(b) $s_0 = 1$, $s_1 = 10$, $w = 2$ and $\omega = .2$.

30.20. This problem is concerned with deriving the solution of the two point boundary value problem for the spring-mass system

$$\begin{cases} s'' + w^2 s = 0, & 0 \le t \le \pi/(2w), \\ s(0) = s_0, \ s(\pi/(2w)) = s_1, \end{cases} \tag{30.34}$$

which corresponds to observing the position of the mass at times $t = 0$ and $t = \pi/(2w)$ and then predicting how it behaves for the rest of the time.

1. Show that $s(t) = A\sin(\omega t) + B\cos(\omega t)$ satisfies the differential equation in (30.34) for any constants A and B.

2. Determine the values of A and B from the values of s at $t = 0$ and $t = \pi/(2\omega)$.

3. Show that the solution determined in (1) and (2) is the only solution by carrying out the following steps:

   (a) Assume there are two solutions s and r and show that $\varepsilon = s - r$ satisfies

   $$\begin{cases} \varepsilon''(t) + \omega^2\varepsilon(t) = 0, & 0 \le t \le \pi/(2\omega), \\ \varepsilon(0) = 0, \ \varepsilon(\pi/(2\omega)) = 0. \end{cases}$$

   (b) Define an energy function E for ε and show that $E'(t) = 0$. Conclude that $E(t) = E_0^2$, a nonnegative constant, for all t.

   (c) From the equation $E(t) = E_0^2$, derive a differential equation

   $$\varepsilon'(t) = \sqrt{E_0^2 - \omega^2\varepsilon(t)^2}. \tag{30.35}$$

   (d) Solve (30.35) by using separation of variables and show the solution is $\varepsilon(t) = \frac{E_0}{\omega}\sin(\omega t + C)$ for some constant C.

   (e) Show that ε must be zero for all t by using the values at $t = 0$ and $t = \pi/(2\omega)$.

31
Fixed Point Iteration and Newton's Method

We conclude the discussion of calculus by applying the idea of the linearization to root and fixed point problems for functions. We have approached these kinds of problems by constructing approximation methods that produce a sequence of iterates $\{x_i\}$ that converge to the root or fixed point $\bar{x}$. The first method we studied was the Bisection Algorithm in Chapter 13, which has the property that the error of the iterates x_i decreases by a factor of $1/2$ after each step. Later in Chapter 15, we considered the Fixed Point Iteration. One motivation was that many models naturally result in fixed point problems. But an equally important motivation is that we are interested in finding methods in which the error of the iterates decreases more quickly than the Bisection Algorithm. We saw that it is possible to do this in the sense that the error of the iterates of some Fixed Point Iterations decrease by a factor smaller than $1/2$ after each step.

In this chapter, we continue the search for approximation methods for root and fixed point problems that converge quickly. The primary method introduced in this chapter is called Newton's method. Most modern techniques for solving root problems use some form of Newton's method at the heart of the algorithm.

31.1   Linearization and the Fixed Point Iteration

The first step is to use the linearization in the analysis of fixed point methods. In Chapter 15, we show that if g is a contraction map on an interval

$I = [a, b]$, and in particular that the Lipschitz constant of g on I is $L < 1$, then the Fixed Point Iteration for g converges. By Theorem 19.1, if g is uniformly strongly differentiable on I with $|g'(x)| < L$ for all x in I, then it is Lipschitz continuous with constant L, so we conclude immediately that the following theorem holds.

Theorem 31.1 *If there is an interval I and a constant $L < 1$ such that g is uniformly strongly differentiable on I and has the properties*

$$g : I \to I \qquad\qquad (31.1)$$

$$|g'(x)| \le L < 1 \text{ for all } x \text{ in } I, \qquad\qquad (31.2)$$

then the sequence $\{x_i\}$ generated by the Fixed Point Iteration starting with any point x_0 in I converges to the unique fixed point $\bar{x}$ of g in I.

EXAMPLE 31.1. Consider the fixed point problem $g(x) = \cos(x) = x$ on the interval $I = [0, \pi/3]$. Because $-\sin(x) = D\cos(x) \le 0$ for $0 \le x \le \pi/3$, cos is strictly decreasing on I. Therefore to show $g : I \to I$, it suffices to check that g evaluated at the endpoints of the interval I are inside I. Since $\cos(0) = 1$, while $\cos(\pi/3) = 1/2$, $g : I \to I$. Furthermore, $0 \le |g'(x)| = \sin(x) \le \sqrt{3}/2 < 1$ for x in I. Hence, the Fixed Point Iteration converges to a unique fixed point $\bar{x}$ in I. Starting with $x_0 = 0$, we find that the iterates agree to 15 places after 91 iterations, and $x_{91} = 0.739085133215161\cdots$.

31.2   Global Convergence and Local Behavior

Theorems 15.1 and 31.1 have some nice qualities. For one thing, we do not need to know that there is a fixed point to use the theorems.[1] If we can find an interval on which g satisfies the properties, e.g., (31.1) and (31.2), the theorems imply that there is a unique fixed point that can be approximated to any desired accuracy using the fixed point iterates $\{x_i\}$ starting with any initial value in the interval. In particular, the initial value does not have to be close to $\bar{x}$. These theorems are examples of **global** convergence results.

One disadvantage of these two theorems is that it can be quite difficult to find an interval I on which g has the required properties.[2] For example, we have carefully chosen functions g that either monotonically increase or decrease in the examples because it is much easier to verify that $g : I \to I$.

---

[1]In one dimension, we can usually use a graph to tell if there is a fixed point, but it is not so easy in higher dimensions. Analogs of all of the fixed point and Newton's method theorems hold for higher dimensions.

[2]Which you may know from trying some of the problems.

In general, checking this property leads to more root problems for g' in order to find maximum and minimum values of g on $I!$

Another disadvantage of these two theorems is that they can seriously overestimate the factor by which the errors decrease after each step of the Fixed Point Iteration when the iterates are close to the fixed point.

EXAMPLE 31.2. We consider the fixed point problem for

$$g(x) = x + \frac{9}{20}e^{-2(x-1/2)} - \frac{9}{20}e \tag{31.3}$$

on the interval $I = [.5, 10]$. The fixed point is $\bar{x} = 1$. We plot g and g' in Fig. 31.1 By inspection, g satisfies (31.1) and (31.2). However,

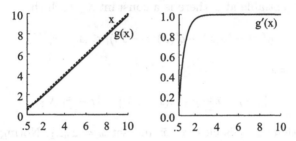

FIGURE 31.1. On the left we plot g in (31.3) together with $y = x$ and on the right we plot $g'(x)$.

$g'(1) \approx .669$ is substantially smaller than $g'(x)$ for most x in $[.5, 10]$. For example, $g'(3) \approx .994$ while $L = g'(10) \approx .999999995$. Theorem 31.1 predicts that the error of the iterates $|x_i - 1|$ decreases by a factor of L at each step, which is of course extremely slow. In Fig. 31.2, we plot the errors $\{|x_i - 1|\}$ and the ratios $\{|x_i - 1|/|x_{i-1} - 1|\}$ for the Fixed Point Iteration beginning with $x_0 = 10$. The errors decrease very slowly

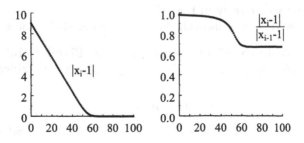

FIGURE 31.2. On the left we plot the errors $\{|x_i - 1|\}$ and on the right the ratios $\{|x_i - 1|/|x_{i-1} - 1|\}$ for the Fixed Point Iteration for g in (31.3) beginning with $x_0 = 10$.

at first but as the iterates get close to 1, the errors begin to decrease much more quickly.

The issue arises when using the Lipschitz constant or the maximum value of g on the entire interval I, because it is too crude a measure of how g behaves if I is relatively large and the iterates are close to the root or fixed point. In Example 31.2, the constant L accurately predicts how the errors decrease for iterates far from $\bar{x}$ but is not accurate when the iterates are close to $\bar{x}$. In other words, the **local behavior** of the iterates may be much better than the global behavior of the Fixed Point Iteration on the entire interval.

A way to get a more accurate analysis of the Fixed Point Iteration when the iterates are close to $\bar{x}$ is to use the linearization of g at $\bar{x}$. Since g is strongly differentiable at $\bar{x}$ there is a constant $\mathcal{K}_{\bar{x}}$ such that

$$|g(x) - (g(\bar{x}) + g'(\bar{x})(x - \bar{x}))| \le |x - \bar{x}|^2 \mathcal{K}_{\bar{x}}$$

or since $g(\bar{x}) = \bar{x}$,

$$|g(x) - \bar{x} - g'(\bar{x})(x - \bar{x}))| \le |x - \bar{x}|^2 \mathcal{K}_{\bar{x}}. \tag{31.4}$$

We assume that x_{n-1} is close to $\bar{x}$ and set $x = x_{n-1}$. Noting that $x_n = g(x_{n-1})$, we get

$$|x_n - \bar{x} - g'(\bar{x})(x_{n-1} - \bar{x})| \le |x_{n-1} - \bar{x}|^2 \mathcal{K}_{\bar{x}}. \tag{31.5}$$

Therefore for x_{n-1} close to $\bar{x}$,

$$x_n - \bar{x} \approx g'(\bar{x})(x_{n-1} - \bar{x}). \tag{31.6}$$

In other words, the error decreases approximately by a factor of $g'(\bar{x})$ when the iterates are close to $\bar{x}$. It is the size of $g'(\bar{x})$ that ultimately determines how the error decreases, not the maximum value of $|g'|$ on the entire interval I. This can be seen clearly in Fig. 31.2.

We can make this analysis of the local convergence behavior more precise.

Theorem 31.2 Local Convergence for the Fixed Point Iteration
If $\bar{x}$ is a solution of $g(x) = x$ and g is uniformly strongly differentiable at $\bar{x}$ and

$$|g'(\bar{x})| < 1, \tag{31.7}$$

then the Fixed Point Iteration converges to $\bar{x}$ for all initial values x_0 sufficiently close to $\bar{x}$. Furthermore,

$$\lim_{n \to \infty} \frac{x_n - \bar{x}}{x_{n-1} - \bar{x}} = g'(\bar{x}). \tag{31.8}$$

It is a good idea to compare Theorem 31.2 with Theorem 31.1 closely. To use Theorem 31.2, we have to know that the fixed point exists and its approximate value so we can verify (31.7), whereas we can verify (31.1) and (31.2) without knowing there is a fixed point. On the other hand, (31.7) is usually easier to check when we do have a rough estimate of the fixed point.

EXAMPLE 31.3.   We apply the theorem to the fixed point problem $g(x) = \log(x+2) = x$. In Fig. 31.3, we show plots of g and g'. From the plot of g we see that $\bar{x}$ is between 1 and 1.5 and $|g'(\bar{x})| \leq .4$. The Fixed Point Iteration beginning with $x_0 = 1$ gives $x_{27} \approx 1.146193220620577$ and all the subsequent iterates agree with x_{27}.

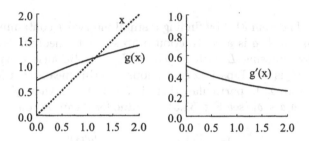

FIGURE 31.3. On the left we plot $g(x) = \log(x+2)$ together with $y = x$ and on the right we plot $g'(x)$.

Theorem 31.1 guarantees that the Fixed Point Iteration converges for any initial value in x_0 even for values that are far away from $\bar{x}$. Theorem 31.2 requires an initial value x_0 that is close to $\bar{x}$ and moreover does not say how close, so again some information about $\bar{x}$ is required. For this reason, we call Theorem 31.2 a **local** convergence result.

EXAMPLE 31.4. The fixed point of $g(x) = (x-1)^3 + .9x + .1$ is $\bar{x} = 1$. It is easy to check that $g'(\bar{x}) = .9 < 1$. Experimentally, the Fixed Point Iteration converges for x_0 in $[.69, 1.31]$ but diverges rapidly for values outside this interval.

On the other hand, the estimate on how quickly the error decreases (31.8) can be very accurate.

EXAMPLE 31.5.   In Fig. 31.4, we plot the errors $\{|x_i - 1|\}$ and the ratios $\{|x_i - 1|/|x_{i-1} - 1|\}$ for the Fixed Point Iteration in Example 31.3 beginning with $x_0 = 1$. The errors begin decreasing by more or less a constant factor after the first few iterations. We find that $|g'(\bar{x})| \approx .3178444$, while the error of x_{21} is roughly $.3178446$ times the error of x_{20}. Both of these values are not far from the crude estimate of $.4$ that are obtained by inspecting the plot of g in Fig. 31.3.

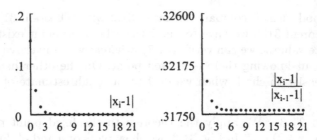

FIGURE 31.4. On the left we plot the errors $\{|x_i - 1|\}$ and on the right the ratios $\{|x_i - 1|/|x_{i-1} - 1|\}$ for the Fixed Point Iteration for $\log(2 + x)$ beginning with $x_0 = 1$.

We prove Theorem 31.2 by finding a small interval I containing $\bar{x}$ as the midpoint on which g is a contraction map so that Theorem 31.1 applies. We begin by choosing L with $|g'(\bar{x})| < L < 1$. Because g' is Lipschitz continuous, $|g'|$ is also Lipschitz continuous and this means that $|g'(x)| \leq L$ for all x close to $\bar{x}$. In particular, there is a $\delta > 0$ such that $|g'(x)| \leq L$ for x in $I = [\bar{x} - \delta, \bar{x} + \delta]$ (see Fig. 31.5). A value for δ can be computed. If the

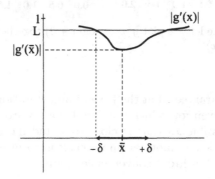

FIGURE 31.5. $|g'(x)| \leq L < 1$ for x in $I = [\bar{x} - \delta, \bar{x} + \delta]$.

Lipschitz constant of g' is K for x near $\bar{x}$, then $|g'(x) - g'(\bar{x})| \leq K|x - \bar{x}|$. To guarantee that $|g'(x)| \leq L$, we can use

$$|g'(x)| \leq |g'(\bar{x})| + K|x - \bar{x}| \leq L,$$

which gives

$$|x - \bar{x}| \leq \frac{L - |g'(\bar{x})|}{K} = \delta.$$

Now we have an interval I on which (31.2) is satisfied, so we only have to check (31.1) and we can use Theorem 31.1. The interval I is simply the set of points x such that $|x - \bar{x}| \leq \delta$, so if x is in I we need to show that

$|g(x) - \bar{x}| \le \delta$. But $|g(x) - \bar{x}| = |g(x) - g(\bar{x})|$, and Theorem 19.1 implies

$$|g(x) - g(\bar{x})| \le L|x - \bar{x}| \le L\delta. \tag{31.9}$$

Since $L < 1$, this shows that (31.1) holds.

To show that (31.8) is true, we divide both sides of (31.5) by $|x_{n-1} - \bar{x}|$ to get

$$\left| \frac{x_n - \bar{x}}{x_{n-1} - \bar{x}} - g'(\bar{x}) \right| \le |x_{n-1} - \bar{x}|\mathcal{K}_{\bar{x}}.$$

(If $x_{n-1} = \bar{x}$ there is nothing to prove.) We conclude that (31.8) is true by taking the limit as n goes to infinity.

By this argument, the Fixed Point Iteration is guaranteed to converge for any x_0 in the interval I constructed in Fig. 31.5. Therefore, the smaller this interval I, the closer x_0 has to be to $\bar{x}$ and consequently the better we have to know the location of $\bar{x}$ before we start the Fixed Point Iteration. The size of I depends on the distance between $|g'(\bar{x})|$ and 1 and how $|g'(x)|$ behaves for x near $\bar{x}$. If $g'(\bar{x})$ is close to 1 and rises steeply as x moves away from $\bar{x}$, then I has to be chosen very small. We illustrate in Fig. 31.6. On the other hand, if $|g'(x)|$ decreases in value as x moves away from $\bar{x}$, then I can be chosen to be large.

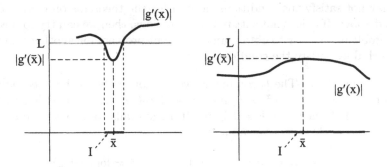

FIGURE 31.6. The size of I depends on how $|g'(x)|$ behaves for x near $\bar{x}$. On the left, $|g'(x)|$ rises steeply from $|g'(\bar{x})|$ and I is small. On the right, $|g'(x)|$ is always less than L and I can be chosen large.

EXAMPLE 31.6. The fixed point of

$$g(x) = .9 + 1.9x - \frac{1}{10}\tan^{-1}(10(x-1))$$

with

$$g'(x) = 1.9 - \frac{1}{1 + 100(x-1)^2} \tag{31.10}$$

is $\bar{x} = 1$. We plot $g'(x)$ in Fig. 31.7. From the plot, we can guarantee con-

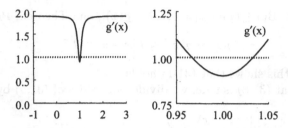

FIGURE 31.7. Plots of $g'(x)$ defined in (31.10). The plot on the right is a closeup for x in $[.95, 1.05]$.

vergence for intervals approximately like $[.97, 1.03]$, or smaller. Experimentally, the Fixed Point Iteration converges to $\bar{x}$ for x_0 in $[.96, 1.06]$ but diverges for $x_0 = .939$ and $x_0 = 1.061$.

EXAMPLE 31.7. The fixed point for $g(x) = \frac{1}{2}\cos(x)$ is $\bar{x} \approx 0.450183611294874$. Now $|g'(x)| = \frac{1}{2}|\sin(x)| \leq \frac{1}{2}$ for all x and experimentally the Fixed Point Iteration converges to $\bar{x}$ for any initial value x_0.

In general when solving a fixed point problem for a given function, g may or may not satisfy the conditions needed for the theorems on convergence stated above. If g does not satisfy the conditions, then we can try to rewrite the problem $g(x) = x$ to obtain an equivalent problem for another function $\tilde{g}$ which does satisfy the conditions.

EXAMPLE 31.8. The theorems above do not apply to the fixed point problem for $g(x) = e^x - 2$ since $\bar{x} \approx 1.146193220621$ and $|g'(\bar{x})| \approx 3.146 > 1$. In fact, the Fixed Point Iteration fails to converge to $\bar{x}$ for any $x_0 \neq \bar{x}$. However, we can solve

$$e^{\bar{x}} - 2 = \bar{x} \iff e^{\bar{x}} = \bar{x} + 2 \iff \bar{x} = \log(\bar{x} + 2)$$

and we can apply Theorem 31.2 to $g(x) = \log(x + 2)$, as we show in Example 31.3.

31.3   High Order Convergence

In the previous section, we found that the errors of the fixed point iterates decrease by approximately a factor of $|g'(\bar{x})|$ after each step when $g'(\bar{x}) \neq 0$. Since smaller values of $|g'(\bar{x})|$ mean the errors decrease more with each step, it is natural to consider what happens when $|g'(\bar{x})| = 0$. In this case, (31.4) reduces to

$$|g(x) - \bar{x}| \leq |x - \bar{x}|^2 \mathcal{K}_{\bar{x}}.$$

If we substitute $x = x_{n-1}$ and $x_n = g(x_{n-1})$, we get

$$|x_n - \bar{x}| \leq \mathcal{K}_{\bar{x}} |x_{n-1} - \bar{x}|^2. \tag{31.11}$$

Induction yields

$$|x_n - \bar{x}| \leq \mathcal{K}_{\bar{x}} \left(\mathcal{K}_{\bar{x}} |x_{n-2} - \bar{x}|^2 \right)^2 \leq \cdots \leq \left(\mathcal{K}_{\bar{x}} |x_0 - \bar{x}| \right)^{2^{n-1}} |x_0 - \bar{x}|.$$

Therefore, the iteration converges to $\bar{x}$ for all x_0 with $\mathcal{K}_{\bar{x}} |x_0 - \bar{x}| < 1$, i.e., for all initial values x_0 that are sufficiently close to $\bar{x}$. Moreover, (31.11) can be written as

$$\mathcal{K}_{\bar{x}} |x_n - \bar{x}| \leq \left(\mathcal{K}_{\bar{x}} |x_{n-1} - \bar{x}| \right)^2.$$

If $|x_{n-1} - \bar{x}| \leq 10^{-p - \log(\mathcal{K}_{\bar{x}})}$ for some p, i.e., if x_{n-1} agrees with $\bar{x}$ to at least $p + \log(\mathcal{K}_{\bar{x}})$ digits, then $|x_n - \bar{x}| \leq 10^{-2p - \log(\mathcal{K}_{\bar{x}})}$ or x_n agrees to about $2p + \log(\mathcal{K}_{\bar{x}})$ digits. In other words, when $g'(\bar{x}) = 0$, the fixed point iterate x_n has roughly twice as many accurate digits as the previous iterate x_{n-1}. This is extremely fast convergence! We say that the sequence $\{x_i\}$ converges at a **second order rate** or at a **quadratic rate**.

EXAMPLE 31.9. The fixed point for

$$g(x) = \frac{2(x+1)}{5 + 4x + x^2} \tag{31.12}$$

is $\bar{x} = \sqrt{2} - 1 \approx 0.414213562373095$. It is straightforward to verify that $g'(\bar{x}) = 0$. In Fig. 31.8, we list the first few fixed point iterates along with the errors. The convergence is very fast compared to the previous

i	x_i	$|x_i - \bar{x}|$
0	0.000000000000000	0.414213562373095
1	0.400000000000000	0.014213562373095
2	0.414201183431953	$1.237894114247684 \times 10^{-5}$
3	0.414213562363800	$9.295619829430279 \times 10^{-12}$
4	0.414213562373095	0.000000000000000

FIGURE 31.8. The first few fixed point iterates and associated errors for g in (31.12).

examples of Fixed Point Iterations, and counting digits, we find that x_i has roughly twice as many accurate digits as x_{i-1}.

It is possible to find Fixed Point Iterations that gain accuracy even more quickly than a second order rate. To cover all possibilities, we give a more precise definition of the **order of convergence** of a sequence $\{x_n\}$ with $\lim_{n \to \infty} x_n = \bar{x}$. We say that $\{x_n\}$ **converges to order** p, or converges at a *p*th **order rate**, if given $p > 0$ there are constants $C > 0$ and $N > 0$ such that

$$|x_n - \bar{x}| \leq C|x_{n-1} - \bar{x}|^p \text{ for all } n \geq N. \tag{31.13}$$

Higher order convergence means faster convergence in the sense that the errors decrease more quickly for each iteration.

EXAMPLE 31.10. The Fixed Point Iterations for $g_1(x) = \frac{1}{2}x$, $g_2(x) = \frac{1}{2}x^2$, $g_3(x) = \frac{1}{2}x^3$, and $g_4(x) = \frac{1}{2}x^4$ converge with order 1, 2, 3, and 4, respectively, on the interval $I = [-1, 1]$ to $\bar{x} = 0$. We list the first few iterates for each in Fig. 31.9. The differences in the rate are clear.

i	$\frac{1}{2}x$	$\frac{1}{2}x^2$	$\frac{1}{2}x^3$	$\frac{1}{2}x^4$
0	1	1	1	1
1	.5	.5	.5	.5
2	.25	.125	.0625	.03125
3	.125	.0078125	.000122$\cdots$	.00000047$\cdots$

FIGURE 31.9. The first few fixed point iterates for the indicated g.

Note that this definition allows for the possibility that the error of the first few iterates might not decrease as quickly as the rest, which makes sense based on the discussions above, especially Example 31.2.

When the order of convergence is $p = 1$, then the iteration only converges if $C < 1$. C is called the **convergence factor** when the convergence is first order and it is customary to compare the rate of convergence of first order convergent sequences by comparing the relative sizes of the convergence factors.

EXAMPLE 31.11. The Fixed Point Iteration for $g_1(x) = \frac{1}{4}x$ converges more quickly than the Fixed Point Iteration for $g_2(x) = \frac{1}{2}(x)$.

When the rate of convergence is higher than 1, the value of C determines how close the initial value x_0 has to be to the fixed point $\bar{x}$ to get the Fixed Point Iteration to converge. The larger C is, the closer x_0 has to be to $\bar{x}$.

EXAMPLE 31.12. The Fixed Point Iteration for $g_1(x) = \frac{1}{2}x^2$ converges to the fixed point $\bar{x} = 0$ for all initial values $|x_0| < 2$ since this means that

$$\frac{1}{2}x_0^2 = \frac{1}{2}|x_0| \times |x_0| < |x_0|.$$

It diverges for any initial value with $|x_0| > 2$. In contrast, the Fixed Point Iteration for $g_2(x) = 4x^2$ converges to $\bar{x} = 0$ for any initial value $|x_0| < \frac{1}{4}$ and diverges for $|x_0| > \frac{1}{4}$.

Theorem 31.2 guarantees that the Fixed Point Iteration converges at a first order rate when $0 < |g'(\bar{x})| < 1$, while the discussion above implies the following theorem is true.

Theorem 31.3 *If $\bar{x}$ is a solution of $g(x) = x$ and g is strongly differentiable at $\bar{x}$ and*

$$|g'(\bar{x})| = 0, \tag{31.14}$$

then the Fixed Point Iteration converges at least at a second order rate to $\bar{x}$ for all initial values x_0 sufficiently close to $\bar{x}$.

31.4 Newton's Method

As mentioned, one motivation for introducing the fixed point problem was to find faster methods for solving root problems. We now use Theorem 31.3 to construct a method for solving root problems that is second order convergent called Newton's method.

One of the simplest ways to rewrite a root problem $f(x) = 0$ as a fixed point problem $g(x) = x$ is to choose

$$g(x) = x - \alpha f(x)$$

where α is a non-zero constant. Based on Theorem 31.3, it is natural to try to choose α so that $g'(\bar{x}) = 0$ and thereby gain second order convergence. Assuming f is strongly differentiable and computing gives

$$g'(x) = 1 - \alpha f'(x),$$

which means that α should be chosen so

$$\alpha = \frac{1}{f'(\bar{x})}$$

and

$$g(x) = x - \frac{f(x)}{f'(\bar{x})}.$$

The Fixed Point Iteration looks like

$$x_i = x_{i-1} - \frac{f(x_{i-1})}{f'(\bar{x})}$$

and converges quadratically. Unfortunately to use this Fixed Point Iteration, we need to know the value of $\bar{x}$.

To try to find a practical method, we use the more sophisticated approach of writing

$$g(x) = x - \alpha(x)f(x)$$

where $\alpha(x)$ is a nonzero strongly differentiable function. Now

$$g'(x) = 1 - \alpha'(x)f(x) - \alpha(x)f'(x).$$

Substituting $\bar{x}$ and using $f(\bar{x}) = 0$, we find that $\alpha(x)$ needs to have the property

$$\alpha(\bar{x}) = \frac{1}{f'(\bar{x})}.$$

One way to get α to have this property at $\bar{x}$ is simply to choose

$$\alpha(x) = \frac{1}{f'(x)}$$

for *all* x. In other words, we compute the Fixed Point Iteration for

$$g(x) = x - \frac{f(x)}{f'(x)}.$$

This gives:

Algorithm 31.1 Newton's method Choose x_0 and for $i = 1, 2, \cdots$, set

$$x_i = x_{i-1} - \frac{f(x_{i-1})}{f'(x_{i-1})}. \tag{31.15}$$

Checking, we differentiate to find

$$g'(x) = \frac{f(x)f''(x)}{(f'(x))^2}$$

so $g'(\bar{x}) = 0$ (because $f(\bar{x}) = 0$) provided that $f'(\bar{x}) \neq 0$ and $f''(\bar{x})$ is defined. Moreover, $g'(x)$ is Lipschitz continuous for x near $\bar{x}$ provided $f'(\bar{x}) \neq 0$ and $f''(x)$ is Lipschitz continuous near $\bar{x}$. If these conditions hold, then Theorem 31.3 implies:

Theorem 31.4 Local Convergence for Newton's method *If $\bar{x}$ is a solution of $f(x) = 0$, where f has a Lipschitz continuous second derivative in an interval containing $\bar{x}$, and if*

$$|f'(\bar{x})| \neq 0, \tag{31.16}$$

then Newton's method (31.15) converges at least at a second order rate to $\bar{x}$ for all initial values x_0 sufficiently close to $\bar{x}$.

EXAMPLE 31.13. Consider the root problem for

$$f(x) = \frac{1}{2+x} - x.$$

Newton's method is the Fixed Point Iteration for

$$g(x) = x - \frac{\frac{1}{2+x} - x}{\frac{-1}{(2+x)^2} - 1} = \frac{2(x+1)}{5 + 4x + x^2},$$

which we recognize from (31.12). Here, $f'(x) \neq 0$. The first few Newton iterates are shown in Fig. 31.8.

EXAMPLE 31.14. The Fixed Point Iteration for $g(x) = \log(x+2)$ discussed in Example 31.3 converges at a linear rate and took 27 iterations to get 15 digits of accuracy. In order to get a second order Fixed Point Iteration, we first rewrite the fixed point problem as a root problem and then apply Newton's method to get a better fixed point problem. We set $f(x) = x - \log(x+2)$ so $f(\bar{x}) = 0$ if and only if $g(\bar{x}) = \bar{x}$. Then

$$f'(x) = 1 - \frac{1}{x+2};$$

so $f'(x) = 0$ only at $x = -1$, which is not $\bar{x}$. Therefore, we can apply Newton's method, which is the Fixed Point Iteration for

$$g(x) = x - \frac{x - \log(x+2)}{1 - \frac{1}{x+2}} = \frac{-x + (2+x)\log(2+x)}{1+x}.$$

We show the results in Fig. 31.10

i	x_i
0	1.000000000000000
1	1.147918433002165
2	1.146193440797909
3	1.146193220620586
4	1.146193220620583
5	1.146193220620583

FIGURE 31.10. The first Newton iterates for $f(x) = x - \log(x+2)$.

It is important to keep in mind that Theorem 31.4 only guarantees that Newton's method converges when the initial value x_0 is sufficiently close to $\bar{x}$. Otherwise, the results of using Newton's method are very unpredictable.

EXAMPLE 31.15. We apply Newton's method to $f(x) = (x-2)(x-1)x(x+.5)(x+1.5)$ without trouble because $f'(x) \neq 0$ at any of the roots of f. We compute 21 Newton iterations for f starting with 5000 equally spaced initial values in $[-3, 3]$, then we record the last value computed by Newton's method. We plot the resulting pairs of points in Fig. 31.11. Each of the roots is contained in an interval in which all initial values produce convergence to the root. But outside these intervals the behavior of the iteration is unpredictable with nearby initial values converging to different roots.

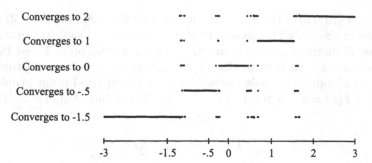

FIGURE 31.11. This plot shows the roots of $f(x) = (x-2)(x-1)x(x+.5)(x+1.5)$ found by Newton's method for 5000 equally spaced initial guesses in $[-3, 3]$. The horizontal position of the points shows the location of the initial guess and the vertical position indicates the twenty first Newton iterate.

31.5   Some Interpretations and History of Newton's Method

We begin this section by giving a couple of interpretations of Newton's method.

The first interpretation is geometric and is a good way to remember the formula for Newton's method. The idea is drawn in Fig. 31.12 Given the

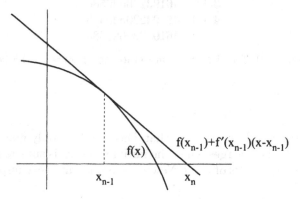

FIGURE 31.12. An illustration of one step of Newton's method from x_{n-1} to x_n.

value x_{n-1}, we would like to trace the graph of f to the point where it crosses the x-axis, which is $\bar{x}$. But this is difficult to do when f is non-linear. The idea is to replace f by a linear approximation, for which we can compute a root easily, and then compute the root of the linear approximation. The linear approximation is often called a **linear model** for f. We can choose different linear approximations, but a natural choice is to use the linearization of f at the current iteration point x_{i-1}. That is presumably the closest value we know to $\bar{x}$.

The linearization of f at x_{n-1} is

$$f(x) \approx f(x_{n-1}) + f'(x_{n-1})(x - x_{n-1}).$$

To approximate the root of $f(x) = 0$, we compute the root x_n of the linearization

$$f(x_{n-1}) + f'(x_{n-1})(x - x_{n-1}) = 0,$$

which is

$$x_n = x_{n-1} - \frac{f(x_{n-1})}{f'(x_{n-1})}.$$

This is precisely the definition of Newton's method!

The second interpretation is particularly useful when we consider root problems in higher dimensions. The Fundamental Theorem says that

$$f(x) = f(x_{n-1}) + \int_{x_{n-1}}^{x} f'(s)\,ds$$

for any x. Therefore, another way to interpret $\bar{x}$ is that it is the number such that

$$f(x_{n-1}) + \int_{x_{n-1}}^{\bar{x}} f'(s)\,ds = 0.$$

Since we don't know $\bar{x}$, we can't evaluate this integral. A way to get around this is to use the rectangle rule approximation,

$$\int_{x_{n-1}}^{x} f'(s)\,ds \approx f'(x_{n-1})(x - x_{n-1})$$

for any x close to x_{n-1}. We define x_n to be the number such that

$$f(x_{n-1}) + f'(x_{n-1})(x_n - x_{n-1}) = 0.$$

This also gives Newton's method.

The history of Newton and Newton-like methods is long and complicated (see Ypma [21] for a detailed account). Methods related to the Bisection Algorithm and Newton's method were known as far back as the Babylonians and there was continued development down to Newton's time, especially in Arabia and China. Newton knew about various techniques and was particularly influenced by the work of Viéte.[3] Newton derived his method using algebraic arguments, not calculus, Moreover, he only explained his method

---

[3]François Viète (France, 1540–1603) was never a professional mathematician, yet made some early important contributions to algebra, geometry, solving equations, and trigonometry as well as writing several text. He also worked to crack Spanish secret codes for King Henry IV.

in the context of solving for the roots of polynomials, though he did use it to solve for a root of a non-polynomial function. Newton's description and implementation of his method were very complicated, which greatly limited its accessibility.

Wallis published the first printed description of Newton's method, essentially following Newton's explanation. Later, Raphson[4] published a book in which he described a method for solving polynomial equations that was equivalent to the method invented by Newton. His account and implementation were much simpler, however, and Raphson considered his method to be new. Like Newton, Raphson did not use calculus in his derivation.

The first person to use calculus to describe Newton's method was Simpson.[5] After that, Lagrange published a description that used the modern f' notation for the derivative, and somewhat later, Fourier published an influential book in which he described Newton's method in modern form and called it "Newton's method." But, we see that a more accurate name might be "Viète-Newton-Raphson-Simpson's method".

31.6   What Is the Error in an Approximate Root?

Naturally, we are interested in the **error**

$$|x_n - \bar{x}|$$

of x_n. However, $\bar{x}$ is usually unknown so the error cannot be computed in general. Instead, we have to estimate the error in some fashion.

One quantity that can be computed is $f(x_n)$, which is called the **residual** of x_n. This is a measure of how well x_n solves $f(x) = 0$ since the residual of the true root $\bar{x}$ is zero, i.e., $f(\bar{x}) = 0$. The question is how to connect the computable residual to the unknown error. We obtain an estimate assuming that f is uniformly strongly differentiable in an interval containing $\bar{x}$. This means there is a constant $\mathcal{K}$ such that for x_n sufficiently close to $\bar{x}$,

$$|f(\bar{x}) - (f(x_n) + f'(x_n)(\bar{x} - x_n))| \le |\bar{x} - x_n|^2 \mathcal{K}.$$

Since $f(\bar{x}) = 0$, we conclude that for x_n is close to $\bar{x}$,

$$0 \approx f(x_n) + f'(x_n)(\bar{x} - x_n)$$

---

[4]The English mathematician Joseph Raphson (1648–1715) was one of the few people to enjoy close access to Newton's work. Raphson wrote several books that described many of Newton's results. Raphson published a version of Newton's method long before Newton got around to doing so.

[5]The English mathematician Thomas Simpson (1710–1761) was an interesting character. He wrote several books and taught as a wandering lecturer in London coffee houses. Rather remarkably, Simpson developed Newton's method for a system and used it to maximize a function of several variables. Simpson is best remembered for his work on interpolation and numerical integration.

or

$$(\bar{x} - x_n) \approx -f'(x_n)^{-1} f(x_n). \tag{31.17}$$

The approximation (31.17) says that the error $\bar{x} - x_n$ is proportional to the residual $f(x_n)$ with constant $f'(x_n)^{-1}$ when x_n is close to $\bar{x}$.

One consequence is that if $|f'(x_n)|$ is very small, then the error may be large even though the residual is very small. In this case the process of computing the root $\bar{x}$ is said to be **ill-conditioned**.

EXAMPLE 31.16. We apply Newton's method to $f(x) = (x-2)^2 - 10^{-15}x$ with root $\bar{x} \approx 1.00000003162278$. Here $f'(\bar{x}) = 10^{-15}$, so that $f'(x_n)$ is very small for all x_n close to $\bar{x}$. We plot the errors and residuals versus iteration in Fig. 31.13. The residuals become small considerably

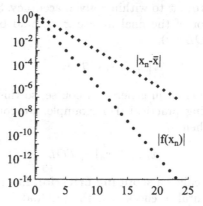

FIGURE 31.13. Plots of the residuals • and errors ◆ versus iteration number for Newton's method applied to $f(x) = (x-2)^2 - 10^{-15}x$ with initial value $x_0 = 2$.

faster than the errors.

Note that by the definition of Newton's method,

$$x_{n+1} = f(x_n) - f(x_n)/f'(x_n);$$

so (31.17) implies

$$|x_n - \bar{x}| \approx |x_{n+1} - x_n|. \tag{31.18}$$

In other words, to get an estimate of the error of x_n, we can compute an extra step of Newton's method to get x_{n+1} and then compute $|x_{n+1} - x_n|$.

EXAMPLE 31.17. We apply Newton's method to $f(x) = x^2 - 2$ and show the error and error estimate (31.18) in Fig. 31.14. The error estimate does a pretty good job.

i	$|x_i - \bar{x}|$	$|x_{i+1} - x_i|$
0	.586	.5
1	.086	.083
2	2.453×10^{-3}	2.451×10^{-3}
3	2.124×10^{-6}	2.124×10^{-6}
4	1.595×10^{-12}	1.595×10^{-12}
5	0	0

FIGURE 31.14. The error and error estimate for Newton's method for $f(x) = x^2 - 2$ with $x_0 = 2$.

An issue closely related to estimating the error is the question of when to stop the iteration. In many situations, the goal of solving a root problem is to approximate the root $\bar{x}$ to within a given accuracy. So ideally, we might specify that the error of the final iterate x_n should be less than a given **error tolerance** $TOL > 0$,

$$|x_n - \bar{x}| \le TOL.$$

We do not know the error in general, of course, so this ideal goal must be replaced by something practical. For example, we could use (31.18) and stop the iteration when

$$|x_{n+1} - x_n| \le TOL. \tag{31.19}$$

Condition (31.19) is called a **stopping criterion** for the iteration. In some cases, it is more natural to check that the residual is sufficiently small. In other words, the iteration is stopped when

$$|f(x_n)| \le TOL. \tag{31.20}$$

31.7   Globally Convergent Methods

The theory only guarantees that Newton's method converges when x_0 is sufficiently close to $\bar{x}$. Moreover, it turns out to be the only way to guarantee convergence in practice as well. The question is how to find good initial values for Newton's method.

One solution is based on the observation that there are iterative methods that converge without requiring a starting value x_0 to be close to $\bar{x}$. For example, recall that the Bisection Algorithm converges to the fixed point starting with an interval of any size as long as the function changes sign at the endpoint values. The Bisection Algorithm is a **globally** convergent method, as opposed to Newton's method, which is **locally** convergent. The problem with globally convergent methods is that they invariably tend to be slow in converging; i.e., they converge at first order.

The idea is to use an algorithm that combines a globally convergent method with a locally convergent method in such a way that the fast local method is used when it is working well, otherwise the slow-but-sure globally convergent method is used. Such a method is sometimes called a **hybrid-Newton method.** It is good to keep in mind that in general it is impossible to guarantee that an iterative approximation method for computing roots approximates a desired root. In other words, there are no truly fast, globally convergent methods. The best we can do is to devise methods that generally compute *some* root starting from almost all initial values. We describe a *simplification* of such a method below.[6]

For example, we could construct an algorithm that uses the Bisection Algorithm until the endpoints are sufficiently close and then switches to Newton's method. The problem is to determine when the switch to Newton's method should take place: there are no natural criteria for deciding. Instead, we try to construct methods that switch from a globally convergent method to Newton's method automatically.

A popular approach is based on considering the Newton "change,"

$$-\frac{f(x_{n-1})}{f'(x_{n-1})}, \tag{31.21}$$

which is added to x_{n-1} to get x_n as determining both a direction and a distance (see Fig. 31.15). Newton's method has the property that $f(x)$ always decreases initially for x, changing value away from x_{n-1} in the direction indicated by (31.21), i.e., toward x_n. Unfortunately if Newton's method indicates a big change from x_{n-1} to x_n, then it is possible for $|f(x_n)| > |f(x_{n-1})|$ (see Fig. 31.15). This is counterproductive since we are trying to make $f(\cdot)$ smaller.

The idea is to accept the computed Newton iterate x_n when $|f(x_n)| < |f(x_{n-1})|$ and to reject the iterate otherwise. When an iterate is rejected, a bisection search is performed on the interval $[x_{n-1}, x_n]$ with the object of finding a x^* between x_{n-1} and x_n such that $|f(x^*)| < |f(x_{n-1})|$. First, we set $x^* = (x_n + x_{n-1})/2$ and check $|f(x^*)| < |f(x_{n-1})|$. If this is true, we set $x_n = x^*$ and proceed forward to compute the next Newton step. If this is false, we set $x_n = x^*$ and return to take another bisection step. In the Newton step shown in Fig. 31.15, the algorithm rejects x_n and takes one step of the bisection search before setting $x_n = x^*$ and continuing with the Newton iteration. The algorithm is:

Algorithm 31.2 hybrid-Newton method

given f, f', x_0

---

[6]The analysis of this method is quite complicated and moreover requires some slight modifications of the algorithm we have written down, so we do not give that and instead refer to Dennis and Schnabel [9].

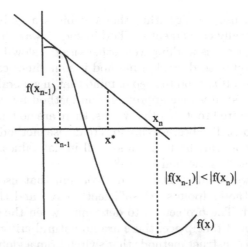

FIGURE 31.15. The Newton step determines both the distance and the direction for obtaining x_n from x_{n-1}. $f(x)$ decreases initially as x changes value from x_{n-1} in the direction of x_n however $|f(x_n)| > |f(x_{n-1})|$ is possible if the change $x_{n-1} \to x_n$ is large.

for $n = 1, 2, 3, \cdots$

    compute $x_n = x_{n-1} - f(x_{n-1})/f'(x_{n-1})$

    if $|f(x_n)| > |f(x_{n-1})|$

        set $x_n = (x_n + x_{n-1})/2$

        while $|f(x_n)| > |f(x_{n-1})|$

            set $x_n = (x_n + x_{n-1})/2$

decide if another Newton iteration is needed .

EXAMPLE 31.18. We implement this algorithm in $MATLAB^{©}$ and run it on the function $f(x) = \tan^{-1}(x - 1)$ with root $\bar{x} = 1$ using $x_0 = 6$. The output of the program is shown in Fig. 31.16 In this example, in fact, Newton's method actually diverges when $x_0 = 6$. However the hybrid-Newton method switches to the safe but slow global method until x_n is close enough to 1 for Newton's method to converge.

31.8   When Good Derivatives Are Hard to Find

In many situations, computing the derivative of f is undesirable or even impossible. For example, f' can be very hard to compute, especially in higher dimensions, and moreover evaluating f' involves computing additional function values, which can be very expensive in terms of computing time. Also, it often happens that f is known only through a set of values

i	method	x_i	$f(x_i)$
0	initial value	6.000000000000000	1.373400766945016
1	Newton iterate	−29.708419940570410	−1.538243471665299
	bisection search	−11.854209970285200	−1.493157178586551
	bisection search	−2.927104985142602	−1.321454920055357
2	Newton iterate	18.774030640348380	1.514593719328215
	bisection search	7.923462827602890	1.427351950092280
	bisection search	2.498178921230144	0.982232920022955
3	Newton iterate	−0.688715155697761	−1.036156901891317
	bisection search	0.904731882766191	−0.094981458393522
4	Newton iterate	1.000575394221151	0.000575394157651
5	Newton iterate	0.999999999873000	−0.000000000127000
6	Newton iterate	1.000000000000000	0.000000000000000

FIGURE 31.16. Results of the hybrid Newton method applied to $f(x) = \tan^{-1}(x - 1)$.

measured during some computational or experimental procedure, so there is no function to differentiate.

In Section 31.5, we interpreted Newton's method as a process of replacing the function f by its linearization at the current iterate and using the linearization to compute the next iterate. To avoid the derivative of f, the basic idea is to use a different linear approximation to the function f. A simple example is to use a secant line passing through the current iterate x_{n-1} and some nearby point $x_{n-1} + h_n$ where h_n is a suitably chosen small number. The slope of this secant line is

$$m_n = \frac{f(x_{n-1} + h_n) - f(x_{n-1})}{h_n}, \qquad (31.22)$$

and the linear approximation to f is

$$f(x) \approx f(x_{n-1}) + m_n(x - x_{n-1}).$$

The new iterate is found by computing the root of the linear approximation and we find

$$x_n = x_{n-1} - \frac{f(x_{n-1})}{m_n}. \qquad (31.23)$$

We illustrate in Fig. 31.17. A Newton method that uses a linear approximation to a function other than the linearization is sometimes called a **quasi-Newton method**.

The important question is whether the new method works, and in particular, if it converges at a second order rate. It turns out that, neglecting round-off error, if h_n tends to zero at the same rate that $x_n - \bar{x}$ tends to zero, and in particular, if there is a constant $c > 0$ such that

$$h_n \leq c|x_{n-1} - \bar{x}| \qquad (31.24)$$

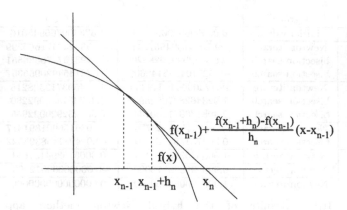

FIGURE 31.17. An illustration of one step of a quasi-Newton method from x_{n-1} to x_n with step h_n.

for all sufficiently large n, or equivalently if there is a constant $\tilde{c}$ such that

$$h_n \leq \tilde{c}|f(x_{n-1})|, \tag{31.25}$$

then the quasi-Newton method does converge at a second order rate for all x_0 sufficiently close to $\bar{x}$. In practice this works at first but as x_{n-1} gets closer to $\bar{x}$, the effects of round-off on the subtraction $f(x_{n-1}+h_n)-f(x_{n-1})$ cause a lot of trouble due to significant cancelation of the leading decimals, owing to the closeness of $x_{n-1}+h_n$ and x_{n-1}. So as a rule of thumb, h_n is never reduced lower than some constant times $\sqrt{u}$, where u is the machine number.

One disadvantage of general quasi-Newton methods is that f must be evaluated at two points, namely, x_{n-1} and $x_{n-1}+h_n$, in order to compute x_n. This motivates the choice $h_n = -(x_{n-1} - x_{n-2})$, which gives

$$m_n = \frac{f(x_{n-1}) - f(x_{n-2})}{x_{n-1} - x_{n-2}} \tag{31.26}$$

and

$$x_n = x_{n-1} - \frac{f(x_{n-1})}{m_n}. \tag{31.27}$$

This is called the **secant method**. We illustrate in Fig. 31.18. Note that the secant method requires two initial values x_0 and x_1. It turns out that the secant method converges with order $(1 + \sqrt{5})/2 \approx 1.6$.

EXAMPLE 31.19. Using the function $f(x) = x^2 - 2$, we compare Newton's method and a quasi-Newton method using $x_0 = 1$ and the secant method with $x_0 = 0$ and $x_1 = 1$ in Fig. 31.19.

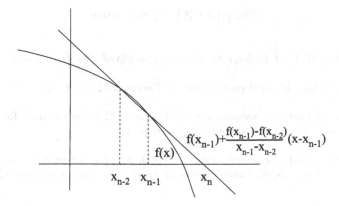

FIGURE 31.18. An illustration of one step of the secant method for computing x_n.

i	Newton's method	quasi-Newton method	secant method
0	1.000000000000000	1.000000000000000	0.000000000000000
1	1.500000000000000	1.400000000000000	1.000000000000000
2	1.416666666666667	1.437500000000000	2.000000000000000
3	1.414215686274510	1.414855072463768	1.333333333333334
4	1.414213562374690	1.414214117144937	1.400000000000000
5	1.414213562373095	1.414213562373512	1.414634146341463
6	1.414213562373095	1.414213562373095	1.414211438474870
7	1.414213562373095	1.414213562373095	1.414213562057320
8	1.414213562373095	1.414213562373095	1.414213562373095

FIGURE 31.19. Results of Newton's method, a quasi-Newton method with $h_n = |f(x_{n-1})|$, and the secant method applied to $f(x) = x^2 - 1$.

31.9   Unanswered Questions

We have given only brief descriptions of the construction of globally convergent hybrid Newton's methods and quasi-Newton methods in which finite differences are used instead of derivatives. We have not proved that these ideas work, nor discussed important practical details. This is an interesting and complicated subject. See Dennis and Schnabel [9] for more details.

Chapter 31   Problems

Problems 31.1–31.10 have to do with the Fixed Point Iteration.

31.1. Prove that the fixed point found in Theorem 31.1 is unique.

31.2. Verify that the assumptions of Theorem 31.1 are satisfied for $g(x) = 1/(2+x)$ on $I5t = [0, 1]$. Compute the fixed point.

31.3. Verify that the assumptions of Theorem 31.1 are satisfied for $g(x) = .5 \tan(x)$ on $I = [-.5, .5]$. Compute the Fixed Point Iteration starting with $x_0 = .5$.

31.4. Find an interval I such that Theorem 31.1 applies for computing the fixed point of $g(x) = 1 - \log(1 + e^{-x})$. Compute the fixed point.

31.5. Make a chart contrasting the assumptions and conclusions of Theorem 31.1 and Theorem 31.2.

31.6. Perform the computations displayed in Example 31.2.

31.7. Verify the claims in Example 31.4.

31.8. Find the fixed points of the following functions by using a graph to verify the assumptions of Theorem 31.2 and finding an initial value x_0 close enough to $\bar{x}$ to give convergence and then computing enough iterations to guarantee 3 digits of accuracy for each fixed point $\bar{x}$. Note that you may have to rewrite the fixed point problem $g(x) = x$ for each fixed point $\bar{x}$ in order to use Theorem 31.2!

$$\begin{array}{ll}
\text{(a) } g(x) = 2 + 2\cos(x/4) & \text{(b) } g(x) = x^3 - x^2 - 1 \\
\text{(c) } g(x) = \log(2 + x^2) & \text{(d) } g(x) = x^6 - 1 \\
\text{(e) } g(x) = 2e^{-x} & \text{(f) } g(x) = x^3.
\end{array}$$

31.9. Set

$$c_1 = .4/64, \quad c_2 = .99999, \quad c_3 = 4 - 4c_2 + \frac{2}{15}c_1$$

and

$$g(x) = c_3 + c_2 x - c_1 \left(\frac{1}{20}x^5 - x^4 + \frac{16}{3}x^3 \right).$$

(a) Verify that Theorem 31.2 applies. (b) Compute 30 fixed point iterations beginning with $x_0 = 0$ and verify that x_i is converging to the fixed point $\bar{x} = 4$. (c) Make a plot of the ratios

$$\frac{|x_i - \bar{x}|}{|x_{i-1} - \bar{x}|}.$$

Explain the results of the graph by using a plot of $|g'(x)|$.

31.10. Rewrite the root problem $f(x) = x^2 - 3$ as a fixed point problem by setting $g(x) = x + c(x^2 - 3)$ and finding a value of c that allows the use of Theorem 31.2.

Problems 31.11–31.17 have to do with high order convergence in the Fixed Point Iteration.

31.11. Estimate the order of convergence, and in the case of first order convergence, the convergence factor for the Fixed Point Iterations computed in Problem 31.8.

31.12. (a) Find the values of c for which you can guarantee that the Fixed Point Iteration for $g(x) = 2 - (1+c)x + cx^3$ converges to $\bar{x} = 1$. (b) What value of c gives second order convergence?

31.13. In each of the following cases, determine if the Fixed Point Iteration converges to the indicated fixed point, and if it does converge, determine the order of convergence and when the order is one the convergence factor:

    (a) $g(x) = x + 9/x^2 - 1$, $\bar{x} = 3$

    (b) $g(x) = \frac{2}{3}x + \frac{1}{x^2}$, $\bar{x} = 3^{1/3}$

    (c) $g(x) = 6/(1+x)$, $\bar{x} = 2$.

31.14. (a) Experimentally verify the claims made about the intervals on which the Fixed Point Iteration converges in Example 31.6 and Example 31.7. (b) Give an explanation for the observation that the interval of convergence in Example 31.6 found experimentally is larger than the region predicted by the analysis. *Hint:* what can be overestimated in the estimates in (31.9).

31.15. (a) Verify theoretically that the Fixed Point Iteration for

$$g(x) = \frac{1}{2}\left(x + \frac{a}{x}\right)$$

where $\bar{x} = \sqrt{a}$ converges quadratically. (b) Try to say something about which initial values guarantee convergence for $a = 3$ by computing some Fixed Point Iterations.

31.16. (a) Show analytically that the Fixed Point Iteration for

$$g(x) = \frac{x(x^2 + 3a)}{3x^2 + a}$$

is third order convergent for computing $\bar{x} = \sqrt{a}$. (b) Compute a few iterations for $a = 2$ and $x_0 = 1$. How many digits of accuracy are gained with each iteration?

31.17. (a) Verify the claims about the convergence rate for the functions in Example 31.10. (b) Verify the claims about convergence for the functions in Example 31.12.

Problems 31.18–31.22 have to do with Newton's method.

31.18. Find the fixed points of the following functions by using Newton's method. Compare to rate of convergence for each computation to the rates obtained for the fixed point computations in Problem 31.8.

(a) $f(x) = 2 + 2\cos(x/4) - x$ (b) $f(x) = x^3 - x^2 - x - 1$
(c) $f(x) = \log(2 + x^2) - x$ (d) $f(x) = x^6 - x - 1$
(e) $f(x) = 2e^{-x} - x$ (f) $f(x) = x^3 - x$.

31.19. Use Newton's method to compute all the roots of $f(x) = x^5 + 3x^4 - 3x^3 - 5x^2 + 5x - 1$.

31.20. Use Newton's method to compute the smallest positive root of $f(x) = \cos(x) + \sin(x)^2(50x)$.

31.21. Use Newton's method to compute the root $\bar{x} = 0$ of the function

$$f(x) = \begin{cases} \sqrt{x}, & x \geq 0, \\ -\sqrt{-x}, & x < 0. \end{cases}$$

Does the method converge? If so, is it converging at second order? Explain your answer.

31.22. Apply Newton's method to $f(x) = x^3 - x$ starting with $x_0 = 1/\sqrt{5}$. Is the method converging? Explain your answer using a plot of $f(x)$.

Problems 31.23–31.25 have to do with error estimation and stopping criteria.

31.23. Modify the code used to solve the root problems in Problem 31.18 to output an error estimate for each x_n using (31.18) as well as computing the real error when the root $\bar{x}$ is input into the program. Rerun the computations you performed in Problem 31.18 and compare the error estimates to the errors as in Fig. 31.14. If you do not have the exact root for a problem, take the value of x_N for a large N as an approximation for $\bar{x}$ and then compare errors for n not too large compared to N.

31.24. Modify the code used to solve the root problems in Problem 31.18 so that it uses a choice of either (31.19) or (31.20) to stop the iteration.

31.25. (a) Derive an approximate relation between the residual $g(x) - x$ of a fixed point problem for g and the error of the fixed point iterate $x_n - \bar{x}$. (b) Devise two stopping criteria for a Fixed Point Iteration. (c) Revise your fixed point code to make use of (a) and (b).

Problems 31.26–31.28 have to do with modifications of Newton's method.

31.26. (a) Implement Algorithm 31.2, making sure the code indicates when a bisection search is performed. (b) Apply your code to the problems in Problem 31.18 compute above and note if any bisection searches are used. (c) Apply your code to Problem 31.19. (d) Apply your code to Problem 31.21. (e) Apply your code to Problem 31.22 with $x_0 < 1/\sqrt{5}$, $x_0 = 1/\sqrt{5}$, and $x_0 > 1/\sqrt{5}$.

31.27. (a) Use a plot to explain why Newton's method diverges for the root problem in Example 31.18. (b) On your plot, illustrate the first Newton iterate and the step-by-step results of the subsequent bisection search.

31.28. (a) Write a code that simultaneously implements Newton's method, a quasi-Newton method with step $h_n = |f(x_{n-1})|$, and the secant method. (b) Apply your code to the problems in Problem 31.18 compute above and compare the convergence of the methods.

In Problems 31.29 and 31.30, we apply Newton's methods in situations in which the conditions that guarantee convergence do not hold.

31.29. Use Newton's method to compute the root $\bar{x} = 1$ of $f(x) = x^4 - 3x^2 + 2x$. Is the method converging quadratically? *Hint:* You can test this by plotting $|x_n - 1|/|x_{n-1} - 1|$ for $n = 1, 2, \cdots$.

31.30. Assume that $f(x)$ has the form $f(x) = (x - \bar{x})^2 h(x)$ where h is a differentiable function with $h(\bar{x}) \neq 0$. (a) Verify that $f'(\bar{x}) = 0$ but $f''(\bar{x}) \neq 0$. (b) Show that Newton's method applied to $f(x)$ converges to $\bar{x}$ at a linear rate and compute the convergence factor.

Calculus Quagmires

There are two major controversies associated with the creation of calculus. Both controversies involved generations of mathematicians and had a strong effect on the development of mathematics. Both troubles yield relevant lessons for modern scientists and mathematicians.

One controversy remains well known, even today, and that is the debate over whether Leibniz or Newton invented calculus first. The debate began a little while after Leibniz and Newton published their results. First, their immediate associates argued; later, Leibniz and Newton themselves became involved as did the following generation or two of mathematicians. The controversy had a strongly negative effect on the pace of development of mathematics because it kept the British mathematicians isolated from the continental European mathematicians for generations.

Yet in retrospect, this debate is entirely meaningless in particular and in general terms. In fact, Leibniz and Newton made most of their calculus discoveries independently and within a few years of each other. Newton made most of his discoveries somewhat before Leibniz, but Leibniz published his results before Newton. Given this, it is simply unworthy of gentlefolk to argue pointlessly about priority.

Perhaps more importantly, this debate is meaningless in general terms. Neither Leibniz nor Newton "invented" calculus. Rather, they contributed substantial progress to the development of calculus, which had begun with the ancient Greeks and continued through to the rigorous construction of the real numbers long after Leibniz and Newton had passed away.

The results of Leibniz and Newton were founded on a substantial body of work on calculus. The ancient Greeks knew about infinite series and a form of integration (see Chapter 27). The study of series continued right up to the time of Leibniz and Newton, and series played a key role in their arguments. The generations of mathematicians immediately before Leibniz and Newton worked directly on the central problems of calculus, such as computing tangents and integration. The calculus textbook of Barrows,[1] who was Newton's predecessor at Cambridge, contained material on finding tangents to curves; differentiation of products and quotients; derivatives of monomials; implicit differentiation; integration including the change of variable formula in a definite integral; and computing the lengths of curves, all described from a geometric point of view. Wallis wrote a similar textbook from an algebraic point of view. Both Leibniz and Newton were fully aware of the prior work on calculus.

Yet, calculus was not a unified subject before Leibniz and Newton. Rather, it consisted of unconnected results for specific functions. The tremendous achievement of Leibniz and Newton was to synthesize these results into a general method and, in particular, to recognize the close relationship between differentiation and integration. They also demonstrated the power of this method by a making a remarkable set of scientific applications.

The development of a general tool for analysis unleashed a revolution in mathematics and science. Since this began with Leibniz and Newton, it is natural to consider them as the inventors of calculus. Yet, to yield to the deplorable human tendency toward the "cult of personality"[2] is to take a very simplistic view of how science and mathematics are conducted. Progress in science and mathematics is a communal affair. Spectacular achievements like those of Leibniz and Newton mark a special kind of genius, yet they are no more than part of the general human march toward understanding nature.

The second controversy was significantly more substantial and important for the development of modern mathematics. This controversy arose because the calculus of Leibniz and Newton was not mathematically rigorous. The central issue was the meaning of *limits*, which were not treated in a mathematically precise way.

This controversy did not take the form of a debate between proponents and opponents of calculus. Rather, it took the form of a serious self-examination by mathematicians. Leibniz and Newton themselves, as well as the subsequent generations of mathematicians, were aware that there were essential holes in the mathematical foundations of calculus. They, and their immediate successors like the Bernoulli's, Dirichlet, Euler, and Lagrange, worried a great deal about putting calculus on a rigorous basis. This fueled

---

[1] Isaac Barrows (1630–1677) held the Lucasian Chair in mathematics at Cambridge University before resigning so Newton could take his place. Barrows wrote very influential texts in geometry, an early form of calculus, and optics.

[2] With ill effects that are seen throughout business, politics, religion, and science.

the ultimately successful efforts of Bolzano, Cantor, Cauchy, Dedekind, and Weierstrass.

However, the power of calculus for describing the physical world and the overwhelming computational and experimental evidence that calculus was correct gave early analysts the confidence to continue developing and using calculus even while they struggled with the lack of rigor. It is important to realize that the revolution in science and mathematics that began in the seventeenth century rested partly on analytic techniques that remained largely unproved as correct until the beginning of this century.[3]

Yet, the very success in applying mathematics to understanding the physical world and the great advances made in mathematics made the need for establishing a rigorous foundation all the more urgent. In fact, there were many incorrect scientific explanations based on faulty mathematics[4] that were subsequently discarded. The gradual progress in establishing rigorous mathematical analysis was important not only mathematically, but scientifically.

In short, the effort to put analysis on a rigorous foundation was conducted by mathematicians who worked in both pure and applied mathematics and were motivated by both mathematical and scientific concerns. For this reason, it is surprising and dismaying to learn that large gulfs have formed between pure and applied and between rigorous and experimental mathematics in this century. Indeed, today a large swath of mathematicians consider these areas to be separate disciplines.

This development can be understood partly on historical grounds. The deep self-examination that lead to rigorous analysis established mathematics as fundamentally different than the other sciences. The standard for mathematical truth rests ultimately on consistent and correct proof rather than on experimental evidence. We believe that correct mathematics is provably correct and that we do not fully understand mathematics until it has been proven true.[5] In contrast, the other sciences rest on experimental evidence and the truths in science cannot be proved to be true in a mathematical sense.

It is the interpretation of "ultimately" in the preceding description that is the source of argument and controversy among mathematicians. Does "ultimately" imply immediacy, i.e., that mathematical arguments are not *mathematics* until they are in the form of rigorous proofs? Or do we in-

---

[3]This trend continues today. For example, theoretical physicists studying subjects like quantum mechanics are using exploratory mathematics that lies far beyond what has been proved to be true.

[4]The cavalier treatment of the convergence of infinite series had particularly misleading consequences.

[5]This is not to say that we understand mathematics just because it has been proven to be true. A proof may not or may only partly explain why a fact is true.

terpret "ultimately" as meaning proof is the final goal, but we accept that there are valid and useful mathematical arguments that are not yet proofs?

Understanding the historical roots of the gulfs between pure and applied and between rigorous and experimental mathematics does not excuse the ignorance and prejudice that keep them in place. *Nothing good for either mathematics or science results from these divisions.* But there are many bad consequences. Principally, the vast majority of research involving mathematics is conducted by non-mathematicians studying scientific and engineering problems. Mathematicians who confine themselves only to mathematics that is provably true, limit themselves to a small part of the mathematical world, consequently losing a rich source of mathematical problems and intuition. At the same time, scientists and engineers in many fields sorely miss the expertise of mathematicians who could help guide them in the exploratory world of applied mathematics.

Part III

You Want Analysis? We've Got Your Analysis Right Here.

32
Notions of Continuity and Differentiability

In the third part of this book, we look more deeply into the properties of functions. We begin in this chapter by considering different ways to define continuity and differentiability and the relations between the different notions. Up to this point, we have employed somewhat restrictive notions of continuity and differentiability in order to make it possible to use constructive arguments to prove major theorems. By considering weaker notions of these concepts, we include more functions in the discussion and also discover some important properties. However, we lose the possibility of using constructive analysis in many cases.

Beginning with this chapter, the discussion takes on a decidedly theoretical flavor and requires more sophistication[1] to read. But, a mastery of the material in this part opens up the doors to the entire world of analysis.

32.1   A General Notion of Continuity

Recall that the intent in defining Lipschitz continuity was to classify a function as varying smoothly in the sense that small changes in input lead to small changes in output. The Lipschitz continuous condition $|f(x)-f(y)| \le L|x-y|$ quantifies the maximum amount a function's value can change for a given change in input. We based the notion of Lipschitz continuity on the behavior of linear functions.

---

[1] Translation: patience and frustration.

But Lipschitz continuity is not the most general way to express the idea that f should vary smoothly.

EXAMPLE 32.1.   Consider $x^{1/3}$, which is Lipschitz continuous on any bounded interval that is bounded away from 0. Checking the Lipschitz condition at 0 gives

$$|x^{1/3} - 0^{1/3}| = |x|^{1/3}.$$

For any constant L,

$$|x|^{1/3} > L|x| \text{ for all } x \text{ sufficiently small;} \qquad (32.1)$$

hence $x^{1/3}$ cannot be Lipschitz continuous on any interval that contains 0 or has 0 as an endpoint.

On the other hand, $|x|^{1/3}$ can be made as close to 0 as desired by making $|x|$ small. So $x^{1/3}$ does vary smoothly as x passes by 0. We can see this from the plot Fig. 32.1.

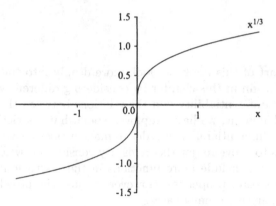

FIGURE 32.1. Plot of $x^{1/3}$.

We make a general definition of continuity that covers such cases.[2] We say that f is **continuous** at $\bar{x}$ if given any sufficiently small $\epsilon > 0$ there is a $\delta > 0$ such that

$$|f(x) - f(\bar{x})| < \epsilon \text{ for all } x \text{ with } |x - \bar{x}| < \delta.$$

In words, this says that the change in value of $f(x)$ from $f(\bar{x})$ can be made arbitrarily small by taking x sufficiently close to $\bar{x}$. Note that $f(x)$ needs to be defined for all x sufficiently close to $\bar{x}$. Note also that $\delta = \delta_{\bar{x},\epsilon}$ usually depends on both $\bar{x}$ and ϵ.

---

[2] Bolzano, Cauchy, and Weierstrass all used this notion of continuity. The notation is due to Weierstrass.

EXAMPLE 32.2. We show that x^2 is continuous at 1. Given $\epsilon > 0$, we want to show $|x^2 - 1| < \epsilon$ for all x close to 1. Now $|x^2 - 1| = |x+1||x-1|$ so restricting x to $[0, 2]$ means that $|x^2 - 1| \leq 3|x - 1|$. Hence if we further restrict x so that $3|x - 1| < \epsilon$, then

$$|x^2 - 1| \leq 3|x - 1| < \epsilon.$$

We can do this by choosing $\delta = \epsilon/3$.

EXAMPLE 32.3. We show that $x^{1/3}$ is continuous at 0. Given $\epsilon > 0$ we want to achieve $|x^{1/3} - 0^{1/3}| = |x|^{1/3} < \epsilon$ for all x close to 1. Now this is true if $\left(|x|^{1/3}\right)^3 < \epsilon^3$ or if $|x| < \epsilon^3 = \delta$.

The last example shows that continuity is somehow a "weaker" property than Lipschitz continuity.

32.2 Properties of Continuous Functions

The general properties of continuous functions follow almost immediately from previous discussion. This is clear after the following theorem, which presents an alternative formulation of continuity.

Theorem 32.1 f *is continuous at* $\bar{x}$ *if and only if* $f(\bar{x})$ *is defined and* $\lim_{x \to \bar{x}} f(x) = f(\bar{x})$.

We leave it as a problem (Problem 32.4) to show this using Theorem 12.5.

EXAMPLE 32.4. The function $f(x) = (x^2 - 1)/(x - 1)$ has a limit at $\bar{x} = 1$ but is not continuous there because $f(1)$ is undefined.

From this result, it follows that:

Theorem 32.2 *Suppose* f *and* g *are continuous at* $\bar{x}$ *and* c *is a number. Then* $f + cg$ *and* fg *are continuous at* $\bar{x}$. *If* $g(\bar{x}) \neq 0$, *then* f/g *is also continuous at* $\bar{x}$.

We leave the proof of this as a problem (Problem 32.5).

Finally, we also leave as a problem (Problem 32.6) to prove a result about the composition of continuous functions.

Theorem 32.3 *If* g *is continuous at* $\bar{x}$ *with* $\bar{y} = g(\bar{x})$ *and* f *is continuous at* $\bar{y}$, *then* $f \circ g$ *is continuous at* $\bar{x}$.

32.3 Continuity on an Interval

As before, we say that a function is **continuous** on an interval I if it is continuous at every point in I.

EXAMPLE 32.5. We show that x^2 is continuous on $(-\infty, \infty)$. Choose a real number $\bar{x}$. Given $\epsilon > 0$, we want to achieve $|x^2 - \bar{x}^2| < \epsilon$ for all x close to $\bar{x}$. Now $|x^2 - \bar{x}^2| = |x + \bar{x}||x - \bar{x}|$ so if we restrict x to $[\bar{x} - 1, \bar{x} + 1]$, then $|x^2 - \bar{x}^2| \leq (2|\bar{x}| + 1)|x - \bar{x}|$. Hence, if we further restrict x so that $(2|\bar{x}| + 1)|x - \bar{x}| < \epsilon$, then

$$|x^2 - \bar{x}^2| \leq (2|\bar{x}| + 1)|x - \bar{x}| < \epsilon.$$

We can do this by choosing $\delta = \epsilon/(2|\bar{x}| + 1)$.

EXAMPLE 32.6. The step function $I(t)$ is discontinuous at $t = 0$ and $t = 1$ but is continuous on $(-\infty, 0)$, $(0, 1)$, and $(1, \infty)$.

EXAMPLE 32.7. The function defined on $[0, 1]$ by

$$Q(x) = \begin{cases} 0, & x \text{ irrational,} \\ 1, & x \text{ rational,} \end{cases}$$

is discontinuous at every point in $[0, 1]$. Let x be any point in $[0, 1]$, then Q takes on the values 0 and 1 for points arbitrarily close to x.

We explore some properties of functions that are continuous on an interval. First, it is straightforward to alter the proof of the Bisection Algorithm so that the theorem applies to continuous functions. We leave the details as a problem (Problem 32.10). We obtain:

Theorem 32.4 Bolzano's Theorem *If f is continuous in an interval $[a, b]$ and $f(a)$ and $f(b)$ have opposite signs, then f has at least one root in (a, b) and the Bisection Algorithm starting with $x_0 = a$ and $X_0 = b$ converges to a root of f in (a, b)*

The Intermediate Value Theorem follows immediately.

Theorem 32.5 Intermediate Value Theorem *Suppose that f is continuous on an interval $[a, b]$. Then for every d between $f(a)$ and $f(b)$ there is at least one point c between a and b such that $f(c) = d$.*

Another interesting fact is that unlike Lipschitz continuity, continuity carries over to an inverse function without qualification, We give the proof of the following theorem as an exercise (Problem 32.11).

Theorem 32.6 Inverse Function Theorem *Let f be a continuous monotone function on $[a, b]$ with $\alpha = f(a)$ and $\beta = f(b)$. Then f has a continuous monotone inverse function defined on $[\alpha, \beta]$. For any x in (α, β), the value of f^{-1} can be computed by applying the Bisection Algorithm to compute the root y of $f(y) - x = 0$ starting on the interval $[a, b]$.*

EXAMPLE 32.8. The function x^3 is Lipschitz continuous and continuous on $[0, 1]$, but its inverse function $x^{1/3}$ is only continuous on $[0, 1]$ and not Lipschitz continuous.

We have two notions of continuous behavior on an interval: continuity and Lipschitz continuity. Lipschitz continuity is apparently the "stronger" notion because Lipschitz continuity implies continuity and in particular Lipschitz continuous functions can vary less abruptly than merely continuous functions. We spend the rest of this section investigating the ways in which the Lipschitz condition is more restrictive.

Actually, there are two ways in which the Lipschitz condition further restricts the behavior of a continuous function. First, it insures that the continuous behavior is uniform across the interval. To be more precise, we recall that if f is continuous on an interval I, then given $\bar{x}$ in I and $\epsilon > 0$, we can find a $\delta_{\bar{x},\epsilon}$ such that for all x in I with $|x - \bar{x}| < \delta_{\bar{x},\epsilon}$, $|f(x) - f(\bar{x})| < \epsilon$. This is *not* a uniform notion of continuity on I because $\delta_{\bar{x},\epsilon}$ depends on $\bar{x}$ as well as ϵ.

EXAMPLE 32.9. Recall that in Example 32.5, we showed x^2 is continuous on $(-\infty, \infty)$ with modulus of continuity $\delta_{\bar{x},\epsilon} = \epsilon/(2|\bar{x}| + 1)$ for any x and $\epsilon > 0$.

EXAMPLE 32.10. We show the function $1/x$ is continuous on $(0, \infty)$. Choosing x, $\bar{x}$ in $(0, \infty)$, we compute

$$\left| \frac{1}{x} - \frac{1}{\bar{x}} \right| = \frac{|x - \bar{x}|}{|x||\bar{x}|}.$$

Things are a little complicated in this example because the right-hand side depends on x as well as $\bar{x}$ and $|x - \bar{x}|$. We can get rid of the dependence on x by assuming that x is restricted so $|x - \bar{x}| < |\bar{x}|/2$, or in particular so $x > \bar{x}/2$. Then

$$\left| \frac{1}{x} - \frac{1}{\bar{x}} \right| = \frac{|x - \bar{x}|}{|x||\bar{x}|} \leq \frac{2|x - \bar{x}|}{|\bar{x}|^2}.$$

Given $\epsilon > 0$, if we assume in addition that $|x - \bar{x}| < |\bar{x}|^2 \epsilon/2$, then

$$\left| \frac{1}{x} - \frac{1}{\bar{x}} \right| \leq \frac{2|x - \bar{x}|}{|\bar{x}|^2} < \epsilon.$$

Hence, given any $\epsilon > 0$,

$$\left| \frac{1}{x} - \frac{1}{\bar{x}} \right| < \epsilon$$

for all $x > 0$ satisfying

$$|x - \bar{x}| < \delta_{\bar{x},\epsilon} = \min\left\{ |\bar{x}|/2, |\bar{x}|^2 \epsilon/2 \right\} = \frac{|\bar{x}|}{2} \min\left\{ 1, |\bar{x}|\epsilon \right\}.$$

In the case when f is continuous on an interval I and we can find a $\delta_{\bar{x},\epsilon} = \delta_\epsilon$ as above that is *independent* of x in I, we say that f is uniformly continuous on I. More precisely, f is **uniformly continuous** on an interval I if for every $\epsilon > 0$ there is a $\delta > 0$ such that $|f(x) - f(y)| < \epsilon$ for all x and y in I with $|x - y| < \epsilon$. The important point is that the degree to which f can vary with a given change in input is the same regardless of the location of the input. The notion of uniform continuity was originally formulated by Heine.[3]

EXAMPLE 32.11. A linear function $f(x) = ax + b$ is uniformly continuous on $(-\infty, \infty)$.

EXAMPLE 32.12. We show that x^2 is uniformly continuous on any bounded interval $[a, b]$. Indeed, Example 32.5 shows this is true since $2|\bar{x}| + 1 \leq 2\max\{|a|, |b|\} + 1$ for any $\bar{x}$ in $[a, b]$. Note that x^2 is *not* uniformly continuous on $(-\infty, \infty)$.

EXAMPLE 32.13. The function x^{-1} is uniformly continuous on any interval $[a, b]$ with $a > 0$. Indeed from Example 32.10, given any $\epsilon > 0$ if we choose $\delta = \epsilon a^2$, then for x and y in $[a, b]$ with $|x - y| < \delta$, we have

$$\left| \frac{1}{x} - \frac{1}{y} \right| < \epsilon.$$

However, x^{-1} is not uniformly continuous on any interval that has 0 as an endpoint (see Fig. 32.2).

A function f that is Lipschitz continuous on an interval I is certainly uniformly continuous on I since

$$|f(x) - f(y)| \leq L|x - y| < \epsilon$$

provided $|x - y| < \delta = \epsilon / L$. But, Lipschitz continuity is more restrictive than uniform continuity in general because it places a limit on how quickly a function can change with a change in input by requiring a linear relationship between ϵ and δ. For general continuous functions, the dependence of δ on ϵ can be more complicated. For example, we can make a useful generalization of the notion of Lipschitz continuity due to Hölder[4] by assuming a power relationship between δ and ϵ. We say that f is **Hölder continuous** on an interval I if there are constants L and $\alpha > 0$ such that

$$|f(x) - f(y)| \leq L|x - y|^{\alpha} \text{ for all } x, y \text{ in } I.$$

---

[3]Heinrich Eduard Heine (1821–1881) was a German mathematician. He discovered some important results in analysis, including some fundamental properties of sets of real numbers.

[4]Otto Ludwig Hölder (1859–1937) was a German mathematician whose main contributions were in group theory. However, he was interested in Fourier series and discovered the important inequality that bears his name.

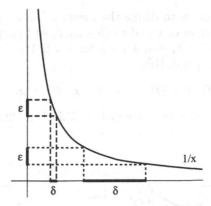

FIGURE 32.2. The function $1/x$ is not uniformly continuous on any interval that has 0 as an endpoint. The plot of $1/x$ indicates that a change in the value of $1/x$ of size ϵ requires a smaller change of size δ in the value of x for smaller values of x.

We call α the **Hölder exponent** of f. A Lipschitz continuous function is Hölder continuous with exponent 1. A Hölder continuous function on I is uniformly continuous on I since

$$|f(x) - f(y)| \le L|x - y|^\alpha < \epsilon$$

as long as $|x - y| < \delta = L^{-1/\alpha}\epsilon^{1/\alpha}$.

EXAMPLE 32.14.   As an example, we verify that $x^{1/2}$ is Hölder continuous with exponent $1/2$ in $[0, \infty)$. For x and y in $[0, \infty)$, $|\sqrt{x} - \sqrt{y}| \le |\sqrt{x} + \sqrt{y}|$. Multiplying by $|\sqrt{x} - \sqrt{y}|$, we get $|\sqrt{x} - \sqrt{y}|^2 \le |x - y|$, or in other words $|\sqrt{x} - \sqrt{y}| \le |x - y|^{1/2}$. Recall that $x^{1/2}$ is not Lipschitz continuous on $[0, \infty)$.

Some of the nice properties of Lipschitz continuous functions result from the fact they are uniformly continuous. For example, recall that a function that is Lipschitz continuous on a bounded interval is bounded. In fact, the boundedness is a consequence of being uniformly continuous. Suppose that f is uniformly continuous on a bounded interval I. We consider the case that $I = [a, b]$ is closed and leave the case when I is open as Problem 32.20. The proof is based on constructing a mesh on $[a, b]$. Set $x_0 = a$ and fix $\epsilon > 0$. By assumption, there is a $\delta > 0$, *independent* of x_0, such that $|f(x) - f(x_0)| < \epsilon$ for all $x_0 \le x < x_0 + \delta$. This means that

$$|f(x)| \le |f(x_0)| + \epsilon \text{ for } x_0 \le x \le x_0 + \delta.$$

We set $x_1 = x_0 + \delta$ and the same argument shows that

$$|f(x)| \le |f(x_1)| + \epsilon \text{ for } x_1 \le x \le x_1 + \delta.$$

We continue this process to define the mesh. If we set n to be the integer that is just greater than or equal to $(b-a)/\delta$, so that $(n-1)\delta < b \le n\delta$, and define a mesh with $x_i = a + i \times \delta$ for $i = 0, 1, \cdots, n-1$ and $x_n = b$. Note that $|x_n - x_{n-1}| \le \delta$. Now,

$$|f(x)| \le |f(x_i)| + \epsilon \text{ for } x_i \le x \le x_{i+1}$$

for $i = 0, 1, \cdots n-1$. We illustrate in Fig. 32.3. But this means that $|f(x)|$

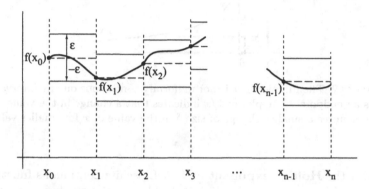

FIGURE 32.3. Illustration of the proof of Theorem 32.7.

is bounded by the maximum of $\{|f(x_0)| + \epsilon, \cdots, |f(x_{n-1})| + \epsilon\}$, which exists since this is a finite set of numbers. Once the case of an open interval is considered (Problem 32.20), we have proved the following theorem:

Theorem 32.7 *A function that is uniformly continuous on a bounded interval is bounded on the interval.*

It is essential for this proof that the interval is bounded and that the function is uniformly continuous.

EXAMPLE 32.15.   The function $f(x) = 1/x$ is continuous on $(0,1)$ but not uniformly continuous. Neither is it bounded.

EXAMPLE 32.16.   The function $y = 2x$ is uniformly continuous on $(-\infty, \infty)$ but is not bounded.

We explore more properties of uniformly continuous functions below.

32.4   Differentiability and Strong Differentiability

We have defined both strong differentiability and differentiability of a function f. These are not equivalent. The definitions imply that if f is strongly differentiable at $\bar{x}$, it is necessarily differentiable. However, differentiability does not imply strong differentiability.

EXAMPLE 32.17. The function $f(x) = x^{4/3}$ is differentiable at 0 since $f'(x) = \frac{4}{3}x^{1/3}$ but is not strongly differentiable. For if we try to compute the error of the linearization, we get

$$\left| x^{4/3} - (0 + 0(x-0)) \right| = |x|^{4/3}.$$

Given any $L > 0$, $|x|^{4/3} > L|x|^2$ for all x sufficiently small; hence, the error of the linearization fails to be sufficiently small.

So it is interesting to compare and contrast the two kinds of differentiability.

If f is strongly differentiable at a point $\bar{x}$, then it is "Lipschitz continuous" at $\bar{x}$ in the sense of Theorem 16.1. Similarly if f is differentiable at $\bar{x}$, then it is continuous at $\bar{x}$. In fact, this follows immediately from the definition because in order for

$$\lim_{x \to \bar{x}} \frac{f(x) - f(\bar{x})}{x - \bar{x}} = f'(\bar{x})$$

to converge, $\lim_{x \to \bar{x}} f(x) - f(\bar{x}) = 0$, which means f is continuous at $\bar{x}$.

Next, we consider the smoothness of the derivative of a function that is differentiable on an interval. It turns out that *neither being differentiable or strongly differentiable on an interval is sufficient to guarantee that f' is continuous*. We explain why with an example.

EXAMPLE 32.18. Consider the continuous function

$$f(x) = \begin{cases} x^2 \sin(1/x), & x \neq 0, \\ 0, & x = 0. \end{cases} \tag{32.2}$$

We plot f in Fig. 32.4.

For $x \neq 0$, $f'(x) = -\cos(1/x) + 2x\sin(1/x)$. While defined for $x \neq 0$, $\lim_{x \to 0} f'(x)$ is undefined because $-\cos(1/x)$ oscillates faster and faster, taking on all values between -1 and 1, as x decreases.

However if we compute the derivative at 0 using the definition,

$$f'(0) = \lim_{x \to 0} \frac{x^2 \sin(1/x) - 0}{x - 0} = \lim_{x \to 0} x \sin(1/x) = 0,$$

we find it is defined and $f'(0) = 0$. Moreover, it is straightforward to show that f is strongly differentiable at each x, including 0. Hence, f is strongly differentiable on any interval containing 0 but f' is not continuous on such an interval.

On the other hand, we also proved above that a function that is uniformly strongly differentiable on an interval has a Lipschitz continuous derivative on the interval. So the uniformity must convey some extra smoothness on the derivative. To understand this, we add uniformity to the definition of

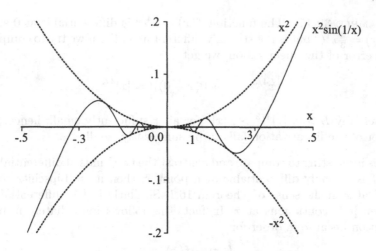

FIGURE 32.4. Plot of f defined in (32.2).

differentiability. We say that f is **uniformly differentiable** on an interval I if given any $\epsilon > 0$ there is a $\delta > 0$ such that

$$\left| \frac{f(y) - f(x)}{y - x} - f'(x) \right| < \epsilon \text{ for all } x \text{ and } y \text{ in } I \text{ with } |x - y| < \delta.$$

This is the direct analog of the definition of uniform continuity. Note that if f is uniformly strongly differentiable on an interval, then it is uniformly differentiable on the interval, but the converse is generally not true.

EXAMPLE 32.19. From the discussion above, it follows that x^2 is uniformly differentiable on any bounded interval $[a, b]$.

EXAMPLE 32.20. The function $1/x$ is differentiable on $(0, 1)$ but is not uniformly differentiable on $(0, 1)$.

We assume that f is uniformly differentiable on an interval I and compute the change in f' between two points x and y in I,

$$|f'(y) - f'(x)| = \left| f'(y) - \frac{f(y) - f(x)}{y - x} + \frac{f(y) - f(x)}{y - x} - f'(x) \right|$$
$$\leq \left| f'(y) - \frac{f(y) - f(x)}{y - x} \right| + \left| \frac{f(y) - f(x)}{y - x} - f'(x) \right|.$$

By assumption, given any $\epsilon > 0$ we can find a $\delta > 0$ such that each of the two quantities on the right are smaller than $\epsilon/2$ for any x and y in I with $|x - y| < \delta$. This proves the following:

Theorem 32.8 *A uniformly differentiable function on an interval has a uniformly continuous derivative on the interval.*

Note that this theorem does not guarantee that f' is Lipschitz continuous.

Some of the other nice properties of uniformly strongly differentiable functions result from the uniformity. One of the most important is that the Mean Value Theorem holds for uniformly differentiable functions.

Theorem 32.9 Mean Value Theorem   *Suppose that f is uniformly differentiable on an interval $[a, b]$. There is at least one point c in $[a, b]$ such that*

$$\frac{f(b) - f(a)}{b - a} = f'(c).$$

The proof of this using an algorithm for approximating the point c is nearly the same as the proof for uniformly strongly differentiable functions. We give this proof as an exercise (Problem 32.22).

Recall that uniformly strongly differentiable functions are also Lipschitz continuous. This is also a consequence of the uniformity. Suppose that f is uniformly differentiable on a closed interval $[a, b]$. The Mean Value Theorem says that for any two points x and y in $[a, b]$, there is a point c in $[a, b]$ such that

$$|f(x) - f(y)| = |f'(c)|\,|x - y|.$$

Now f' is uniformly continuous on $[a, b]$. Theorem 32.7 implies that it is bounded, i.e., there is an M such that $|f'(x)| \leq M$ for all x in $[a, b]$. Hence, $|f(x) - f(y)| \leq M|x - y|$ for all x and y in $[a, b]$.

We summarize the results on the smoothness of a differentiable function as a theorem.

Theorem 32.10 *A function that is differentiable on an interval is continuous on the interval. A function that is uniformly differentiable on a closed interval is Lipschitz continuous on the interval.*

32.5   Weierstrass' Principle and Uniform Continuity

In the discussion above, we saw that uniformity on an interval is a strong condition with many good consequences. Therefore, the following theorem, due to Dirichlet, is quite remarkable.

Theorem 32.11 Principle of Uniform Continuity   *A function that is continuous in a closed, bounded interval is uniformly continuous in that interval.*

One good thing about this theorem is that it is often much easier to show a function is continuous at each point in an interval than to show it is uniformly continuous on the interval. But it is important to note that the interval *must* be closed.

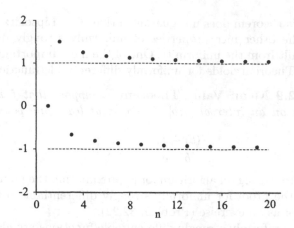

FIGURE 32.5. Plot of some values of $\{(-1)^n + 1/n\}$.

EXAMPLE 32.21. The function $f(x) = 1/x$ is continuous on $(0,1)$ but is not uniformly continuous on $(0,1)$ (or on $[0,1]$ for that matter, but then f is not defined at 0).

The proof of this theorem uses a remarkable and important fact about sequences of numbers called Weierstrass' principle. The Weierstrass principle is concerned with sequences that do not necessarily converge but nonetheless have some kind of regular behavior.

EXAMPLE 32.22. The sequence $\{n^2\}$ does not converge because the terms grow without bound as n increases; i.e., it diverges to infinity.

EXAMPLE 32.23. The sequence $\{(-1)^n + 1/n\} = \{0, 3/2, -2/3, 5/4, -4/5, 7/6, -6/7, 9/8, -8/9, 11/10, \cdots\}$ also fails to converge since the terms fail to approach a single number (see Fig. 32.5). On the other hand, the odd terms $\{0, -2/3, -4/5, -6/7, -8/9, , \cdots\}$ do approach -1 while the even terms $\{3/2, 5/4, 7/6, 9/8, 11/10, \cdots\}$ approach 1.

EXAMPLE 32.24. The sequence $\{\sin(n)\}$ clearly does not converge. What exactly happens is hard to determine (see Fig. 32.6).

We want to distinguish sequences that have the property that some part converges to a limit. Given a sequence $\{x_n\}_{n=1}^{\infty}$, a **subsequence** is a sequence of the form $\{x_{n_k}\}_{k=1}^{\infty} = \{x_{n_1}, x_{n_2}, \cdots\}$ where $n_1 < n_2 < n_3 < \cdots$ is a subset of the natural numbers. We also say that a subsequence is obtained by **extracting** an infinite number of terms from the sequence $\{x_n\}$. We say that x is a **limit point** of a sequence $\{x_n\}$ if we can extract a subsequence $\{x_{n_k}\}$ that converges to x in the usual sense.

EXAMPLE 32.25. The sequence $\{n^2\}$ has no limit points because any subsequence grows without bound.

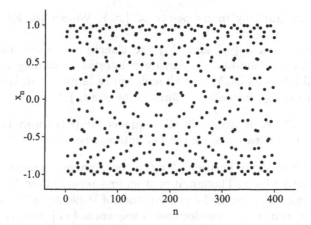

FIGURE 32.6. Plot of some values of $\{\sin(n)\}$.

EXAMPLE 32.26. The sequence $\{(-1)^n + 1/n\} = \{0, 3/2, -2/3, 5/4,$ $-4/5, 7/6, -6/7, 9/8, -8/9, 11/10, \cdots\}$ has the two limit points -1 and 1 (see Fig. 32.5). The subsequence obtained by taking the odd terms converges to -1 and the subsequence obtained by taking the even terms converges to 1.

Another way to characterize a limit point is the following observation.

Theorem 32.12 *A point x is a limit point of $\{x_n\}$ if and only if every open interval that contains x contains a term in $\{x_n\}$ distinct from x.*

Proving this is a good exercise (Problem 32.25).

Note that a limit point x of $\{x_n\}$ is not necessarily a limit of the sequence $\{x_n\}$, which indeed may not even converge as in the example above. On the other hand, if $\{x_n\}$ converges to a limit x, then x is necessarily a limit point of $\{x_n\}$, and, in fact, every subsequence of $\{x_n\}$ necessarily converges to x. We leave the proof as Problem 32.23.

Weierstrass' principle talks about the existence of limit points:

Theorem 32.13 Weierstrass' Principle   *Every bounded sequence has at least one limit point, i.e., has at least one convergent subsequence.*

The proof is a modification of the argument used for convergence of the Bisection Algorithm. Since the sequence $\{x_n\}$ is bounded, the elements are contained in a bounded interval $[y_1, Y_1]$. Starting with this interval, we construct a sequence of nested intervals $[y_i, Y_i]$ with $|Y_i - y_i| = \frac{1}{2}|Y_{i-1} - y_{i-1}|$ each of which contains infinitely many points of $\{x_n\}$. We know that $[y_1, Y_1]$ contains infinitely many points of $\{x_n\}$. Assume that we have $[y_{i-1}, Y_{i-1}]$ containing infinitely many points of $\{x_n\}$. Define the midpoint $m_{i-1} = (Y_{i-1} + y_{i-1})/2$. At least one of the two intervals $[y_{i-1}, m_{i-1}]$ or $[m_{i-1}, Y_{i-1}]$

must contain infinitely many points of $\{x_n\}$. We set $[y_i, Y_i]$ to be that interval.

Now the sequences $\{y_i\}$ and $\{Y_i\}$ are Cauchy sequences and therefore converge to a common limit x. Any open interval containing x necessarily contains all intervals $[y_i, Y_i]$ for all sufficiently large i, and hence contain infinitely many x_n. Hence x is a limit point of $\{x_n\}$.[5]

EXAMPLE 32.27. Remarkably, the sequence $\{\sin(n)\}$ shown in Fig. 32.6 must contain at least one convergent subsequence.

Like the Mean Value Theorem, Weierstrass' principle is useful for proving many interesting facts. Though, it is often first recast into other equivalent forms. For example, one useful consequence of Weierstrass' principle is the following theorem about non-decreasing sequences $\{x_n\}$ with $x_1 \leq x_2 \leq \cdots$ and non-increasing sequences $\{x_n\}$ with $x_1 \geq x_2 \geq \cdots$.

Theorem 32.14 *A bounded non-decreasing or non-increasing sequence $\{x_n\}$ converges to a limit.*

We leave the proof as an exercise (Problem 32.26).

Another important consequence has to do with the existence of "tightest" bounds on a bounded set of numbers. Recall that in Chapter 8, we defined a set of numbers A to be **bounded** with size less than or equal to $b - a$ if the numbers in the set are contained in a finite interval $[a, b]$, and if we can find the *smallest* interval $[a, b]$ with this property, then we define the **size** of A to be $|A| = b - a$. This begs the question of whether or not the smallest interval that contains a bounded set of numbers can be determined. This has two components: determining a and b. We call any number greater than the members in a set an **upper bound** on the set and likewise any number smaller than the members in a set a **lower bound** on the set. In mathematical terms, a is a lower bound for a set A if $a \leq x$ for all x in A, and b is an upper bound for a set A if $x \leq b$ for all x in B. If there is a smallest upper bound on a set A, we call this number the **least upper bound** or **supremum** of A and write it as $\sup A$. Likewise, if there is a largest lower bound on a set A, we call this number the **greatest lower bound** or **infimum** of A and write $\inf A$. Mathematically,

$$\text{lower bounds } x \leq \inf A \leq y \text{ in } A \leq \sup A \leq \text{ upper bounds } z.$$

---

[5]This proof of Weierstrass' principle is rather deceptive in the sense that it does not give an algorithm for computing the limit point despite its close resemblance to proof for the Bisection Algorithm. The critical difference between the two proofs is the decision process for choosing the subintervals. The point is that we cannot verify that a given interval contains an infinite number of terms of a sequence by counting sequentially. We always have to stop at some point in practice and at that point we do not know how many might be left: a finite number or an infinite number. Recall that we pointed out that a strict constructivist would raise a similar objection to the definition of a strict inequality.

If we can find inf A and sup A, then the size of A is simply $|A| = \sup A -$ inf A. Note that if either sup A or inf A exists, then it must either be a member of A or the limit of some sequence of points in A. For example, consider sup A. If it is not in A, then we must be able to find numbers in A arbitrarily close to sup A, since otherwise we could get upper bounds of A smaller, then sup A. This means in particular, we can find a sequence of numbers $\{x_n\}$ in A with $|\sup A - x_n| < 1/n$ and so $\{x_n\}$ converges to sup A.

EXAMPLE 32.28. The set of numbers $A = \{1/n\}$ for natural numbers n is bounded with sup $A = 1$ being a member of A and inf $A = 0$ not in A.

When sup A is in A, we call sup A the **maximum** of A and write sup $A =$ max A. Likewise when inf A is in A, we call inf A the **minimum** of A and write inf $A = \min A$.

Weierstrass' principle implies a result originally proved by Bolzano:

Theorem 32.15 Least Upper Bound Principle *A bounded set of numbers has a greatest lower bound and a least upper bound.*

We prove the existence of the least upper bound and leave the greatest lower bound as an exercise. Assuming that A is a bounded set, we define b_n to be the smallest rational upper bound on A with denominator 2^n. Note that only a finite number of numerators have to be checked to determine b_n. Then for any x in A, we have

$$x \leq b_{n+1} \leq b_n \leq b_1 \text{ for all } n \geq 1.$$

The sequence $\{b_n\}$ is therefore bounded and non-increasing and thus must have a limit b. Now b must be an upper bound for A because otherwise some of the b_n must fail to be upper bounds since the b_n become arbitrarily close to b. Moreover, b must be less than or equal to any other upper bound by the construction of $\{b_n\}$.

With these theorems in hand, we turn to proving some facts about continuous functions. To begin with, we use Weierstrass' principle to prove Theorem 32.11. Suppose that f is continuous in the bounded interval $[a, b]$. If f is not uniformly continuous, then there is an $\epsilon > 0$ such that there exist points x and y in $[a, b]$ arbitrarily close together yet $|f(x) - f(y)| \geq \epsilon$. In particular, given any natural number n we can find points x_n and y_n in $[a, b]$ for which $|f(x_n) - f(y_n)| \geq \epsilon$ and $|x_n - y_n| < 1/n$.

The sequence $\{x_n\}$ is bounded since it is contained in $[a, b]$ and Theorem 32.13 implies that it contains a subsequence $\{x_{n_k}\}$ that converges to a limit point x. We claim that x must be contained in $[a, b]$. Suppose, for example, that $x > b$. Then $|x - x_{n_k}| \geq |x - b|$ for any term x_{n_k} and hence $\{x_{n_k}\}$ could not converge to x.[6] Because the terms in $\{y_n\}$ become

---

[6]It is essential for the interval $[a, b]$ to be closed for this to be true.

arbitrarily close to the terms in $\{x_n\}$ as n increases, there is a subsequence $\{y_{m_k}\}$ of $\{y_n\}$ that also converges to x.

Since f is continuous on $[a, b]$, we must have

$$\lim_{k \to \infty} f(x_{n_k}) = \lim_{k \to \infty} f(y_{m_k}) = f(x).$$

But this contradicts the assumption that $|f(x_n) - f(y_n)| \geq \epsilon$ for all n.

By Theorem 32.7, it follows that a continuous function on a closed interval is bounded. But we can say even more. Suppose that f is continuous on the interval $[a, b]$. Then the set of numbers $A = \{f(x), a \leq x \leq b\}$ is bounded. It therefore has a least upper bound $M = \sup A$ and a greatest lower bound $m = \inf A$. By the comment above, M and m are the limits of some sequences of points in A. For example, there is a sequence $\{x_n\}$ with x_n in $[a, b]$ such that $\lim_{n \to \infty} f(x_n) = M$. But, $\{x_n\}$, being bounded, must contain a convergent subsequence $\{x_{n_k}\}$ converging to a limit point x, which is in $[a, b]$. By the continuity of f, $f(x) = \lim_{k \to \infty} f(x_{n_k}) = M$. In short, f actually takes the value $\sup_{a \leq x \leq b} f(x)$ at some point in $[a, b]$. Likewise, f also takes the value $\inf_{a \leq x \leq b} f(x)$ at some point in $[a, b]$. When f takes the value $\sup_{x \text{ in } A} f(x)$ we call this the **maximum value** of f on A. Likewise, when f takes the value $\inf_{x \text{ in } A} f(x)$ we call this the **minimum value** of f on A. We have proved a result originally due to Weierstrass:

Theorem 32.16 Extremum Principle for a Continuous Function
A continuous function on a closed interval is bounded and attains its maximum and minimum values at some points in the interval.

Using this result, we can prove another version of the Mean Value Theorem that applies to a larger class of functions, albeit at the expense of a proof that does not give a computational algorithm. The theorem is:

Theorem 32.17 Mean Value Theorem *Suppose f is continuous on an interval $[a, b]$ and differentiable on (a, b). Then there is a point c in (a, b) such that*

$$f'(c) = \frac{f(b) - f(a)}{b - a}.$$

It is a good idea to compare this theorem to Theorem 21.1. As before, this theorem follows from the nonconstructive form of Rolle's theorem (Problem 32.28).

Theorem 32.18 Rolle's Theorem *Suppose g is continuous on an interval $[a, b]$, differentiable on (a, b), and $g(a) = g(b) = 0$. Then there is a point c in (a, b) such that $g'(c) = 0$.*

We prove Theorem 32.18. Since g is continuous on $[a, b]$, it attains its maximum M and minimum m values at some points in $[a, b]$. Since $g(a) = 0$, we must have $m \leq 0 \leq M$. Now if $m = M$, it follows that $g(x) = 0$ for all

x and so $g'(x) = 0$ for all x and we are done. So we suppose that $M > 0$ and we let c be the point in (a, b) with $g(c) = M$.[7]

Since $g(x) \leq g(c) = M$ for all x in $[a, b]$, we have $g(x) - g(c) \leq 0$ for all x in $[a, b]$. Thus

$$\frac{g(x) - g(c)}{x - c} \begin{cases} \geq 0, & x < c, \\ \leq 0, & x > c. \end{cases}$$

Letting x approach c first from below and then from above, we conclude that $g'(c) \leq 0$ and $g'(c) \geq 0$. Hence, $g'(c) = 0$ as desired.

32.6   Some Differentiability Equivalences

We conclude by using these results to find equivalences between the various notions of differentiability.

Above, we show that if f is uniformly differentiable on an interval, then f' is uniformly continuous on the interval. Now, we assume that f is differentiable on an interval I and moreover f' is uniformly continuous on I. For x and y in I, $x \neq y$, we compute the error of the linearization

$$|f(x) - (f(y) + f'(y)(x - y))| = |f(x) - f(y) - f'(y)(x - y)|.$$

By Theorem 32.17, there is a point c between x and y such that

$$|f(x) - (f(y) + f'(y)(x - y))| = |f'(c)(x - y) - f'(y)(x - y)|$$
$$= |f'(c) - f'(y)|\,|x - y|.$$

Division yields

$$\left| \frac{f(x) - f(y)}{x - y} - f'(y) \right| = |f'(c) - f'(y)|.$$

Because f' is uniformly continuous on I, for any $\epsilon > 0$ there is a $\delta > 0$ such that

$$\left| \frac{f(x) - f(y)}{x - y} - f'(y) \right| = |f'(c) - f'(y)| < \epsilon$$

for all x and y in I with $|x - y| < \delta$.[8] This shows that f is uniformly differentiable and we have proved the following extension of Theorem 32.8:

Theorem 32.19 *A function is uniformly differentiable on an interval if and only if it is differentiable on the interval and the derivative is uniformly continuous on the interval.*

---

[7]Note that $c \neq a$ or b.

[8]This uses the fact that c is between x and y.

As we said above, differentiability on an interval does not imply strong differentiability on the interval. But we can prove the following equivalence.

Theorem 32.20 *A function is uniformly strongly differentiable on an interval if and only if it is uniformly differentiable on the interval and its derivative is Lipschitz continuous on the interval.*

We already know that if a function f is uniformly strongly differentiable, then it is uniformly differentiable and its derivative is Lipschitz continuous. The converse is a consequence of the Mean Value Theorem. Suppose the Lipschitz constant of f' is K. We compute the error of the linearization at a point $\bar{x}$ in $[a, b]$,

$$|f(x) - (f(\bar{x}) + f'(\bar{x})(x - \bar{x}))| = |f(x) - f(\bar{x}) - f'(\bar{x})(x - \bar{x})|.$$

By the Mean Value Theorem, there is a point c between x and $\bar{x}$ such that

$$|f(x) - (f(\bar{x}) + f'(\bar{x})(x - \bar{x}))| = |f'(c)(x - \bar{x}) - f'(\bar{x})(x - \bar{x})|$$
$$= |f'(c) - f'(\bar{x})|\,|x - \bar{x}|.$$

The Lipschitz condition implies

$$|f(x) - (f(\bar{x}) + f'(\bar{x})(x - \bar{x}))| \le K|x - \bar{x}|^2.$$

This gives the conclusion.

Chapter 32 Problems

Problems 32.1–32.9 are concerned with the general notion of continuity.

32.1. Prove the claim about (32.1).

32.2. Prove that $\sqrt{x}$ is continuous at 0 and at 1.

32.3. Prove that x^3 is continuous at 2.

32.4. Prove Theorem 32.1.

32.5. Prove Theorem 32.2.

32.6. Prove Theorem 32.3.

32.7. Prove that $\sqrt{x}$ is continuous on $[0, \infty)$.

32.8. Prove that x^3 is continuous on $(-\infty, \infty)$.

32.9. Suppose f is continuous on an interval $[a, b]$ and only takes on rational values. Prove f is constant.

Problems 32.10–32.20 have to do with continuity on an interval.

32.10. Prove Theorem 32.4.

32.11. Prove Theorem 32.6.

32.12. Verify the claims in Example 32.11.

32.13. Verify the claims in Example 32.12.

32.14. Verify the claims in Example 32.13.

32.15. Consider $f(x) = 1/x$ on $(0, 1)$. Given any $\epsilon > 0$ show there are points x and y in $(0, 1)$ arbitrarily close such that $|f(x) - f(y)| \geq \epsilon$. *Hint:* Choose $\delta > 0$. Set $x = \delta/4$ and $y = 3\delta/4$. Show that $|f(x) - f(y)| \geq \epsilon$ for all δ sufficiently small.

32.16. Prove that x^3 is continuous on any bounded interval.

32.17. Prove that the function

$$f(x) = \begin{cases} \dfrac{1}{\log_2 |x|}, & x \neq 0, \\ 0, & x = 0, \end{cases}$$

is continuous but not Hölder continuous on $[-.5, .5]$. *Hint:* Consider what happens at zero using the sequence $x_n = 2^{-n/\alpha}$ for $\alpha > 0$.

32.18. Assume that $f(x)$ is defined on $(-\infty, \infty)$ and that

$$|f(x) - f(y)| \le (y - x)^2$$

for all x and y. Prove that f is constant.

32.19. State and prove a theorem about the uniform continuity of the composition of two uniformly continuous functions.

32.20. Finish the proof of Theorem 32.7 by showing that it also holds for an interval $I = (a, b)$. *Hint:* The problem now is defining the initial mesh point x_0, which cannot be a. By assumption, given $\epsilon > 0$ there is a $\delta > 0$ such that $|f(x) - f(y)| < \epsilon$ for all x and y in I with $|x - y| < \delta$. Choose x_0 to be a point in $(a, a + \delta)$. Now define the rest of the mesh accordingly. Take care about defining x_n.

Problems 32.21 and 32.22 have to do with differentiability.

32.21. Show that f defined in (32.2) is strongly differentiable at every x.

32.22. Modify the proof of Theorem 21.1 to prove Theorem 32.9. *Hint:* The crucial point is to prove that if g is uniformly differentiable on an interval $[a, b]$ containing a point y and $g'(y) > 0$ there is a point $\tilde{y} > y$ such that $g(\tilde{y}) > g(y)$. Similarly, there is a point $\tilde{y} < y$ with $g(\tilde{y}) < g(y)$ and corresponding results when $g'(y) < 0$. To show this, consider that for any $\epsilon > 0$ there is a $\delta > 0$ such that for all $\tilde{y} > y$ and $|\tilde{y} - y| < \delta$,

$$\left| \frac{g(\tilde{y}) - g(y)}{\tilde{y} - y} - g'(y) \right| < \epsilon.$$

From this deduce that for all such $\tilde{y}$,

$$(g'(y) - \epsilon)(\tilde{y} - y) + g(y) < g(\tilde{y}) < (g'(y) + \epsilon)(\tilde{y} - y) + g(y).$$

Now choose $\epsilon = g'(y)/2$ and draw the conclusion.

Problems 32.23–32.30 have to do with Weierstrass' principle and related results.

32.23. Prove that if $\{x_n\}$ converges to a limit x, then x is necessarily a limit point of $\{x_n\}$, and, in fact, every subsequence of $\{x_n\}$ necessarily converges to x.

32.24. Draw the sequence of intervals generated by the proof of Weierstrass' principle for the sequence $\{1 - 1/n\}$, $n \ge 1$.

32.25. Prove Theorem 32.12.

32.26. Use Theorem 32.13 to prove Theorem 32.14.

32.27. Prove that a bounded set of numbers has a greatest lower bound.

32.28. Show that Theorem 32.17 follows from Theorem 32.18.

32.29. Assume that f is continuous on $[a, b]$ and differentiable on (a, b). Prove that $f'(c) = 0$ at any point c in (a, b) where f attains is maximum or minimum value.

32.30. Prove the following theorem:

Theorem 32.21 Integral Mean Value Theorem   *Suppose f is continuous on $[a, b]$. There is a point c in (a, b) such that*

$$\int_a^b f(x)\, dx = f(c)(b - a).$$

Explain that this means that a continuous function takes on its average value at least once on an interval.

Problems 32.31–32.35 contain miscellaneous results for continuous functions.

32.31. Suppose that g is differentiable on $(-\infty, \infty)$ with a bounded derivative; i.e., there is a constant M such that $|g'(x)| \leq M$ for all x. Prove that for all sufficiently small $\epsilon > 0$, the function $f(x) = x + \epsilon g(x)$ is invertible.

32.32. Suppose that f is a continuous function on an interval $[a, b]$. Show that there is a function g that is continuous on $(-\infty, \infty)$ such that $g(x) = f(x)$ for $a \leq x \leq b$. Such a function is called a **continuous extension** of f. Show by example that the result is false if the interval (a, b) is open instead.

32.33. A function f defined on an interval (a, b) is **convex** if for any x and y in (a, b)

$$f(sx + (1 - s)y) \leq sf(x) + (1 - s)f(y) \text{ for all } 0 < s < 1.$$

Prove that a convex function is continuous.

32.34. Prove that a monotone function f on an interval $[a, b]$ that takes on every value between $f(a)$ and $f(b)$ as x varies between a and b at least once is continuous. Why is monotonicity necessary?

32.35. Assume f is a continuous function on $[0, 1]$ such that $0 \leq f(x) \leq 1$ for $0 \leq x \leq 1$. Prove that $f(x) = x$ for at least one x in $[0, 1]$.

33
Sequences of Functions

When studying the solution of differential equations, we emphasized that the solution of a differential equation can rarely be written down as an explicit formula in terms of known functions. Instead, we try to approximate the solution using a sequence of relatively simple functions that converge to the solution. Indeed, this was precisely the approach used to solve $y'(x) = f(x)$ on $[a, b]$[1] for x in $[a, b]$, in constructing a Cauchy sequence of functions $\{Y_N\}$ that converges uniformly to y. An important step in this process was proving that y "inherits" some useful properties from the sequence $\{Y_N\}$, such as being Lipschitz continuous.

In this chapter, we study the question of inherited properties for abstract sequences of functions that converge to a limit. In other words, if $\{f_n\}$ converges to a function f, what properties of the functions f_n carry over to the limit f? In general, we consider the sequence $\{f_n\}$ as a sequence of successively more accurate approximations of the limit f, as in the case of integration, but without specifying how the approximating sequence is constructed.[2]

Following the trend in this part of the book, we begin by generalizing the notion of convergence of a sequence of functions. Suppose $\{f_n\}$ is a sequence of functions on an interval I such that the sequence of *numbers* $\{f_n(x)\}$ converges for each x in I. We define the **limit** of $\{f_n\}$ to be the

---

[1]Which is equivalent to computing $\int_a^x f(s)\, ds$.

[2]Recall that we adopted a similarly abstract attitude when discussing the properties of real numbers.

function f with the value

$$f(x) = \lim_{n \to \infty} f_n(x) \text{ for each } x \text{ in } I.$$

More precisely, for any x in I and $\epsilon > 0$, there is a N such that for $n > N$,

$$|f(x) - f_n(x)| < \epsilon.$$

We say that $\{f_n\}$ **converges pointwise** to f on I. Note that pointwise convergence is not the same as uniform convergence. Recall that a sequence of functions $\{f_n\}$ **converges uniformly** to f on an interval I if for every $\epsilon > 0$ there is an N such that for all x in I,

$$|f_n(x) - f(x)| < \epsilon \text{ for } n \geq N.$$

EXAMPLE 33.1. The sequence $\{x^n\}_{n=1}^{\infty}$ converges pointwise on $[0, 1]$ but not uniformly. In fact,

$$\lim_{n \to \infty} x^n = \begin{cases} 0, & 0 \leq x < 1, \\ 1, & x = 1. \end{cases}$$

However, given any n, there is an x in $[0, 1)$ with x^n arbitrarily close to 1 (see Fig. 33.1). So x^n cannot converge uniformly to 0 for $0 \leq x < 1$.

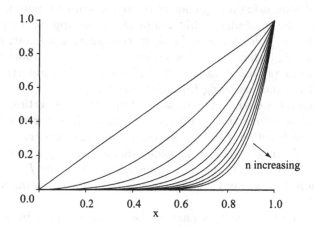

FIGURE 33.1. Plots of the functions $\{x^n\}$ for $n = 1, 2, \cdots, 10$.

In pointwise convergence, the sequences of numbers $\{f_n(x)\}$ may converge at different rates for each x.

We are particularly concerned with the inheritance of continuity, differentiability, and integrability from a convergent sequence of functions. In each of these cases, the question can be rephrased in terms of exchanging the order of two limit processes. As an example, consider continuity. By Theorem 32.1, if each element in $\{f_n\}$ is continuous at $\bar{x}$, then

$$\lim_{x \to \bar{x}} f_n(x) = f_n(\bar{x}) \text{ for each } n.$$

If $\{f_n\}$ converges to f, then to show that f is continuous at $\bar{x}$, i.e to show that

$$\lim_{x \to \bar{x}} f(x) = f(\bar{x}),$$

we have to verify that

$$\lim_{x \to \bar{x}} f(x) = \lim_{x \to \bar{x}} \lim_{n \to \infty} f_n(x) = \lim_{n \to \infty} \lim_{x \to \bar{x}} f_n(x) = \lim_{n \to \infty} f_n(\bar{x}). \qquad (33.1)$$

The question is therefore equivalent to whether the interchange in the order of the limits in the middle of (33.1) is justified.

While intuition might suggest that taking multiple limits "should" be independent of order, in fact this is a place where intuition can be very misleading. *In general, order does matter when taking multiple limits.* A simple example shows why.

EXAMPLE 33.2. Consider the sequence with double indices

$$\left\{\frac{m}{n+m}\right\}_{n=1,\, m=1}^{\infty}$$

First we compute

$$\lim_{m \to \infty} \lim_{n \to \infty} \frac{m}{n+m} = \lim_{m \to \infty} 0 = 0.$$

Yet,

$$\lim_{n \to \infty} \lim_{m \to \infty} \frac{m}{n+m} = \lim_{n \to \infty} 1 = 1.$$

It is easy to understand the differences in the limits by listing the sequence first in order of increasing m while fixing n and then in order of increasing n while fixing m.

It is not surprising that some additional assumptions on a convergent sequence of functions are required to guarantee that the limit inherits particular properties.

33.1   Uniform Convergence and Continuity

In fact, Example 33.1 already shows that continuity is not preserved in general. In that case, the functions in $\{x^n\}$ are all uniformly continuous functions on $[0, 1]$; nonetheless, they converge to a discontinuous function, which is 0 for $0 \leq x < 1$ and 1 at $x = 1$.

However, if the sequence $\{f_n\}$ converges *uniformly*, then continuity is preserved. In particular, we prove the following theorem:

Theorem 33.1 *Suppose* $\{f_n\}$ *is a sequence of continuous functions on an interval I that converges uniformly to f on I. Then f is continuous on I.*

Following the discussion above, for each $\bar{x}$ in I,

$$\lim_{n \to \infty} f_n(\bar{x}) = f(\bar{x}) \text{ and } \lim_{x \to \bar{x}} f_n(x) = f_n(\bar{x}),$$

while we need to show that

$$\lim_{x \to \bar{x}} f(x) = f(\bar{x}).$$

We estimate[3]

$$|f(x) - f(\bar{x})| = |f(x) - f_n(x) + f_n(x) - f_n(\bar{x}) + f_n(\bar{x}) - f(\bar{x})|$$
$$\leq |f(x) - f_n(x)| + |f_n(x) - f_n(\bar{x})| + |f_n(\bar{x}) - f(\bar{x})|.$$

By the uniform convergence, for any $\epsilon > 0$, there is an $N > 0$ such that $n \geq N$ implies

$$|f(x) - f_n(x)| < \epsilon/3 \text{ and } |f_n(\bar{x}) - f(\bar{x})| < \epsilon/3 \text{ for all } x, \bar{x} \text{ in } I.$$

For any $n \geq N$, there is a $\delta > 0$ such that if $|x - \bar{x}| < \delta$ and x is in I, then

$$|f_n(x) - f_n(\bar{x})| < \epsilon/3.$$

Hence, given any $\epsilon > 0$ there is a $\delta > 0$ such that if $|x - \bar{x}| < \delta$ and x is in I, then

$$|f(x) - f(\bar{x})| < \epsilon/3 + \epsilon/3 + \epsilon/3 = \epsilon.$$

It is important to note that while uniform convergence is enough to guarantee that the limit of continuous functions is continuous, it is *not* necessary.

EXAMPLE 33.3. Consider the sequence $\{nxe^{-nx}\}$. The first few terms are shown in Fig. 33.2. For each x, $f_n(x) \to 0$ as $n \to \infty$. Yet $f_n(1/n) = e^{-1}$ for all n, hence the convergence cannot be uniform. So this is a sequence of continuous functions that converge pointwise to a continuous function, but not uniformly.

---

[3]The strategy here is to take advantage of the facts that $f_n(x)$ approaches $f(x)$ for each x and $f_n(x)$ approaches $f_n(\bar{x})$ as x approaches $\bar{x}$ for each n.

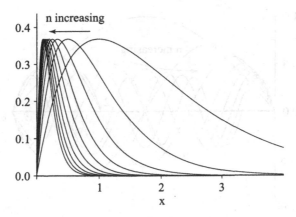

FIGURE 33.2. Plots of the functions $\{nxe^{-nx}\}$ for $n = 1, 2, \cdots, 10$.

33.2 Uniform Convergence and Differentiability

We next consider a sequence of functions $\{f_n\}$ that are differentiable on an interval I which converges to a function f and try to determine if f is differentiable and if $\{f_n'\}$ converges to f'. Since f is differentiable at $\bar{x}$ if

$$\lim_{x \to \bar{x}} \frac{f(x) - f(\bar{x})}{x - \bar{x}} = f'(\bar{x})$$

converges, this question is equivalent to the question of whether the following equality is true:

$$\lim_{x \to \bar{x}} \lim_{n \to \infty} \frac{f_n(x) - f_n(\bar{x})}{x - \bar{x}} = \lim_{n \to \infty} \lim_{x \to \bar{x}} \frac{f_n(x) - f_n(\bar{x})}{x - \bar{x}}.$$

Again it is necessary to assume something more than mere pointwise convergence in general.

EXAMPLE 33.4. The sequence $\{x^n\}$ on $[0, 1]$ consists of strongly differentiable functions, yet the limit is discontinuous at 1, so it is certainly not differentiable there.

However, simply adding uniform convergence is not sufficient either.

EXAMPLE 33.5. Consider the sequence of functions $\{\sin(nx)/\sqrt{n}\}$. The first few terms are shown in Fig. 33.3. This sequence converges uniformly to $f(x) = 0$ on any interval since

$$\left| \frac{\sin(nx)}{\sqrt{n}} - 0 \right| \le \frac{1}{\sqrt{n}} \text{ for all } x.$$

Moreover, f is differentiable and $f'(x) = 0$ for all x. Yet, $f_n'(x) = \sqrt{n} \cos(nx)$ and $\{f_n'\}$ does not converge for most values of x.

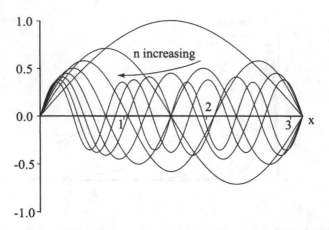

FIGURE 33.3. Plots of the functions $\{\sin(nx)/\sqrt{n}\}$ for $n = 1, 2, \cdots, 8$.

We prove the following theorem:

Theorem 33.2 *Suppose that $\{f_n\}$ is a sequence of functions with continuous derivatives $\{f_n'\}$ on $[a, b]$, $\{f_n(\bar{x})\}$ converges for some $\bar{x}$ in $[a, b]$, and $\{f_n'\}$ converges uniformly on $[a, b]$. Then $\{f_n\}$ converges uniformly to a differentiable function f on $[a, b]$ and $\{f_n'\}$ converges uniformly to f'.*

Note that the convergence of the sequence of derivatives $\{f_n'\}$ of terms in a sequence $\{f_n\}$ is not sufficient to guarantee the sequence itself converges.

EXAMPLE 33.6.   Consider the sequence $\{n + x/n\}$, which does not converge. The sequence of derivatives $\{1/n\}$ converges uniformly to zero.

We begin by showing that under the assumptions of Theorem 33.2, the sequence $\{f_n\}$ converges uniformly on $[a, b]$. For indices n, m, we estimate

$$|f_n(x) - f_m(x)| = |f_n(x) - f_m(x) - (f_n(\bar{x}) - f_m(\bar{x})) + (f_n(\bar{x}) - f_m(\bar{x}))|$$
$$\leq |f_n(x) - f_m(x) - (f_n(\bar{x}) - f_m(\bar{x}))| + |(f_n(\bar{x}) - f_m(\bar{x}))|$$

The Mean Value Theorem 32.17 applied to the function $f_n(x) - f_m(x)$ implies there is a c between x and $\bar{x}$ such that

$$f_n(x) - f_m(x) - (f_n(\bar{x}) - f_m(\bar{x})) = (f_n'(c) - f_m'(c))(x - \bar{x}).$$

By the uniform convergence of $\{f_n'\}$, for any $\epsilon > 0$, there is an N_1 such that for $n > N_1$

$$|f_n'(c) - f_m'(c)| < \frac{\epsilon}{2(b - a)} \text{ for any } c \text{ in } [a, b].$$

Since $|x - \bar{x}| \leq b - a$, for any $\epsilon > 0$ there is an N_1 such that for $n > N_1$,

$$|f_n(x) - f_m(x) - (f_n(\bar{x}) - f_m(\bar{x}))| < \frac{\epsilon}{2(b - a)}(b - a) = \frac{\epsilon}{2}$$

for all x and $\bar{x}$ in $[a, b]$.

On the other hand, since $\{f_n(\bar{x})\}$ converges, given any $\epsilon > 0$ there is an N_2 such that for $n, m > N_2$,

$$|(f_n(\bar{x}) - f_m(\bar{x}))| < \frac{\epsilon}{2}.$$

Thus, given any $\epsilon > 0$, there is a $N = \max\{N_1, N_2\}$ such that for $n, m > N$,

$$|f_n(x) - f_m(x)| < \frac{\epsilon}{2} + \frac{\epsilon}{2} = \epsilon \text{ for all } x \text{ in } [a, b].$$

Theorem 25.2 implies that $\{f_n\}$ converges uniformly.

Now $\{f_n\}$ converges uniformly to some function f and $\{f_n'\}$ converges uniformly to some function $\tilde{f}$ on $[a, b]$. We want to show that f is differentiable and $f' = \tilde{f}$.

Fixing $\bar{x}$ in $[a, b]$, consider the sequence

$$\left\{ \frac{f_n(x) - f_n(\bar{x})}{x - \bar{x}} \right\} \tag{33.2}$$

defined for x in $[a, b]$, $x \neq \bar{x}$. Note that

$$\lim_{x \to \bar{x}} \frac{f_n(x) - f_n(\bar{x})}{x - \bar{x}} = f_n'(\bar{x}).$$

We show this sequence converges uniformly by estimating for indices n, m.

$$\left| \frac{f_n(x) - f_n(\bar{x})}{x - \bar{x}} - \frac{f_m(x) - f_m(\bar{x})}{x - \bar{x}} \right| = \left| \frac{f_n(x) - f_m(x) - (f_n(\bar{x}) - f_m(\bar{x}))}{x - \bar{x}} \right|.$$

Using the Mean Value Theorem as above, there is a c between x and $\bar{x}$ such that

$$\left| \frac{f_n(x) - f_n(\bar{x})}{x - \bar{x}} - \frac{f_m(x) - f_m(\bar{x})}{x - \bar{x}} \right| = \left| \frac{(f_n'(c) - f_m'(c))(x - \bar{x})}{x - \bar{x}} \right|$$

$$= |f_n'(c) - f_m'(c)|.$$

By the uniform convergence of $\{f_n'\}$, given any $\epsilon > 0$, there is an N such that for $n, m > N$,

$$\left| \frac{f_n(x) - f_n(\bar{x})}{x - \bar{x}} - \frac{f_m(x) - f_m(\bar{x})}{x - \bar{x}} \right| < \epsilon \text{ for all } x \neq \bar{x} \text{ in } [a, b].$$

Hence, the sequence in (33.2) converges uniformly for x in $[a, b]$, $x \neq \bar{x}$. The limit is

$$\lim_{n \to \infty} \frac{f_n(x) - f_n(\bar{x})}{x - \bar{x}} = \frac{f(x) - f(\bar{x})}{x - \bar{x}} \text{ for } x \neq \bar{x}.$$

Now we estimate

$$\left| \frac{f(x) - f(\bar{x})}{x - \bar{x}} - f_n'(x) \right|$$

$$\leq \left| \frac{f(x) - f(\bar{x})}{x - \bar{x}} - \frac{f_n(x) - f_n(\bar{x})}{x - \bar{x}} \right| + \left| \frac{f_n(x) - f_n(\bar{x})}{x - \bar{x}} - f_n'(\bar{x}) \right|.$$

Given any $\epsilon > 0$, there is an N_1 such that for $n > N_1$,

$$\left| \frac{f(x) - f(\bar{x})}{x - \bar{x}} - \frac{f_n(x) - f_n(\bar{x})}{x - \bar{x}} \right| < \frac{\epsilon}{4} \text{ for all } x \neq \bar{x} \text{ in } [a, b].$$

As for the second term, the Mean Value Theorem 32.17 implies there is a c between x and $\bar{x}$ such that

$$\left| \frac{f_n(x) - f_n(\bar{x})}{x - \bar{x}} - f_n'(\bar{x}) \right| = \left| \frac{f_n'(c)(x - \bar{x})}{x - \bar{x}} - f_n'(\bar{x}) \right|$$
$$= |f_n'(c) - f_n'(x)|.$$

Now we estimate

$$|f_n'(c) - f_n'(x)| \leq |f_n'(c) - \tilde{f}(c)| + |\tilde{f}(c) - \tilde{f}(x)| + |\tilde{f}(x) - f_n'(x)|.$$

By the uniform convergence of $\{f_n'\}$, for any $\epsilon > 0$ there is an N_2 such that for $n > N_2$,

$$|f_n'(c) - \tilde{f}(c)| < \frac{\epsilon}{4} \text{ and } |f_n'(x) - \tilde{f}(x)| < \frac{\epsilon}{4} \text{ for any } x, c \text{ in } [a, b].$$

Moreover, Theorem 33.1 implies that $\tilde{f}$ is continuous and there is a $\delta > 0$ such that for all x in $[a, b]$ with $|x - \bar{x}| < \delta$,

$$|\tilde{f}(c) - \tilde{f}(x)| < \frac{\epsilon}{4},$$

since c is between x and $\bar{x}$. We conclude that given any $\epsilon > 0$ there is a $\delta > 0$ and a $N = \max\{N_1, N_2\}$ such that for all $n > N$ and x in $[a, b]$, $x \neq \bar{x}$, $|x - \bar{x}| < \delta$,

$$\left| \frac{f(x) - f(\bar{x})}{x - \bar{x}} - f_n'(x) \right| < \frac{\epsilon}{4} + \frac{\epsilon}{4} + \frac{\epsilon}{4} + \frac{\epsilon}{4} = \epsilon.$$

Passing to the limit as $n \to \infty$, we conclude that

$$\left| \frac{f(x) - f(\bar{x})}{x - \bar{x}} - \tilde{f}(x) \right| < \epsilon$$

for any $\epsilon > 0$ and x sufficiently close to $\bar{x}$. This proves the theorem.

33.3   Uniform Convergence and Integrability

We finally consider a sequence of functions $\{f_n\}$ that are continuous on an interval $[a, b]$ which converges to a function f. Each f_n as well as f is integrable on $[a, b]$ since all the functions are continuous. The question is whether the integrals of $\{f_n\}$ converge to the integral of f. When f is integrable on $[a, b]$, then

$$\int_a^b f(x)\, dx = \lim_{N \to \infty} \sum_{i=1}^{2^N} f(x_{N,i-1}) \Delta x_N,$$

where for each N, $\Delta x_N = (b - a)/2^N$ and $x_{N,i} = a + i \times \Delta x_N$ for $i = 0, 1, \cdots, 2^N$. Hence, the question can be rephrased as whether the following equality holds:

$$\lim_{N \to \infty} \lim_{n \to \infty} \sum_{i=1}^{2^N} f_n(x_{N,i-1}) \Delta x_N = \lim_{n \to \infty} \lim_{N \to \infty} \sum_{i=1}^{2^N} f_n(x_{N,i-1}) \Delta x_N.$$

Again, something more than pointwise convergence is needed.

EXAMPLE 33.7. Consider the sequence $\{f_n(x)\} = \{nx(1 - x^2)^n\}$. A few terms are shown in Fig. 33.4. For $0 < x \le 1$,

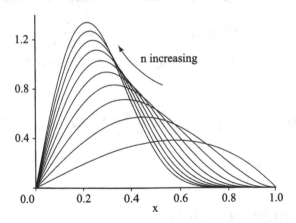

FIGURE 33.4. Plots of the functions $\{nx(1 - x^2)^n\}$ for $n = 1, 2, \cdots, 10$.

$$\lim_{n \to \infty} nx(1 - x^2)^n = 0$$

since $|1 - x^2| < 1$. Also $f_n(0) = 0$ for all n, and we conclude that $f_n \to f = 0$ as $n \to \infty$. However, the convergence is not uniform.

Moreover, it is easy to verify that

$$\int_0^1 nx(1 - x^2)^n \, dx = \frac{n}{2(n + 1)}$$

and therefore

$$\lim_{n \to \infty} \int_0^1 f_n(x) \, dx = \frac{1}{2} \neq 0 = \int_0^1 \lim_{n \to \infty} f_n(x) \, dx.$$

We prove:

Theorem 33.3 *Suppose that $\{f_n\}$ is a sequence of continuous functions on $[a, b]$ that converges uniformly to f on $[a, b]$. Then*

$$\int_a^b f(x) \, dx = \lim_{n \to \infty} \int_a^b f_n(x) \, dx.$$

This is not difficult. With the mesh notation above,

$$\left| \int_a^b f(x) \, dx - \int_a^b f_n(x) \, dx \right| = \lim_{N \to \infty} \left| \sum_{i=1}^{2^N} (f(x_{N,i-1}) - f_n(x_{N,i-1})) \Delta x_N \right|.$$

By the uniform convergence, given any $\epsilon > 0$, there is a N such that $n > N$ implies

$$|f(x_{N,i-1}) - f_n(x_{N,i-1})| < \frac{\epsilon}{b - a}$$

for all $1 \leq i \leq 2^N$. Hence, for any $\epsilon > 0$, there is an N_1 such that for $n > N_1$,

$$\left| \int_a^b f(x) \, dx - \int_a^b f_n(x) \, dx \right| < \lim_{N \to \infty} \frac{\epsilon}{b - a} \sum_{i=1}^{2^N} \Delta x_N = \frac{\epsilon}{b - a}(b - a) = \epsilon.$$

This proves the theorem.

33.4   Unanswered Questions

The material on integration in Section 33.3 points to some serious short-comings in the Riemann theory of integration. Example 33.7 shows that integration and taking limits are not interchangeable; i.e., $\int \lim f_n \, dx$ does not necessarily equal $\lim \int f_n \, dx$. In fact, a sequence of integrable functions $\{f_n\}$ that converges on an interval does not necessarily converge to an integrable function. Another shortcoming of the Riemann theory is that integration is defined only on intervals, though in practice we may need to integrate over more complicated sets of numbers. This occurs frequently in probability theory, for example.

There are several alternative definitions of integration that fix up the shortcomings of the Riemann integral while reducing to the Riemann integral on continuous functions on intervals. Perhaps the most prominent alternative theory is due to Lebesgue.[4] See Rudin [19] for an introduction to the Lebesgue integral.

---

[4]Henri Léon Lebesgue (1875–1941) was a French mathematician. He is best known for his creation of measure theory and the Lebesgue theory of integration, which have had a profound impact on analysis. He also made important contributions to Fourier analysis, potential theory, and topology.

Chapter 33   Problems

33.1. Let $\{s_{ij}\}_{i,j=1}^{\infty}$ have the following properties: $\lim_{j\to\infty} s_{ij} = S_i$ exists for each i and $\lim_{i\to\infty} s_{ij} = U_j$ converges *uniformly* for all j. Show that $\lim_{i\to\infty} S_i$ exists and $\lim_{i\to\infty} S_i = \lim_{j\to\infty} U_j$. In other words, $\lim_{i\to\infty} \lim_{j\to\infty} s_{ij} = \lim_{j\to\infty} \lim_{i\to\infty} s_{ij}$. Discuss Example 33.2 in context with this result.

33.2. Compute the limit of $\{1/(1+x^{2n})\}$. Is the limit continuous? Determine if the convergence is uniform.

33.3. Suppose $f(x)$ is a continuous function on $[0,1]$ with $f(1) = 0$. Show that $\{f(x)x^n\}$ converges uniformly to 0.

33.4. Suppose that f is a uniformly continuous function on $(-\infty,\infty)$. For each natural number $n > 0$ define $f_n(x) = f(x + 1/n)$. Show that $\{f_n\}$ converges uniformly on $(-\infty,\infty)$.

33.5. For $\alpha > 0$, $0 \le x \le 1$, and integers $n \ge 2$, define

$$f_n(x) = \begin{cases} xn^{\alpha}, & 0 \le x \le 1/n, \\ \left(\frac{2}{n} - x\right)n^{\alpha}, & 1/n \le x \le 2/n, \\ 0, & 2/n \le x \le 1. \end{cases}$$

(a) Show that f_n is continuous. (b) Show that $\lim_{n\to infty} f_n = 0$. (c) Decide if the convergence is uniform or not depending on the value of α.

33.6. For integers $n > 1$ and $x \ge 0$, define

$$f_n(x) = \begin{cases} 0, & 0 \le x \le 1/(n+1), \\ \sin^2(\pi/x), & 1/(n+1) \le x \le 1/n, \\ 0, & 1/n \le x. \end{cases}$$

Show that $\{f_n\}$ converges to a continuous function but not uniformly.

33.7. Suppose that $\{f_n\}$ converges uniformly to f on $[a,b]$ and g is a function on $[a,b]$. Find conditions on g that guarantee that $\{gf_n\}$ converges uniformly to gf on $[a,b]$.

33.8. (a) Suppose that $\{f_n\}$ and $\{g_n\}$ converge uniformly to f and g for x in an interval I and c is a number. Prove that $\{f_n + g_n\}$ and $\{cf_n\}$ converge uniformly and determine the limits. (b) Suppose in addition that $\{f_n\}$ and $\{g_n\}$ are sequences of bounded functions and show that $\{f_ng_n\}$ converges uniformly on I.

33.9. Prove that the limit of a uniformly convergent sequence of functions that are uniformly continuous on an interval I is itself uniformly continuous.

33.10. Construct sequences $\{f_n\}$ and $\{g_n\}$ that converge uniformly on an interval I such that $\{f_ng_n\}$ converges on I but not uniformly.

33.11. Prove that a uniformly convergent sequence of bounded functions $\{f_n\}$ on an interval I is uniformly bounded; i.e., there is an M such that $|f_n(x)| \le M$ for all n and x.

33.12. *(Hard)* Remove the assumption that f_n' is continuous in Theorem 33.2.

33.13. (a) Show that the sequence $\{x/(1 + nx^2)\}$ converges uniformly to a function f. (b) Show that $f'(x) = \lim_{n\to\infty} x/(1 + nx^2)$ for $x \ne 0$ but not for $x = 0$.

33.14. Define $f_n(x) = n^2 x e^{-nx}$ for natural numbers $n > 0$ and real x. Prove that $\{f_n\}$ converges pointwise to 0 and $\{f_n'\}$ also converges pointwise to 0, but $\{f_n'\}$ does not converge uniformly.

33.15. For natural numbers $n > 0$ and real x, define

$$f_n(x) = \begin{cases} 1/n, & |x| \le 1/n, \\ |x|, & |x| \ge 1/n. \end{cases}$$

Prove that $\{f_n\}$ converges to $|x|$ uniformly on $(-\infty, \infty)$. Note that each f_n is differentiable at $x = 0$, but the limit $|x|$ is not.

33.16. Let $\{f_n\}$ be a sequence of continuous functions that converges uniformly to a function f for x in a set of numbers S. Prove that

$$\lim_{n\to\infty} f_n(x_n) = f(x)$$

for every sequence of points $\{x_n\}$ in S such that $x_n \to x$ and x is in S. Is the converse true?

33.17. *(Hard)* Suppose that $\{f_n\}$ is a sequence of continuous functions on a bounded interval $[a, b]$ that converges to a continuous function f. Prove that if $\{f_n(x)\}$ is monotone increasing or monotone decreasing to $f(x)$ for each x, then $\{f_n\}$ actually converges uniformly to f on $[a, b]$. *Hint:* Assume the sequence is decreasing. If the claim is not true, then there is an $\epsilon > 0$ such that for each n there is a natural number m_n and a point x_n in $[a, b]$ with $f_{m_n}(x_n) > f(x_n) + \epsilon$. Get a contradiction.

34
Relaxing Integration

No, this chapter does not contain the secret to integration that frees every mathematics student to find peace with the world. Rather we apply the ideas about functions in Chapter 32 and Chapter 33 to relax the assumptions we used to prove that integration works in Chapter 25. In particular, we show that merely continuous functions can be integrated and we use much more general meshes to compute the integral. We conclude this chapter by applying these ideas to define and compute the length of a curve.

34.1 Continuous Functions

The analysis of integration in Chapter 25 depends on the assumption of Lipschitz continuity of the integrand. We show here that what is actually important in that analysis is the uniform continuity of the integrand that follows automatically from the assumption of Lipschitz continuity. Likewise in Chapter 32, we saw that many nice properties of the Lipschitz continuous functions are due to the uniform nature of Lipschitz continuity. The analysis in this section follows the analysis in Chapter 25 closely, which means it is filled with tedious details. A reasonable way to approach the material is just to compare the proofs in this section and in Chapter 25 to see the differences needed to handle uniform rather than Lipschitz continuity.

Note that it suffices to assume continuity, as opposed to uniform continuity, because of Theorem 32.11, which says that a function that is continuous on a closed, bounded interval is uniformly continuous on the interval. So we assume the function f is continuous on the bounded interval $[a, b]$, and we want to show that the initial value problem

$$\begin{cases} y'(x) = f(x), & a < x \leq b, \\ y(a) = 0. \end{cases} \tag{34.1}$$

has a unique solution, which we write as

$$y(x) = \int_a^x f(s) \, ds,$$

that can be approximated to any desired degree of accuracy. Recall that once we have solved (34.1) with initial value 0, we can easily solve a problem with an arbitrary initial value y_0.

Even though f may not be Lipschitz continuous, continuity suffices to define the same approximate solution Y_N used above. We create a **mesh** of equally spaced points $\{x_{N,i}\}$ in $[a, b]$ by setting $\Delta x_N = (b - a)/2^N$ for a natural number N and $x_{N,i} = a + i \times \Delta x_N$ for $i = 0, 1, \cdots, 2^N$. Note in particular that $x_{N,0} = a$ and $x_{N,2^N} = b$. Solving the approximate initial value problems (25.3), interval by interval, we compute the approximate solution Y_N such that for $x_{N,n-1} \leq x < x_{N,n}$,

$$Y_N(x) = \sum_{i=1}^{n-1} f(x_{N,i-1})\Delta x_N + f(x_{N,n-1})(x - x_{N,n-1}), \tag{34.2}$$

while the nodal value of Y_N at $x_{N,n}$ is

$$Y_{N,n} = Y_N(x_{N,n}) = \sum_{i=1}^{n} f(x_{N,i-1})\Delta x_N. \tag{34.3}$$

We just have to show that $\{Y_N\}$ forms a Cauchy sequence that converges to a unique function that satisfies (34.1).

Using the same notation as before, we choose natural numbers $M \geq N$ and we define $\mu(i)$ to be the set of indices j such that $[x_{M,j-1}, x_{M,j}]$ is contained in $[x_{N,i-1}, x_{N,i}]$ (see Fig. 34.1). We can write

$$Y_M(x_{N,n}) = \sum_{i=1}^{n} \sum_{j \text{ in } \mu(i)} f(x_{M,j-1})\Delta x_M$$

and

$$Y_N(x_{N,n}) = \sum_{i=1}^{n} f(x_{N,i-1})\Delta x_N = \sum_{i=1}^{n} \sum_{j \text{ in } \mu(i)} f(x_{N,i-1})\Delta x_M.$$

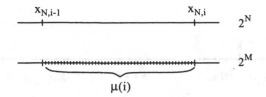

FIGURE 34.1. The definition of $\mu(i)$.

Estimating, we find

$$|Y_M(x_{N,n}) - Y_N(x_{N,n})| \leq \sum_{i=1}^{n} \sum_{j \text{ in } \mu(i)} |f(x_{M,j-1}) - f(x_{N,i-1})| \Delta x_M.$$

By uniform continuity, for any $\epsilon > 0$ there is a $\delta > 0$ such that $|f(x_{M,j-1}) - f(x_{N,i-1})| < \epsilon$ if $|x_{M,j-1} - x_{N,i-1}| < \delta$. Since

$$|x_{M,j-1} - x_{N,i-1}| < |x_{N,i} - x_{N,i-1}| \text{ for } j \text{ in } \mu(i),$$

given any $\delta > 0$ there is an $\bar{N}$ such that for $N > \bar{N}$, $|x_{M,j-1} - x_{N,i-1}| < \delta$ for $1 \leq i \leq 2^N$ and j in $\mu(i)$. We conclude that for any $\epsilon > 0$, there is an $\bar{N}$ such that for $M \geq N > \bar{N}$,

$$|Y_M(x_{N,n}) - Y_N(x_{N,n})| \leq \epsilon \sum_{i=1}^{n} \sum_{j \text{ in } \mu(i)} \Delta x_M = \epsilon(x_{N,n} - x_{N,0}) \leq \epsilon(b-a).^{1}$$

So we can make the difference between Y_M and Y_N arbitrarily small at nodes by taking $M \geq N$ sufficiently large. A similar argument works for the values of x in between the nodes. Hence, $\{Y_N\}$ is a uniform Cauchy sequence that converges uniformly, $\lim_{N \to \infty} Y_N(x) = y(x)$, for $a \leq x \leq b$.

Using a similar argument, we can show that Y_N is continuous on $[a, b]$ for each N and therefore the limit function y is continuous on $[a, b]$ by Theorem 33.1. We have to show that y is differentiable and satisfies (34.1). Note that $y(a) = 0$.

As in Chapter 25, given $\bar{x}$ and $x > \bar{x}$ in $[a, b]$, for each N we choose m_N so that $x_{N,m_N-1} < \bar{x} \leq x_{N,m_N}$ and n_N so that $x_{N,n_N-1} < x \leq x_{N,n_N}$ (see Fig. 34.2). By this choice,

$$x - \bar{x} = (x - x_{N,n_N-1}) + \sum_{j=m_N}^{n_N-1} \Delta x_N - (\bar{x} - x_{N,n_N-1})$$

and

$$\lim_{N \to \infty} x_{N,m_N-1} = \lim_{N \to \infty} x_{N,m_N} = \bar{x} \text{ and } \lim_{N \to \infty} x_{N,n_N-1} = \lim_{N \to \infty} x_{N,n_N} = x.$$

---

[1]Compare this result to (25.12).

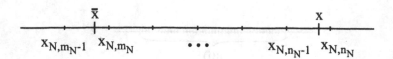

FIGURE 34.2. The choice of m_N and n_N.

Moreover,

$$Y_N(\bar{x}) = Y_N(x_{N,m_N-1}) + f(x_{N,m_N-1})(\bar{x} - x_{N,m_N-1})$$

and

$$Y_N(x) = Y_N(x_{N,m_N-1}) + \sum_{j=m_N}^{n_N-1} f(x_{N,j-1})\Delta x_N$$
$$+ f(x_{N,n_N-1})(x - x_{N,n_N-1}).$$

Subtraction gives

$$Y_N(x) - Y_N(\bar{x}) = f(x_{N,n_N-1})(x - x_{N,n_N-1}) + \sum_{j=m_N}^{n_N-1} f(x_{N,j-1})\Delta x_N$$
$$- f(x_{N,m_N-1})(\bar{x} - x_{N,m_N-1}).$$

We can rewrite this as

$$Y_N(x) - Y_N(\bar{x}) = f(\bar{x})(x - \bar{x})$$
$$+ (f(x_{N,n_N-1}) - f(\bar{x}))(x - x_{N,n_N-1})$$
$$+ \sum_{j=m_N}^{n_N-1} (f(x_{N,j-1}) - f(\bar{x}))\Delta x_N$$
$$- (f(x_{N,m_N-1}) - f(\bar{x}))(\bar{x} - x_{N,m_N-1}),$$

and so

$$|Y_N(x) - Y_N(\bar{x}) - f(\bar{x})(x - \bar{x})|$$
$$\leq |f(x_{N,n_N-1}) - f(\bar{x})|\,|x - x_{N,n_N-1}|$$
$$+ \sum_{j=m_N}^{n_N-1} |f(x_{N,j-1}) - f(\bar{x})|\Delta x_N$$
$$+ |f(x_{N,m_N-1}) - f(\bar{x})|\,|\bar{x} - x_{N,m_N-1}|. \quad (34.4)$$

So far the analysis has followed steps (25.15)–(25.19) precisely. Now, however, we estimate in (34.4) using the uniform continuity of f. Given $\epsilon > 0$, there is a $\delta > 0$ such that $|f(y) - f(\bar{x})| < \epsilon$ for all y with $|y - \bar{x}| < \delta$. Given this δ, we assume that $|x - \bar{x}| < \delta/2$ and choose $\bar{N}$ so that $\Delta x_N < \delta/2$ for all $N > \bar{N}$. We have $|x_{M,n_M-1} - \bar{x}| < \delta/2$, $|x_{N,i} - \bar{x}| \leq |x - \bar{x}| < \delta/2$ for $n_N \leq i \leq m_N - 1$, and $|x_{N,m_N} - \bar{x}| \leq |x - \bar{x}| + |x_{N,m_N} - x| < \delta$. Hence,

$$|Y_N(x) - Y_N(\bar{x}) - f(\bar{x})(x - \bar{x})| < 3\epsilon|x_{N,n_N} - x_{N,m_N-1}|.$$

Letting $N \to \infty$, we get

$$|y(x) - (y(\bar{x}) + f(\bar{x})(x - \bar{x}))| < 3\epsilon|x - \bar{x}|.^2 \qquad (34.5)$$

It is straightforward to treat the cases when $\bar{x} > x$ and $\bar{x} = a$ or b. Hence, for any $\epsilon > 0$, there is a $\delta > 0$ such that

$$\left| \frac{y(x) - y(\bar{x})}{x - \bar{x}} - f(\bar{x}) \right| \leq 3\epsilon$$

for all $x \neq \bar{x}$ with $|x - \bar{x}| < \delta/2$. So y is differentiable at $\bar{x}$ and $y'(\bar{x}) = f(\bar{x})$ for $a \leq \bar{x} \leq b$.

We summarize this analysis as two theorems.[3]

Theorem 34.1 Fundamental Theorem of Calculus *If f is a continuous function on $[a, b]$, there is a unique solution y of (34.1) that is approximated by the function*

$$Y_N(x) = \sum_{i=1}^{n-1} f(x_{N,i-1})\Delta x_N + f(x_{N,n-1})(x - x_{N,n-1}),$$

where $\Delta x_N = (b - a)/2^N$ for a natural number N, $x_{N,i} = a + i \times \Delta x_N$ for $i = 0, 1, \cdots, 2^N$, and $x_{N,n-1} < x \leq x_{N,n}$. The approximation is uniformly accurate in the sense that given any $\epsilon > 0$ there is a $\bar{N}$ such that for all $N > \bar{N}$,

$$|y(x) - Y_N(x)| \leq (b - a)\epsilon \text{ for } a \leq x \leq b. \qquad (34.6)$$

Rewritten as a result for integration, this theorem is:

Theorem 34.2 Fundamental Theorem of Calculus *If f is a continuous function on $[a, b]$, then*

$$\int_a^b f(x)\, dx$$

---

[2]Compare this to (25.20).

[3]Compare these results to Theorem 25.4 and Theorem 25.5.

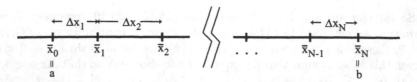

FIGURE 34.3. A nonuniform mesh for $[a, b]$.

exists and given any $\epsilon > 0$ there is a $\bar{N}$ such that for $N > \bar{N}$,

$$\left| \int_a^b f(x)\, dx - \sum_{i=1}^{2^N} f(x_{N,i-1}) \Delta x_N \right| \leq (b-a)\epsilon,$$

where $\Delta x_N = (b-a)/2^N$ for a natural number N and $x_{N,i} = a + i \times \Delta x_N$ for $i = 0, 1, \cdots, 2^N$.

34.2   General Meshes

Next we relax the assumptions on the mesh used to compute the approximate solution Y_N to (34.1). We want to allow the sizes of the subintervals to vary and to allow a choice of interpolation points inside the subintervals. The main difficulty is figuring out how to compare approximations computed on two different meshes when the meshes are no longer "nested."

The main reason to use more general meshes is computational, so we restrict the discussion to the computation of the definite integral $\int_a^b f(x)\, dx$, where f is a continuous function on $[a, b]$. From the results in Section 34.1, we know the integral is well-defined and can be approximated to any desired degree of accuracy using uniform, nested meshes. We want to show that it can also be approximated using more general meshes.

We partition $[a, b]$ using a **mesh** $\mathcal{T}_N$ determined by a set of $N+1$ **nodes**

$$\mathcal{T}_N = \{\bar{x}_{\mathcal{T}_N,0}, \bar{x}_{\mathcal{T}_N,1}, \cdots, \bar{x}_{\mathcal{T}_N,N}\} = \{\bar{x}_0, \bar{x}_1, \cdots, \bar{x}_N\}$$

with

$$a = \bar{x}_0 < \bar{x}_1 < \cdots < \bar{x}_N = b.$$

Unless absolutely necessary, we drop the subscript denoting the particular mesh $\mathcal{T}_N$ under discussion from any mesh-related quantity. Since the subintervals $[\bar{x}_{n-1}, \bar{x}_n]$ vary in length, we let $\Delta x_n = \bar{x}_n - \bar{x}_{n-1}$ (see Fig. 34.3). To measure the "fineness" of $\mathcal{T}_N$, we use the size of the largest subinterval,

$$\Delta_{\mathcal{T}_N} = \max_{1 \leq n \leq N} \Delta x_n,$$

which we call the **mesh size**.

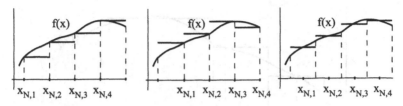

FIGURE 34.4. Three piecewise constant interpolants of f. The interpolation points x_n in $[\bar{x}_{n-1}, \bar{x}_n]$ are, respectively, $x_n = \bar{x}_{n-1}$, $x_n = \bar{x}_n$, and $x_n = (\bar{x}_{n-1} + \bar{x}_n)/2$.

Finally, we choose an interpolation point x_n in each subinterval $[\bar{x}_{n-1}, \bar{x}_n]$, $1 \leq n \leq N$. In the previous discussions, we used $x_n = \bar{x}_{n-1}$. In Fig. 34.4, we show three different interpolants of a function on a uniform mesh where the interpolation points occupy the same position in the subintervals. We can even vary the position of the interpolation points, as illustrated in Fig. 34.5.

We construct the approximate solution $Y_{\mathcal{T}_N, N} = Y_N$ to (34.1) interval by interval as before. On $[\bar{x}_0, \bar{x}_1]$, we compute Y_N solving

$$\begin{cases} Y'_N = f(x_1), & \bar{x}_0 \leq x \leq \bar{x}_1, \\ Y_N(\bar{x}_0) = 0, \end{cases}$$

to get $Y_N(x) = f(x_1)(x - \bar{x}_0)$ for $\bar{x}_0 \leq x \leq \bar{x}_1$ with nodal values $Y_{N,0} = Y(\bar{x}_0) = 0$ and $Y_{N,1} = Y_N(\bar{x}_1) = f(x_1)\Delta x_1$. Then given the nodal value $Y_{N,n-1}$, we solve

$$\begin{cases} Y'_N = f(x_n), & \bar{x}_{n-1} \leq x \leq \bar{x}_n, \\ Y_N(\bar{x}_{n-1}) = Y_{N,n-1}. \end{cases}$$

Continuing for $1 \leq n \leq N$, we get the final formula for the approximation of $\int_a^b f(x)\, dx$,

$$Y_{N,N} = \sum_{n=1}^{N} f(x_n)\Delta x_n. \tag{34.7}$$

We interpret this result in terms of area under the curve f in Fig. 34.6.

EXAMPLE 34.1. We repeat Example 25.4 using different interpolation points. The problem is to compute the integral approximation for $f(x) = x$ on $[0, 1]$ using a uniform mesh with $N + 1$ nodes. We have $\Delta x_N = 1/N$ and nodes $\bar{x}_n = n/N$, $0 \leq n \leq N$. If we use the interpolation point $x_n = \bar{x}_n$ on $[x_{n-1}, x_n]$, then

$$Y_{N,n} = \sum_{i=1}^{n} \frac{i}{N} \times \frac{1}{N} = \frac{1}{N^2}\frac{n(n+1)}{2},$$

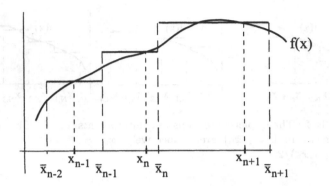

FIGURE 34.5. An interpolant of f on a nonuniform mesh where the interpolation points in different subintervals are chosen in different locations.

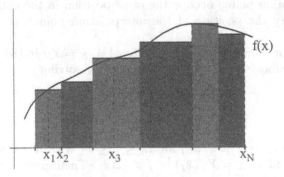

FIGURE 34.6. The area underneath the piecewise constant interpolant of f. We alternate the shading to distinguish contributions from neighboring rectangles.

and

$$Y_{N,N} = Y_N(1) = \frac{1}{2} + \frac{1}{2N}.$$

Using $x_n = (\bar{x}_n + \bar{x}_{n-1})/2$ instead gives

$$Y_{N,n} = \sum_{i=1}^{n} \frac{1}{2}\left(\frac{i}{N} + \frac{i-1}{N}\right)\frac{1}{N} = \frac{1}{N^2}\frac{n^2}{2}$$

and

$$Y_{N,N} = Y_N(1) = \frac{1}{2}.$$

Notice that this formula gives the exact answer for any N!

We want to show that there is a unique number such that if $\{\mathcal{T}_N\}_{N=1}^{\infty}$ is a sequence of meshes with $\Delta_{\mathcal{T}_N} \to 0$, the sum (34.7) converges to this number. Since one such sequence is made up of the uniform, nesting meshes used above, this number must be $\int_a^b f(x)\,dx$.

First, we compare Y_M and Y_N with $M > N$ where $\mathcal{T}_M$ is a **refinement** of $\mathcal{T}_N$, which means that all the nodes in $\mathcal{T}_N$ are nodes of $\mathcal{T}_M$. To avoid double subscripts, we denote $\mathcal{T}_M = \{\bar{y}_0, \bar{y}_1, \cdots, \bar{y}_M\}$, choose corresponding interpolation points $\{y_i\}$, and let

$$Y_{M,M} = \sum_{m=1}^{M} f(y_m)\Delta y_m,$$

where $\Delta y_m = \bar{y}_m - \bar{y}_{m-1}, 1 \leq m \leq M$.

Any two consecutive nodes $\bar{x}_{n-1}, \bar{x}_n$ in $\mathcal{T}_N$ are in $\mathcal{T}_M$ so there are integers i and j such that

$$\bar{x}_{n-1} = \bar{y}_{i-1} \text{ and } \bar{x}_n = \bar{y}_j.$$

We illustrate in Fig. 34.7. We compare the contributions to the approximate

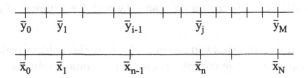

FIGURE 34.7. The nested meshes $\mathcal{T}_M$ and $\mathcal{T}_N$.

integrals from $[\bar{x}_{n-1}, \bar{x}_n]$, which are

$$\sum_{l=i}^{j} f(y_l)(\bar{y}_l - \bar{y}_{l-1})$$

and

$$f(x_n)(\bar{x}_n - \bar{x}_{n-1}) = \sum_{l=i}^{j} f(x_n)(\bar{y}_l - \bar{y}_{l-1}).$$

Given $\epsilon > 0$, there is a $\delta > 0$ such that $|f(y_l) - f(x_n)| < \epsilon$ for all l with $|y_l - x_n| < \delta$. Since $|y_l - x_n| \leq |\bar{x}_n - \bar{x}_{n-1}|$ for $i \leq l \leq j$, the condition holds provided $\Delta_{\mathcal{T}_N} < \delta$. If this is true, we get

$$\left| \sum_{l=i}^{j} (f(y_l) - f(x_n))(\bar{y}_l - \bar{y}_{l-1}) \right| \leq \sum_{l=i}^{j} \epsilon(\bar{y}_l - \bar{y}_{l-1}) \leq \epsilon(\bar{x}_n - \bar{x}_{n-1}).$$

Adding, we conclude that for any $\epsilon > 0$ there is a $\delta > 0$ such that for all meshes $\mathcal{T}_M$ and $\mathcal{T}_N$ where $\mathcal{T}_M$ is a refinement of $\mathcal{T}_N$ and $\Delta_{\mathcal{T}_N} < \delta$,

$$|Y_{M,M} - Y_{N,N}| \leq \epsilon \sum_{n=1}^{N} (\bar{x}_n - \bar{x}_{n-1}) = \epsilon(b - a).$$

We have shown that the difference between the final nodal values of two approximations computed on nested meshes can be made arbitrarily small by insuring that the meshes are sufficiently **refined**, i.e., that the corresponding mesh sizes are sufficiently small.

We finally want to remove the assumption that the meshes are nested. We let $\mathcal{T}_N$ and $\mathcal{T}_M$ be any two meshes. To compare the corresponding approximations, we use the mesh $\mathcal{T}_{N+M}$ constructed by taking the union of the nodes in $\mathcal{T}_N$ and $\mathcal{T}_M$.[4] We illustrate in Fig. 34.8. Without being

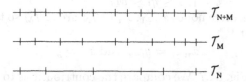

FIGURE 34.8. Two meshes $\mathcal{T}_M$ and $\mathcal{T}_N$ and their "union" $\mathcal{T}_{N+M}$.

precise, we choose interpolation points $\{z_i\}$ in the subintervals of $\mathcal{T}_{N+M}$ and let Y_{N+M} denote the corresponding approximate solution.

Now we estimate

$$|Y_{M,M} - Y_{N,N}| \leq |Y_{M,M} - Y_{N+M,N+M}| + |Y_{N+M,N+M} - Y_{N,N}|,$$

where $Y_{N+M,N+M}$ is the final nodal value of Y_{N+M}. By the results above, given any $\epsilon > 0$ there is a $\delta > 0$ such that if $\Delta_{\mathcal{T}_M} < \delta$ and $\Delta_{\mathcal{T}_N} < \delta$,

$$|Y_{M,M} - Y_{N,N}| < 2\epsilon(b-a).$$

Hence, the difference between the final nodal values of approximate solutions computed on two different meshes can be made arbitrarily small by insuring that the meshes are sufficiently refined.

We let $\{\mathcal{T}_N\}$ be a sequence of meshes and $\{Y_N\}$ the corresponding approximate solutions where $\Delta_{\mathcal{T}_N} \to 0$ as $N \to \infty$. In particular, given any $\delta > 0$ there is a $\bar{N}$ such that $\Delta_{\mathcal{T}_N} < \delta$ for all $N > \bar{N}$. Thus for any $\epsilon > 0$ there is a $\bar{N}$ such that

$$|Y_{N,N} - Y_{M,M}| < 2\epsilon(b-a) \text{ for } M > \bar{N} \text{ and } N > \bar{N}.$$

Hence, $\{Y_{N,N}\}$ is a Cauchy sequence and $\lim_{N\to\infty} Y_{N,N} = Y$ exists. We claim this limit is independent of the sequence of meshes and interpolation points.

---

[4]Note we are abusing notation because $\mathcal{T}_{N+M}$ probably has fewer than $N+M+2 = N+1+M+1$ nodes. But we don't have to be precise because we do not use $\mathcal{T}_{N,M}$ to compute an approximation. We only need its existence and the fact that it is a refinement of both $\mathcal{T}_N$ and $\mathcal{T}_M$.

We let $\{\bar{\mathcal{T}}_N\}$ denote another sequence of meshes with $\Delta_{\bar{\mathcal{T}}_N} \to 0$ and $\{\bar{Y}_N\}$ denote the corresponding approximations with $\lim_{N \to \infty} \bar{Y}_{N,N} = \bar{Y}$. Given any $\epsilon > 0$ there is a $\bar{N}$ such that

$$|Y_{N,N} - \bar{Y}_{N,N}| < 2\epsilon(b - a) \text{ for } N > \bar{N}.$$

Hence, $|Y - \bar{Y}| < 2\epsilon(b-a)$ for any $\epsilon > 0$, i.e., $Y = \bar{Y}$. Since we could choose the sequence of uniform meshes used above, we must have $Y = \int_a^b f(x)\,dx$. We summarize as a theorem.[5]

Theorem 34.3 Fundamental Theorem of Calculus *If f is a continuous function on $[a, b]$ and $\{\mathcal{T}_N\}$ is a sequence of meshes on $[a, b]$ with $\Delta_{\mathcal{T}_N} \to 0$ as $N \to \infty$, then*

$$\int_a^b f(x)\,dx$$

exists and given any $\epsilon > 0$ there is a $\bar{N}$ such that for $N > \bar{N}$,

$$\left| \int_a^b f(x)\,dx - \sum_{n=1}^N f(x_n)\Delta x_n \right| \leq (b - a)\epsilon,$$

where $\{x_n\}$ are the set of interpolation points for $\mathcal{T}_N$ and $\{\Delta x_n\}$ are the sizes of the subintervals.

34.3 Application to Computing the Length of a Curve

In Chapter 27, we discussed two applications of integration, namely, defining and computing the area underneath a curve and the average value of a function. In this section, we use integration to define and compute the length of the curve formed by the graph of a function (see Fig. 34.9). The length of a curve is important in physical applications. For example, we are often interested in the total distance traveled by a particle that is constrained to move along a certain path that can be described as the graph of a function. As in the other applications of integration, we have a strong geometric intuition that the length of a curve is well-defined.[6]

This problem is a good application of the ideas in this chapter. Sometimes when doing analysis, we are forced to use certain subintervals and/or interpolation points. This happens when defining the length of a curve for example.

---

[5]The sixth and final version!
[6]But we require an analytic definition in order to achieve inner peace.

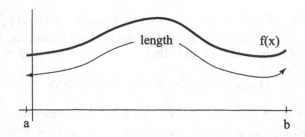

FIGURE 34.9. The length of the curve formed by the graph of a function.

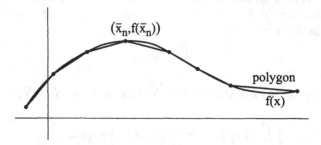

FIGURE 34.10. The length of a polygon approximating the graph of a function by interpolation.

To define the length of a curve f on an interval $[a, b]$, we use the idea behind integration by first defining the length of a polygonal approximation to the curve. We choose a mesh $\mathcal{T}_N = \{\bar{x}_0, \bar{x}_1, \cdots, \bar{x}_N\}$, with $a = \bar{x}_0 < \bar{x}_1 < \cdots < \bar{x}_N = b$ and subinterval lengths $\{\Delta x_n\}$, and consider the length of the polygon joining the points $\{(\bar{x}_0, f(\bar{x}_0)), (\bar{x}_1, f(\bar{x}_1)), \cdots, (\bar{x}_N, f(\bar{x}_N))\}$ (see Fig. 34.10).

By Pythagorean's theorem (see Fig. 34.11), the distance between $(\bar{x}_{n-1}, f(\bar{x}_{n-1}))$ and $(\bar{x}_n, f(\bar{x}_n))$ is

$$\sqrt{(\bar{x}_n - \bar{x}_{n-1})^2 + (f(\bar{x}_n) - f(\bar{x}_{n-1}))^2}.$$

By the Mean Value Theorem 32.17, there is a point x_n in $[\bar{x}_{n-1}, \bar{x}_n]$ such that

$$\sqrt{(\bar{x}_n - \bar{x}_{n-1})^2 + (f(\bar{x}_n) - f(\bar{x}_{n-1}))^2}$$
$$= \sqrt{(\bar{x}_n - \bar{x}_{n-1})^2 + (f'(x_n)(\bar{x}_n - \bar{x}_{n-1}))^2}$$
$$= \sqrt{1 + (f'(x_n))^2} \Delta x_n.$$

Summing all the lengths of all the straight segments of the polygon yields

$$\sum_{n=1}^{N} \sqrt{1 + (f'(x_n))^2} \Delta x_n. \qquad (34.8)$$

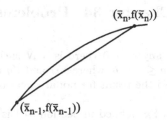

FIGURE 34.11. The length of one segment of the polygon approximating the graph of a function by interpolation.

If f' is continuous, then the Fundamental Theorem 34.3 implies that the sum in (34.8) converges to a definite limit as the mesh is refined. This limit, which we define to be the **length of the curve defined by f from a to b**, is

$$\int_a^b \sqrt{1 + f'(x)^2}\, dx = \lim_{\Delta_{T_N} \to 0} \sum_{n=1}^N \sqrt{1 + (f'(x_n))^2}\, \Delta x_n.$$

Note that in the sum (34.8), we do *not* have control of the location of the interpolation points $\{x_n\}$.

EXAMPLE 34.2. We compute the length of the curve of $f(x) = 2x^{3/2}$ from $x = 0$ to $x = 1$. Since $f'(x) = 3x^{1/2}$, we compute

$$\int_0^1 \sqrt{1 + (3x^{1/2})^2}\, dx = \int_0^1 \sqrt{1 + 9x}\, dx$$

$$= \frac{1}{9} \int_1^{10} \sqrt{u}\, du = \frac{2}{27}(10^{3/2} - 1)$$

There are actually very few functions for which the length of the corresponding curve can be computed analytically. The sum (34.8) is impractical for computing approximations of the length with the interpolation points $\{x_i\}$ given by the Mean Value Theorem. However, we simply choose different interpolation points when we want to use (34.8) in practice.

Chapter 34   Problems

34.1. Prove that given any $\epsilon > 0$ there is a $\bar{N}$ such that if $M \geq N > \bar{N}$, $|Y_M(x) - Y_N(x)| < \epsilon$ for $a \leq x \leq b$, where Y_M and Y_N are the functions defined in Section 34.1. We proved the result for nodal values of x.

34.2. Prove the function Y_N defined in Section 34.1 is continuous on $[a, b]$.

34.3. Show (34.5) when $\bar{x} > x$ and $\bar{x} = a$ or b.

34.4. Explain why (34.6) is valid.

Problems 34.5–34.7 have to do with approximate integration on general meshes.

34.5. Verify (34.7).

34.6. (a) Repeat Problem 25.5 and Problem 25.6 using $x_n = \bar{x}_n$ and a uniform mesh with $N + 1$ nodes. (b) Repeat Problem 25.5 and Problem 25.6 using $x_n = (\bar{x}_n + \bar{x}_{n-1})/2$ and a uniform mesh with $N + 1$ nodes.

34.7. Let $\{\mathcal{T}_N\}$ be a set of meshes for $[a, b]$ with $\Delta_{\mathcal{T}_N} \to 0$ as $N \to \infty$ and for a mesh $\mathcal{T}_N$ with nodes $\{\bar{x}_0, \bar{x}_1, \cdots, \bar{x}_N\}$, where $\bar{x}_0 = a < \bar{x}_1 < \cdots < \bar{x}_N = b$ and $\Delta x_n = \bar{x}_n - \bar{x}_{n-1}$ for $1 \leq n \leq N$, let x_n and y_n be points in $[\bar{x}_{n-1}, \bar{x}_n]$ for $1 \leq n \leq N$. Assume f and g are continuous functions in $[a, b]$. Show that

$$\lim_{\Delta_{\mathcal{T}_N} \to 0} \sum_{n=1}^{N} f(x_n)g(y_n)\Delta x_n = \int_a^b f(x)g(x)\, dx.$$

Interpret this result in terms of computing weighted averages of functions. *Hint:* Consider

$$\sum_{n=1}^{N} f(x_n)(g(y_n) - g(x_n))\Delta x_n.$$

Problem 34.8 gives another way to establish the existence of the integral which, in particular, applies to functions that are not necessarily continuous.

34.8. Let f be a function on a finite interval $[a, b]$ that is bounded, i.e., there is a number M such that $|f(x)| \leq M$ for $a \leq x \leq b$. For a mesh $\mathcal{T}_N$ on $[a, b]$ with nodes $\{x_0, x_1, \cdots, x_N\}$, where $x_0 = a < x_1 < \cdots < x_N = b$ and $\Delta x_n = x_n - x_{n-1}$ for $1 \leq n \leq N$, let M_n be the least upper bound of f on $[x_{n-1}, x_n]$ and m_n be the greatest lower bound of f on $[x_{n-1}, x_n]$. Both of these bounds exist because f is bounded on $[a, b]$. The **upper sum** of f on $\mathcal{T}_N$ is

$$U_N = \sum_{n=1}^{N} M_n\Delta x_n,$$

while the **lower sum** of f on $\mathcal{T}_N$ is

$$L_N = \sum_{n=1}^{N} m_n \Delta x_n.$$

(a) Prove that $U_N \geq L_N$.

(b) Show that if a mesh is refined by the addition of nodes, then the upper sum on the new mesh is either the same or smaller than the upper sum on the old mesh and likewise the lower sum is either the same or greater than the lower sum on the old mesh.

(c) The **upper Darboux integral** of f, denoted by $\mathcal{M}$, is the greatest lower bound of all upper sums U_N of f for all meshes. Likewise, the **lower Darboux integral** of f, denoted by $\mathcal{L}$, is the least upper bound of all lower sums L_N of f for all meshes. If the upper and lower Darboux integrals of f exist and are equal, we call the common value the **Darboux integral** of f.[7] Prove that if f is continuous, then the upper and lower Darboux integrals of f exist and are equal and prove that the resulting Darboux integral is equal to the usual integral of f.

(d) Note that the concept of the Darboux integral applies to functions f that are merely defined and bounded on $[a, b]$; i.e., the functions may not be continuous. Hence, we have given a definition of integrability that does not depend on assuming continuity. Part (c) shows that this definition agrees with the usual definition when the integrand is continuous. As an example to show that the new definition is more general, prove that the Darboux integral of a monotone bounded, though not necessarily continuous, function exists. *Hint:* It is possible to get an explicit formula for the Darboux integrals of f on uniform meshes.

(e) Either compute the Darboux integral of the step function $I(x)$ ($I(x) = 1$ for $0 \leq x \leq 1$ and 0 for all other x) on $[-1, 2]$ or prove it does not exist. Do the values of $I(x)$ at $x = 0$ and 1 make a difference?

(f) Find an example of a function that does not have a Darboux integral.

Problems 34.9–34.13 involve computing the length of a curve.

34.9. Compute the length of the curve $f(x) = x$ on $[0, 1]$ both geometrically and using integration.

34.10. Compute the length of $f(x) = \frac{1}{3}(x^2 + 2)^{3/2}$ from 0 to 2.

34.11. Compute the length of $f(x) = (4 - x^{2/3})^{3/2}$ from 0 to 8.

34.12. Compute the length of $f(x) = \frac{1}{6}x^3 + \frac{1}{2}x^{-1}$ from 1 to 3.

34.13. Compute the length of $f(x)$ from 1 to 2 where $f(x)$ is any solution of the differential equation $y' = (x^4 - 1)^{1/2}$.

---

[7]Named after the French mathematician Jean Gaston Darboux (1842–1917), who made important contributions to analysis and differential geometry. He was highly honored for his work during his lifetime.

35

Delicate Limits and Gross Behavior

In this chapter, we examine some "delicate limits" of functions. In particular, we have so far avoided taking limits of functions as the inputs tend to infinity and avoided considering functions that increase or decrease without bound as the inputs tend to some limit. In other words, we have largely avoided discussing limits of functions when infinity is involved. However, knowing how a function behaves as the inputs increase in size or knowing that a function increases without bound as the inputs tend to some limit is often important in practice. So we begin by extending the idea of a limit to cover both situations in a way that is consistent with the usual "finite" limits. After that, we derive a useful tool called L'Hôpital's rule for computing limits in situations that threaten to involve infinity. Finally, we introduce some language that is very useful for discussing the rate at which a function increases or decreases in value.

35.1 Functions and Infinity

We consider first "taking a limit at ∞." We say that the **function** f **converges to** L **at** ∞, and write

$$\lim_{x \to \infty} f(x) = L,$$

if the number L has the property that given any $\epsilon > 0$ there is an m such that

$$|f(x) - L| < \epsilon \text{ for all } x > m.$$

In words, $f(x)$ approaches L as x increases in size. Likewise, a **function f converges to L at** $-\infty$, and we write

$$\lim_{x \to -\infty} f(x) = L,$$

if the number L has the property that given any $\epsilon > 0$ there is an m such that

$$|f(x) - L| < \epsilon \text{ for all } x < m.$$

In words, $f(x)$ approaches L as x decreases in size.

EXAMPLE 35.1.   We show that

$$\lim_{x \to \infty} \left(1 + \frac{1}{x}\right) = 1.$$

Given $\epsilon > 0$, then

$$\left|\left(1 + \frac{1}{x}\right) - 1\right| = \left|\frac{1}{x}\right| < \epsilon$$

if $x > 1/\epsilon = m$.

EXAMPLE 35.2.   The plot of sin (see Fig. 35.1), makes it clear that $\lim_{x \to \infty} \sin(x)$ is undefined. Analytically, given any number y in $[-1, 1]$ there are arbitrarily large x with $\sin(x) = y$.

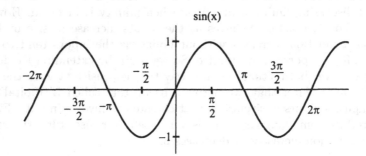

sin(x)

FIGURE 35.1. Plot of sin.

Note that we can think of the limit at ∞ as a one-sided limit from the left, i.e.,

$$\lim_{x \to \infty} f(x) = \lim_{x \uparrow \infty} f(x),$$

and likewise the limit at $-\infty$ as a one-sided limit from the right, i.e.,

$$\lim_{x \to -\infty} f(x) = \lim_{x \downarrow \infty} f(x).$$

Next, we define "infinite limits." We say that f **converges to ∞ at a number** a, and write

$$\lim_{x \to a} f(x) = \infty,$$

if for every M there is a $\delta > 0$ such that

$$f(x) > M \text{ for all } x \text{ with } 0 < |x - a| < \delta.$$

In words, $f(x)$ can be made arbitrarily large by taking x sufficiently close to a. Likewise, f **converges to $-\infty$ at a number** a,

$$\lim_{x \to a} f(x) = -\infty,$$

if for every M there is a $\delta > 0$ such that

$$f(x) < M \text{ for all } x \text{ with } 0 < |x - a| < \delta.$$

EXAMPLE 35.3. We show that $\lim_{x \to 0} x^{-2} = \infty$. Given any $M > 0$, then

$$\frac{1}{x^2} > M$$

for all x with $x^2 < 1/M$ or $|x| < 1/\sqrt{M} = \delta$.

This definition can be trickier to apply than it might appear. Consider the two functions plotted in Fig. 35.2. Neither function is continuous at a.

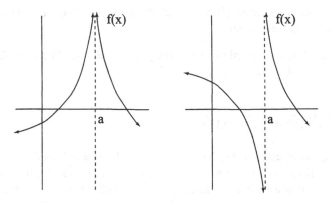

FIGURE 35.2. Plot of two functions that become large in magnitude as x approaches a.

However, the function on the left converges to ∞ as x approaches a, while the function on the right does *not*, because it behaves differently on either side of a. We often have to consider one-sided limits to evaluate an infinite limit. Thus,

$$\lim_{x \to a} f(x) = \infty \text{ if and only if } \lim_{x \uparrow a} f(x) = \infty \text{ and } \lim_{x \downarrow a} f(x) = \infty,$$

with the obvious definition of the one-sided limits.

EXAMPLE 35.4.   We show that $\lim_{x \to 0} x^{-1}$ is undefined. For if $x > 0$, then given any $M > 0$, $1/x > M$ as long as $0 < x < 1/M = \delta$. Yet, if $x < 0$, then given any $M < 0$, $1/x < M$ as long as $1/M = \delta < x < 0$. Hence, $\lim_{x \uparrow 0} x^{-1} = -\infty$ and $\lim_{x \downarrow 0} x^{-1} = \infty$.

Of course, these different definitions can be combined. We say that f **converges to ∞ at ∞**, and write

$$\lim_{x \to \infty} f(x) = \infty,$$

if for every M there is an m such that

$$f(x) > M \text{ for all } x > m$$

and f **converges to $-\infty$ at ∞**, and write

$$\lim_{x \to \infty} f(x) = -\infty,$$

if for every M there is an m such that

$$f(x) < M \text{ for all } x > m.$$

We leave it as an exercise (Problem 35.2) to define f converging to ∞ as x approaches $-\infty$ and so on.

EXAMPLE 35.5.   We show that $\lim_{x \to \infty} x^2 = \infty$. For any $M > 0$, $x^2 > M$ for all x with $x > \sqrt{M} = N$.

35.2   L'Hôpital's Rule

The examples treated so far have been straightforward applications of the definitions. However as is the case with the usual "finite" limits, limits that involve or threaten to involve infinity can be difficult to evaluate. We have dealt with some examples already. For example when differentiating sin, we encountered the limit

$$\lim_{x \to 0} \frac{\sin(x)}{x}.$$

This is a difficult limit because both the numerator and denominator tend to zero and it is unclear what their ratio does. We dealt with this by doing some complicated geometry. Another example of a relatively difficult limit is

$$\lim_{x \to \infty} \frac{\log(x^3 + 1)}{\log(x^2 + 5x)}.$$

In this case, both the numerator and the denominator increase without bound and it is unclear what their ratio does.

These are both examples of what are called **indeterminate forms**. Indeterminate forms include limits of ratios of functions in which the numerator and denominator both tend to zero or both tend to plus or minus infinity. In an abuse of notation, these two cases are often labeled "0/0" and "∞/∞," though these two expressions are actually meaningless. Indeterminate forms also include limits of a product of two functions where one function tends to zero and the other increases without bound and the difference of two functions both of which tend to plus or minus infinity. These are labeled "0 · ∞" and "∞ − ∞," respectively. Other indeterminate forms include "∞⁰," "1^∞," and "0⁰" with the obvious interpretations. We discuss specific examples below.

In this section, we state and prove L'Hôpital's rule, which is an often-useful tool for evaluating indeterminate forms.[1] Because it applies to several different situations, the statement of the general result is neither easy to read or understand. So we first give some motivation in the simplest case.

Suppose that f and g are differentiable functions in an open interval containing a with $f(a) = g(a) = 0$ and that we want to compute

$$\lim_{x \to a} \frac{f(x)}{g(x)}.$$

We can rewrite this limit as

$$\lim_{x \to a} \frac{f(x)}{g(x)} = \lim_{h \to 0} \frac{f(a + h)}{g(a + h)}.$$

Now for h small,

$$f(x + h) \approx f(a) + hf'(a) = hf'(a)$$
$$g(x + h) \approx g(a) + hg'(a) = hg'(a).$$

Hence, if $f'(a)/g'(a)$ is defined, then for h small,

$$\frac{f(x + h)}{g(x + h)} \approx \frac{f'(a)}{g'(a)},$$

---

[1]This result is named after the French mathematician Guillaume Francois Antoine Marquis de L'Hôpital (1661–1704). L'Hôpital paid Johann Bernoulli for private lessons on the calculus of Leibniz and the right to use some of Bernoulli's results in his textbook, which was the first textbook on differential calculus. L'Hôpital's rule was almost certainly discovered by Johann Bernoulli, though L'Hôpital himself was a reasonable mathematician.

suggesting that

$$\lim_{x \to a} \frac{f(x)}{g(x)} = \frac{f'(a)}{g'(a)}.$$

This is essentially L'Hôpital's rule, which in general replaces the limit of the ratio of two functions by the limit of the ratio of their derivatives in certain circumstances.

Theorem 35.1 L'Hôpital's Rule *Suppose f and g are differentiable functions on (a, b) and $g'(x) \neq 0$ for all x in (a, b), where $-\infty \leq a < b \leq \infty$.*

1. *Assume that*

$$\lim_{x \downarrow a} \frac{f'(x)}{g'(x)} = A$$

   exists, where A may be finite or infinite. If (a) $\lim_{x \downarrow a} f(x) = 0$ and $\lim_{x \downarrow a} g(x) = 0$ or (b) $\lim_{x \downarrow a} g(x) = \pm\infty$, then

$$\lim_{x \downarrow a} \frac{f(x)}{g(x)} = A.$$

2. *Assume that*

$$\lim_{x \uparrow b} \frac{f'(x)}{g'(x)} = B$$

   exists, where B may be finite or infinite. If (a) $\lim_{x \uparrow b} f(x) = 0$ and $\lim_{x \uparrow b} g(x) = 0$ or (b) $\lim_{x \uparrow b} g(x) = \pm\infty$, then

$$\lim_{x \uparrow b} \frac{f(x)}{g(x)} = B.$$

This statement of L'Hôpital's rule is framed in terms of one-sided limits. If we want to apply this to evaluate $\lim_{x \to a} f(x)$ where a is finite, then we rewrite the limit as the common value of the one-sided limits from the left and right of a.

EXAMPLE 35.6. To compute

$$\lim_{x \to 0} \frac{\sin(x)}{x},$$

note that $\sin(x)$ and x are differentiable everywhere, while $(x)' = 1 \neq 0$ for any x. Since $(\sin(x))' = \cos(x)$,

$$\lim_{x \downarrow 0} \frac{\cos(x)}{1} = \lim_{x \uparrow 0} \frac{\cos(x)}{1} = 1.$$

We conclude that

$$\lim_{x \to 0} \frac{\sin(x)}{x} = 1.$$

EXAMPLE 35.7. To compute

$$\lim_{x \to \infty} \frac{\log(x^3 + 1)}{\log(x^2 + 5x)},$$

note that the numerator and denominator are differentiable, while $(\log(x^2 + 5x))' = (2x+5)/(x^2+5x) \neq 0$ for $x > 0$ and moreover $\lim_{x \to \infty} \log(x^2 + 5x) = \infty$. We compute

$$\frac{(\log(x^3 + 1))'}{(\log(x^2 + 5x))'} = \frac{\frac{3x^2}{x^3+1}}{\frac{2x+5}{x^2+5x}} = \frac{3x^4 + 15x^3}{2x^4 + 5x^3 + 2x + 5}.$$

Using a trick developed for rational functions, we conclude that

$$\lim_{x \to \infty} \frac{3x^4 + 15x^3}{2x^4 + 5x^3 + 2x + 5} = \lim_{x \to \infty} \frac{(x^{-4})(3x^4 + 15x^3)}{(x^{-4})(2x^4 + 5x^3 + 2x + 5)}$$

$$= \lim_{x \to \infty} \frac{3 + 15x^{-1}}{2 + 5x^{-1} + 2x^{-3} + 5x^{-4}} = \frac{3}{2}. \quad (35.1)$$

However, we can also show (35.1) by repeated application of L'Hôpital's rule. This is just

$$\lim_{x \to \infty} \frac{3x^4 + 15x^3}{2x^4 + 5x^3 + 2x + 5} = \lim_{x \to \infty} \frac{12x^3 + 45x^2}{8x^3 + 15x^2 + 2}$$

$$= \lim_{x \to \infty} \frac{36x^2 + 90x}{24x^2 + 30x} = \lim_{x \to \infty} \frac{36x + 90}{24x + 30}$$

$$= \lim_{x \to \infty} \frac{36}{24} = \frac{3}{2},$$

where at each step the assumptions of the theorem hold as long as the new limit exists.

We prove case 1 and leave case 2 as an exercise (Problem 35.8). The proof is based on a generalization of the Mean Value Theorem 32.17.

Theorem 35.2 Generalized Mean Value Theorem *If f and g are continuous functions in $[a, b]$ and differentiable in (a, b), then there is a c in (a, b) such that*

$$(f(b) - f(a))g'(c) = (g(b) - g(a))f'(c). \quad (35.2)$$

We give the proof of this as an exercise (Problem 35.4). Note that the standard non-constructive Mean Value Theorem 32.17 follows from this result by taking $g = x$.

We begin by assuming (a) holds and treat three cases, beginning with $-\infty < A < \infty$. By definition, for any $\epsilon > 0$ there is an $m > a$ such that

$$\left| A - \frac{f'(t)}{g'(t)} \right| < \epsilon \text{ for } a < t < m.$$

Choosing w and x with $a < w < x < m$, then because $g'(s) \neq 0$ for any s in (a, b), Theorem 35.2 implies there is a t in (w, x) with

$$\frac{f(x) - f(w)}{g(x) - g(w)} = \frac{f'(t)}{g'(t)}.$$

This means that

$$\left| A - \frac{f(x) - f(w)}{g(x) - g(w)} \right| < \epsilon.$$

Letting $w \to a$, we conclude that for any $\epsilon > 0$ there is an $m > a$ such that

$$\left| A - \frac{f(x)}{g(x)} \right| < \epsilon \text{ for } a < x < m.$$

This proves the result.

For the second case, we assume $A = -\infty$. By definition, given any $M > A$ there is an $m > a$ such that

$$A < \frac{f'(t)}{g'(t)} < M \text{ for } a < t < m.$$

Now repeating the argument above, we conclude that for any $M > A$ there is an $m > a$ such that

$$A < \frac{f(x)}{g(x)} < M \text{ for } a < x < m. \tag{35.3}$$

We treat $A = \infty$ is a similar way. We leave the details of both arguments as exercises (Problems 35.5 and 35.6).

Now we consider assumption (b), breaking the proof down into two cases, beginning with $-\infty \leq A < \infty$. We show the result using one-sided limits assuming that $g(x) \to \infty$. The case $g(x) \to -\infty$ follows directly.

Given any $M > A$, we choose $\tilde{M}$ such that $A < \tilde{M} < M$. By definition, there is an $m > a$ such that

$$\frac{f'(t)}{g'(t)} < \tilde{M} < M \text{ for } a < t < m.$$

If $a < x < y < m$, then there is a t in (x, y) such that

$$\frac{f(x) - f(y)}{g(x) - g(y)} = \frac{f(y) - f(x)}{g(y) - g(x)} = \frac{f'(t)}{g'(t)} < \tilde{M}. \tag{35.4}$$

Since $g(x)$ increases without bound as x approaches a, by reducing m if necessary, $g(x) > 0$ and $g(x) > g(y)$ for all $a < x < y < m$. We multiply (35.4) by $(g(x) - g(y))/g(x) > 0$ to get

$$\frac{f(x) - f(y)}{g(x)} < \tilde{M} \frac{g(x) - g(y)}{g(x)} = \tilde{M} - \tilde{M} \frac{g(y)}{g(x)}$$

or

$$\frac{f(x)}{g(x)} < \tilde{M} - \tilde{M}\frac{g(y)}{g(x)} + \frac{f(y)}{g(x)}.$$

Since $g(x) \to \infty$ as $x \downarrow a$, for fixed y and all x sufficiently close to a,

$$\tilde{M} - \tilde{M}\frac{g(y)}{g(x)} + \frac{f(y)}{g(x)} < M.$$

Hence, we conclude that for any $M > A$ there is an $m > a$ such that

$$\frac{f(x)}{g(x)} < M \text{ for } a < x < m. \tag{35.5}$$

Now if $A = -\infty$, then the desired result has been proved. Otherwise, if $-\infty < A \le \infty$, we use a very similar argument (Problem 35.7) to show that for any $N < A$ there is an $n > a$ such that

$$N < \frac{f(x)}{g(x)} \text{ for } a < x < n. \tag{35.6}$$

This proves the result when $A = \infty$ directly, and together (35.5) and (35.6) prove the result when $-\infty < A < \infty$.

We mentioned that the other indeterminate forms are treated by rewriting them as the indeterminate forms "0/0" or "∞/∞" and then using L'Hôpital's rule if necessary. We conclude this section by giving some examples.

EXAMPLE 35.8. An example of the indeterminate form "$\infty - \infty$" is

$$\lim_{x \to \infty} x - \sqrt{x^2 + 1}.$$

To evaluate this limit, we first write

$$x - \sqrt{x^2 + 1} = \left(x - \sqrt{x^2 + 1}\right)\frac{x + \sqrt{x^2 + 1}}{x + \sqrt{x^2 + 1}}$$
$$= \frac{x^2 - (x^2 + 1)}{x + \sqrt{x^2 + 1}} = -\frac{1}{x + \sqrt{x^2 + 1}}.$$

Since

$$\lim_{x \to \infty} \frac{1}{x + \sqrt{x^2 + 1}} = 0$$

we conclude that

$$\lim_{x \to \infty} x - \sqrt{x^2 + 1} = 0.$$

EXAMPLE 35.9. When considering the exponential function, we encountered the indeterminate form "1^∞" in the form

$$L = \lim_{x \to 0} (1 + x)^{1/x}.$$

Since the logarithm is a monotone increasing, continuous function,

$$\log(L) = \log\left(\lim_{x \to 0}(1+x)^{1/x}\right) = \lim_{x \to 0} \log\left((1+x)^{1/x}\right),$$

provided the second limit exists. Now

$$\log\left((1+x)^{1/x}\right) = \frac{\log(1+x)}{x}.$$

We apply L'Hôpital's rule to evaluate

$$\lim_{x \to 0} \frac{\log(1+x)}{x} = \lim_{x \to 0} \frac{(1+x)^{-1}}{1} = 1.$$

We conclude that $\log(L) = 1$ or $L = e$.

35.3   Orders of Magnitude

We saw that L'Hôpital's rule is a useful device for comparing the rates at which two functions change values when both increase without bound or both tend to zero as the input changes. In this section, we introduce some useful language for comparing the rates of growth of two functions.

Consider two functions $f(x)$ and $g(x)$ that both tend to ∞ as x tends to ∞. We say that f **becomes infinite at a higher order (rate) than** g if

$$\lim_{x \to \infty} \left| \frac{f(x)}{g(x)} \right| = \infty.$$

This means that while both $|f|$ and $|g|$ increase without bound as x increases, $|f|$ increases more quickly.

EXAMPLE 35.10. Clearly, x^3 becomes infinite at a higher order than x^2.

EXAMPLE 35.11. In Chapter 29, we proved that for any p, $\exp(x)$ becomes larger than x^p for sufficiently large x, and likewise x^p eventually becomes larger than $\log(x)$. We can make these comparisons more precise.

First, we show that for any natural number n,

$$\lim_{x \to \infty} \frac{e^x}{x^n} = \infty,$$

by an inductive application of L'Hôpital's rule,

$$\lim_{x \to \infty} \frac{e^x}{x^n} = \lim_{x \to \infty} \frac{e^x}{nx^{n-1}} = \lim_{x \to \infty} \frac{e^x}{n(n-1)x^{n-2}}$$

$$= \cdots \lim_{x \to \infty} \frac{e^x}{n(n-1)\cdots 1} = \infty. \quad (35.7)$$

Now for any $p > 0$, let n be the largest natural number less than or equal to p. Then for $x > 1$,

$$\frac{e^x}{x^p} \geq \frac{e^x}{x^p} \frac{x^p}{x^n} = \frac{e^x}{x^n}. \quad (35.8)$$

Hence for any $p > 0$,

$$\lim_{x \to \infty} \frac{e^x}{x^n} = \infty \quad (35.9)$$

and therefore $\exp(x)$ becomes infinite at a higher order than x^p for any $p > 0$. Using this, it is straightforward to show that x^p becomes infinite to higher order than $\log(x)$ for any $p > 0$.

Likewise, if $f(x)$ and $g(x)$ tend to infinity as $x \to \infty$, we say that f **becomes infinite at a lower order (rate) than** g if

$$\lim_{x \to \infty} \left| \frac{f(x)}{g(x)} \right| = 0.$$

Finally, we say that f **and** g **become infinite at the same order** if there are constants $c_1 < c_2$ such that

$$c_1 < \left| \frac{f(x)}{g(x)} \right| < c_2$$

for all sufficiently large x.

EXAMPLE 35.12. In Example 35.7, we showed that $\log(x^3 + 1)$ and $\log(x^2 + 5x)$ grow at the same order of magnitude.

Note that in the last case, the ratio of the functions need not tend to a limit.

To complete the language, we say that x^p has an **order of magnitude** p as $x \to \infty$. Any function that becomes infinite at the same order as x^p also has an order of magnitude p. The examples above show that $\exp(x)$ has an order of magnitude larger than any $p > 0$ as $x \to \infty$, while $\log(x)$ has a lower order of magnitude than any $p > 0$.

While this notation is often useful, it cannot be used to compare the rates of growth of any two functions.

EXAMPLE 35.13. We cannot compare the functions $x^2 \sin^2(x) + x + 1$ and $x^2 \cos^2(x) + 1$ using orders of magnitude. The ratio of these two functions neither remains between two constants nor tends to zero or infinity.

But on the other hand, it is not meant to compare all functions. For example, knowing how one of the functions in Example 35.13 behaves gives no useful information about how the other function behaves.

We can apply the idea of orders of magnitude to compare the rates that two functions decrease to zero by using a change of variables, namely,

$$\lim_{x \downarrow 0} f(x) = \lim_{y \to \infty} f(1/y).$$

To compare two functions $f(x)$ and $g(x)$ as $x \downarrow 0$, we compare $f(1/y)$ and $g(1/y)$ as $y \to \infty$. It follows that f **vanishes at a higher order than** g if

$$\lim_{x \downarrow 0} \left| \frac{f(x)}{g(x)} \right| = 0.$$

We say that f **vanishes at a lower order than** g if

$$\lim_{x \downarrow 0} \left| \frac{f(x)}{g(x)} \right| = \infty.$$

Finally, we say that f **and** g **vanish at the same order** if there are constants $c_1 < c_2$ such that

$$c_1 < \left| \frac{f(x)}{g(x)} \right| < c_2$$

for all sufficiently small x.

Consequently, we say that x^p for $p > 0$ decreases or vanishes with order of magnitude p as $x \downarrow 0$, and we assign orders of magnitude to general functions depending on how they compare to x^p.

EXAMPLE 35.14. In Example 35.11, we showed that $\log(x)$ vanishes with a lower order of magnitude than x^p for any $p > 0$ and likewise $\exp(-1/x)$ vanishes with a higher order of magnitude.

EXAMPLE 35.15. The function $x^p + x$ vanishes at the same order as x if $p > 1$ and at the same order as x^p if $0 < p < 1$.

There is a succinct notation for discussing orders of magnitudes, called the "big O" and "little o" notation, which is due to Landau.[2] If the function f is of lower order of magnitude than g, we write

$$f = o(g)$$

---

[2]The German mathematician Edmund Georg Hermann Landau (1877–1938) wrote many papers in number theory, making fundamental contributions to analytic number theory in particular.

and say that f is "little o" of g. This means that

$$\frac{f}{g} \to 0$$

when $x \to \infty$ or $x \downarrow 0$ as is relevant. We can abbreviate the results above as simply

$$x^p = \mathbf{o}(x^q) \text{ for } p < q \text{ as } x \to \infty$$
$$\log(x) = \mathbf{o}(x^p) \text{ for } p > 0 \text{ as } x \to \infty$$
$$x^p = \mathbf{o}(e^x) \text{ for } p > 0 \text{ as } x \to \infty$$
$$x^p = \mathbf{o}(x^q) \text{ for } p > q \text{ as } x \downarrow 0$$
$$\log(x) = \mathbf{o}(1/x^p) \text{ for } p > 0 \text{ as } x \downarrow 0$$
$$e^{-1/x} = \mathbf{o}(x^p) \text{ for } p > 0 \text{ as } x \downarrow 0.$$

The "big O" notation is used to indicate that one function is, at most, the same order of magnitude as the other. We say that $f = \mathbf{O}(g)$ if there are constants c_1 and c_2 such that

$$c_1 < \left| \frac{f(x)}{g(x)} \right| < c_2$$

for all relevant values of x.

EXAMPLE 35.16.

$$\sqrt{x} = \mathbf{O}(\sqrt{x+1}) \text{ for all } x \geq 0$$
$$\sqrt{x} = \mathbf{O}(x) \text{ for all } x \geq 1$$
$$\log(x) = \mathbf{O}(x) \text{ for all } x \geq 1$$
$$x = \mathbf{O}(\sin(x)) \text{ for all } x \leq 1.$$

Chapter 35   Problems

In Problem 35.1, we ask you to use the definitions to compute the indicated limits. The rest of the problems involving the computation of limits should be tackled using L'Hôpital's rule.

35.1. Compute the following limits using the definition, or show they are not defined,

$$\text{(a) } \lim_{x \to \infty} x^5 \qquad\qquad \text{(b) } \lim_{x \to \infty} 2/(x+1)$$

$$\text{(c) } \lim_{x \to 3} 1/(x-3)^4 \qquad \text{(d) } \lim_{x \to 1} x/(x-1) \,.$$

35.2. Write down a definition for a function to converge to ∞ as $x \to -\infty$.

35.3. Compute the following limits:

$$\text{(a) } \lim_{x \to 1} \frac{x^3 + x^2 - 2x}{x^3 - x^2 + x - 1} \qquad \text{(b) } \lim_{x \to 0} \frac{x - \sin(x)}{x^3}$$

$$\text{(c) } \lim_{x \to 0} \frac{1 - \cos(x)}{x^2} \qquad \text{(d) } \lim_{x \to \infty} \frac{x^4 - 3000x + 1}{x^2 + 2x + 4}$$

$$\text{(e) } \lim_{x \to \infty} \frac{\log(x - 1)}{\log(x^2 - 1)} \qquad \text{(f) } \lim_{x \to \infty} \frac{x \log(x)}{(x + 1)^2}.$$

35.4. Prove Theorem 35.2. *Hint:* Consider the function

$$h(x) = \big(f(b) - f(a)\big)g(x) - \big(g(b) - g(a)\big)f(x) - \big(f(b)g(a) - f(a)g(b)\big).$$

35.5. Verify (35.3).

35.6. Carry out the proof of case 1. of Theorem 35.1 under assumption (a) with $A = \infty$.

35.7. Carry out the proof of case 1. of Theorem 35.1 under assumption (b) with $-\infty < A \le \infty$.

35.8. Carry out the proof of case 2. of Theorem 35.1.

35.9. Compute the following limits:

$$\text{(a) } \lim_{x \to 0} \big(\sin(x) \log(x)\big) \qquad \text{(b) } \lim_{x \to 1} \left(\frac{1}{\log(x)} - \frac{x}{x - 1} \right)$$

$$\text{(c) } \lim_{x \to \infty} x^{1/x} \qquad\qquad \text{(d) } \lim_{x \downarrow 0} \big(\log(1 + x)\big)^x.$$

35.10. (a) Show that Newton's method applied to a differentiable function $f(x)$ with $f(\bar{x}) = f'(\bar{x}) = 0$, where $\bar{x}$ is the root of f in question, converges at a linear rate. *Hint:* Use L'Hôpital's rule to show that $\lim_{x \to \bar{x}} g'(x) = 1/2$ where $g(x) = x - f(x)/f'(x)$. (b) What is the rate of convergence of the following variant of Newton's method in the case of a double root: $g(x) = x - 2f(x)/f'(x)$?

35.11. Verify the details in (35.7).

35.12. Verify (35.8) and show this implies (35.9).

35.13. Show that x^p becomes infinite to higher order than $\log(x)$ for any $p > 0$.

35.14. Verify the claims in Example 35.13.

35.15. Show that $\sin(x) = o(x)$ as $x \to \infty$.

35.16. Verify the equations in Example 35.16.

35.17. Translate the results in Problem 35.3 into the "big O" and "little o" notation.

35.18. Let $f(x)$ be a differentiable function on an open interval containing 0 and suppose that both $f(0) = f'(0) = 0$. Show that f vanishes to higher order than x at $x \to 0$.

36
The Weierstrass Approximation Theorem

Recall that the fundamental idea underlying the construction of the real numbers is approximation by the simpler rational numbers. Firstly, numbers are often determined as the unknown roots of some equation and when we cannot solve the equation explicitly, as is most often the case, then we must compute approximate solutions. But even if we write down a real number symbolically, like $\sqrt{2}$, for example, we cannot specify its numerical value completely in general. In this case, we approximate the real number to any desired accuracy using rational numbers with finite decimal expansions.

The situation for functions is completely analogous. In general, functions that are specified as the solutions of differential equations cannot be written down explicitly in terms of known functions. Instead, we must look for good approximations. Moreover, most of the functions that we can write down, i.e., those involving exp, log, sin, and so on, are "complicated" in the sense that they take on real values that cannot be written down explicitly. To use these functions in practical computations, we must resort to using good approximations of their values. Put it this way; when we press the e^x key on a calculator, we do not get e^x, rather we get a good approximation.

This raises one of the fundamental problems of analysis, which is figuring out how to approximate a given function using simpler functions. In this chapter, we begin the study of this problem by proving a fundamental result which says that any continuous function can be approximated arbitrarily well by polynomials. This is an important result because polynomials are relatively simple. In particular, a polynomial is specified completely by a finite set of coefficients. In other words, the relatively simple polynomials play the same role with respect to continuous functions that rational numbers play with real numbers.

The result is due to Weierstrass and it states:

Theorem 36.1 Weierstrass Approximation Theorem *Assume that f is continuous on a closed bounded interval I. Given any $\epsilon > 0$, there is a polynomial P_n with sufficiently high degree n such that*

$$|f(x) - P_n(x)| < \epsilon \text{ for } a \le x \le b. \tag{36.1}$$

There are many different proofs of this result, but in keeping with our constructivist tendencies, we present a constructive proof based on Bernstein[1] polynomials. The motivation for this approach rests in probability theory. We do not have space in this book to develop probability theory, but we describe the connection in an intuitive way. Later in Chapter 37 and Chapter 38, we investigate other polynomial approximations of functions that arise from different considerations.

Before beginning, we note that it suffices to prove Theorem 36.1 for the interval $[0, 1]$. The reason is that the arbitrary interval $a \le y \le b$ is mapped to $0 \le x \le 1$ by $x = (a - y)/(a - b)$ and vice versa by $y = (b - a)x + a$. If g is continuous on $[a, b]$, then $f(x) = g((b - a)x + a)$ is continuous on $[0, 1]$. If the polynomial P_n of degree n approximates f to within ϵ on $[0, 1]$, then the polynomial $\tilde{P}_n(y) = P_n((a - y)/(a - b))$ of degree n approximates $g(y)$ to within ϵ on $[a, b]$.

36.1   The Binomial Expansion

One ingredient needed to construct the polynomial approximations is an important formula called the binomial expansion. For natural numbers $0 \le m \le n$, we define the **binomial coefficient** $\binom{n}{m}$, or n **choose** m, by

$$\binom{n}{m} = \frac{n!}{m!(n - m)!}.$$

EXAMPLE 36.1.

$$\binom{4}{2} = \frac{4!}{2!2!} = 6, \quad \binom{6}{1} = \frac{6!}{1!5!} = 6, \quad \binom{3}{0} = \frac{3!}{3!0!} = 1$$

We can interpret n choose m as the number of distinct subsets with m elements that can be chosen from a set of n objects, or the number of combinations of n objects taken m at a time.

---

[1]The Russian mathematician Sergi Natanovich Bernstein (1880–1968) studied in France before returning to Russia to work. He proved significant results in approximation theory and probability.

EXAMPLE 36.2. We compute the probability $\mathcal{P}$ of getting an ace of diamonds in a poker hand of 5 cards chosen at random from a standard deck of 52 cards. Recall the formula

$$\mathcal{P}(\text{event}) = \text{probability of an event}$$

$$= \frac{\text{number of outcomes in the event}}{\text{total number of possible outcomes}}$$

that holds if all outcomes are equally likely. The total number of 5 card poker hands is $\binom{52}{5}$. Obtaining a "good" hand amounts to choosing any 4 cards from the remaining 51 cards after getting an ace of diamonds. So there are $\binom{51}{4}$ good hands. This means

$$\mathcal{P} = \frac{\binom{51}{4}}{\binom{52}{5}} = \frac{51!}{4!47!} \frac{5!47!}{52!} = \frac{5}{52}.$$

It is straightforward (Problem 36.3) to show the following identities,

$$\binom{n}{m} = \binom{n}{n-m}, \quad \binom{n}{1} = \binom{n}{n-1}, \quad \binom{n}{n} = \binom{n}{0} = 1. \qquad (36.2)$$

An important application of the binomial coefficient is the following theorem.

Theorem 36.2 Binomial Expansion   *For any natural number* n,

$$(a+b)^n = \sum_{m=0}^{n} \binom{n}{m} a^m b^{n-m}. \qquad (36.3)$$

EXAMPLE 36.3.

$$(a+b)^2 = a^2 + 2ab + b^2$$
$$(a+b)^3 = a^3 + 3a^2b + 3ab^2 + b^3$$
$$(a+b)^4 = a^4 + 4a^3b + 6a^2b^2 + 4ab^3 + b^4$$

The proof is by induction. For $n = 1$,

$$(a+b)^1 = a + b = \binom{1}{0}a + \binom{1}{1}b.$$

We assume the formula is true for $n - 1$, so that

$$(a+b)^{n-1} = \sum_{m=0}^{n-1} \binom{n-1}{m} a^m b^{n-1-m},$$

and prove it holds for n.

We multiply out

$$(a + b)^n = (a + b)(a + b)^{n-1}$$

$$= \sum_{m=0}^{n-1} \binom{n-1}{m} a^{m+1} b^{n-1-m} + \sum_{m=0}^{n-1} \binom{n-1}{m} a^m b^{n-m}.$$

Now changing variables in the sum,

$$\sum_{m=0}^{n-1} \binom{n-1}{m} a^{m+1} b^{n-1-m} = \sum_{m=1}^{n-1} \binom{n-1}{m-1} a^m b^{n-m} + a^n b^0,$$

while

$$\sum_{m=0}^{n-1} \binom{n-1}{m} a^m b^{n-m} = a^0 b^n + \sum_{m=1}^{n-1} \binom{n-1}{m} a^m b^{n-m}.$$

Hence,

$$(a + b)^n = a^0 b^n + \sum_{m=1}^{n-1} \left(\binom{n-1}{m-1} + \binom{n-1}{m} \right) a^m b^{n-m} + a^n b^0. \quad (36.4)$$

It is a good exercise (Problem 36.5) to show that

$$\binom{n-1}{m-1} + \binom{n-1}{m} = \binom{n}{m}. \quad (36.5)$$

Using this in (36.4) proves the result.

We use the binomial expansion to drive two other useful formulas. We differentiate both sides of

$$(x + b)^n = \sum_{m=0}^{n} \binom{n}{m} x^m b^{n-m} \quad (36.6)$$

to get

$$n(x + b)^{n-1} = \sum_{m=0}^{n} m \binom{n}{m} x^{m-1} b^{n-m}.$$

Setting $x = a$ and multiplying through by a/n,

$$a(a + b)^{n-1} = \sum_{m=0}^{n} \frac{m}{n} \binom{n}{m} a^m b^{n-m}. \quad (36.7)$$

Differentiating (36.6) twice (Problem 36.6) gives

$$\left(1 - \frac{1}{n} \right) a^2 (a + b)^{n-2} = \sum_{m=0}^{n} \left(\frac{m^2}{n^2} - \frac{m}{n^2} \right) \binom{n}{m} a^m b^{n-m}. \quad (36.8)$$

36.2 The Law of Large Numbers

The approximating polynomials used to prove Theorem 36.1 are constructed by taking linear combinations of more elementary polynomials called binomial polynomials. In this section, we explore the properties of the binomial polynomials and their connection to probability.

We set $a = x$ and $b = 1 - x$ in the binomial expansion (36.3) to get

$$1 = (x + (1 - x))^n = \sum_{m=0}^{n} \binom{n}{m} x^m (1 - x)^{n-m}. \tag{36.9}$$

We define the $m + 1$ **binomial polynomials** of degree n as the terms in the expansion, so

$$p_{n,m}(x) = \binom{n}{m} x^m (1 - x)^{n-m}, \quad m = 0, 1, \cdots, n.$$

EXAMPLE 36.4.

$$p_{2,0}(x) = \binom{2}{0} x^0 (1 - x)^2 = (1 - x)^2$$

$$p_{2,1}(x) = \binom{2}{1} x^1 (1 - x)^1 = 2x(1 - x)$$

$$p_{2,2}(x) = \binom{2}{2} x^2 (1 - x)^0 = x^2$$

If $0 \le x \le 1$ is the probability of an event E, then $p_{n,m}(x)$ is the probability that E occurs exactly m times in n independent trials.

EXAMPLE 36.5. In particular, consider tossing an coin with probability x that a head (H) occurs and, correspondingly, probability $1 - x$ that a tail (T) occurs. The coin is "unfair" if $x \ne 1/2$. The probability of the occurrence of a particular sequence of n tosses containing m heads, e.g.,

$$\underbrace{HTTHHTHTHTTHHHTHTHTTT\cdots T}_{m \text{ heads in } n \text{ tosses}},$$

is $x^m (1 - x)^{m-n}$ by the multiplication rule for probabilities. There are $\binom{n}{m}$ sequences of n tosses with exactly m heads. By the addition rule for probabilities, $p_{n,m}(x)$ is the probability of getting exactly m heads in n tosses.

The binomial polynomials have several useful properties, some of which follow directly from the connection to probability. For example, we interpret

$$\sum_{m=0}^{n} p_{n,m}(x) = 1 \tag{36.10}$$

as saying that event E with probability x occurs either exactly $0, 1, \cdots,$ or n times in n independent trials with probability 1. Since $p_{n,m}(x) \geq 0$ for $0 \leq x \leq 1$, (36.10) implies that $0 \leq p_{n,m}(x) \leq 1$ for $0 \leq x \leq 1$, as it must since it is a probability.

A couple more useful properties: (36.7) implies

$$\sum_{m=0}^{n} m p_{n,m}(x) = nx \tag{36.11}$$

and (36.8) implies

$$\sum_{m=0}^{n} m^2 p_{n,m}(x) = (n^2 - n)x^2 + nx. \tag{36.12}$$

An important use of the binomial polynomials is an application to the Law of Large Numbers. Suppose we have an event E that has probability x of occurring, such as the unfair coin from Example 36.5. But suppose we don't know the probability. How might we determine x? If we conduct a single trial, e.g., flip the coin once, we might see event E or might not. One trial does not give much information for determining x. However, if we conduct a large number $n \gg 1$ of trials, then intuition suggests that E should occur approximately nx times out of n trials, at least "most of the time."

EXAMPLE 36.6. The connection between the probability of occurrence in one trial and the frequency of occurrence in many trials is not completely straightforward to determine. Consider coin tossing again. If we flip a fair coin $100,000$ times, we expect to see *around* $50,000$ heads *most of the time*. Of course, we could be very unlucky and get all tails. But the probability of this occurring is

$$\left(\frac{1}{2}\right)^{100000} \approx 10^{-30103}.$$

On the other hand, it is also unlikely that we will see heads in exactly half of the tosses. In fact, one can show that the probability of getting heads exactly half of the time is approximately $1/\sqrt{\pi n}$ for n large, and therefore also goes to zero as n increases.

A Law of Large Numbers encapsulates in some way the intuitive connection between the probability of an event occurring in one trial and the frequency that the event occurs in a large number of trials. A mathematical expression of this intuition is a little tricky to state, however, as we saw in Example 36.6. We prove the following version that is originally due to Jacob Bernoulli.

Theorem 36.3 Law of Large Numbers *Assume that event E occurs with probability x and let m denote the number of times E occurs in n trials. Let $\epsilon > 0$ and $\delta > 0$ be given. The probability that m/n differs from x by less than δ is greater than $1 - \epsilon$, i.e.,*

$$P\left(\left|\frac{m}{n} - x\right| < \delta\right) > 1 - \epsilon, \tag{36.13}$$

for all n sufficiently large.

Note that we can choose $\epsilon > 0$ and $\delta > 0$ arbitrarily small at the cost of making n possibly very large, hence the name of the theorem. Also note that while this result says that it is likely that event E will occur approximately xn times in n trials, *it does not say that event E will occur exactly xn times in n trials nor does it say that event E must occur approximately xn times in n trials.* Thus, this result does not contradict the computations in Example 36.6.

Phrased in terms of the binomial polynomials, we want to show that given $\epsilon, \delta > 0$,

$$\sum_{\substack{0 \le m \le n \\ \left|\frac{m}{n} - x\right| < \delta}} p_{n,m}(x) > 1 - \epsilon \tag{36.14}$$

for n sufficiently large.

Consider the complementary sum

$$\sum_{\substack{0 \le m \le n \\ \left|\frac{m}{n} - x\right| \ge \delta}} p_{n,m}(x) = 1 - \sum_{\substack{0 \le m \le n \\ \left|\frac{m}{n} - x\right| < \delta}} p_{n,m}(x),$$

which we estimate simply as

$$\sum_{\substack{0 \le m \le n \\ \left|\frac{m}{n} - x\right| \ge \delta}} p_{n,m}(x) \le \frac{1}{\delta^2} \sum_{\substack{0 \le m \le n \\ \left|\frac{m}{n} - x\right| \ge \delta}} \left(\frac{m}{n} - x\right)^2 p_{n,m}(x) \le \frac{1}{n^2\delta^2} S_n$$

where

$$S_n = \sum_{m=0}^{n} (m - nx)^2 p_{n,m}(x)$$

$$= \sum_{m=0}^{n} m^2 p_{n,m}(x) - 2nx \sum_{m=0}^{n} m p_{n,m}(x) + n^2 x^2 \sum_{m=0}^{n} p_{n,m}(x). \tag{36.15}$$

Using (36.10), (36.11), and (36.12), we find S_n simplifies (Problem 36.9) to $S_n = nx(1-x)$. Since $x(1-x) \le 1/4$ for $0 \le x \le 1$, $S_n \le n/4$. Therefore,

$$\sum_{\substack{0 \le m \le n \\ \left|\frac{m}{n} - x\right| \ge \delta}} p_{n,m}(x) \le \frac{1}{4n\delta^2} \tag{36.16}$$

and

$$\sum_{\substack{0\leq m\leq n \\ \left|\frac{m}{n}-x\right|<\delta}} p_{n,m}(x) \geq 1 - \frac{1}{4n\delta^2}.$$

In particular, for fixed ϵ, $\delta > 0$, we can insure that $(4n\delta^2)^{-1} < \epsilon$ by choosing $n > 1/(4\delta^2\epsilon)$.

36.3   The Modulus of Continuity

In order to prove a strong version of Theorem 36.1, we introduce a useful generalization of Lipschitz continuity.

First note that by Theorem 32.11, the continuous function f on $[a, b]$ in Theorem 36.1 is actually uniformly continuous on $[a, b]$. That is given $\epsilon > 0$ there is a $\delta > 0$ such that $|f(x) - f(y)| < \epsilon$ for all x, y in $[a, b]$ with $|x-y| < \delta$.[2] Now a Lipschitz continuous function f with constant L is uniformly continuous because $|f(x) - f(y)| \leq L|x - y| < \epsilon$ for all x, y with $|x - y| < \delta = \epsilon/L$. On the other hand, uniformly continuous functions are not necessarily Lipschitz continuous. They do, however, satisfy a generalization of the condition that defines Lipschitz continuity called the modulus of continuity.

The generalization is based on the observation that if f is uniformly continuous on a closed, bounded interval $I = [a, b]$, then for any $\delta > 0$, the set of numbers

$$\{|f(x) - f(y)| \text{ with } x, y \text{ in } I, |x - y| < \delta\} \tag{36.17}$$

is bounded. Otherwise, f could not be uniformly continuous (Problem 36.10). But, Theorem 32.15 then implies that the set of numbers (36.17) has a least upper bound. Turning this around, we define the **modulus of continuity** $\omega(f, \delta)$ of a general function f on a general interval I by

$$\omega(f, \delta) = \sup_{\substack{x,y \text{ in } I \\ |x-y|<\delta}} \{|f(x) - f(y)|\}.$$

Note that $\omega(f, \delta) = \infty$ if the set (36.17) is not bounded. We can guarantee that $\omega(f, \delta)$ is finite if f is uniformly continuous and I is a closed interval, but if f is not uniformly continuous and/or I is open or unbounded, then $\omega(f, \delta)$ might be infinite.

EXAMPLE 36.7.   We know x^2 is uniformly continuous on $[0, 1]$. Now consider the difference $|x^2 - y^2| = |x - y| |x + y|$, where $|x - y| < \delta$.

---

[2]Uniformity refers to the fact that δ can be chosen independently of x and y.

The values of $|x - y|$ increases monotonically from 0 to δ, while the corresponding largest values of $|x + y|$ decrease monotonically from 2 to $2 - \delta$. The largest value of their product occurs when $|x - y| = \delta$ so that $\omega(x^2, \delta) = 2\delta - \delta^2$.

EXAMPLE 36.8. $\omega(x^{-1}, \delta)$ on $(0, 1)$ is infinite.

EXAMPLE 36.9. $\omega(\sin(x^{-1}), \delta) = 2$ on $(0, 1)$ since for any $\delta > 0$ we can find x and y within δ of 0, and hence within δ of each other, such that $\sin(x^{-1}) = 1$ and $\sin(y^{-1}) = -1$.

Note that the functions in Example 36.8 and Example 36.9 are not uniformly continuous on the indicated intervals. In fact, if f is uniformly continuous on $[a, b]$, then $\omega(f, \delta) \to 0$ as $\delta \to 0$ (Problem 36.14).

If f is Lipschitz continuous on $[a, b]$ with constant L, then $\omega(f, \delta) \le L\delta$. In this sense, the modulus of continuity is a generalization of the idea of Lipschitz continuity.

36.4   The Bernstein Polynomials

To construct the approximating polynomial, we partition $[0, 1]$ by a uniform mesh with $n + 1$ nodes

$$x_m = \frac{m}{n}, \quad m = 0, \cdots, n.$$

The **Bernstein polynomial** of degree n for f on $[0, 1]$ is

$$B_n(f, x) = B_n(x) = \sum_{m=0}^{n} f(x_m) p_{n,m}(x). \tag{36.18}$$

Note that the degree of B_n is at most n.

The reason that the Bernstein polynomials become increasingly accurate approximations as the degree n increases is rather intuitive. The formula for $B_n(x)$ decomposes into two sums,

$$B_n(x) = \sum_{x_m \approx x} f(x_m) p_{n,m}(x) + \sum_{|x_m - x| \text{ large}} f(x_m) p_{n,m}(x).$$

The first sum converges to $f(x)$ as n becomes large, since we can find nodes $x_m = m/n$ arbitrarily close to x by taking n large.[3] The second sum converges to zero by the Law of Large Numbers. This is exactly what we prove below.

Before stating a convergence result, we consider a couple of examples.

---

[3] Recall that any real number can be approximated arbitrarily well by rational numbers.

EXAMPLE 36.10. The Bernstein polynomial B_n for x^2 on $[0,1]$ with $n \geq 2$ is given by

$$B_n(x) = \sum_{m=0}^{n} \left(\frac{m}{n}\right)^2 p_{n,m}(x).$$

By (36.12), this means

$$B_n(x) = \left(1 - \frac{1}{n}\right) x^2 + \frac{1}{n} x = x^2 + \frac{1}{n} x(1-x).$$

We see that $B_n(x^2, x) \neq x^2$ and in fact the error

$$|x^2 - B_n(x)| = \frac{1}{n} x(1-x)$$

decreases like $1/n$ as n increases.

EXAMPLE 36.11. We compute B_1, B_2, and B_3 for $f(x) = e^x$ on $[0,1]$,

$$B_1(x) = e^0(1-x) + e^1 x = (1-x) + ex$$
$$B_2(x) = (1-x)^2 + 2e^{1/2}x(1-x) + ex^2$$
$$B_3(x) = (1-x)^3 + 3e^{1/2}x(1-x)^2 + 3e^{2/3}x^2(1-x) + ex^3.$$

We plot these functions in Fig. 36.1.

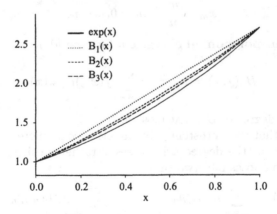

FIGURE 36.1. The first three Bernstein polynomials for e^x.

We prove:

Theorem 36.4 Bernstein Approximation Theorem *Let f be a continuous function on $[0,1]$ and $n \geq 1$ a natural number. Then*

$$|f(x) - B_n(f,x)| \leq \frac{9}{4}\omega(f, n^{-1/2}). \tag{36.19}$$

If f is Lipschitz continuous with constant L, then

$$|f(x) - B_n(f, x)| \leq \frac{9}{4} L n^{-1/2}. \tag{36.20}$$

Theorem 36.1 follows immediately since for $\epsilon > 0$, we simply choose n sufficiently large so that

$$|f(x) - B_n(f, x)| \leq \frac{9}{4} \omega(f, n^{-1/2}) < \epsilon.$$

Using (36.10), we write the error as a sum involving the differences between $f(x)$ and the values of f at the nodes:

$$f(x) - B_n(x) = \sum_{m=0}^{n} f(x) p_{n,m}(x) - \sum_{m=0}^{n} f(x_m) p_{n,m}(x)$$

$$= \sum_{m=0}^{n} (f(x) - f(x_m)) p_{n,m}(x)$$

We expect that the differences $f(x) - f(x_m)$ should be small when x is close to x_m by the continuity of f. To take advantage of this, for $\delta > 0$, we split the sum into two parts

$$f(x) - B_n(x) = \sum_{\substack{0 \leq m \leq n \\ |x - x_m| < \delta}} (f(x) - f(x_m)) p_{n,m}(x)$$

$$+ \sum_{\substack{0 \leq m \leq n \\ |x - x_m| \geq \delta}} (f(x) - f(x_m)) p_{n,m}(x). \tag{36.21}$$

The first sum is small by the continuity of f, since

$$\left| \sum_{\substack{0 \leq m \leq n \\ |x - x_m| < \delta}} (f(x) - f(x_m)) p_{n,m}(x) \right| \leq \sum_{\substack{0 \leq m \leq n \\ |x - x_m| < \delta}} |f(x) - f(x_m)| p_{n,m}(x)$$

$$\leq \omega(f, \delta) \sum_{\substack{0 \leq m \leq n \\ |x - x_m| < \delta}} p_{n,m}(x)$$

$$\leq \omega(f, \delta) \sum_{m=0}^{n} p_{n,m}(x) = \omega(f, \delta).$$

We can get a crude bound on the second sum in (36.21) easily. Since f is continuous on $[0, 1]$ there is a constant C such that $|f(x)| \leq C$ for $0 \leq x \leq 1$. Therefore,

$$\sum_{\substack{0 \leq m \leq n \\ |x - x_m| \geq \delta}} (f(x) - f(x_m)) p_{n,m}(x) \leq 2C \sum_{\substack{0 \leq m \leq n \\ |x - x_m| \geq \delta}} p_{n,m}(x) \leq \frac{C}{n \delta^2}$$

by (36.16). So we can make the second sum as small as desired by taking n large.

To get a sharper estimate on the second sum in (36.21), we use a trick similar to that used to prove Theorem 19.1. We let M be the largest integer less than or equal to $|x - x_m|/\delta$ and choose M uniformly spaced points $y_1, y_2, \cdots, y_M$ in the interval spanned by x and x_m so that each of the resulting $M + 1$ intervals have length $|x - x_m|/(M + 1) < \delta$.

Now, we can write

$$f(x) - f(x_m) = (f(x) - f(y_1)) + (f(y_1) - f(y_2)) + \cdots$$
$$+ (f(y_M) - f(x_m)).$$

Therefore,

$$|f(x) - f(x_m)| \leq (M + 1)\omega(f, \delta) \leq \left(1 + \frac{|x - x_m|}{\delta}\right)\omega(f, \delta).$$

We use this to estimate the second sum in (36.21),

$$\left|\sum_{\substack{0 \leq m \leq n \\ |x - x_m| \geq \delta}} (f(x) - f(x_m))p_{n,m}(x)\right|$$

$$\leq \omega(f, \delta)\left(\sum_{\substack{0 \leq m \leq n \\ |x - x_m| \geq \delta}} p_{n,m}(x) + \frac{1}{\delta}\sum_{\substack{0 \leq m \leq n \\ |x - x_m| \geq \delta}} |x - x_m|p_{n,m}(x)\right).$$

Using the fact that $|x - x_m|/\delta = M \geq 1$,

$$\left|\sum_{\substack{0 \leq m \leq n \\ |x - x_m| \geq \delta}} (f(x) - f(x_m))p_{n,m}(x)\right|$$

$$\leq \omega(f, \delta)\left(\sum_{\substack{0 \leq m \leq n \\ |x - x_m| \geq \delta}} p_{n,m}(x) + \frac{1}{\delta^2}\sum_{\substack{0 \leq m \leq n \\ |x - x_m| \geq \delta}} (x - x_m)^2 p_{n,m}(x)\right)$$

$$\leq \omega(f, \delta)\left(\sum_{m=0}^{n} p_{n,m}(x) + \frac{1}{\delta^2}\sum_{m=0}^{n} (x - x_m)^2 p_{n,m}(x)\right)$$

$$\leq \omega(f, \delta)\left(1 + \frac{1}{4n\delta^2}\right)$$

by (36.11) and (36.12). So

$$\left|\sum_{\substack{0 \leq m \leq n \\ |x - x_m| \geq \delta}} (f(x) - f(x_m))p_{n,m}(x)\right| \leq \omega(f, \delta)\left(1 + \frac{1}{4n\delta^2}\right).$$

Putting the estimates on the sums back into (36.21),

$$|f(x) - B_n(x)| \leq \omega(f, \delta) \left(2 + \frac{1}{4n\delta^2}\right).$$

Setting $\delta = n^{-1/2}$ proves the theorem.

36.5   Accuracy and Convergence

We can interpret Theorem 36.4 as saying that the Bernstein polynomials $\{B_n(f, x)\}$ converge uniformly to $f(x)$ on $[0, 1]$ as $n \to \infty$. In other words, the errors of the Bernstein polynomials B_n for a given function f on $[0, 1]$ tend to zero as n increases. This is a strong property; unfortunately, the price is that the convergence is very slow in general.

EXAMPLE 36.12.   To demonstrate how slowly the Bernstein polynomials can converge, we plot the Bernstein polynomial of degree 4 for $\sin(\pi x)$ on $[0, 1]$ in Fig. 36.2.

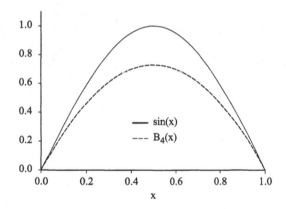

FIGURE 36.2. A plot of the Bernstein polynomial $B_4(x)$ for $\sin(\pi x)$.

If the error bound in (36.19) is accurate, i.e.,

$$|f(x) - B_n(x)| \approx \frac{9}{4}\omega(f, n^{-1/2}) \approx Cn^{-1/2} \text{ for some constant } C,$$

then we have to increase n by a factor of 100 in order to see an improvement of 10 (one additional digit of accuracy) in the error. This follows because from the computation

$$\frac{|f(x) - B_{n_1}(x)|}{|f(x) - B_{n_2}(x)|} \approx \frac{n_1^{-1/2}}{n_2^{-1/2}} = 10^{-1}$$

we need $n_2 = 100n_1$.

The error can decrease more quickly in some cases. Above, we saw that the error for x^2 decreases like $1/n$. But even this is relatively slow compared to some other polynomial approximations and for this reason the Bernstein polynomials are not often encountered in practice.

36.6   Unanswered Questions

We have shown that continuous functions can be approximated by polynomials. But we have not really explained why polynomials are well-suited for approximating functions. In other words, what are the properties of polynomials that make them good approximations? Are there other sets of functions that have similar approximation properties? Atkinson [2], Isaacson and Keller [15], and Rudin [19] have interesting material on these topics.

Chapter 36   Problems

36.1. Evaluate $\binom{8}{3}$.

36.2. Explain the claim that $\binom{n}{m}$ gives the number of ways that n objects can be arranged in groups of m.

36.3. Prove (36.2).

36.4. Expand $(a+b)^6$.

36.5. Prove (36.5).

36.6. Prove (36.8).

36.7. Verify (36.12).

36.8. Determine a formula for the probability of getting exactly $n/2$ heads when tossing a fair coin n times, where n is even. Make a plot of the formula for a n in the range of 1 to 100 and test the claim that it approaches $\sqrt{\pi n}$ for n large.

36.9. Prove that S_n defined in (36.15) is equal to $S_n = nx(1-x)$.

Problems 36.10–36.15 have to do with the modulus of continuity. Several of the proofs in this book could be generalized by using the modulus of continuity instead of Lipschitz continuity.

36.10. Prove that if f is uniformly continuous on $[a, b]$, then for any $\delta > 0$ the set of numbers (36.17) is bounded.

36.11. Evaluate

(a) $\omega(x^2, \delta)$ on $[0, 2]$ (b) $\omega(1/x, \delta)$ on $[1, 2]$ (b) $\omega(\log(x), \delta)$ on $[1, 2]$.

36.12. Verify Example 36.8.

36.13. Verify Example 36.9.

36.14. Prove that if f is uniformly continuous on $[a, b]$, then $\omega(f, \delta) \to 0$ as $\delta \to 0$.

36.15. Prove that if f has a continuous derivative on $[a, b]$, then $\omega(f, \delta) \le \max_{[a,b]} |f'|\delta$.

Computing Bernstein polynomial approximations can be tedious. You might want to use MAPLE © , for example, to do Problems 36.16–36.21.

36.16. Compute formulas for $p_{3,m}$, $m = 0, 1, 2, 3$.

36.17. Verify the computations in Example 36.11.

36.18. Compute the Bernstein polynomials for x on $[0, 1]$.

36.19. Compute and plot the Bernstein polynomials for $\exp(x)$ on $[1, 3]$ of degree 1, 2, and 3.

36.20. (a) Compute a summation formula for the Bernstein polynomial for x^3 on $[0, 1]$ for degree ≥ 3. (b) Find an explicit formula for the Bernstein polynomial from (a) that does not involve summation. (c) Write down a formula for the error.

36.21. Compute and plot the Bernstein polynomials for $\sin(\pi x)$ on $[0, 1]$ of degree 1, 2, 3, and 4.

We have shown that the Bernstein polynomials approximate a differentiable function, which is continuous of course, uniformly well. In Problem 36.22, we ask you to show that the derivative of the function is also approximated by the derivatives of the function's Bernstein polynomials.

36.22. If $f(x)$ has a continuous first derivative in $[0, 1]$, prove that the derivatives of the Bernstein polynomials $\{P_n'(f, x)\}$ converge uniformly to $f'(x)$ on $[0, 1]$.

Hint: First, verify the formulas

$$p_{n,m}' = n(p_{n-1,m-1} - p_{n-1,m}) \text{ for } m = 1, \cdots, m - 1$$

$$p_{n,n}' = np_{n-1,n-1}, \quad p_{n,0}' = -np_{n-1,0}.$$

Then find a summation formula for the error $f'(x) - P_n'(x)$ and rearrange the sum in terms of $p_{n-1,m}$ for $m = 0, 1, \cdots, n - 1$.

36.23. If f is continuous on $[0, 1]$ and if

$$\int_0^1 f(x)x^n \, dx = 0 \text{ for } n = 0, 1, 2, \cdots,$$

, then prove that $f(x) = 0$ for $0 \leq x \leq 1$. *Hint:* This says that the integral of the product of f and *any* polynomial is zero. Use Theorem 36.1 to first prove that

$$\int_0^1 f^2(x) \, dx = 0.$$

*We say that the real numbers $\mathbb{R}$ are **separable** because any real number can be approximated to arbitrary accuracy by a rational number. The analogous property holds for the space of continuous functions on a closed, bounded interval, which is the content of the theorem we ask you to prove in Problem 36.24.*

36.24. Prove the following extension of the Weierstrass Approximation Theorem:

Theorem 36.5 *Assume that f is continuous on a closed bounded interval I. Given any $\epsilon > 0$, there is a polynomial P_n with rational coefficients with finite decimal expansions and of sufficiently high degree n such that*

$$|f(x) - P_n(x)| < \epsilon \text{ for } a \leq x \leq b.$$

Hint: Use Theorem 36.1 to first get an approximate polynomial and then analyze the effect of replacing its coefficients by rational approximations.

37
The Taylor Polynomial

In Chapter 36, we saw that a continuous function can be approximated arbitrarily well by polynomials. Unfortunately, the Bernstein polynomials used to prove this result converge slowly. Consequently, obtaining even a moderately accurate approximation of a function can require a high degree polynomial and many function values.

This motivates the search for other methods for computing polynomial approximations of a given function. We begin this search by generalizing the idea of the linearization of a function at a point introduced in Chapter 16. Recall that the linearization of a function at a point is a linear polynomial that has the same value of the function at a point and whose graph is tangent to the graph of the function. We show how to generalize this idea to find a polynomial approximation of a function of arbitrary degree given that the function is sufficiently smooth.

37.1 A Quadratic Approximation

As motivation, we start by extending the idea of linearization to compute a quadratic approximation of a function. Namely, we look for a quadratic polynomial of a function f near $\bar{x}$ with the property that its error is cubic in $|x - \bar{x}|$. Note for $|x - \bar{x}|$ small, the error of the quadratic approximation is smaller than the error of the linear approximation.

Mathematically, we look for an approximation satisfying

$$\left| f(x) - \left(f(\bar{x}) + m_1(x - \bar{x}) + m_2(x - \bar{x})^2 \right) \right| \leq |x - \bar{x}|^3 \mathcal{K}_{\bar{x}}, \qquad (37.1)$$

for x close to $\bar{x}$, where m_1, m_2, and $\mathcal{K}_{\bar{x}}$ are constants . For $|x - \bar{x}|$ small, $|x - \bar{x}|^3 |\mathcal{K}_{\bar{x}}|$ is much smaller than either $m_1 |x - \bar{x}|$ or $m_2 |x - \bar{x}|^2$.

EXAMPLE 37.1. We compute the quadratic approximation to $1/x$ at $\bar{x} = 1$. We look for m_1, m_2, and $\mathcal{K}_1$ so that

$$\left| \frac{1}{x} - \left(1 + m_1(x - 1) + m_2(x - 1)^2 \right) \right| \le |x - 1|^3 \mathcal{K}_1$$

for x near 1. We expect to have to keep x away from 0. We simplify the expression on the left and pull out a common factor to get

$$\left| \frac{1}{x} - \left(1 + m_1(x - 1) + m_2(x - 1)^2 \right) \right| = |x - 1| \left| \frac{1}{x} + m_1 + m_2(x - 1)) \right|.$$

So $\lim_{x \to 1} \left| \frac{1}{x} - \left(1 + m_1(x - 1) + m_2(x - 1)^2 \right) \right| = 0$ for any m_1 and m_2. Extending the ideas behind (16.6) and (16.7), we also want

$$\lim_{x \to 1} \frac{\left| \frac{1}{x} - \left(1 + m_1(x - 1) + m_2(x - 1)^2 \right) \right|}{|x - 1|}$$

$$= \lim_{x \to 1} \left| \frac{1}{x} + m_1 + m_2(x - 1) \right| = 0.$$

This forces $m_1 = -1$, which is the derivative of $1/x$ at 1. Next we want

$$\lim_{x \to 1} \frac{\left| \frac{1}{x} - \left(1 - (x - 1) + m_2(x - 1)^2 \right) \right|}{|x - 1|^2} = \lim_{x \to 1} \left| \frac{1}{x} - m_2(x - 1) \right| = 0,$$

forcing $m_2 = 1$. Using the same kinds of computations, we estimate the error as

$$\left| \frac{1}{x} - \left(1 + m_1(x - 1) + m_2(x - 1)^2 \right) \right| \le |x - 1|^3 \frac{1}{x}. \tag{37.2}$$

We can bound $1/x$ by some $\mathcal{K}_1$ provided x is restricted to an interval I_1 containing 1 and bounded away from zero.

We conclude that the quadratic approximation to $1/x$ at 1 is

$$\frac{1}{x} \approx 1 - (x - 1) + (x - 1)^2$$

for all $x \ne 0$. We plot the approximation in Fig. 37.1.

37.2   Taylor's Representation of a Polynomial

While the idea behind linearization can be extended to compute a polynomial approximation of arbitrary degree as above, the procedure is clumsy.[1]

---

[1] And that is being polite about it.

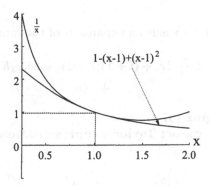

FIGURE 37.1. The quadratic approximation $1-(x-1)+(x-1)^2$ of $1/x$ at $\bar{x}=1$. Compare to the linear approximation shown in Fig. 16.11.

We are after a formula for the approximation that is "easy" to compute as well as a useful expression for the error of the approximation.

We begin by considering the case of a polynomial of degree n,

$$p(x) = c_0 + c_1 x + \cdots + c_n x^n,$$

where $c_0, c_1, \cdots, c_n$ are some numbers. In the spirit of linearization near a point a, we want to rewrite $p(x)$ to better reflect its behavior for values of x near a. We suppose that $x = a + h$ where h is small, and substitute to find

$$p(a+h) = c_0 + c_1(a+h) + \cdots + c_n(a+h)^n.$$

Now expanding all the powers and collecting terms, we see there are coefficients $\bar{c}_0, \cdots, \bar{c}_n$ depending on $c_0, \cdots, c_n$ and a such that

$$p(a+h) = \bar{c}_0 + \bar{c}_1 h + \cdots + \bar{c}_n h^n. \tag{37.3}$$

The natural question is what are the values of $\{\bar{c}_i\}$? Taylor's representation of a polynomial is the realization that

$$\bar{c}_i = \frac{1}{i!} p^{(i)}(a) \text{ for } 0 \le i \le nm, \tag{37.4}$$

where $p^{(i)} = d^i p/dx^i$. For $i = 0$, substituting $h = 0$ into (37.3) gives

$$p(a) = \bar{c}_0.$$

The proof of (37.4) in the general case is based on the observation that if two functions satisfy $f(x) = g(x)$ for all x in an open interval, then all of their derivatives are equal in the interval. Differentiating both sides of (37.3) with respect to h, the Chain Rule implies

$$p'(a+h) = \bar{c}_1 + 2\bar{c}_2 h + 3\bar{c}_3 h^2 + \cdots + n\bar{c}_n h^{n-1}.$$

At $h = 0$, $p'(a) = \bar{c}_1$.

Differentiating i times yields an expansion of the form

$$p^{(i)}(a + h) = i(i - 1) \cdots 1\bar{c}_i + (i + 1)(i - 1) \cdots 2\bar{c}_{i+1}h + \cdots$$
$$+ n(n - 1) \cdots (n - i + 1)\bar{c}_n h^{n-i+1}.$$

Substituting $h = 0$ gives (37.4).

For later use, we write out **Taylor's representation of a polynomial** in compact form

$$p(a + h) = p(a) + \frac{p'(a)}{1!}h + \frac{p^{(2)}(a)}{2!}h^2 + \cdots + \frac{p^{(n)}(a)}{n!}h^n. \qquad (37.5)$$

EXAMPLE 37.2. We compute the Taylor representation of $p(x) = x - 2x^3$ near $a = 1$. We have

$$p(1) = -1, \; p'(1) = 1 - 6 \times 1^2 = -5, \; p^{(2)}(1) = -12 \times 1, \; p^{(3)} = -12,$$

so

$$p(1 + h) = -1 - 5h - 12h^2 - 12h^3.$$

37.3   The Taylor Polynomial for a General Function

The formula (37.5) can be applied to an arbitrary function that has a sufficient number of derivatives. Of course if the function is not a polynomial, we cannot expect to get equality in the expansion. Given a function f that has n continuous derivatives in an open interval containing the point a, the **Taylor polynomial** of degree n at a of f is

$$T_n(a, x) = T_n(f, a, x) = f(a) + f'(a)(x - a) + \frac{f^{(2)}(a)}{2!}(x - a)^2 + \cdots$$
$$+ \frac{f^{(n)}(a)}{n!}(x - a)^n. \quad (37.6)$$

Note that this is (37.5) with $h = x - a$ usually considered to be small.

We define the **remainder of the Taylor polynomial** of f to be the error

$$R_n(a, x) = R_n(f, a, x) = f(x) - T_n(f, a, x), \qquad (37.7)$$

so

$$f(x) = T_n(f, a, x) + R_n(f, a, x).$$

EXAMPLE 37.3. We compute the Taylor polynomial of $f(x) = \log(x)$ of degree n at $a = 1$.

n	$f^{(n)}(x)$	$f^{(n)}(a)$
0	$\log(x)$	0
1	x^{-1}	1
2	$-x^{-2}$	-1
3	$2x^{-3}$	2
4	$-6x^{-4}$	-6
$\vdots$	$\vdots$	$\vdots$
n	$(-1)^{n+1}(n-1)!x^{-n}$	$(-1)^{n+1}(n-1)!$

So

$$T_n(\log(x), 1, x) = 0 + (x-1) - \frac{(x-1)^2}{2} + \frac{(x-1)^3}{3}$$
$$- \frac{(x-1)^4}{4} + \cdots + \frac{(-1)^{n+1}(x-1)^n}{n} \quad (37.8)$$

or

$$T_n(\log(x), 1, x) = \sum_{i=1}^{n} \frac{(-1)^{i+1}(x-1)^i}{i}. \quad (37.9)$$

EXAMPLE 37.4. We compute the Taylor polynomial of $f(x) = e^x$ of degree n at $a = 0$.

n	$f^{(n)}(x)$	$f^{(n)}(a)$
0	e^x	1
1	e^x	1
2	e^x	1
3	e^x	1
$\vdots$	$\vdots$	$\vdots$
n	e^x	1

So

$$T_n(e^x, 0, x) = 1 + x + \frac{2^2}{2!} + \frac{x^3}{3!} + \cdots + \frac{x^n}{n!}. \quad (37.10)$$

or

$$T_n(e^x, 0, x) = \sum_{i=0}^{n} \frac{x^i}{i!}. \quad (37.11)$$

EXAMPLE 37.5. We compute the Taylor polynomial of $f(x) = \sin(x)$ of degree $2n - 1$ for n even and $2n + 1$ for n odd at $a = 0$.

n	$f^{(n)}(x)$	$f^{(n)}(a)$
0	$\sin(x)$	0
1	$\cos(x)$	1
2	$-\sin(x)$	0
3	$-\cos(x)$	-1
4	$\sin(x)$	0
5	$\cos(x)$	1
$\vdots$	$\vdots$	$\vdots$
n even	$(-1)^{n/2}\sin(x)$	0
n odd	$(-1)^{(n-1)/2}\cos(x)$	$(-1)^{(n-1)/2}$

So

$$T_n(\sin(x), 0, x) = 0 + x + 0 - \frac{x^3}{3!} + 0 + \frac{x^5}{5!} + \cdots \qquad (37.12)$$

or

$$T_n(\sin(x), 0, x) = \sum_{i=0}^{\tilde{n}} \frac{(-1)^i x^{2i+1}}{(2i+1)!}, \quad \tilde{n} = \begin{cases} n-1, & n \text{ even,} \\ n, & n \text{ odd.} \end{cases} \qquad (37.13)$$

37.4   The Error of the Taylor Polynomial

Since the Taylor polynomial of an arbitrary function is unlikely to be exact, it is important to analyze the remainder $R_n(f, a, x)$ so as to understand when it is small. We do this by deriving a couple of exact formulas for $R_n(f, a, x)$ under the assumption that f has $n+1$ continuous derivatives, i.e., at least one more than the degree of the Taylor polynomial we are using.

First, we consider the error of the linearization of a function. Recall that if f is strongly differentiable at a, then there is an open interval I_a containing a and a constant $\mathcal{K}_a$ such that

$$\left| f(x) - \left(f(a) + f'(a)(x-a)\right) \right| \leq (x-a)^2 \mathcal{K}_a \quad \text{for all } x \text{ in } I_a. \qquad (37.14)$$

We show that if f has two continuous derivatives in an open interval containing a and x, then (37.14) holds automatically.

In this case,

$$R_1(a, x) = f(x) - \left(f(a) + f'(a)(x-a)\right).$$

Now we fix x and consider a as the variable. Since $f(a)$ and $f'(a)$ are differentiable, $R_1(a, x)$ is differentiable with respect to a, and using the Product Rule,

$$R_1'(a, x) = \frac{d}{da} R_1(a, x) = f'(a) - f'(a) + f''(a)(x-a),$$

or

$$R_1'(a, x) = f''(a)(x - a).$$

At $a = x$, the remainder $R_1(x, x) = 0$. Hence,

$$R_1(a, x) = R_1(a, x) - R_1(x, x) = \int_x^a R_1'(s, x)\, ds = -\int_a^x R_1'(s, x)\, ds,$$

and so

$$R_1(a, x) = \int_a^x (x - s) f''(s)\, ds. \tag{37.15}$$

This is the **integral form of the remainder of Taylor's polyomial**.

There is another formula for the remainder that is very useful in practice. We can derive it from (37.15) using the following Mean Value Theorem for integration, which generalizes Problem 32.30.

Theorem 37.1 Generalized Integral Mean Value Theorem *Suppose that $f(x)$ and $\omega(x)$ are continuous on $[a, b]$ and moreover $\omega(x) \geq 0$ for $a \leq x \leq b$ and $\int_a^b \omega(x)\, dx > 0$. Then there is a point c in $[a, b]$ such that*

$$\int_a^b f(x)\omega(x)\, dx = f(c) \int_a^b \omega(x)\, dx. \tag{37.16}$$

Recall that

$$\bar{f} = \frac{\int_a^b f(x)\omega(x)\, dx}{\int_a^b \omega(x)\, dx}$$

is the weighted average of f with respect to ω over $[a, b]$. So Theorem 37.1 says that f takes on its weighted average $\bar{f}$ at some point c provided the weight function is positive.

The theorem follows immediately from the Intermediate Value Theorem 32.5 if $\bar{f}$ is between the minimum m and maximum M values of f on $[a, b]$. Recall that f attains its minimum and maximum values on $[a, b]$ because it is continuous. The Intermediate Value Theorem says that f takes on all the values, including $\bar{f}$, between m and M at least once in $[a, b]$.

If we multiply the inequality

$$m \leq f(x) \leq M$$

by the *nonnegative* number $\omega(x)$, we get

$$m\omega(x) \leq f(x)\omega(x) \leq M\omega(x) \text{ for all } a \leq x \leq b.$$

Integration gives

$$m \int_a^b \omega(x)\, dx \leq \int_a^b f(x)\omega(x)\, dx \leq M \int_a^b \omega(x)\, dx,$$

or by dividing,

$$m \leq \bar{f} \leq M.$$

Now consider (37.15) when $x > a$. This means that $x - s \geq 0$ for $a \leq s \leq x$ and we can apply Theorem 37.1 to assert that there is a point $a \leq c \leq b$ such that

$$R_1(a, x) = f''(c) \int_a^b (x - s)\, ds = \frac{f''(c)}{2}(x - a)^2. \qquad (37.17)$$

This is called the **Lagrange form of the remainder of the Taylor polynomial**. It is easy to show that (37.17) also holds when $x < a$.

Note that (37.17) implies (37.14) with $\mathcal{K}_a = |f''(c)|/2$.

EXAMPLE 37.6. We estimate the remainder for the linear Taylor polynomial for e^x at $a = 0$. By (37.17), there is a c between a and x such that

$$|e^x - (1 + x)| = |R_1(0, x)| = \frac{e^c}{2}x^2.$$

The fact that c is unknown is annoying, but we can replace e^c by an upper bound. For example if $x < 0$, then $e^c < e^0$ so $|R_1(0, x)| \leq x^2/2$, and if $0 < x < 1$, then $R_1(0, x) \leq ex^2/2$.

We can apply exactly the same argument to estimate $R_n(a, x)$ assuming that $f^{(n+1)}$ is continuous at a. We start with the definition

$$R_n(a, x) = f(x) - f(a) - f'(a)(x - a) - \frac{f^{(2)}(a)}{2!}(x - a)^2 - \cdots$$
$$- \frac{f^{(n)}(a)}{n!}(x - a)^n.$$

We fix x and consider a as the variable. By assumption, $R_n(a, x)$, like the right-hand side, has at least one continuous derivative with respect to a. Differentiating with respect to a,

$$R_n'(a, x) = -f'(a) + f'(a) - f''(a)(x - a) + f''(a)(x - a) + \cdots$$
$$- \frac{f^{(n)}(a)}{(n-1)!}(x - a)^{n-1} + \frac{f^{(n)}(a)}{(n-1)!}(x - a)^{n-1}$$
$$- \frac{f^{(n+1)}(a)}{n!}(x - a)^n$$
$$= -\frac{f^{(n+1)}(a)}{n!}(x - a)^n.$$

As before $R_n(x, x) = 0$, so

$$R_n(a, x) = -\int_a^x R_n'(s, x)\, ds = \int_a^x \frac{f^{(n+1)}(s)}{n!}(s - a)^n\, ds.$$

Also as before, we can apply Theorem 37.1 to conclude there is a point c between a and x such that

$$R_n(a, x) = f^{(n+1)}(c) = \int_a^x \frac{(s-a)^n}{n!} \, ds = \frac{f^{(n+1)}(c)}{(n+1)!}(x-a)^{n+1}.$$

We summarize these results as a theorem.

Theorem 37.2 Error Formulas for Taylor Polynomials *Suppose f has $n+1$ continuous derivatives in an open interval I containing a point a. Then for all x in I,*

$$f(x) = \sum_{i=0}^n \frac{f^{(i)}(a)}{i!}(x-a)^i + R_n(a, x),$$

where

$$R_n(a, x) = \int_a^x \frac{f^{(n+1)}(s)}{n!}(s-a)^n \, ds. \tag{37.18}$$

There is a point c between a and x such that

$$R_n(a, x) = \frac{f^{(n+1)}(c)}{(n+1)!}(x-a)^{n+1}. \tag{37.19}$$

EXAMPLE 37.7. We compute the remainder for the Taylor polynomial for $\log(x)$ around $a = 1$ computed in Example 37.3. Since $f^{(n+1)}(x) = (-1)^n n! x^{-n-1}$,

$$R_n(1, x) = (-1)^n \int_1^x s^{-n-1}(x-s)^n \, ds$$

or

$$R_n(1, x) = \frac{(-1)^n c^{-n-1}}{n+1}(x-1)^{n+1} \text{ for some } c \text{ between } 1 \text{ and } x.$$

EXAMPLE 37.8. We compute the remainder for the Taylor polynomial for e^x around $a = 0$ computed in Example 37.4. Since $f^{(n+1)}(x) = e^x$,

$$R_n(0, x) = \frac{1}{n!} \int_1^x e^s (x-s)^n \, ds$$

or

$$R_n(0, x) = \frac{e^c}{(n+1)!} x^{n+1} \text{ for some } c \text{ between } 0 \text{ and } x.$$

EXAMPLE 37.9. We compute the remainder for the Taylor polynomial for $\sin(x)$ around $a = 0$ computed in Example 37.5. When n is even, $f^{(n+1)}(x) = (-1)^{n/2}\cos(x)$ and

$$R_n(0, x) = \frac{(-1)^{n/2}}{n!} \int_1^x \cos(s)(x-s)^n \, ds$$

or

$$R_n(0, x) = \frac{(-1)^{n/2} \cos(c)}{(n+1)!} x^{n+1} \text{ for some } c \text{ between 0 and } x.$$

When n is odd, $f^{(n+1)}(x) = (-1)^{(n+1)/2} \sin(x)$ and

$$R_n(0, x) = \frac{(-1)^{(n+1)/2}}{n!} \int_1^x \sin(s)(x-s)^n \, ds$$

or

$$R_n(0, x) = \frac{(-1)^{(n+1)/2} \sin(c)}{(n+1)!} x^{n+1} \text{ for some } c \text{ between 0 and } x.$$

37.5   Another Point of View

The importance of Taylor's Theorem 37.2 for analysis cannot be overstated. It is therefore a good idea to understand the result on as many levels as possible. In this section, we give an alternate derivation of the Lagrange formula for the remainder that shows that Taylor's theorem is a generalization of the Mean Value Theorem.

Indeed when $n = 0$, (37.19) implies

$$f(x) = f(a) + f'(c)(x-a)$$

or

$$\frac{f(x) - f(a)}{x - a} = f'(c)$$

for some c between a and x. This is nothing more than the Mean Value Theorem 32.17.

For a fixed $x \neq a$, let r be the number defined by

$$f(x) = T_n(x) + r(x-a)^{n+1}.$$

We want to show that $r = f^{(n+1)}(c)/(n+1)!$ for some c between a and x.

We set

$$g(t) = f(t) - T_n(t) - r(t-a)^{n+1}$$

for t between a and x. Because the degree of T_n is n, it follows that

$$g^{(n+1)}(t) = f^{(n+1)}(t) - (n+1)!r$$

for all t between a and x. If there is a c between a and x such that $g^{(n+1)}(c) = 0$, then we are done.

By construction of the Taylor polynomial, $g(a) = 0$. By the choice of M, $g(x) = 0$ as well. Hence by the Mean Value Theorem 32.17, there is a

c_1 between a and x such that $g'(c_1) = 0$. Now $g'(a) = 0$, hence another application of the Mean Value Theorem shows there is a number c_2 between a and c_1 such that $g''(c_2) = 0$. Then we can repeat the argument to find c_3 in (a, c_2) with $g^{(3)}(c_3) = 0$. In fact, we can repeat this argument $n + 1$ times in fact to conclude there is a number $c = c_{n+1}$ between a and c_n, hence between a and x, such that $g^{(n+1)}(c) = 0$.

37.6   Accuracy and Convergence

A convergence result for the Taylor polynomial would say that the Taylor polynomials T_n for a given function f on a given interval $[a, b]$ become more accurate as n increases. This is certainly desirable because it takes more computational work to compute T_n. Consider the Lagrange form of the remainder (37.19),

$$\frac{|f^{(n+1)}(c)||x - a|^{n+1}}{(n+1)!}.$$

The denominator $(n+1)!$ increases very quickly with n, so that really helps out. If $|x - a| < 1$, then $|x - a|^{n+1}$ is also small for sufficiently large n. In fact, it decreases in size exponentially as n increases. On the other hand, if $|x - a| > 1$, then $|x - a|^{n+1}$ increases exponentially as n increases. We have to balance both of these factors against the size of $|f^{(n+1)}(c)|$. If f for example has the nice property that all of its derivatives are uniformly bounded, then taking more derivatives doesn't hurt anything. If, however, successive derivatives of f increase in size, then this has a negative impact on the size of the remainder.

EXAMPLE 37.10. We estimate the maximum size of the remainder for the Taylor polynomial for $\log(x)$ at $a = 1$ computed in Example 37.7. Taking absolute values,

$$|R_n| = \frac{1}{n+1}\left(\frac{|x - 1|}{c}\right)^{n+1}.$$

If $1 \le c \le x$, then $1 \ge c^{-1} \ge x^{-1}$, so

$$x - 1 \ge \frac{x - 1}{c} \ge \frac{x - 1}{x}$$

and

$$\left(1 - \frac{1}{x}\right)^{n+1} \le \left(\frac{x - 1}{c}\right)^{n+1} \le (x - 1)^{n+1}.$$

If $1 \le x < 2$, the remainder must decrease exponentially as n increases. If $x = 2$, the remainder decreases at least like $1/(n + 1)$. If $x > 2$, the remainder can increase exponentially as n increases.

In a similar way, we can show that if $.5 < x < 1$, the remainder must decreases exponentially as n increases. If $x = .5$, the remainder must decrease like $1/(n + 1)$. If $x < .5$, the remainder can increase exponentially.

EXAMPLE 37.11. We estimate the remainder of the Taylor polynomial for e^x at $a = 0$ computed in Example 37.8. Now

$$|R_n| \leq \frac{e^x}{(n + 1)!} |x|^{n+1}.$$

From this it follows that for any fixed x, $\lim_{n \to \infty} |R_n| = 0$. Given x, let $\tilde{n}$ be the largest integer smaller than x. Then

$$\frac{|x|^{n+1}}{(n + 1)!} = \frac{|x|}{1} \frac{|x|}{2} \frac{|x|}{3} \cdots \frac{|x|}{\tilde{n}} \times \frac{|x|}{\tilde{n} + 1} \cdots \frac{|x|}{n + 1}.$$

Now

$$\frac{|x|}{1} \frac{|x|}{2} \frac{|x|}{3} \cdots \frac{|x|}{\tilde{n}} \leq |x|^{\tilde{n}},$$

while

$$\frac{|x|}{i} \leq 1, \quad \tilde{n} + 1 \leq i \leq n.$$

Hence,

$$\frac{|x|^{n+1}}{(n + 1)!} \leq \frac{|x|^{\tilde{n}+1}}{n + 1}.$$

Since $\tilde{n}$ is fixed, this shows that $|R_n| \to 0$ as $n \to \infty$.

In general, determining the size of the remainder of the Taylor polynomial can be very complicated. As a rule, we can only expect $T_n(f, a, x)$ to be a good approximation of $f(x)$ if $f^{(n+1)}$ is continuous near a and $|x - a|$ is sufficiently small, and conversely, we can expect the error to get larger as $|x - a|$ increases.

EXAMPLE 37.12. The Taylor polynomial of degree $2n$ for $1/(1 + x^2)$ at $a = 0$ is

$$T_{2n}(x) = 1 - x^2 + x^4 - x^6 + \cdots \pm x^{2n}. \tag{37.20}$$

We plot a few of the Taylor polynomials in Fig. 37.2. All of the approximations are accurate for x near $a = 0$, but become inaccurate quickly as x moves away from 0. In fact, if we tried to plot these approximation on the interval $[0, 2]$, we would have to use a vertical scale in the hundreds because the Taylor polynomials become so large.

Note that if for some n, $f^{(n+1)}(x)$ fails to exist at $x = a$ or at some nearby point, then this can severely impact the accuracy of the Taylor polynomial T_n.

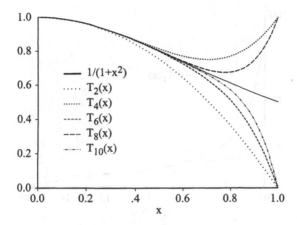

FIGURE 37.2. Some Taylor polynomials for $1/(1 + x^2)$ at $a = 0$.

EXAMPLE 37.13. We compute the Taylor polynomial T_1 of $f(x) = x^{-1}$ at $a = .1$. Here, of course, $f(x)$ and its derivatives are undefined at $x = 0$.

n	$f^{(n)}(x)$	$f^{(n)}(a)$
0	x^{-1}	10
1	$-x^{-2}$	-100
2	$2x^{-3}$	

So
$$T_1(x^{-1}, .1, x) = 10 - 100(x - .1)$$
with remainder
$$R_1(.1, x) = c^{-3}(x - .1)^3 \text{ for some } c \text{ between } x \text{ and } .1$$

If $0 < x < .1$, then the remainder can be very large! We can see this reflected in the plot of T_1 shown in Fig. 37.3. The error of T_1 increases very rapidly as $x < .1$ decreases.

So as far as a convergence result goes, if f has uniformly bounded derivatives of any order on a sufficiently small interval $[a, b]$, then the error of T_n on $[a, b]$ decreases to 0 as n increases. This is not a very satisfactory result. In particular, requiring f to have uniformly bounded derivatives of any order is very restrictive. In general, we cannot expect such a convergence result to hold.

We conclude with an observation on the cost of computing the Taylor polynomial of a function. In the case that f has uniformly bounded derivatives near a and $|x - a|$ is small, the accuracy of the Taylor polynomials of f at a is truly fantastic. The cost of this accuracy is a great deal of information about the function f at the point a, namely, the values of f and

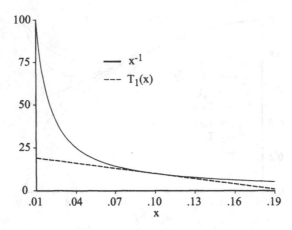

FIGURE 37.3. The Taylor polynomial T_1 for x^{-1} at $a = .1$.

its derivatives at a. Compare this to the Bernstein polynomial of degree $n + 1$, which requires only $n + 1$ values of f at $n + 1$ points. In other words, computing a Bernstein polynomial requires evaluating *one* function, while computing a Taylor polynomial requires evaluating *many* functions.

37.7   Unanswered Questions

In Examples 37.10 and 37.11, we saw that indeed the Taylor polynomials for those functions become more accurate as the degree increases. Yet for other functions, this is not true. This raises the natural question: what kind of functions have the property that their Taylor polynomials become increasingly accurate approximations as the degree increases? Another way to phrase this question is, what kind of functions have a convergent Taylor series, obtained as the limit of the sequence of Taylor polynomials as the degree increases?[2] It turns out that this question lies at the heart of the analysis of smooth functions and is the starting point for what is called the theory of analytic functions. These issues are discussed in detail in the subject of complex analysis, which could be described roughly as the calculus of complex-valued functions of complex-valued variables (see Ahlfors [1] for more details).

---

[2]We can think of Taylor polynomials as the partial sums of the Taylor series.

37.8   Some History of Taylor Polynomials

The history of Taylor series and polynomials is difficult to sort out. Taylor series for specific functions were known well before Leibniz and Newton and those two made heavy use of such expansions in deriving calculus. Taylor[3] is credited with writing down one of the earliest general formulations of Taylor series, perhaps because this appeared in a heavily influential textbook he wrote. Leibniz and Johann Bernoulli also derived the Taylor series independently around the same time, while Maclaurin[4] described a special case that is sometimes given his name. Lagrange gave the first formula for the remainder of the finite order Taylor polynomial and also first emphasized the significance of Taylor series and polynomials for analysis.

---

[3]Brook Taylor (1685–1731) was an English mathematician. He made fundamental discoveries in physics and astronomy, founded the theory of finite differences, and discovered integration by parts and Taylor series. Moreover, he wrote two influential textbooks. Unfortunately, the latter part of his life was marred by several personal tragedies.

[4]Colin Maclaurin (1698–1746) was a Scottish mathematician who was personally close to Newton. Maclaurin made fundamental contributions to calculus, geometry, and physics, as well as helping to establish the foundations of actuarial science. Maclaurin wrote the first general textbook describing Newton's results in calculus. He was also very concerned about the foundations of calculus and attempted to put Newton's results on a more rigorous foundation in his book.

Chapter 37 Problems

37.1. Verify (37.2).

37.2. Prove that $m_1 = f'(\bar{x})$ in the quadratic approximation (37.1).

37.3. Compute the quadratic approximation for $(x + 2)^2$ at $\bar{x} = 1$. *Hint:* Look for a lazy way to do this.

37.4. Compute Taylor's representation for the polynomial $p(x) = 3 + x - 2x^2 + 4x^4$ near $a = 1$ and $a = -2$.

You have to use some clever induction to find a formula for the general Taylor polynomial for a given function, as you discover in Problem 37.5.

37.5. Compute the Taylor polynomials of degree n of the following functions at the indicated points a:

    (a) $\log(x + 1)$, $a = 0$ (b) e^{2x}, $a = 0$ (c) e^x, $a = 1$

    (d) $\sin(2x)$, $a = 0$ (e) $\cos(x)$, $a = 0$ (f) $\cos(x)$, $a = \pi/2$.

The point of Problems 37.6–37.9 is that it is also possible to "cheat" and find Taylor polynomials for a complicated function by using the Taylor polynomials of simpler functions.

37.6. Compute the first three nonzero terms in the Taylor polynomial for $\sin^2(x)$ at $a = 0$ by squaring a Taylor polynomial for $\sin(x)$.

37.7. Compute the Taylor polynomial of degree n for $\sin^{-1}(x)$ at $a = 0$ by using the formula,

$$\sin^{-1}(x) = \int_0^x \frac{dt}{\sqrt{1 - t^2}}.$$

37.8. Compute the Taylor polynomial of degree n for

$$\int_0^x e^{-s^2}\, ds$$

at $a = 0$.

37.9. Derive (37.20) by using the formula for the Taylor polynomial and by using polynomial long division.

In Problems 37.10–37.12, we ask you to find bounds on the errors of Taylor polynomials.

37.10. Estimate the maximum size of the remainder of the Taylor polynomial of degree n for $\log(x)$ at $a = 1$ in the case that $0 < x < 1$ following the argument in Example 37.10.

37.11. Estimate the maximum size of the remainder of the Taylor polynomial of degree n for $\sin(x)$ at $a = 0$. *Hint:* See Example 37.11.

37.12. Estimate the sizes of the remainders for the Taylor polynomials computed in Problem 37.5.

37.13. *Uniqueness of a Taylor polynomial.* Suppose we have an expansion

$$f(x) = a_0 + a_1 x + \cdots + a_n x^n + R_n(x),$$

where $a_0, \cdots, a_n$ are constants, R_n is n times continuously differentiable, and $R_n(x)/x^n \to 0$ as $x \to 0$. Show that

$$a_k = \frac{f^{(k)}(0)}{k!}, \quad k = 0, \cdots, n.$$

In Problems 37.14–37.15, we ask you to find different ways to derive formulas for the remainder of the Taylor polynomial.

37.14. (a) Suppose $g(h)$ has continuous derivatives of order $n+1$ for $0 \le h \le H$. If $g(0) = g'(0) = \cdots = g^{(n)}(0) = 0$ while $|g^{(n+1)}(h)| \le M$ for $0 \le h \le H$, show

$$|g^{(n)}(h)| \le Mh, \ |g^{(n-1)}(h)| \le \frac{Mh^2}{2!}, \ \cdots, |g(h)| \le \frac{Mh^n}{n!}, \quad 0 \le h \le H.$$

(b) Assume f is smooth on $a \le x \le b$. Apply (a) to $g(h) = R_n(f, a, a + h) = f(a + h) - T_n(f, a, a + h)$ to obtain a rough estimate of the remainder of the Taylor polynomial for f.

37.15. Derive the integral formula for the remainder R_n by applying integration by parts repeatedly to

$$f(a + h) - f(a) = \int_0^h f'(x + s)\, ds.$$

In Problems 37.16 and 37.17, we present two applications of Taylor polynomials.

37.16. Suppose f has three continuous derivatives on $[a, b]$. Prove that

$$\lim_{h \to 0} \frac{f(x + h) - 2f(x) + f(x - h)}{h^2} = f''(x)$$

for all $a < x < b$.

37.17. Suppose $f^{(2)}$ is continuous on $[a, b]$ and $f''(x) \ge 0$ for $a \le x \le b$. Show that for any $\bar{x}$ in $[a, b]$, $f(x)$ is never smaller than the value of the tangent line of f at $\bar{x}$.

38

Polynomial Interpolation

So far we have described two ways to approximate a given function f by polynomials. The Bernstein polynomial approximation uses values of f at $n + 1$ equally spaced points in an interval to produce an approximating polynomial of degree $n + 1$. The Taylor polynomial approximation uses the $n + 1$ values of f and its first n derivatives at a common point to produce an approximating polynomial of degree n. However, both of these approximations have disadvantages. While uniformly accurate on an interval, the Bernstein polynomial can require very many values of f to obtain even moderate accuracy, while the Taylor polynomial requires f *and* its derivatives and, moreover, can only be expected to be accurate near one point.

So we are still motivated to look for other polynomial approximations of a function. We introduce another approach in this chapter called interpolation. The **(polynomial) interpolation problem** for a function f on an interval $[a, b]$ is to find a polynomial p that agrees with f at $n + 1$ points $a = x_0 < x_1 < \cdots < x_n = b$, called the **interpolation nodes**. In other words, $p(x_i) = f(x_i)$, $i = 0, 1, \cdots, n$. The polynomial p is called the **polynomial interpolant** of f and is said to **interpolate** f at the nodes.

The polynomial interpolation problem is natural from a couple of viewpoints. First, we use values of the function f in an interval to compute the polynomial, as does the Bernstein polynomial approximation, but we specify the polynomial is *exact* at the nodes, which follows the idea behind the Taylor polynomial. Second, the interpolation problem arises naturally when conducting physical experiments. In many situations, we know theoretically that two quantities y and x are related by an unknown function

$y = f(x)$ but in the laboratory we are only able to measure values of the function at specific points $x_0, x_1, \cdots, x_n$.

EXAMPLE 38.1.   After ignition, the temperature inside an engine increases as a smooth function of the time since ignition. Experimentally, we can record the temperature perhaps every few seconds manually, and every few tenths of a second electronically.

In this situation, we would like to find a smooth function through the points $x_0, x_1, \cdots, x_n$ to use in place of f, which is unknown, so we can predict values between the nodes or perhaps differentiate to get a instantaneous rate of change or integrate to get the average change. The interpolation problem is one way to do this.

38.1   Existence and Uniqueness

We start by showing that there is a unique solution to the polynomial interpolation problem in a specific sense. In particular given $x_0 < x_1 < \cdots < x_n$, we show there is a unique polynomial p_n of degree n such that

$$p_n(x_i) = f(x_i), \quad 0 \le i \le n.$$

Recall that a polynomial of degree n is specified by $n + 1$ coefficients,

$$p(x) = a_0 + a_1 x + \cdots + a_n x^n. \tag{38.1}$$

In the context of computing a polynomial approximation of a function, we call these coefficients the **degrees of freedom** of the polynomial approximation because we are free to choose the coefficients in order to compute the approximation. We claim there is a unique polynomial interpolant with the property that the number of degrees of freedom is the same as the number of function values.

We first prove that there is at least one polynomial interpolant of degree n and then that there is only one. To do this, we use a special way to write polynomials of degree n that is different than the standard form (38.1).

We call the monomials $\{1, x, \cdots, x^n\}$ a **basis** for the set of polynomials of degree n and less because *every* polynomial of degree n and less can be written as a linear combination of these monomials in a *unique* way, as in (38.1).[1] We call $\{a_0, \cdots, a_n\}$ the **coefficients** of p in (38.1) with respect to the basis $\{1, x, \cdots, x^n\}$.

It turns out that there are many sets of basis polynomials for the set of polynomials of degree n and less.

---

[1] Recall that we prove this in Example 7.7.

EXAMPLE 38.2. For example, 1 and x are the standard basis of monomials for the linear polynomials. But 1 and $x+2$ are another basis. For if $p(x)$ is a linear polynomial with $p(x) = a_0 \times 1 + a_1 \times x$, we can also write

$$p(x) = a_0 + a_1 x = a_0 1 + a_1(x+2) - 2a_1$$
$$= (a_0 - 2a_1) \times 1 + a_1 \times (x+2)$$

Thus, we have written $p(x)$ as a linear combination of 1 and $x+2$. Moreover, and this is very important, the coefficients are unique because the original coefficients a_0 and a_1 are unique.

EXAMPLE 38.3. It is a good exercise to prove that the set of polynomials

$$1, (x-c), (x-c)^2, \cdots (x-c)^n$$

is a basis for the polynomials of degree n and less, where c is any fixed number.[2] The basis of monomials is just the special case with $c = 0$.

But not every set of polynomials is necessarily a basis.

EXAMPLE 38.4. The polynomials $\{1, x + x^2\}$ are *not* a basis for the polynomials of degree 2 and less because there are not enough functions. For example, we cannot write x as a linear combination of 1 and $x+x^2$.

EXAMPLE 38.5. The set of polynomials $\{1, x, 2+x, x^2\}$ is *not* a basis for the polynomials of degree 2 and less because there are too many functions and consequently uniqueness is lost. For example, we can write $2x + 1$ in two different ways,

$$2x + 1 = 1 \times 1 + 2 \times x = -3 \times 1 + 2 \times (2+x).$$

We use a particular basis for the polynomials of degree n and less to solve the polynomial interpolation problem. Given the points $x_0, x_1, \cdots, x_n$, the **Lagrange basis** $\{l_{n,0}(x), l_{n,1}(x), \cdots, l_{n,n}(x)\}$ for the polynomials of degree n and less is defined like this; for $0 \le i \le n$,

$$l_{n,i}(x)$$
$$= \frac{(x-x_0)(x-x_1)\cdots(x-x_{i-1})(x-x_{i+1})\cdots(x-x_n)}{(x_i-x_0)(x_i-x_1)\cdots(x_i-x_{i-1})(x_i-x_{i+1})\cdots(x_i-x_n)}. \quad (38.2)$$

Note that $l_{n,i}$ has degree n for every i and also that

$$l_{n,i}(x_j) = \begin{cases} 1, & i = j, \\ 0, & i \ne j, \end{cases} \text{ for } 0 \le i, j \le n. \quad (38.3)$$

In other words, a Lagrange basis function is 1 at one particular node and 0 at all the rest.

---

[2]Recall that we use these polynomials to define the Taylor polynomial.

EXAMPLE 38.6. The Lagrange basis for the polynomials of degree 1 and less is

$$\{l_{1,0}(x), l_{1,1}(x)\} = \left\{ \frac{x - x_1}{x_0 - x_1}, \frac{x - x_0}{x_1 - x_0} \right\}.$$

We plot these functions in Fig. 38.1.

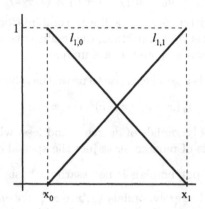

FIGURE 38.1. The Lagrange basis for the polynomials of degree 1 and less.

EXAMPLE 38.7. The Lagrange basis functions for the polynomials of degree 2 is

$$\{l_{2,0}(x), l_{2,1}(x), l_{2,2}(x)\}$$
$$= \left\{ \frac{(x - x_1)(x - x_2)}{(x_0 - x_1)(x_0 - x_2)}, \frac{(x - x_0)(x - x_2)}{(x_1 - x_0)(x_1 - x_2)}, \frac{(x - x_0)(x - x_1)}{(x_2 - x_0)(x_2 - x_1)} \right\}.$$

We plot the basis corresponding to the nodes x_0, $x_1 = (x_0 + x_2)/2$, and x_2 in Fig. 38.2.

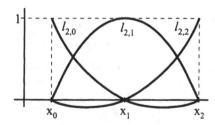

FIGURE 38.2. The Lagrange basis for the polynomials of degree 2 and less with nodes x_0, $x_1 = (x_0 + x_2)/2$, and x_2.

By the way, we have not proved that the Lagrange basis is indeed a basis, but that follows as soon as we show the interpolant is unique.

Because of (38.3), we can solve the interpolation problem quickly. In fact, the polynomial p_n of degree n that interpolates f at the nodes $x_0, \cdots, x_n$ is simply

$$p_n(x) = f(x_0)\, l_{n,0}(x) + f(x_1)\, l_{n,1}(x) + \cdots + f(x_n)\, l_{n,n}(x).$$

We just have to check that for $0 \le i \le n$,

$$
\begin{aligned}
p_n(x_i) &= f(x_0)\, l_{n,0}(x_i) + f(x_1)\, l_{n,1}(x_i) + \cdots + f(x_i)\, l_{n,i}(x_i) \\
&\qquad + f(x_{i+1})\, l_{n,i+1}(x_{i+1}) + \cdots + f(x_n)\, l_{n,n}(x_i) \\
&= f(x_0) \times 0 + f(x_1) \times 0 + \cdots + f(x_i) \times 1 \\
&\qquad + f(x_{i+1}) \times 0 + \cdots + f(x_n) \times 0 \\
&= f(x_i).
\end{aligned}
$$

EXAMPLE 38.8. The linear polynomial that interpolates e^x at $x = 0$ and $x = 1$ is

$$p_1(x) = e^0\, \frac{x-1}{0-1} + e^1\, \frac{x-0}{1-0} = 1 + (e^1 - 1)x.$$

EXAMPLE 38.9. The quadratic polynomial that passes through $(0,1)$, $(1,0)$, and $(2,2)$ is

$$p_2(x) = 1 \times \frac{(x-1)(x-2)}{2} + 0 \times \frac{(x-0)(x-2)}{1} + 2 \times \frac{(x-0)(x-1)}{2}.$$

So we have found at least one polynomial interpolant of degree n for f with respect to the nodes $x_0, \cdots, x_n$. We just have to show there is only one such polynomial. This follows in fact from elementary properties of polynomials. Recall that if a polynomial $p(x)$ has a root r, then $(x - r)$ divides evenly, i.e., without remainder, into $p(x)$. Suppose there are two polynomials $p_n(x)$ and $q_n(x)$ of degree n that agree at $x_0, \cdots, x_n$. Their difference $d_n(x) = p_n(x) - q_n(x)$ is a polynomial of degree m, $0 \le m \le n$, with $n+1$ distinct roots $x_0, \cdots, x_n$. Using polynomial division for the roots $x_0, \cdots, x_{m-1}$, we can write

$$d_n(x) = c(x - x_0)(x - x_1) \cdots (x - x_{m-1})$$

for some constant c. But if we substitute x_m,

$$d_n(x_m) = 0 = c(x_m - x_0)(x_m - x_1) \cdots (x_m - x_{m-1}).$$

Since $x_m - x_0 \ne 0, \cdots, x_m - x_{m-1} \ne 0$, $c = 0$. In other words, $d_n(x) = 0$ for all x and therefore $p_n = q_n$.

This also implies that the Lagrange basis is indeed a basis. For given distinct points $x_0, \cdots, x_n$, every polynomial p of degree n and less can be written *uniquely* as

$$p(x) = p(x_0)\, l_{n,0}(x) + p(x_1)\, l_{n,1}(x) + \cdots + p(x_n)\, l_{n,n}(x).$$

We can interpret this result as the fact that the interpolating polynomial of degree n for a polynomial P of degree m with $m \leq n$ is simply P, i.e., $P = p_n$ for any polynomial P with $\deg(P) \leq n$.

EXAMPLE 38.10. Recall that the Bernstein polynomial of degree $n \geq 2$ for x^2 on $[0, 1]$ is $B_n(x) = x^2 + \frac{1}{n}x(1-x)$. The interpolating polynomial of degree $n \geq 2$ for x^2 on $[0, 1]$ is $p_n(x) = x^2$.

We summarize this discussion as a theorem:

Theorem 38.1 Existence of the Polynomial Interpolant *There is a unique polynomial $p_n(x)$ of degree n or less that takes on the $n+1$ values $f_0, f_1, \cdots, f_n$ at the $n+1$ distinct points $x_0, x_1, \cdots, x_n$, respectively. The polynomial p_n is given by*

$$p_n(x) = f_0\, l_{n,0}(x) + f_1\, l_{n,1}(x) + \cdots + f_n\, l_{n,n}(x). \tag{38.4}$$

38.2   The Error of a Polynomial Interpolant

In the case that the interpolating polynomial p_n is computed using $n+1$ values of a function $f(x_0), f(x_1), \cdots, f(x_n)$, it is natural to wonder about the size of the error

$$e(x) = f(x) - p_n(x)$$

at *any* point x in $[a, b]$. We know that $e(x_0) = e(x_1) = \cdots e(x_n) = 0$, the question is what happens in between? If we assume that f has $n+1$ continuous derivatives in $[a, b]$, then it is possible to obtain a precise formula for the error.

To do this, we use yet another generalization of Rolle's Theorem 32.18.

Theorem 38.2 General Order Rolle's Theorem *If $f(x)$ has continuous derivatives of order $n+1$ in $[a, b]$ and vanishes at $n+2$ distinct points in $[a, b]$ there is a point c in (a, b) such that $f^{(n+1)}(c) = 0$.*

We use induction to prove this. Suppose the zeroes of f are $x_0 < x_1 < \cdots < x_{n+1}$. Since $f(x_i) = f(x_{i+1})$ for $0 \leq i \leq n$, Rolle's Theorem 32.18 implies there is a point in (x_i, x_{i+1}) at which f' is zero for each such i. In other words, f' is zero at $n+1$ distinct points in $[a, b]$. The same argument now applies to show that there are n distinct points in $[a, b]$ at which $f'' = (f')'$ is zero. Induction proves the theorem.

Now consider the function

$$E(x) = e(x) - K(x - x_0)(x - x_1) \cdots (x - x_n)$$

for a constant K. First we note that $E(x)$ vanishes at the $n + 1$ distinct points $x_0, \cdots, x_n$. Moreover if y is any point in (a, b) distinct from $x_0, \cdots, x_n$, we can choose K so that $E(y) = 0$. Namely, we set

$$K = \frac{e(y)}{(y - x_0) \cdots (y - x_n)}.$$

Hence, $E(x)$ vanishes at the $n+2$ distinct points $x_0, \cdots, x_n, y$ and moreover has $n + 1$ continuous derivatives by assumption. Thus, there is a c in (a, b) such that $E^{(n+1)}(c) = 0$.

Since $E^{(n+1)}(x) = f^{(n+1)}(x) - K(n + 1)!$, we conclude that $f^{(n+1)}(c) - K(n + 1)! = 0$ or

$$K = \frac{f^{(n+1)}(c)}{(n + 1)!}.$$

Now we set $y = x$ for some x in $[a, b]$. Using $E(x) = 0$, we conclude the following theorem:

Theorem 38.3 Error Formula for Interpolation   *Assume $f(x)$ has $n + 1$ continuous derivatives on $[a, b]$ and $a = x_0 < x_1 < \cdots < x_n = b$ are $n + 1$ distinct nodes. There is a point c in (a, b) such that*

$$e(x) = \frac{f^{(n+1)}(c)}{(n + 1)!} (x - x_0) \cdots (x - x_n). \tag{38.5}$$

EXAMPLE 38.11. We compute quadratic interpolants to $\sin(x)$ and $\cos(x)$ on $[0, \pi]$ using $x_0 = 0$, $x_1 = \pi/2$, and $x_2 = \pi$. We get

$$\sin(x) \approx p_2(x) = \frac{4}{\pi^2} x(x - \pi)$$

and

$$\cos(x) \approx p_2(x) = 1 - \frac{2}{\pi} x.$$

For the errors, we note that $\sin^{(3)}(c) = -\cos(c)$ and $\cos^{(3)}(c) = \sin(c)$. Therefore the error of the error of the approximation for sin satisfies

$$e(x) = \frac{-\cos(c)}{6} x(x - \pi/2)(x - \pi),$$

while the error of the approximation for cos satisfies

$$e(x) = \frac{\sin(c)}{6} x(x - \pi/2)(x - \pi).$$

We plot the approximations and their errors in Fig. 38.3.

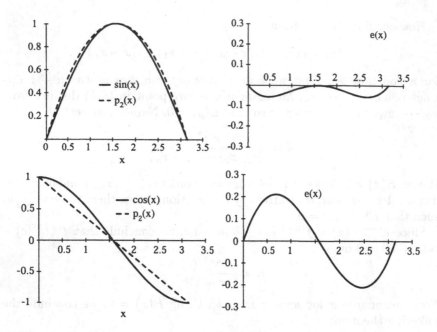

FIGURE 38.3. Plots of the quadratic interpolating polynomials for sin and cos on $[0, \pi]$ along with their errors.

38.3   Accuracy and Convergence

Theorem 38.3 gives an exact formula for the error at each point x. But there are some disadvantages to this result. First, even though we only use values of f to compute the polynomial interpolant, we require pointwise values of the $n + 1^{st}$ derivative of f to estimate the error. This is difficult to obtain at best, and in the case of experimentally determined data, impossible to get. Second it gives a bound at *each* point x in the interval $[a, b]$. In other words, we have to evaluate the estimate for each x where we wish to know the error.

The standard way to deal with these two issues is to derive an error *bound* from the estimate (38.5). A bound is typically bigger than the error but, being cruder, typically requires less information. It is a good exercise to prove the following bound using Theorem 38.3.

Theorem 38.4 Error Bound for Interpolation *Assume $f(x)$ has $n+1$ continuous derivatives on $[a, b]$ and $a = x_0 < x_1 < \cdots < x_n = b$ are $n + 1$ distinct nodes. Assume further that $|f^{(n+1)}|$ is bounded by a constant M on $[a, b]$. Then*

$$\max_{a \leq x \leq b} |e(x)| \leq \frac{M}{(n + 1)!} \max_{a \leq x \leq b} |x - x_0| \cdots |x - x_n|. \qquad (38.6)$$

This gives a bound on the maximum value of the error on $[a, b]$, hence we only need to compute the bound once, even if we want the error at several points, and it only requires knowledge of some bound on $f^{(n+1)}$ rather than pointwise values.

EXAMPLE 38.12. In Example 38.11, we computed the quadratic interpolating polynomials for cos and sin on $[0, \pi]$. Since $|\cos(c)| \leq 1$ and $|\sin(c)| \leq 1$ for any c, we obtain a bound on the error for both polynomials of

$$\max |e| \leq \frac{1}{6} \max_{[0,\pi]} |x||x - \pi/2||x - \pi| \leq 1/4.$$

In fact, the error of the interpolating polynomial for sin is much smaller while the bound is not too far off for the error of the polynomial for cos.

EXAMPLE 38.13. Consider the interpolating polynomial p_n of e^x using $n+1$ equally distributed nodes in $[0, 1]$. Now $d^{n+1}e^x/dx^{n+1} = e^x$ while $\max_{[0,1]} |x - x_0| \cdots |x - x_n| \leq 1^{n+1}$. We conclude that

$$\max_{0 \leq x \leq 1} |e(x)| \leq \frac{e^1}{(n+1)!}.$$

Note that this implies that the error of p_n tends to zero as n increases.

It is useful to consider some special general cases.

EXAMPLE 38.14. Consider the constant interpolant $p_0(x) = f(a)$ for all $a \leq x \leq b$. The bound (38.6) is

$$\max_{a \leq x \leq b} |e(x)| \leq (b - a) \max_{a \leq x \leq b} |f'(x)|. \tag{38.7}$$

EXAMPLE 38.15. Consider the linear interpolant $p_1(x)$ using the nodes $x_0 = a$ and $x_1 = b$. The bound (38.6) is

$$\max_{a \leq x \leq b} |e(x)| \leq \frac{1}{8}(b - a)^2 \max_{a \leq x \leq b} |f''(x)|. \tag{38.8}$$

To get (38.8), we compute

$$\max_{a \leq x \leq b} |x - a||x - b|.$$

The maximum of this occurs at the maximum or minimum of $q(x) = (x - a)(x - b)$. The extreme values of this function occur either where $q'(x) = 0$ or $x = a$ or $x = b$. The last two give $q(a) = q(b) = 0$ while $q'(x) = 2x - (a+b) = 0$ at $x = (a+b)/2$, and $q((a+b)/2) = (b-a)^2/4$. Putting this in (38.6) gives (38.8).

EXAMPLE 38.16. Consider the quadratic interpolant $p_2(x)$ using the nodes $x_0 = a$, $x_1 = (a+b)/2$, and $x_1 = b$. The bound (38.6) is

$$\max_{a \leq x \leq b} |e(x)| \leq \frac{\sqrt{3}}{216} (b-a)^3 \max_{a \leq x \leq b} |f^{(3)}(x)|. \tag{38.9}$$

From (38.6), we conclude that if all of f's derivatives are uniformly bounded by a constant M and if $b - a$ is small, then increasing the order of the interpolating polynomial leads to a smaller error. But if $b - a$ is large, then (38.6) does not imply the error is small. In fact, the factor

$$(x - x_0)(x - x_1) \cdots (x - x_n)$$

in (38.5) is zero when x is equal to one of the nodes, but if x is not equal to any node and $b - a$ is large, then we have to expect some of the terms to be large and some to be small. When $b - a$ is large, it is difficult to make general statements about the error of the polynomial interpolant even when the function has uniformly bounded derivatives of any order.

In fact, the error of the interpolant can be very large even for moderate n, as the next example shows.

EXAMPLE 38.17. We interpolate $f(x) = e^{-8x^2}$ at nine equally spaced nodes $x_0 = -2$, $x_1 = -1.5, \cdots, x_8 = 2$ in $[-2, 2]$. We plot $p_8(x)$ together with $f(x)$ in Fig. 38.4. The error away from the interpolation nodes is

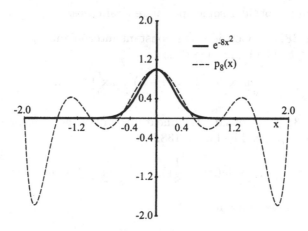

FIGURE 38.4. Plots of the e^{-8x^2} and $p_8(x)$ computed using nodes $x_0 = -2$, $x_1 = -1.5, \cdots, x_8 = 2$.

clearly large. Using $MAPLE^{©}$, we can see that $|f^{(9)}(x)|$ is bounded by 8×10^7 on $[-2, 2]$ while $\max |x + 2| |x + 1.5| \cdots |x - 2| \leq 10$. The error bound (38.8) says that $\max |e(x)| \leq 2205$, which is not much help!

A convergence result for interpolation would imply that the error of the interpolating polynomials p_n for a given function f on a given interval $[a, b]$ tend to zero as n increases. The discussion above implies that such a result holds *if* f has uniformly bounded derivatives of all orders and the interval $[a, b]$ is sufficiently small. But we cannot expect such a convergence result to hold in general. This is similar to the situation for Taylor polynomials.

38.4   A Piecewise Polynomial Interpolant

While we cannot expect to get convergence for the polynomial interpolants of a general function on a large interval as the degree increases, interpolation does provide a way to generate a method of approximating a function that does have nice convergence properties. The idea is to divide the large interval into subintervals and then use piecewise polynomial interpolants with respect to this subdivision. Recall that we first used the idea of piecewise polynomial approximations when studying integration for essentially the same reason.[3]

Given a function $f(x)$ and a set of nodes $a = x_0 < x_1 < \cdots x_n = b$, the **piecewise linear interpolant** of f on $[a, b]$ with respect to the mesh $\{x_0, \cdots, x_n\}$ is the function $P(x)$ that is linear on each subinterval $[x_{i-1}, x_i]$ for $1 \le i \le n$ and satisfies $P(x_i) = f(x_i)$ for $0 \le i \le n$. We illustrate in Fig. 38.5

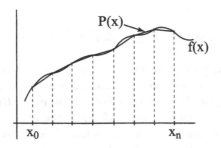

FIGURE 38.5. The piecewise linear interpolant of f.

Some facts follow immediately. First, there is a unique piecewise linear interpolant of f. This follows from the uniqueness of the linear polyno-

---

[3]The subject of piecewise polynomial approximation is extremely important in many applications of mathematics. However, it is better to study that subject after learning about some other kinds of background material, such as linear algebra, that we cannot cover in this book. So we do not present a general discussion of piecewise polynomial approximations. We just consider an especially simple example that nonetheless shows the power of this technique (see Atkinson [2] and Eriksson, Estep, Hansbo, and Johnson [10] for further information).

mial interpolant since the piecewise interpolant is just a linear polynomial interpolant on each subinterval. On $[x_{i-1}, x_i]$, $P(x)$ is given by

$$P(x) = f(x_{i-1})\frac{x - x_i}{x_{i-1} - x_i} + f(x_i)\frac{x - x_{i-1}}{x_i - x_{i-1}}.$$

Moreover, the piecewise linear interpolant of a continuous function is also continuous.

Lastly, we can derive a formula and a bound for the error from Theorem 38.3. Again because the piecewise linear polynomial is just a linear interpolant on each subinterval, if we choose x in $[a, b]$, then x is in $[x_{i-1}, x_i]$ for some i. We apply Theorem 38.3 on $[x_{i-1}, x_i]$ to assert there is a point c in (x_{i-1}, x_i) such that

$$e(x) = \frac{f''(c)}{2}(x - x_{i-1})(x - x_i).$$

If $|f''|$ is bounded by M on $[a, b]$, then we find that if $x_{i-1} \le x \le x_i$,

$$|e(x)| \le \frac{M}{2} \max_{x_{i-1} \le x \le x_i} |x - x_{i-1}||x - x_i| \le \frac{M}{8}(x_i - x_{i-1})^2$$

and

$$\max_{a \le x \le b} |e(x)| \le \frac{M}{8} \max_{1 \le i \le n} (x_i - x_{i-1})^2. \tag{38.10}$$

If the nodes are uniformly spaced so $x_i - x_{i-1} = \Delta x = (b - a)/n$ for $1 \le i \le n$, then

$$\max_{a \le x \le b} |e(x)| \le \frac{M}{8}\Delta x^2 = \frac{M(b - a)}{8n^2}.$$

These bounds imply that the error of the piecewise linear interpolant of a given function on a given interval tends to zero as the mesh sizes tend to zero provided only that f has a bounded second derivative on the interval. If the mesh points are distributed uniformly, then the error decreases like $1/n^2$ where n is the number of mesh points. Recall that the error of the Bernstein polynomials decreases like $1/\sqrt{n}$, which is much slower.

EXAMPLE 38.18. If we use (38.10) to bound the error of the piecewise linear interpolant of e^{-8x^2} on $[-2, 2]$, we find

$$\Delta x = .5 \quad \Longrightarrow \quad \max_{[-2,2]} |e| \le 1$$

$$\Delta x = .25 \quad \Longrightarrow \quad \max_{[-2,2]} |e| \le .25$$

$$\Delta x = .125 \quad \Longrightarrow \quad \max_{[-2,2]} |e| \le .0625$$

$$\vdots \qquad\qquad\qquad \vdots$$

This is a vast improvement over computing higher and higher order polynomial interpolants using the same sets of nodes!

Note that it is considerably cheaper to write down the piecewise linear interpolant of a function using n nodes than the corresponding Bernstein polynomial. We do, however, need to make a decision (i.e., decide on the correct subinterval) in order to evaluate the function.

38.5   Unanswered Questions

We have described three different ways to produce an approximation of a given function and have shown that each method can produce an accurate approximation given the right conditions. Unfortunately, we have also seen that each method has some serious shortcomings in other circumstances. So this leaves open the search for good ways to approximate continuous and smoother functions. This search falls under the rubric of approximation theory, which is a centrally important topic in analysis and numerical analysis.

Among other methods for computing polynomial approximations, we want to mention an important technique called orthogonal projection. This is similar to the idea of polynomial interpolation; however, the polynomial approximation is chosen to have the same weighted average values (see Section 27.2) as the function in question with respect to a certain choice of weights.

We also point out that we could look for approximations using other sets of approximating functions chosen to better "match" the behavior of the given function. For example, in Problem 38.8 we ask you to solve the interpolation problem that uses a set of exponential functions, which would be well suited to approximate an exponentially growing function. On the other hand, if we want to approximate a continuous *periodic* function, it would be natural to seek approximations using sets of trigonometric functions.

Atkinson [2] and Isaacson and Keller [15] contains more material on these subjects.

Chapter 38   Problems

Problems 38.1–38.5 have to do with the idea of a basis for polynomials. We have not developed the theory of vector spaces, so we could not give a complete explanation of the idea of a basis. But if you do these problems, then at least you can develop an intuitive understanding.

38.1. Prove that $\{2 + x, x\}$ is a basis for the linear polynomials.

38.2. Either prove the following sets are a basis for the quadratic polynomials or prove they are not a basis:

(a) $\{1, 2 - x - x^2, x\}$

(b) $\{x^2, x^2 + x, x^2 - 2x\}$

(c) $\{1, 2 - x^2, x + x^2, 1 + x^2\}$.

38.3. Verify the claim in Example 38.3.

38.4. Prove that for any $n \geq 1$,

$$\sum_{i=0}^{n} l_{n,i}(x) = 1 \text{ for all } x.$$

38.5. Draw rough sketches of the Lagrange basis polynomials for the polynomials of degree 3 and less on $[a, b]$ with equally-spaced nodes $x_0 = a, x_1 = a + (b - a)/3, x_2 = a + 2(b - a)/3, x_3 = b$.

38.6. Compute the following polynomial interpolants:

(a) $p_2(x)$ interpolating $\log(x)$ at $\{1, 2, 3\}$

(b) $p_3(x)$ interpolating $\sqrt{x}$ at $\{0, .25, .64, 1\}$

(c) $p_2(x)$ interpolating $\sin(x)$ at $\{0, \pi, 2\pi\}$

(d) $p_3(x)$ passing through points $(-1, 1), (0, 2), (1, -1), (2, 0)$.

To tackle Problems 38.7–38.9, either try to use Theorem 38.1 directly, or use some of the ideas behind in its proof.

38.7. Find the degree of the polynomial passing through the points $(-3, -5)$, $(-2, 0), (-1, -1), (0, -2),$ and $(1, 3)$.

38.8. Prove that the following interpolation problem has a unique solution. Given $f(x)$ and nodes $a = x_0 < x_1 < \cdots x_n = b$, find a function

$$e_n(x) = c_0 + c_1 e^x + c_2 e^{2x} + c_3 e^{3x} + \cdots + c_n e^{nx}$$

such that

$$e_n(x_i) = f(x_i) \text{ for } 0 \leq i \leq n.$$

Hint: Rewrite the problem as a polynomial interpolation problem.

38.9. Consider a rational interpolation problem: Given a function $f(x)$ and distinct nodes $a = x_0 < x_1 < x_2 = b$, find a rational function

$$q(x) = \frac{c_0 + c_1 x}{1 + c_2 x}$$

such that

$$q(x_i) = f(x_i) \text{ for } 0 \leq i \leq 2.$$

Does this problem have a unique solution? This may require some assumptions or restrictions!

38.10. (a) Use Theorem 38.3 to find an expression for the error of the quadratic interpolant of $\log(x)$ at the nodes $\{1, 1.5, 2\}$. (b) Derive a bound on the error using Theorem 38.4.

38.11. Use Theorem 38.3 to find an expression for the error of the polynomial interpolant of degree 4 of e^x at five equally spaced nodes in $[0, 1]$. (b) Derive a bound on the error using Theorem 38.4.

38.12. Assume that $f(x)$ has continuous derivatives of order three and less on $[a, b]$ and let $p_2(x)$ be the quadratic interpolant at the nodes $\{x_0 = a, x_1 = (a + b)/2, x_2 = b\}$.

    (a) Derive a formula for the error $f'(x) - p_2'(x)$ for $a \leq x \leq b$.

    (b) Derive special formulas for the errors $f'(x_i) - p_2'(x_i)$ for $i = 0, 1, 2$.

38.13. Prove Theorem 38.4.

38.14. Consider the third degree Taylor polynomial computed at $a = 1$ and the third degree interpolant with nodes $\{1, 4/3, 5/3, 2\}$ of $\log(x)$.

    (a) Using Theorem 38.4, compare the bounds on the errors of both approximations on $[1, 2]$.

    (b) Plot the errors of the both approximations on $[1, 2]$.

38.15. Verify (38.9).

38.16. For $f(x) = 1/(1 + x^2)$ computing $p_{10}(x)$ using 11 equally spaced nodes in $[-5, 5]$. Plot the both $f(x)$ and $p_n(x)$ in the same plot and also plot the error. *Hint:* This problem should be programmed using for example $MATLAB^{©}$.

38.17. Write a $MATLAB^{©}$ function that returns the piecewise linear interpolant of a user-specified function $f(x)$ defined by a user-specified set of nodes $a = x_0 < x_1 < \cdots < x_n = b$.

38.18. Use a symbolic manipulation package like $MAPLE^{©}$ to verify the computations in Example 38.18.

38.19. Define a piecewise constant interpolant $Q(x)$ of a function $f(x)$ with respect to a set of nodes $a = x_0 < x_1 < \cdots < x_n$ by

$$Q(x) = f(x_i) \text{ for } x_{i-1} \leq x < x_i, \quad 1 \leq i \leq n - 1$$

and
$$Q(x) = f(x_{n-1}) \text{ for } x_{n-1} \leq x \leq x_n.$$

(a) Prove that Q is unique. Is Q continuous in general?

(b) Find an exact expression for the error $e(x) = f(x) - Q(x)$ at any $a \leq x \leq b$ assuming f has a continuous derivative.

(c) Find a bound for $\max_{a \leq x \leq b} |e(x)|$ assuming f has a bounded first derivative. Simplify the bound under the assumption that the mesh points are uniformly distributed.

38.20. Define a piecewise quadratic interpolant $Q(x)$ of a function $f(x)$ with respect to a set of nodes $a = x_0 < x_1 < \cdots < x_n$, where n is even, by grouping the nodes in threes as $\{x_0, x_1, x_2\}, \{x_2, x_3, x_4\}, \cdots, \{x_{n-2}, x_{n-1}, x_n\}$ and computing quadratic interpolants of f on each subinterval $[x_0, x_2], [x_2, x_4], \cdots, [x_{n-2}, x_n]$.

(a) Find a formula for $Q(x)$ for any $a \leq x \leq b$.

(b) Prove that Q is unique and continuous.

(c) Find an exact expression for the error $e(x) = f(x) - Q(x)$ at any $a \leq x \leq b$ assuming that f has a continuous third derivative.

(d) Find a bound for $\max_{a \leq x \leq b} |e(x)|$ assuming f has a bounded third derivative. Simplify the bound under the assumption that the mesh points are uniformly distributed.

(e) Plot the piecewise quadratic and piecewise linear interpolants of $\sin(x)$ on $[0, \pi]$ computed using three equally spaced nodes in $[0, \pi]$.

39
Nonlinear Differential Equations

We finish this introduction to analysis with a brief glimpse into the world of nonlinear differential equations. This is a perfect way to end the book because on one hand, it uses nearly every idea considered so far, and on the other, it is a springboard into the many areas of analysis that remain ahead.

The problem we consider is this: given a point a, an **initial value** y_a, and a function $f(x, y)$, find a point $b > a$ and a function $y(x)$ that is differentiable on $[a, b]$ satisfying the **initial value problem**

$$\begin{cases} y'(x) = f(x, y(x)), & a \leq x \leq b, \\ y(a) = y_a. \end{cases} \tag{39.1}$$

See Chapter 23 for a discussion of the relevance of (39.1) for modeling. We consider the components a, y_a, and $f(x, y)$ as **data** for the initial value problem. In typical applications, they are determined by the model. Specifying a final point b might also be part of the application. But as we see later, we might not be able to find a solution on any arbitrary interval.

EXAMPLE 39.1. In Example 29.1, we propose a simple linear model of population growth, $P'(t) = kP(t)$. Realistically, we might expect that a more complicated relation, $P'(t) = f(P(t))$, holds in most situations. For example, the linear model does not take into account the limitations on population size that exist because of the environment, e.g., the physical size of a petri dish in the case of trapped bacteria.[1] There is also competition for food and mates, and in the case of humans, energy and medicine.

Thus, it is not surprising to find out that the linear population model is only valid for populations that are "small" relative to the available resources. In order to model population growth over long time intervals, we need more realistic models. As a first step, we might consider adding a "competition" term to the linear model that lowers the growth rate when the population becomes large. We could expect that competition is determined by the number of encounters between members of the species being modeled. Statistically, the average number of encounters between two members per unit time is proportional to the population squared. This leads to the initial value problem for the **logistic equation**,

$$\begin{cases} P' = k_1 P - k_2 P^2, & 0 \le t \le b, \\ P(0) = P_a, \end{cases} \tag{39.2}$$

where $k_1 > 0$ and $k_2 > 0$ are constants. This was introduced by the Dutch mathematician Verhulst in 1837 and is the continuous analog of the discrete Verhulst model discussed in Section 4.4.

When modeling populations, k_1 determines the linear birth rate of the species at small populations, while k_2 is determined by the available resources. In general, we expect k_2 to be much smaller than k_1, so when the population P is small, the term $-k_2 P^2$ is negligible compared to $k_1 P$ and the population grows approximately at an exponential rate as predicted by the linear model. However, when the population becomes sufficiently large, then $-k_2 P^2$ is no longer negligible and significantly lowers the rate of growth.

EXAMPLE 39.2. The spruce budworm is a serious threat to the health of balsam firs in the Rockies. One well-studied model for the population is a modification of the logistics equation that takes into account predation by birds. The model has the form

$$\begin{cases} P' = k_1 P - k_2 P^2 - k_3 \frac{P^2}{1+P^2}, & 0 \le t \le b, \\ P(0) = P_a, \end{cases} \tag{39.3}$$

where k_1, k_2, and k_3 are positive constants. The constant k_1 is determined by the linear birth rate of the spruce budworm, while k_2 is determined by the density of foliage. The new predation term $-k_3 P^2/(1+P^2)$ is chosen to mimic the measured predation effects. It approaches a constant rate when the population becomes large, but falls off quickly to

---

[1] Or the state of our poor mistreated Earth in the case of humans.

zero when the population is small as birds tend to move away when food becomes scarce. We plot an example in Fig. 39.1.

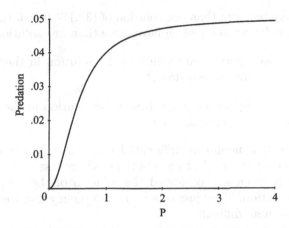

FIGURE 39.1. A plot of the predation term $k_3 P^2/(1+P^2)$ in the spruce budworm model with $k_3 = .05$.

Recall that we derived the theory of integration in Chapter 25 as a way of solving a special case of (39.1), namely,

$$\begin{cases} y'(x) = f(x), & a \le x \le b, \\ y(a) = y_a, \end{cases} \tag{39.4}$$

where f does not depend on the unknown solution y. There is a tremendous difference between (39.1) and (39.4) arising from the possibly nonlinear dependence on the unknown. The analysis of (39.4), as complicated as that is, is consequently much simpler than the analysis of (39.1).

To solve and analyze (39.1), we have to develop tools to handle non-linearity: just as finding numeric roots of nonlinear equations leads to the Bisection Algorithm, fixed point iteration, differentiation, and Newton's method. In fact, all of the issues encountered in the solution of nonlinear equations for numbers are reflected in the solution of nonlinear differential equations.[2]

We describe two approaches to solving (39.1). However, before doing that, we first describe some questions that have to be addressed. To begin with, it is important to realize that the solution of (39.1) can almost never be written down in closed form, i.e., as a combination of the usual functions.[3] This raises four fundamental questions:

---

[2]This has motivated some of the most important and interesting developments in analysis over the last two centuries.

[3]In fact, this claim can be made very precise (see Braun [4]).

1. Does the solution of (39.1) exist? Existence is generally an abstract quality given that we cannot write down the solution.

2. Can there be more than one solution of (39.1)? Recall that nonlinear equations for numbers often have more than one solution.

3. How can we approximate values of the solution in the general case when we cannot write it down?

4. How can we determine properties of the solution in the general case when we cannot write it down.

An analysis of a nonlinear differential equation is generally aimed at addressing one or more of these questions. Of course, we answered the same questions when we considered the solution of the simpler problem (39.4) by integration. They just take on more urgency now because finding the solution is more difficult.

Based on the experience with (39.4), it is not surprising that we have to make assumptions on f in order to answer the questions above. After all, we have only solved (39.4) when $f(x)$ is continuous in x. We likewise assume that $f(x, y)$ behaves continuously with respect to x. But what about its behavior with respect to y?

EXAMPLE 39.3. It turns out that we can solve the logistics equation (39.2) in closed form because it is separable.[4] We rewrite the differential equation using differential notation as

$$\frac{dP}{k_1 P - k_2 P^2} = dt.$$

Integrating both sides, on the right from 0 to t and on the left correspondingly from $P_a = P(0)$ to $P(t)$, we find that $P(t)$ is determined from

$$\int_{P_a}^{P(t)} \frac{dP}{k_1 P - k_2 P^2} = \int_0^t dt = t.$$

Now we use partial fractions to write

$$\frac{1}{k_1 P - k_2 P^2} = \frac{1}{P(k_1 - k_2 P)} = \frac{1/k_1}{P} + \frac{k_2/k_1}{k_1 - k_2 P},$$

so

$$\int_{P_a}^{P(t)} \frac{dP}{k_1 P - k_2 P^2} = \frac{1}{k_2} \int_{P_a}^{P(t)} \frac{dP}{P} + \frac{k_2}{k_1} \int_{P_a}^{P(t)} \frac{dP}{k_1 - k_2 P}.$$

---

[4]It is convenient to have an interesting nonlinear example with a known solution. We use the logistics problem to test our approaches to solving (39.1).

By computing each of the integrals on the right[5] and collecting terms using properties of the logarithm, we find

$$\frac{1}{k_1} \log \left(\frac{P(t)}{P_a} \left| \frac{k_1 - k_2 P_a}{k_1 - k_2 P(t)} \right| \right) = t.$$

It is not hard to show that

$$\frac{k_1 - k_2 P_a}{k_1 - k_2 P(t)} > 0 \text{ for } t > 0,$$

and after some tedious algebra, we obtain

$$P(t) = \frac{k_1 P_a}{k_2 P_a + (k_1 - k_2 P_a)e^{-k_1 t}} \text{ for } t \geq 0. \qquad (39.5)$$

From this formula, we conclude the solution exists for all $t \geq 0$ and that $P(t) \to k_1/k_2$ as $t \to \infty$ for any P_a. If $P_a < k_1/k_2$, then $P(t) < k_1/k_2$ for all t, while if $P_a > k_1/k_2$, then $P(t) > k_1/k_2$ for all t. We plot some solutions in Fig. 39.2.

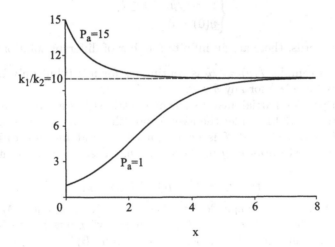

FIGURE 39.2. Two solutions of the logistic equation (39.2) with $k_1 = 1$ and $k_2 = .1$. The solution corresponding to $P_a = 1$ has the characteristic "s"-shape associated with solutions of the logistics problem.

In this example, the function $f(P) = k_1 P - k_2 P^2$ is smooth and there is a solution for all time. Yet, a seemingly small change can lead to completely different behavior.

---

[5]This technique is called **separation of variables**.

EXAMPLE 39.4. It is easy to show that the unique solution of

$$\begin{cases} y'(x) = y^2, & 0 \leq x, \\ y(0) = 1, \end{cases}$$

is $y(x) = 1/(1-x)$ using separation of variables. This function is defined for $0 \leq x < 1$ but "blows up" as x approaches 1.

In the second example, the function $f(y) = y^2$ is also perfectly smooth, yet we do not have a solution for all x.

If we consider functions that are less smooth, other interesting things can happen. We recall Example 23.13.

EXAMPLE 39.5. The functions $y(t) = 0$ for all $t \geq 0$ and

$$y(t) = \begin{cases} 0, & 0 \leq t \leq c, \\ \dfrac{(t-c)^2}{4}, & t \geq c, \end{cases}$$

for any $c \geq 0$ solves

$$\begin{cases} y' = \sqrt{y}, & 0 \leq t, \\ y(0) = 0. \end{cases}$$

In other words, there are an infinite number of different solutions.

In the last example, $f(y) = \sqrt{y}$ is continuous for $y \geq 0$ but only Lipschitz continuous for $y \geq \delta$ for any $\delta > 0$.

We can provide partial answers to the questions 1–4 above for a general investigation of (39.1) under the assumption that there are constants $B > a$ and $M > 0$ such that f is continuous in x and **locally uniformly Lipschitz continuous** in y in the sense that there is a constant $L > 0$ such that

$$|f(x, y_2) - f(x, y_1)| \leq L|y_2 - y_1|$$

for all (x, y) in the rectangle $\bar{\Re}$ described by $a \leq x \leq B$ and $-M + y_a \leq y_1, y_2 \leq y_a + M$ (see Fig. 39.3). The "uniformity" refers to the fact that the Lipschitz constant L is independent of x in $[a, B]$.[6]

These continuity assumptions on f turn out to guarantee existence of a unique solution at least for some $x > a$. It works like this: we can use the Lipschitz condition on f with respect to y as long as $(x, y(x))$ remains in $\bar{\Re}$. But $y(x)$ is continuous and starts with the value y_a at $x = a$, $(x, y(x))$ remains inside $\bar{\Re}$ for at least some $x > a$. We call such existence results **short time existence**.

---

[6]Of course if $f(x, y)$ is Lipschitz continuous for all y uniformly with respect to all x, i.e., f is globally uniformly Lipschitz continuous, then it is locally uniformly Lipschitz continuous on any such rectangle. But we recall that there are few nonlinear functions that are Lipschitz continuous on $\mathbb{R}$.

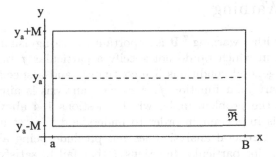

FIGURE 39.3. The rectangle $\bar{\Re} = \{(x, y) : a \leq x \leq B, \, -M + y_a \leq y \leq y_a + M\}$.

Of course, once $(x, y(x))$ leaves $\bar{\Re}$, we have no idea what happens. In Example 39.4, $f(y) = y^2$ is Lipschitz continuous on any finite interval $[y_a - M, y_a + M]$ but not on the entire set of reals. In that problem, the solution eventually becomes larger than any fixed number and the local existence analysis breaks down. Consequently, we can find a solution that exists for x up to 1 but not for all x.

On the other hand, Lipschitz continuity is not required to get a unique solution. But the following example shows that weakening the Lipschitz continuity to a less strong version of Hölder continuity can lead to problems.

EXAMPLE 39.6. Assuming that

$$|f(x, y_2) - f(x, y_1)| \leq L |y_2 - y_1|^\alpha$$

for all (x, y) in a rectangle $\bar{\Re}$, where $0 < \alpha < 1$, is not sufficient to guarantee that a unique solution exists. For example, both $y(x) = 0$ and

$$y(x) = \begin{cases} 0, & 0 \leq x \leq c, \\ \left(\epsilon(x - c)\right)^{1/\epsilon}, & x \geq c, \end{cases}$$

for any $c > 0$, solve

$$\begin{cases} y' = y^{1-\epsilon}, & 0 \leq x, \\ y(0) = 0, \end{cases}$$

for $0 < \epsilon < 1$.

39.1   A Warning

We conclude with a warning.[7] It is important to distinguish a general analysis of (39.1), in which we do not specify a particular f but rather only assume some general conditions like continuity, and a specific analysis of (39.1) for a particular function f. A general analysis is aimed at finding conditions on the problem under which questions 1–4 above can be answered. This is important in order to understand the "structure" of the problem. But a general analysis does *not* preclude being able to answer these questions for particular functions f that fail to satisfy the assumptions of the general analysis. Consider Example 39.4 and Example 39.1. The respective functions $f(y) = y^2$ and $f(P) = k_1 P - k_2 P^2$ are equivalent in *general* terms like being Lipschitz continuous. But they force completely different behavior on the solutions of the respective initial value problems, with the solution corresponding to $f(y) = y^2$ blowing up at $t = 1$, and the solution corresponding to $f(P) = k_1 P - k_2 P^2$ existing for all time.

It is a great mistake to learn the general theory for differential equations and then to believe that this theory covers all the interesting cases or even necessarily has much to do with any particular case.

---

[7]Why is it older people want to warn off younger people from having the same kinds of fun they had when they were young?

Chapter 39 Problems

39.1. Verify the details of Example 39.4.

39.2. Verify the claims in Example 23.13.

39.3. Show that there is more than one solution of

$$\begin{cases} y'(x) = \sin(2x)y^{1/3}, & 0 \le x, \\ y(0) = 0. \end{cases}$$

Hint: $y \equiv 0$ is one solution. Ignore the initial condition and use separation of variables to find others.

39.4. Verify the details of Example 39.3.

39.5. Verify the claims in Example 39.6.

40

The Picard Iteration

In the first approach[1] to solving (40.1),

$$\begin{cases} y'(x) = f(x, y(x)), & a \leq x \leq b, \\ y(a) = y_a, \end{cases} \tag{40.1}$$

we reformulate the problem as a fixed point problem and then use a contraction map to generate a sequence that converges to the solution. This is the direct analog of our approach to solving fixed point problems for numbers.[2] There is one important difference, however: the fixed point we are seeking is now a *function*, not a number.

Based on the discussion in Chapter 39, we assume that there are constants $B > a$ and $M > 0$ such that f is continuous in x and **locally uniformly Lipschitz continuous** in y in the sense that there is a constant $L > 0$ such that

$$|f(x, y_2) - f(x, y_1)| \leq L|y_2 - y_1|$$

for all (x, y) in the rectangle $\bar{\Re}$ described by $a \leq x \leq B$ and $-M + y_a \leq y_1, y_2 \leq y_a + M$ (see Fig. 39.3).

---

[1]Historically, the method of successive approximations, or Picard's iteration, was discovered after Euler's method and infinite series were used to prove existence and uniqueness of solutions.

[2]It is a good idea to review the material in Chapter 15.

40.1   Operators and Spaces of Functions

We begin by describing the basic ingredients for posing a fixed point problem whose solution is a function.

An **operator** is a function whose inputs are functions and whose output consists of functions, i.e., a function on functions. The fixed point problems we study in this chapter are for operators.

The idea of operators may seem like a hard concept- and it is- but we are familiar with a number of examples.

EXAMPLE 40.1.   Any fixed number c is associated naturally with an operator $\mathcal{A}$ defined by $\mathcal{A}(f(x)) = cf(x)$ for any function f. This is a **linear operator** since for any functions f, g and number d, $\mathcal{A}(df(x) + g(x)) = c(df(x) + g(x)) = dcf(x) + cg(x) = d\mathcal{A}(f(x)) + \mathcal{A}(g(x))$.

EXAMPLE 40.2.   Any fixed function $g(x)$ defines an operator $\mathcal{G}$ via composition on the set of functions that have values in the domain of g. Namely, $\mathcal{G}(f(x)) = g(f(x))$. Note that this is not generally defined for all functions. For example when $g(x) = \sqrt{x}$, $\mathcal{G}$ is defined on functions with nonnegative values.

EXAMPLE 40.3.   Differentiation is a linear operator that is defined on functions that have a derivative. Sometimes D is called the **differential operator** and we write $D : f(x) \to f'(x)$.

EXAMPLE 40.4.   Integration is a linear operator defined on the set of continuous functions.

Recall that an important ingredient of a Fixed Point Iteration for computing a fixed point of a function is that the function should map an interval into itself. The operator analog of this property is that the operator should map a set of functions into itself. The fixed point is located in this set of functions. So we have to consider sets, or **spaces**, of functions in order to talk about a fixed point problem for an operator.

There are very many spaces of functions and studying the various examples would be a huge and difficult task. We settle for discussing a particular example that is well suited for our purpose. Given a finite interval $[a, b]$ and a natural number $q \geq 0$, we define the space $C^q([a, b])$ to be the set of continuous functions that have continuous derivatives of order q and lower on $[a, b]$. When $q = 0$, then this is just the space of continuous functions on $[a, b]$.

These spaces have several nice properties. For one thing, if f, g are in $C^q([a, b])$ for $q \geq 0$ and c is a number, then $f + cg$ is in $C^q([a, b])$. The zero function $f(x) \equiv 0$ is also in $C^q([a, b])$ for all q.[3] Another important property

---

[3]In the language of linear algebra, $C^q([a, b])$ is a vector space.

is that $C^q([a, b])$ is contained in $C^{q-1}([a, b])$ for $q \geq 1$. Finally, the space of continuous functions $C^0([a, b])$ is well suited for the purpose of studying the convergence of a Fixed Point Iteration because of the important fact that a uniformly convergent sequence of continuous functions is continuous.[4] In other words, a uniformly convergent sequence of functions in $C^0([a, b])$ converges to a function in $C^0([a, b])$.[5]

However, to use this last property, we must deal with an operator that maps continuous functions into continuous functions. Finding such operators is not straightforward.

EXAMPLE 40.5. The operator A defined by $A(f) = cf$, where c is a number, maps $C^q([a, b])$ into $C^q([a, b])$ for any $q \geq 0$.

EXAMPLE 40.6. If $g(x)$ is continuous and its domain is all real numbers, then the composition operator $G(f) = g \circ f$ maps $C^0([a, b])$ into $C^0([a, b])$.

EXAMPLE 40.7. Differentiation maps $C^q([a, b])$ into $C^{q-1}([a, b])$ for $q \geq 1$.

EXAMPLE 40.8. Integration maps $C^q([a, b])$ into $C^{q+1}([a, b])$ and hence into $C^q([a, b])$ for $q \geq 0$.

40.2   A Fixed Point Problem for a Differential Equation

Example 40.7 suggests that it might be difficult to pose a fixed point problem for a differential equation in a space $C^q([a, b])$ since differentiation does not map $C^q([a, b])$ into itself. We get around this by rewriting the differential equation (40.1) into an equivalent form that does not have this difficulty.

The new formulation is based on the Fundamental Theorem 34.1, which implies in particular that

$$\int_a^x y'(x)\, dx = y(x) - y(a)$$

for any differentiable function y. Hence, any solution of (40.1) also solves the **integral equation**

$$y(x) = y_a + \int_a^x f(s, y(s))\, ds \text{ for } a \leq x \leq b. \qquad (40.2)$$

---

[4]See Theorem 33.1.

[5]The analogous fact for numbers, i.e., that a convergent sequence of numbers in a closed interval converges to a limit in the interval, is crucially important for the previous fixed point results.

We call (40.2) the **integral form**, or **weak form**, of (40.1) because it does not involve an explicit derivative of y.

Conversely, we can take (40.2) as the fundamental problem and look for a function y that is continuous on $[a, b]$ and satisfies (40.2) for $a \leq x \leq b$. Now we can see the meaning of "weak," namely, (40.2) does not require its solution to be differentiable. The solution merely has to be continuous in order that the integral be defined. This follows because f is uniformly Lipschitz continuous in y. Since y is continuous on $[a, b]$, by decreasing b if necessary, we can guarantee that $|y(x) - y_a| \leq M$ for $a \leq x \leq b$. After all, $y(a) = y_a$ and it cannot move too far away for x near a. But this means that $f(x, y(x))$ is continuous on $[a, b]$ since $|f(x_2, y(x_2)) - f(x_1, y(x_1)))| \leq L|y(x_2) - y(x_1)|$ for x_1, x_2 in $[a, b]$. The difference on the right can be made as small as desired by taking x_2 close to x_1. Therefore, the integral in (40.2) is defined for $a \leq x \leq b$.

This raises a fundamental issue: do (40.1) and (40.2) have the same solutions?[6] We have already shown that any solution of (40.1) solves (40.2), so we have to show that a solution of (40.2) solves (40.1). The key observation is that $y(x)$ and therefore $f(x, y(x))$ are continuous on $[a, b]$, which means that

$$y(x) = y_a + \int_a^x f(s, y(s)) \, ds$$

is differentiable for $a \leq x \leq b$. In other words, even though the solution $y(x)$ of (40.2) is only required to be continuous, it is actually differentiable. Moreover,

$$y'(x) = \frac{d}{dx} \int_a^x f(s, y(s)) \, ds = f(x, y(x)) \text{ for } a \leq x \leq b$$

and so y solves (40.1).

From the point of view of formulating a fixed point problem, Example 40.8 explains what is gained by rewriting the differential equation (40.1) as the integral equation (40.2). Namely, if we define the **integral operator** $\mathcal{L}$ on $C^0([a, b])$ by

$$\mathcal{L}(g(x)) = y_a + \int_a^x f(s, g(s)) \, ds, \tag{40.3}$$

then $\mathcal{L}$ maps $C^0([a, b])$ into $C^0([a, b])$ and moreover any fixed point of $\mathcal{L}$ is a solution of (40.2). Thus, we have reformulated the initial value problem (40.1) into the fixed point problem

$$\mathcal{L}(y) = y \tag{40.4}$$

on the set of continuous functions for the integral operator (40.3).

---

[6]Of course, we want to answer yes.

Note that the discussion in this section is an example of *a priori* analysis. We derive properties of the solution *assuming that it exists.* We have yet to show that the solution actually does exist.

40.3   The Banach Contraction Mapping Principle

We state and prove a contraction mapping principle that applies to a fixed point problem for a general operator. In Section 40.4, we show that the fixed point problem (40.4) for a differential equation satisfies the assumptions of this theorem.

The abstract fixed point problem for an operator $\mathcal{A}$ defined on a space of functions S is to find y in S such that $\mathcal{A}(y) = y$. The Fixed Point Iteration is:

Algorithm 40.1 Fixed Point Iteration Choose y_0 in S and for $i = 1$, $2, \cdots$, set

$$y_i = \mathcal{A}(y_{i-1}).\tag{40.5}$$

The following theorem says that the Fixed Point Iteration converges under the right assumptions on S and $\mathcal{A}$.

Theorem 40.1 Banach Contraction Mapping Principle *Let S be a nonempty space of functions defined on an interval $[a, b]$ and $\mathcal{A}$ an operator defined on S satisfying:*

1. Every function g in S is uniformly bounded on $[a, b]$.

2. Every uniform Cauchy sequence in S converges to a limit in S.

3. $\mathcal{A}$ maps S into S.

4. There is a constant $0 < K < 1$ such that

$$\sup_{a \le x \le b} |\mathcal{A}(g(x)) - \mathcal{A}(\tilde{g}(x))| \le K \sup_{a \le x \le b} |g(x) - \tilde{g}(x)|\tag{40.6}$$

for all $g, \tilde{g}$ in S.

Then there is a unique solution $y(x)$ in S of the fixed point problem $\mathcal{A}(y) = y$ and the Fixed Point Iteration $\{y_i\}$ generated by Algorithm 40.1 converges uniformly to y on $[a, b]$ for any y_0 in S.

Assumptions 1 and 2 put conditions on S that guarantee that a sequence of functions in S that converges has a limit in S. If the limit were not in S, then we might have trouble for example applying the operator A to the limit. Assumptions 3 and 4 put conditions on $\mathcal{A}$ that guarantee the Fixed Point Iteration converges. An operator $\mathcal{A}$ satisfying 3 and 4 above is said to be a **contraction map** on S.

The proof of Theorem 40.1 follows the basic plan of the proof of Theorem 15.1. We first show that the Fixed Point Iteration is a uniform Cauchy sequence. We begin by estimating the difference between two successive members. By assumption 3, y_i is in S for all i. Therefore, we can use the contraction assumption 4 inductively to conclude that for $i \geq 2$ and $a \leq x \leq b$,

$$|y_i(x) - y_{i-1}(x)| = |\mathcal{A}(y_{i-1}(x)) - \mathcal{A}(y_{i-2}(x))| \leq K|y_{i-1}(x) - y_{i-2}(x)|$$
$$\leq K^{i-1}|y_1(x) - y_0(x)|.$$

By assumption 1, $|y_1(x) - y_0(x)|$ is uniformly bounded by a constant C. Hence,

$$\sup_{a \leq x \leq b} |y_i(x) - y_{i-1}(x)| \leq CK^{i-1}, \quad 2 \leq i. \tag{40.7}$$

Since $K < 1$, the difference between y_i and y_{i-1} can be made uniformly small by taking i large.

To show that $\{y_i\}$ is a Cauchy sequence, we have to show the same holds for $\sup_{a \leq x \leq b} |y_i(x) - y_j(x)|$ for any $j \geq i$ sufficiently large. For $j > i$, we expand

$$|y_i(x) - y_j(x)| = |y_i(x) - y_{i+1}(x) + y_{i+1}(x) - y_{i+2}(x)$$
$$+ \cdots + y_{j-1}(x) - y_j(x)|$$
$$\leq \sum_{k=i}^{j-1} |y_k(x) - y_{k+1}(x)|.$$

We use (40.7) on each term in the sum to get

$$|y_i(x) - y_j(x)| \leq \sum_{k=i}^{j-1} CK^k = CK^i \frac{1 - K^{j-i}}{1 - K}$$

by the formula for the geometric sum. Since $K < 1$, $1 - K^{j-i} \leq 1$ and therefore

$$|y_i(x) - y_j(x)| \leq \frac{CK^i}{1 - K}, \quad a \leq x \leq b.$$

Since K^i approaches 0 as i increases, $\sup_{a \leq x \leq b} |y_i(x) - y_j(x)|$ with $j \geq i$ can be made as small as desired by taking i large. In other words, $\{y_i\}$ is a uniform Cauchy sequence in S and, by assumption 2, converges to a function y in S.

Next, we have to verify that the limit y of $\{y_i\}$ is a fixed point of $\mathcal{A}$. By the way, since y is in S, it makes sense to write $\mathcal{A}(y)$. By assumption 4,

$$\sup_{a \leq x \leq b} |\mathcal{A}(y(x)) - \mathcal{A}(y_i(x))| \leq K \sup_{a \leq x \leq b} |y(x) - y_i(x)|.$$

Since $\{y_i\}$ converges uniformly to y, $\{\mathcal{A}(y_i)\}$ converges uniformly to $\mathcal{A}(y)$. We take the limit as $i \to \infty$ in $y_i = \mathcal{A}(y_{i-1})$ to conclude $y = \mathcal{A}(y)$.

Finally, we show there is a unique fixed point in S. If y and $\tilde{y}$ are two fixed points, then assumption 4 implies

$$\sup_{a \leq x \leq b} |y(x) - \tilde{y}(x)| = \sup_{a \leq x \leq b} |\mathcal{A}(y(x)) - \mathcal{A}(\tilde{y}(x))| \leq K \sup_{a \leq x \leq b} |y(x) - \tilde{y}(x)|.$$

Since $K < 1$, we must have $y = \tilde{y}$.

40.4   Picard's Iteration

The fixed point iteration Algorithm 40.1 applied to the integral operator $\mathcal{L}$ in (40.3) is called **Picard's iteration** or the **method of successive approximations**. This technique was first used by Liouville[7] to solve a specific second order differential equation and Picard[8] generalized the technique.

Before computing some examples, we first verify that the fixed point problem (40.4) satisfies the assumptions 1–4 of Theorem 40.1 for some b with $a < b < B$. For $a < b < B$ to be chosen, we let $\Re$ denote the "subrectangle" of $\bar{\Re}$ described by $a \leq x \leq b$ and $-M + y_a \leq y_1, y_2 \leq y_a + M$. We take S to be the set of continuous functions on $[a, b]$ whose graphs lie in $\Re$ and $A = \mathcal{L}$ defined by (40.3).

1. Continuous functions are bounded on closed, finite intervals by Theorem 32.16.

2. The limit of a uniform Cauchy sequence of continuous functions is continuous by Theorem 33.1. The graph of the limit of a uniformly convergent sequence of continuous functions whose graphs lie in $\Re$ must also lie in $\Re$.

3. Showing that $\mathcal{L}$ maps S into S requires a little work. Certainly, $\mathcal{L}$ maps continuous functions into continuous functions. We have to verify that it maps a function with a graph in $\Re$ into another function with a graph in $\Re$. The important observation is that $f(x, \cdot)$ is itself uniformly bounded on S. Let z be a function in S. For any $a < b \leq B$ and $a \leq x \leq b$,

$$|f(x, z(x)) - f(x, y_a)| \leq L|z(x) - y_a| \leq LM,$$

---

[7]Joseph Liouville (1809–1882) was a French mathematician. Liouville wrote many papers, making important discoveries in analysis, astronomy, differential equations, differential geometry, and number theory.

[8]Charles Emile Picard (1856–1941) was a French mathematician. He made fundamental discoveries in algebraic geometry, analysis, differential equations, and the theory of functions. He also studied applications in elasticity, electricity, and heat.

since the graph of z lies in $\Re$ by assumption. Therefore,

$$|f(x, z(x))| \le LM + \sup_{a \le x \le b} |f(x, y_a)| = \tilde{M},$$

since f is continuous with respect to x and $\sup_{a \le x \le b} |f(x, y_a)|$ is finite. But then for z in $\mathcal{S}$,

$$|\mathcal{L}(z) - y_a| = \left| \int_a^x f(s, z(s))\, ds \right| \le \tilde{M}(x - a) \le \tilde{M}(b - a). \quad (40.8)$$

This says that the image of the integral operator $\mathcal{L}$ applied to a function in $\mathcal{S}$ is contained in the triangular region inside $\Re$ between the lines with slopes $\pm\tilde{M}$ passing through (a, y_a) (see Fig. 40.1). The desired result holds if we choose $b > a$ so that

$$b \le a + \frac{M}{\tilde{M}} \qquad (40.9)$$

and define $\Re$ accordingly.

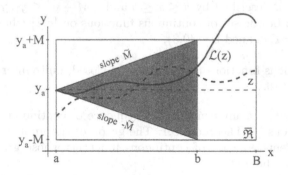

FIGURE 40.1. The image of the integral operator $\mathcal{L}$ applied to a function in $\mathcal{S}$ is contained in the triangular shaded region inside $\bar{\Re}$ between the lines through (a, y_a) with slopes $\pm\tilde{M}$.

4. Let z and $\tilde{z}$ be in $\mathcal{S}$. Then for b in (40.9), the uniform Lipschitz continuity of f implies that

$$|\mathcal{L}(z(x)) - \mathcal{L}(\tilde{z}(x))| = \left| \int_a^x \big(f(s, z(s)) - f(s, \tilde{z}(s)) \big)\, ds \right|$$

$$\le L \int_a^x |z(s) - \tilde{z}(s)|\, ds$$

$$\le L(x - a) \sup_{a \le s \le x} |z(s) - \tilde{z}(s)|.$$

Taking the sup for $a \leq x \leq b$ and decreasing b if necessary to insure $0 < K = L(b-a) < 1$, i.e., choosing

$$b < a + \min\left\{\frac{1}{L}, \frac{M}{\tilde{M}}\right\}, \qquad (40.10)$$

proves the desired result.

We summarize as a theorem.

Theorem 40.2 Picard Existence Theorem *Suppose that $f(x,y)$ is continuous in x and uniformly Lipschitz continuous in y for all (x,y) in the rectangle $\tilde{\mathfrak{R}}$ described by $a \leq x \leq B$ and $-M+y_a \leq y \leq y_a+M$. Then there is a b with $a < b \leq B$ such that (40.1) has a unique solution for $a \leq x \leq b$ and the Picard iteration converges to this solution for any continuous initial iterate whose graph is contained in the rectangle $\mathfrak{R}$ described by $a \leq x \leq b$ and $-M + y_a \leq y \leq y_a + M$.*

We compute some examples.

EXAMPLE 40.9. We compute the Picard iterates for

$$\begin{cases} y' = y, & 0 \leq x, \\ y(0) = 1, \end{cases}$$

which has solution $y(x) = e^x$. In this case, $f(y) = y$ and $f'(y) = 1$, so $L = 1$. If $|y - 1| \leq M$, then $\tilde{M} = 1 + M$. So convergence is guaranteed for all

$$b \leq \frac{M}{1+M}$$

by (40.10). In other words, we can guarantee convergence on $[0, 1)$. Starting with $y_0 = y_a = 1$,

$$y_1(x) = 1 + \int_0^x 1 \, ds = 1 + x$$

and

$$y_2(x) = 1 + \int_0^x (1+s) \, ds = 1 + x + \frac{x^2}{2}.$$

Inductively,

$$y_i(x) = 1 + \int_0^x y_{i-1}(s) \, ds = 1 + \int_0^x \left(1 + s + \cdots + \frac{s^{i-1}}{(i-1)!}\right) ds$$

$$= 1 + s + \cdots + \frac{s^i}{i!}.$$

Therefore, $y_i(x)$ is nothing more than the Taylor polynomial for e^x around 0 of degree i! We have now given a second proof that this Taylor polynomial converges to e^x as the degree increases for $0 \leq x < 1$.

Since we know that for x in any bounded interval, the Taylor polynomial for e^x converges as the degree increases (see Example 37.11), we see that (40.10) may be pessimistic in terms of predicting the length of the interval on which the Picard iteration converges.

EXAMPLE 40.10. We compute the Picard iterates for the logistics equation (39.2) with $k_1 = k_2 = 1$ and $P_a = 1/2$ and solution

$$P(t) = \frac{1}{1 + e^{-t}} \text{ for } t \geq 0.$$

Now $f(P) = P - P^2$. For $|P - 1/2| \leq M$,

$$L = \max_{|P - 1/2| \leq M} |1 - 2P| = |1 - 2(\frac{1}{2} + M)| = 2M.$$

Here $\tilde{M}P_a - P_a^2 + LM = 1/4 + 2M^2$ and by (40.10), we can guarantee convergence for

$$b = \frac{M}{\frac{1}{4} + 2M^2} = \frac{4M}{1 + 8M^2}.$$

The formula for b is concave up and its smallest value occurs when

$$\frac{d}{dM} \frac{4M}{1 + 8M^2} = 0 \implies M = \frac{1}{\sqrt{8}}.$$

Hence, there is convergence at least for x up to $b = 1/\sqrt{2}$.

Starting with $P_0 = P_a = 1/2$, we use $MAPLE^{©}$ to compute

$$P_0 = \frac{1}{2}$$

$$P_1(t) = \frac{1}{2} + \frac{t}{4}$$

$$P_2(t) = \frac{1}{2} + \frac{t}{4} - \frac{t^3}{48}$$

$$P_3(t) = \frac{1}{2} + \frac{t}{4} - \frac{t^3}{48} + \frac{t^5}{480} - \frac{t^7}{16128}$$

We plot these functions along with P in Fig. 40.2. Note that the Picard iterates are very accurate on $[0, 1/\sqrt{2}]$ but are not accurate on much larger intervals.

40.5   Unanswered Questions

We can use Theorem 40.1 to guarantee that a unique solution of (40.1) exists on some short time interval (a, b). Example 39.4 shows that indeed

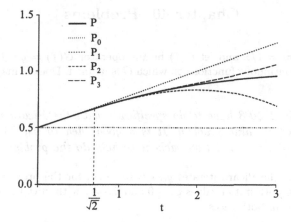

FIGURE 40.2. Plots of the solution $P(t)$ of the logistic equation along with the first four Picard iterates.

there are problems for which a solution exists for some time but there is a time at which the solution "blows up" and ceases to exist. However, in most applications, we require that the solution exist over some given time interval that may not be short and often is very long. Such existence results are called **global existence**. In these cases, an important question is how to show that a solution of (40.1) exists over the required time interval? It is possible to find general conditions on (40.1) that guarantee a solution exists for all time (see Problem 40.9), but these are so onerous that they apply to very few real models. Instead, we generally need to use analysis that is specialized to a particular model and its properties to prove a global existence result.

Chapter 40   Problems

40.1. Let $g(x) = 1/x$ and let $G(f)$ be the operator $G(f) = g \circ f$. Describe a subset of the continuous functions on which G is defined. Does G map this subset into itself?

Problems 40.2–40.8 have to do specifically with the Picard iteration for solving an initial value problem. It is a good idea to use MAPLE© or another symbolic manipulation package to help do the problems.

40.2. Compute the Picard iterates y_i, $i = 0, \cdots, 4$, for the problem $y' = x + y$ for $0 \le x$ using both initial values $y_a = 0$ and $y_a = 1$ with initial iterate $y_0 = y_a$. Plot the results in both cases.

40.3. Compute the Picard iterates y_i, $i = 0, \cdots, 4$, for the problem $y' = y^2$ for $0 \le x$ using the initial value $y_a = 1$ and initial iterate $y_0 = y_a$ and plot the results. Compare to the expected behavior of the true solution.

40.4. Compute the Picard iterates y_i, $i = 0, \cdots, 4$, for the problem $y' = 1 - y^3$ for $0 \le x$ using both initial values $y_a = 0$ and $y_a = 1$ with initial iterate $y_0 = y_a$ in both cases. Plot the results in both cases. Explain the results when $y_a = y_0 = 1$.

40.5. Compute the Picard iterates y_i, $i = 0, \cdots, 4$, for the problem $y' = y^{1/2}$ for $0 \le x$ using the initial value $y_a = 0$ and two different initial iterates $y_0 = 0$ and $y_0 = 1$. Explain your results.

40.6. Compute the Picard iterates y_i, $i = 0, \cdots, 4$, for the problem $y' = y - y^3$ for $0 \le x$ using the initial value $y_a = 1$ and two different initial iterates $y_0 = 1$ and $y_0 = 1 - x$. Plot the results in both cases.

40.7. Find intervals $[a, b]$ on which the Picard iteration is guaranteed to converge via (40.10) for Problem 40.2, Problem 40.3, Problem 40.4, and Problem 40.5.

40.8. Compute the Picard iterates P_i, $i = 0, 1, 2$, for the spruce budworm model (39.3) with $k_1 = 1$, $k_2 = .1$, and $k_3 = .05$ for $0 \le t$ using the initial value $P_a = 1/2$ and initial iterate $P_0 = 1/2$. Plot the results.

40.9. Assume $f(x, y)$ is continuous in x for $a \le x < \infty$, $|f(x, y)| \le \tilde{M}$ for $a \le x < \infty$ and $-\infty < y < \infty$, and $|f(x, y) - f(x, z)| \le L|y - z|$ for $a \le x < \infty$ and $-\infty < y < \infty$. Show the solution of

$$\begin{cases} y' = f(x, y), & a \le x, \\ y(a) = y_a, \end{cases}$$

exists for all $x \ge a$.

Problems 40.10–40.13 are applications of Theorem 40.1 to fixed point problems for various kinds of operators.

40.10. Consider the operator $\mathcal{A}(f) = cf$, where c is a number. Let $\mathcal{S} = C^q([a, b])$. Find conditions on c that guarantees the Fixed Point Iteration converges for y_0 in $\mathcal{S}$. What is the fixed point of $\mathcal{A}$?

40.11. Consider the operator $G(f) = g \circ f$ where $g(x) = x^2/4$. Let $\mathcal{S} =$ the set of continuous functions whose values are between -1 and 1. Verify that Theorem 40.1 applies. What are the fixed points of G? Which fixed point is found by the Fixed Point Iteration for initial iterates in $\mathcal{S}$?

40.12. Does Theorem 40.1 apply to $G(f) = g \circ f$, where $g(x) = \sqrt{x}$ and $\mathcal{S} =$ the set of continuous functions whose values are between 0 and 1?

40.13. (a) Consider the general integral equation: find y in $C^0([a, b])$ such that

$$y(x) = f(x) + \lambda \int_a^x e^{xs} y(s) \, ds, \quad a \le x \le b,$$

where λ is a number and $f(x)$ is a continuous function on $[a, b]$. Define an appropriate operator and then find conditions on b, f, and λ that allow Theorem 40.1 to guarantee a solution exists. (b) Do the same for $C^0([a, b])$ such that

$$y(x) = f(x) + \lambda \int_a^x \sin(x + s) y(s) \, ds, \quad a \le x \le b.$$

(c) Finally, consider the general integral equation

$$y(x) = f(x) + \lambda \int_a^x K(x, s) y(s) \, ds, \quad a \le x \le b.$$

$K(x, s)$ is called the **kernel**. Find conditions on b, f, λ, and $K(x, y)$ that allows Theorem 40.1 to guarantee a solution exists.

Problem 40.14 is an application of Theorem 40.1 to prove an important generalization of the Inverse Function Theorem called the Implicit Function Theorem. The situation is a model in which there is a parameter that can vary, so any solution of the model naturally changes as the parameter changes. The Implicit Function Theorem gives conditions under which we can guarantee that the solution of the model depends continuously on the parameter. To prove the most general result, we have to consider functions of several variables, which we avoid doing in this book. But, the following special case gives the flavor of the theorem.

40.14. Prove the following theorem.

Implicit Function Theorem Let $f(x, y)$ be defined for $a \le x \le b$ and $-\infty < y < \infty$. Suppose $f(x, y)$ is continuous with respect to x, differentiable with respect to y, and furthermore there are constants m, M such that

$$0 < m \le \frac{d}{dy} f(x, y) \le M < \infty$$

for $a \le x \le b$ and $-\infty < y < \infty$. Then

$$f(x, y) = 0$$

has a unique solution $y(x)$ for each x in $[a, b]$ such that $y(x)$ is a continuous function.

Hint: Let $\mathcal{S} = C^0([a, b])$ and consider the operator

$$\mathcal{A}(z(x)) = z(x) - \frac{1}{M} f(x, z(x)).$$

Use Theorem 40.1. By the way, the Mean Value Theorem for f looks like

$$f(x, y) - f(x, \tilde{y}) = \frac{d}{dy} f(x, c)(y - \tilde{y})$$

for some $\tilde{y} < c < y$.

41
The Forward Euler Method

In Chapter 40, we prove that there is a unique solution y of

$$\begin{cases} y'(x) = f(x, y(x)), & a \leq x \leq b, \\ y(a) = y_a, \end{cases} \tag{41.1}$$

by using a Fixed Point Iteration that *theoretically* produces a sequence of increasing accurate approximations to y. In practice, it turns out that computing the Picard iterates requires evaluating a succession of increasingly complicated integrals. Consequently, we cannot compute very many iterates in general. Unfortunately, the Picard iteration may converge very slowly.

So, we are still left with the problem of devising a practical method for approximating the solution of (41.1) to a desired accuracy. In this chapter, we construct such a method that is closely related to the approach used for defining and computing the integral.[1]

41.1   The Forward Euler Method

As in Chapter 40, we use the equivalent integral or weak form of (41.1): find a function y that is continuous on $[a, b]$ and satisfies

$$y(x) = \mathcal{L}(y) = y_a + \int_a^x f(s, y(s))\, ds \text{ for } a \leq x \leq b. \tag{41.2}$$

---

[1]It is a good idea to review Chapters 25 and 34.

In Section 40.2, we discuss the **integral operator** $\mathcal{L}$ and, in particular, prove that (41.2) and (41.1) are equivalent in the sense that a solution of one problem is also a solution of the other problem.

We define an approximate solution by considering $f(x, y(x))$ as a function of x and applying the rectangle quadrature rule to approximate the integral in (41.2). We choose a mesh $\mathcal{T} = \{x_0, x_1, \cdots, x_N\}$ with $a = x_0 < x_1 < \cdots < x_N = b$ and step sizes $\Delta x_n = x_n - x_{n-1}$, $1 \leq n \leq N$. We use the maximum step size $\Delta x_{\mathcal{T}} = \Delta x = \max_{1 \leq n \leq N} \Delta x_n$ as the measure of the "fineness" of the mesh $\mathcal{T}$.

The approximation is a piecewise linear, continuous function with respect to the mesh $\mathcal{T}$. We use capital letters to denote the approximation. The nodal values $\{Y_n\}_{n=0}^{N}$ are determined via the following algorithm.

Algorithm 41.1 Forward Euler Method Set $Y_0 = y_a$ and for $1 \leq n \leq N$,

$$Y_n = Y_0 + \sum_{k=1}^{n} f(x_{k-1}, Y_{k-1})\Delta x_k. \tag{41.3}$$

Correspondingly, for $x_{n-1} \leq x \leq x_n$, the approximation is given by

$$Y(x) = Y_0 + \sum_{k=1}^{n-1} f(x_{k-1}, Y_{k-1})\Delta x_k + f(x_{n-1}, Y_{n-1})(x - x_{n-1}). \tag{41.4}$$

This is called the forward Euler approximation because the left-hand end-point is used to define the integral approximation on each interval $[x_{n-1}, x_n]$. An important consequence is that the equations determining $\{Y_n\}$ are *linear*. In contrast, for example, if we use the right-hand endpoint on each interval, the resulting equations for the nodal values are nonlinear (see Problem 41.6).

EXAMPLE 41.1. We compute the forward Euler approximations for

$$\begin{cases} y' = y, & 0 \leq x, \\ y(0) = 1, \end{cases} \tag{41.5}$$

which has solution $y(x) = e^x$. We use uniform meshes on $[0,1]$ with 5, 10, 20, and 40 intervals and plot the resulting approximations in Fig. 41.1. Compare to the results of the Picard iteration in Example 40.9.

EXAMPLE 41.2. We compute the forward Euler approximations for the logistics equation with $k_1 = k_2 = 1$,

$$\begin{cases} P' = P - P^2, & 0 \leq t, \\ P(0) = 1/2, \end{cases} \tag{41.6}$$

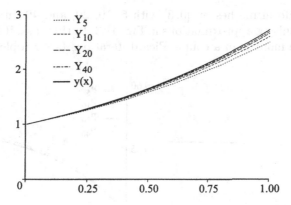

FIGURE 41.1. Plots of forward Euler approximations for (41.5) with 5, 10, 20, and 40 uniformly spaced nodes in $[0, 1]$ along with the true solution.

which has solution $P(t) = 1/(1 + e^{-t})$. We use uniform meshes on $[0, 3]$ with 5, 10, 20, and 40 intervals and plot the resulting approximations in Fig. 41.2. Compare to the results of the Picard iteration in Example 40.10.

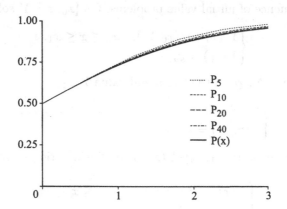

FIGURE 41.2. Plots of forward Euler approximations for (41.6) with 5, 10, 20, and 40 uniformly spaced nodes in $[0, 3]$ along with the true solution.

EXAMPLE 41.3. We compute the forward Euler approximations for the model of the spruce budworm with $k_1 = k_2 = k_3 = 1$,

$$\begin{cases} P' = P - P^2 - P^2/(1 + P^2), & 0 \le t, \\ P(0) = 1/2. \end{cases} \qquad (41.7)$$

We use uniform meshes on $[0, 3]$ with 5, 10, 20, and 40 intervals and plot the resulting approximations in Fig. 41.3. In this case, it is difficult to compute more than a couple Picard iterations (see Problem 40.8).

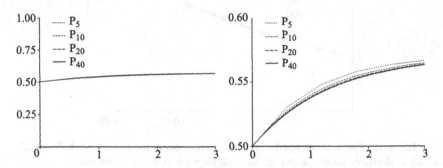

FIGURE 41.3. Plots of forward Euler approximations for (41.7) with 5, 10, 20, and 40 uniformly spaced nodes in $[0, 3]$. On the top, we plot the results on the same scale used in Fig. 41.2 so that the effects of the predation can be seen. On the bottom, we change scale to make the effect of changing the mesh more clear.

We can interpret the forward Euler method as a way of approximately solving a sequence of initial value problems. On $[x_0, x_1]$, Y solves

$$\begin{cases} Y'(x) = f(x_0, Y_0), & x_0 \leq x \leq x_1, \\ Y(x_0) = y_a, \end{cases}$$

and for $1 \leq n \leq N$, given the last nodal value Y_{n-1},

$$\begin{cases} Y'(x) = f(x_{n-1}, Y_{n-1}), & x_{n-1} \leq x \leq x_n, \\ Y(x_{n-1}) = Y_{n-1}. \end{cases}$$

This means that on $[x_{n-1}, x_n]$, $Y(x)$ is the linearization of the function $\tilde{y}$ solving

$$\begin{cases} \tilde{y}'(x) = f(x, \tilde{y}(x)), & x_{n-1} \leq x \leq x_n, \\ \tilde{y}(x_{n-1}) = Y_{n-1}, \end{cases}$$

at x_{n-1} (see Fig. 41.4). $\tilde{y}$ is called the **local solution** on $[x_{n-1}, x_n]$. In Problem 41.7, we ask you to show that $\tilde{y}$ exists. The term "forward" is suggestive of extrapolating the value of $f(x, Y(x))$ from x_{n-1} over $[x_{n-1}, x_n]$.

41.2   Equicontinuity and Arzela's Theorem

We have constructed a collection, or **family**, of potential approximations $\{Y\}$ of y corresponding to all of the possible meshes $\{\mathcal{T}\}$, of which there are

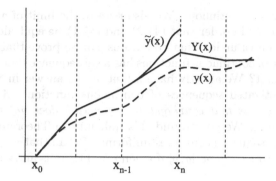

FIGURE 41.4. On $[x_{n-1}, x_n]$, the forward Euler approximation $Y(x)$ is the linearization of the local solution $\tilde{y}$ at x_{n-1}.

very many! We want to show that the solution y exists and that it is possible to approximate y to any desired accuracy using $\{Y\}$. We can formulate these results as follows. Assume that we have a sequence of meshes with the property that the corresponding maximum step sizes tend to zero. Then the corresponding approximations form a uniform Cauchy sequence on $[a, b]$ that converges to a limit, which is the solution y, as $n \to \infty$. This is what we prove.

We used the same strategy to show the rectangle rule approximation converges to the integral of a continuous function in Chapter 25. However, when considering integration, we use a specific sequence of uniform meshes that makes it relatively easy to compare approximate solutions corresponding to different meshes. In this setting, we avoid specifying anything about the sequence of meshes other than the fact that their maximum size tends to zero. We could modify the argument used for integration in Chapter 25 to handle arbitrary meshes (see Problem 41.23). However, we take a different approach in this chapter.

To show that Euler's method converges, we use a theorem about functions that is the analog of Weierstrass' principle (Theorem 32.13), which says that every bounded sequence of *numbers* contains a convergent subsequence. We show that the family of Euler approximations $\{Y\}$ contains a subsequence that converges uniformly to the solution.

We begin by assuming that we have a family of functions $\mathcal{F} = \{g\}$ defined on a finite interval $[a, b]$ that are uniformly bounded. So there is a constant B such that for all g in $\mathcal{F}$,

$$|g(x)| \le B \text{ for } a \le x \le b.$$

This is the analog of having a bounded set of numbers.

However, the intended application to solving the integral equation (41.2) requires more. Namely, the family $\mathcal{F}$ has to consist of continuous functions in order for the integral in (41.2) to be defined. Indeed, the family of Euler approximations are continuous. We also require the limit of a sequence in $\mathcal{F}$ to be continuous in order that (41.2) and (41.1) be equivalent.

So the question is: under what conditions can we prove that a uniformly bounded family of continuous functions has a subsequence that converges to a continuous limit? We already have a hint at the answer in Section 33.1, where we investigated sequences of continuous functions. *A sequence of continuous functions that converges to a function does not have to have a continuous limit.* We got around this difficulty in Theorem 33.1 by assuming that the sequence converges uniformly. Recall that a sequence $\{g_n\}$ converges uniformly to g on $[a, b]$ if given $\epsilon > 0$ there is an N such that for all $n > N$,

$$|g(x) - g_n(x)| < \epsilon \text{ for } a \leq x \leq b.$$

The uniformity refers to the fact that the convergence happens at a minimum rate for all $a \leq x \leq b$. *A sequence of continuous functions that converges uniformly has to converge to a continuous function.*

Hence if we determine conditions under which a uniformly bounded family $\mathcal{F}$ of continuous functions has a uniformly convergent subsequence, then we know that the subsequence has a continuous limit. This requires some additional assumption on $\mathcal{F}$, as the next example illustrates.

EXAMPLE 41.4. Define the family of functions $\mathcal{F} = \{g_n\}$ on $[0, 1]$ by

$$g_n(x) = \frac{x^2}{x^2 + (1 - nx)^2} \quad n = 1, 2, 3, \cdots.$$

Now, $|g_n(x)| \leq 1$ for all $0 \leq x \leq 1$ and $n \geq 1$, so $\mathcal{F}$ is uniformly bounded. All the functions g_n are continuous, and therefore uniformly continuous on $[0, 1]$, and for $0 \leq x \leq 1$,

$$\lim_{n \to \infty} g_n(x) = 0,$$

so $\{g_n\}$ converges to a continuous function. However, $g_n(1/n) = 1$ for $n \geq 1$, hence *no* subsequence of $\{g_n\}$ can converge uniformly.

The sequence in Example 41.4 fails to have a uniformly convergent subsequence even though it converges. The reason is that even though each function in the sequence is uniformly continuous in x, the functions are not uniformly continuous with respect to n. We can avoid this difficulty by assuming a sort of "double" uniform continuity on $\mathcal{F}$. A family $\mathcal{F} = \{g\}$ of functions g defined on $[a, b]$ is **equicontinuous** on $[a, b]$ if for every $\epsilon > 0$ there is a $\delta > 0$ such that

$$|g(x) - g(z)| < \epsilon$$

for all g in $\mathcal{F}$ and x, z in $[a, b]$ with $|x - z| < \delta$. Equicontinuity says the family $\mathcal{F}$ is uniformly continuous in both input and in the functions.

EXAMPLE 41.5. Consider $\mathcal{F} = \{x^2 + x/n\}$ for $n \geq 1$ defined on $[0, 1]$. It is straightforward to show that these functions are Lipschitz continuous uniformly with respect to n. For x, z in $[0, 1]$,

$$\left| x^2 + \frac{x}{n} - \left(z^2 + \frac{z}{n} \right) \right| = \left| (x + z) + \frac{1}{n} \right| |x - z| \leq 2|x - z|.$$

Hence, $\mathcal{F}$ is equicontinuous.

EXAMPLE 41.6. Consider $\mathcal{F} = \{nx^2\}$ for $n \geq 1$ defined on $[0, 1]$. Now for x, z in $[0, 1]$,

$$|nx^2 - nz^2| = n(x + z)|x - z|.$$

To insure that

$$|nx^2 - nz^2| < \epsilon$$

for some $\epsilon > 0$, we have to restrict the size of $|x - z|$ to be smaller than $\epsilon/(n(x + z))$, and for $x, z \neq 0$, this involves n. Hence, $\mathcal{F}$ cannot be equicontinuous.

Assuming that the family $\mathcal{F}$ is uniformly bounded and equicontinuous, it is sufficient to show that it contains a uniformly convergent subsequence. The proof, like the proof of the Weierstrass' principle, appears to be constructive. However, each step requires deciding whether a given region contains the graphs of an infinite number of functions, something that cannot be checked by a computer.

We let B be a uniform bound on the functions in $\mathcal{F}$. Since $\mathcal{F}$ is equicontinuous on $[a, b]$, given any $\epsilon > 0$ there is a $\delta_\epsilon > 0$ such that

$$|g(x) - g(z)| < \epsilon \text{ for } g \text{ in } \mathcal{F}, \ x, z \text{ in } [a, b] \text{ with } |x - z| < \delta_\epsilon.$$

Let $\epsilon_i = B/2^i$ for $i = 1, 2, 3, \cdots$ and let $\delta_i = \delta_{\epsilon_i}$.

Consider the rectangle $\Re$ described by $a \leq x \leq b$ and $-B \leq g \leq B$. We show the rectangle in Fig. 41.5. We construct a sequence of regions $\{b_n\}$ contained in $\Re$ such that b_n is contained in b_{n-1}, b_n spans $[a, b]$ in width, but has decreasing "height", and each b_n contains an infinite number of functions from $\mathcal{F}$. The functions in $\{b_n\}$ form a uniformly convergent Cauchy sequence.

We construct the sequence $\{b_n\}$ inductively. We begin by dividing up $\Re$ into a checkerboard pattern of smaller rectangles of width δ_1, or possibly smaller in the case of rectangles on the right-hand edge, and height ϵ_1. We denote the vertical "strips" of rectangles in a column by $S_1, S_2, \cdots, S_r$.

No function g in $\mathcal{F}$ can have a graph that spans more than two adjacent rectangles in S_1 because of the equicontinuity condition and the choice of δ_1. Therefore, at least two adjacent rectangles in S_1 must contain an infinite number of functions in $\mathcal{F}$. We shaded two such rectangles in S_1.

The graph of any function g in $\mathcal{F}$ that is in the shaded rectangles must pass through one of four adjacent rectangles that overlap the shaded region

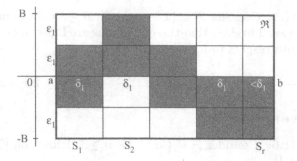

FIGURE 41.5. The first rectangle checkerboard pattern used in the proof of Arzela's theorem. The height of the smaller rectangles is ϵ_1 and the width is δ_1 except possibly for the rectangles on the right-hand edge, which may have smaller width. The shaded band b_1 contains the graphs of an infinite number of functions in S.

in S_2 because of continuity. Again, the graph of such a function cannot span more than two adjacent rectangles. Hence, one of the adjacent pairs of rectangles from among these four must contain an infinite number of functions in $\mathcal{F}$.

Continuing in the same way, we obtain a "band" b_1 of height $2\epsilon_1$ spanning $[a, b]$ containing the graphs of an infinite number of functions in $\mathcal{F}$. We shade an example of such a band in Fig. 41.5. We let $\{g^{(1)}\}$ denote the infinite family of functions from $\mathcal{F}$ whose graphs lie in b_1. We choose a particular member and call it $\bar{g}_1$.

Next, we treat the family $\{g^{(1)}\}$ and the band b_1 in the same way using ϵ_2 and δ_2. See Fig. 41.6. We divide up the rectangles in b_1 into rectangles

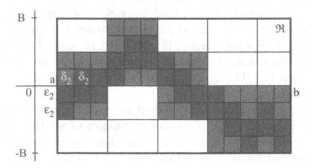

FIGURE 41.6. The second rectangle checkerboard pattern used in the proof of Arzela's theorem. The height of the smaller rectangles is ϵ_2 and the width is δ_2 except possibly for the rectangles on the right-hand edge, which may have smaller width. The shaded band b_2 contains the graphs of an infinite number of functions in $\{g^{(1)}\}$.

of height ϵ_2 and width δ_2, except possibly for rectangles on the right-hand edges of the strips in b_1, which might be smaller in width. Arguing as for b_1, we obtain a band b_2 of height $2\epsilon_2$ contained in b_1 that contains the graphs of infinitely many of the functions in $\{g^{(1)}\}$. We let $\{g^{(2)}\}$ denote the infinite family of functions from $\mathcal{F}$ whose graphs lie in b_2 and $\bar{g}_2$ be a particular member of this family.

Repeating, we get a sequence of bands $\{b_n\}$ such that b_n has height $2\epsilon_n$, width $b - a$, and b_n is contained in b_{n-1}. We also get a sequence of functions $\{\bar{g}_n\}$ from $\mathcal{F}$ with the graph of $\bar{g}_m$ contained in b_n for $m \geq n$. We conclude that $\{\bar{g}_n\}$ is a uniform Cauchy sequence of continuous functions that must therefore converge uniformly to a continuous limit.

We summarize as an important theorem, named after Arzela.[2]

Theorem 41.1 Arzela's Theorem *A uniformly bounded, equicontinuous family of functions on $[a, b]$ contains a subsequence that converges uniformly to a continuous function on $[a, b]$.*

Equicontinuity seems like a rather strong condition. But we can prove a sort of converse to Theorem 41.1. Suppose that we have a uniformly convergent sequence of continuous functions $\{g_n\}$ on $[a, b]$ and suppose $\epsilon > 0$ is given. Since $\{g_n\}$ is a uniform Cauchy sequence, there is an N such that for $n > N$,

$$|g_n(x) - g_N(x)| < \epsilon, \quad a \leq x \leq b.$$

Since continuous functions on a bounded interval are uniformly continuous on the interval by Theorem 32.11, there is a $\delta > 0$ such that for $1 \leq n \leq N$,

$$|g_n(x) - g_n(z)| < \epsilon, \quad x, z \text{ in } [a, b] \text{ and } |x - z| < \delta.$$

Note that even though this looks like the condition for equicontinuity for the family, it is not because we are only guaranteeing the continuity of a *finite* number of the functions, namely, $g_1, g_2, \cdots, g_N$. A finite set of uniformly continuous functions is always equicontinuous (see Problem 41.9).

If $n > N$ and x, z in $[a, b]$ with $|x - z| < \delta$, then

$$|g_n(x) - g_n(z)| \leq |g_n(x) - g_N(x)| + |g_N(x) - g_N(z)| + |g_N(z) - g_n(z)|$$
$$\leq 3\epsilon.$$

The first and last terms on the right are small because of the uniform convergence of g and the middle term on the right is small because of the uniform continuity of g_N. Since we can also make the difference $|g_n(x) - g_n(z)|$ less than ϵ for $n \leq N$, we have proved:

Theorem 41.2 *Suppose $\{g_n\}$ is a sequence of continuous functions on $[a, b]$ that converges uniformly to g on $[a, b]$. Then $\{g_n\}$ is equicontinuous on $[a, b]$.*

---

[2]Cesare Arzela (1847–1912) was an Italian analyst.

This result says that even if the original family $\mathcal{F}$ is not equicontinuous, any subsequence of $\mathcal{F}$ that converges uniformly is equicontinuous.

41.3   Convergence of Euler's Method

We use Arzela's Theorem 41.1 to show that Euler's method converges. We assume that we have a sequence of meshes with the property that the maximum step sizes tend to zero. Then we use Arzela's theorem to conclude that the family of corresponding Euler approximations $\mathcal{F}$ contains a uniform Cauchy subsequence on $[a, b]$ that converges to a continuous limit, which is the solution y, as $N \to \infty$.

Guaranteeing that $\mathcal{F}$ is uniformly bounded and equicontinuous requires assumptions on f. Based on the discussion in Chapter 39, we assume that there are constants $B > a$ and $M > 0$ such that f is continuous in x and **locally uniformly Lipschitz continuous** in y in the sense that there is a constant $L > 0$ such that

$$|f(x, y_2) - f(x, y_1)| \leq L|y_2 - y_1|$$

for all (x, y) in the rectangle $\tilde{\Re}$ described by $a \leq x \leq B$ and $-M + y_a \leq y_1, y_2 \leq y_a + M$ (see Fig. 39.3).

Recall that this implies that $f(x, \cdot)$ is uniformly bounded on the set of continuous functions whose graphs lie in $\tilde{\Re}$. Let z be such a function. Then for any $a < b \leq B$ and $a \leq x \leq b$,

$$|f(x, z(x)) - f(x, y_a)| \leq L|z(x) - y_a| \leq LM,$$

since the graph of z lies in $\tilde{\Re}$ by assumption. Therefore,

$$|f(x, z(x))| \leq LM + \sup_{a \leq x \leq b} |f(x, y_a)| = \tilde{M},$$

since f is continuous with respect to x and therefore $\sup_{a \leq x \leq b} |f(x, y_a)|$ is finite.

We show that the family $\mathcal{F}$ of Euler approximations is uniformly bounded and equicontinuous on an interval $[a, b]$ where $a < b \leq B$ is to be chosen. We let $\Re$ denote the "sub-rectangle" of $\tilde{\Re}$ described by $a \leq x \leq b$ and $-M + y_a \leq y \leq y_a + M$. Evaluating the integral operator $\mathcal{L}$ applied to a continuous function z whose graph is contained in $\Re$, we find

$$|\mathcal{L}(z) - y_a| = \left| \int_a^x f(s, z(s)) \, ds \right| \leq \tilde{M}(x - a) \leq \tilde{M}(b - a).$$

This says that the image of the integral operator $\mathcal{L}$ applied to z is contained in the triangular region inside $\tilde{\Re}$ between the lines with slopes $\pm \tilde{M}$ passing through (a, y_a) (see Fig. 41.7). We choose $b > a$ so that

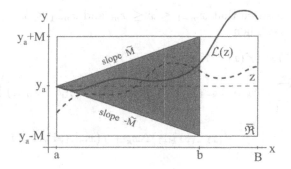

FIGURE 41.7. The image of the integral operator $\mathcal{L}$ applied to a continuous function whose graph is contained in $\tilde{\Re}$ is contained in the triangular shaded region inside $\tilde{\Re}$ between the lines through (a, y_a) with slopes $\pm \tilde{M}$.

$$b \le a + \frac{M}{\tilde{M}}.$$

This means that the image of a continuous function whose graph is in $\Re$ under $\mathcal{L}$ is another continuous function whose graph is in $\Re$.

We choose a mesh from $\{\mathcal{F}\}$ and denote the nodes $\{x_0, x_1, \cdots, x_N\}$ and the corresponding Euler approximation by Y. By (41.4),

$$|Y(x) - Y_0| \le \sum_{k=1}^{n-1} |f(x_{k-1}, Y_{k-1})| \Delta x_k + |f(x_{n-1}, Y_{n-1})|(x - x_{n-1}) \quad (41.8)$$

for $x_{n-1} \le x \le n$, $1 \le n \le N$. First consider $x_0 \le x \le x_1$, where

$$|Y(t) - Y_0| \le |f(x_0, Y_0)||x - x_0| \le \tilde{M}(x - x_0) \le \tilde{M}(b - a) \le M.$$

Assume that we have proved $|Y(x) - Y_0| \le M$ for $x_{n-2} \le x \le x_{n-1}$. Then for $x_{n-1} \le x \le x_n$, (41.8) implies

$$|Y(x) - Y_0| \le \tilde{M} \sum_{k=1}^{n-1} \Delta x_k + \tilde{M}(x - x_{n-1}) = \tilde{M}(x - x_0)$$

$$\le \tilde{M}(b - a) \le M. \quad (41.9)$$

Hence, $\mathcal{F}$ is uniformly bounded on $[a, b]$ with bound M.

Next, we show that $\mathcal{F}$ is equicontinuous. In fact, the Euler approximations in $\mathcal{F}$ are all Lipschitz continuous with constant $\tilde{M}$ uniformly with respect to their respective meshes. Choose a mesh $\mathcal{T}_N$ and denote the nodes $\{x_0, x_1, \cdots, x_N\}$ and the corresponding Euler approximation by Y. Next, choose points $a \le \bar{x} < x \le b$. If x and $\bar{x}$ are in the same subinterval $[x_{n-1}, x_n]$ for $1 \le n \le N$, then immediately

$$|Y(x) - Y(\bar{x})| \le |f(x_{n-1}, Y_{n-1})||x - \bar{x}| \le \tilde{M}|x - \bar{x}|.$$

Otherwise, assume that $x_{m-1} \leq \bar{x} \leq x_m$ and $x_{n-1} \leq x \leq x_n$ for some $1 \leq m < n \leq N$. Then

$$Y(x) - Y(\bar{x}) = (Y(x) - Y_0) - (Y(\bar{x}) - Y_0)$$

$$= \left(\sum_{k=1}^{n-1} f(x_{k-1}, Y_{k-1}) \Delta x_k + f(x_{n-1}, Y_{n-1})(x - x_{n-1}) \right)$$

$$- \left(\sum_{k=1}^{m-1} f(x_{k-1}, Y_{k-1}) \Delta x_k + f(x_{m-1}, Y_{m-1})(\bar{x} - x_{m-1}) \right).$$

$$(41.10)$$

We expand the first sum on the right-hand side as

$$\sum_{k=1}^{m-1} f(x_{k-1}, Y_{k-1}) \Delta x_k$$

$$+ f(x_{m-1}, Y_{m-1})(\bar{x} - x_{m-1}) + f(x_{m-1}, Y_{m-1})(x_m - \bar{x})$$

$$+ \sum_{k=m+1}^{n-1} f(x_{k-1}, Y_{k-1}) \Delta x_k + f(x_{n-1}, Y_{n-1})(x - x_{n-1}),$$

(with the convention that the second sum is empty if m=n-1). So (41.10) means

$$Y(x) - Y(\bar{x}) = f(x_{m-1}, Y_{m-1})(x_m - \bar{x}) + \sum_{k=m+1}^{n-1} f(x_{k-1}, Y_{k-1}) \Delta x_k$$

$$+ f(x_{n-1}, Y_{n-1})(x - x_{n-1}). \quad (41.11)$$

By the uniform bound on f, we conclude that

$$|Y(x) - Y(\bar{x})| \leq \tilde{M} \left((x_m - \bar{x}) + \sum_{k=m+1}^{n-1} \Delta x_k + (x - x_{n-1}) \right) \leq \tilde{M}|x - \bar{x}|.$$

$$(41.12)$$

Hence, the Euler approximations in $\mathcal{F}$ are equicontinuous.

By Arzela's Theorem 41.1, we conclude that $\mathcal{F}$ contains a subsequence of Euler functions that converges to a continuous function $y(x)$. We denote this subsequence by $\{Y^{(n)}\}$ and the corresponding maximum step sizes by $\{\Delta x^{(n)}\}$. We next argue that the limit y is a solution of the differential equation (41.1).

We do this by first showing that the Euler approximations $\{Y^{(n)}\}$ "almost" solve the differential equation. Specifically, we choose $\bar{x}$ in $[a, b]$ and show that for any $\epsilon > 0$,

$$\left| \frac{Y^{(n)}(x) - Y^{(n)}(\bar{x})}{x - \bar{x}} - f(\bar{x}, \bar{y}) \right| < \epsilon \quad (41.13)$$

for $|x - \bar{x}|$ sufficiently small and n sufficiently large, where $\bar{y} = y(\bar{x})$.

We require some information on how $f(\cdot, \cdot)$ behaves as both inputs change simultaneously. Note that $y_a - M \leq y(x) \leq y_a + M$ for $a \leq x \leq b$. We assume that $a < \bar{x} < b$ and $y_a - M < \bar{y} < y_a + M$ and leave the case that $\bar{x}$ or $\bar{y}$ is one of the endpoints of the respective intervals as Problem 41.16.

Now for x in $[a, b]$ and z in $[y_a - M, y_a + M]$,

$$|f(x, z) - f(\bar{x}, \bar{y})| \leq |f(x, z) - f(x, \bar{y})| + |f(x, \bar{y}) - f(\bar{x}, \bar{y})|$$
$$\leq L|z - \bar{y}| + |f(x, \bar{y}) - f(\bar{x}, \bar{y})|.$$

Hence given $\epsilon > 0$, there is a $\delta > 0$ such that

$$|f(x, z) - f(\bar{x}, \bar{y})| < \epsilon \tag{41.14}$$

provided $|x - \bar{x}| < 2\delta$ and $|z - \bar{y}| < 4\tilde{M}\delta$. For δ sufficiently small, this defines a rectangle $\tilde{\Re}$ contained in $\Re$ (see Fig. 41.8).

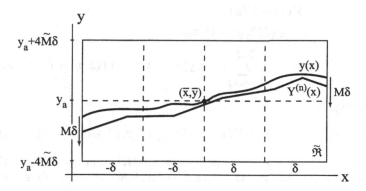

FIGURE 41.8. The rectangle $\tilde{\Re}$ defined by $|x - \bar{x}| < 2\delta$ and $|z - \bar{y}| < 4\tilde{M}\delta$.

We choose N sufficiently large that $n > N$ implies $\Delta x^{(n)} < \delta$ and

$$|y(x) - Y^{(n)}(x)| < \tilde{M}\delta \text{ for } a \leq x \leq b.$$

Then both $(x, Y^{(n)}(x))$ and $(x, y(x))$ are in $\tilde{\Re}$ for $|x - \bar{x}| < 2\delta$. The first claim follows from (41.12), since

$$|Y^{(n)}(x) - \bar{y}| \leq |Y^{(n)}(x) - Y^{(n)}(\bar{x})| + |Y^{(n)}(\bar{x}) - \bar{y}| \leq 2\tilde{M}\delta + \tilde{M}\delta,$$

and the second claim from

$$|y(x) - \bar{y}| \leq |y(x) - Y^{(n)}(x)| + |Y^{(n)}(x) - Y^{(n)}(\bar{x})| + |Y^{(n)}(\bar{x}) - \bar{y}|$$
$$\leq 4\tilde{M}\delta.$$

Assuming that $x_{m-1} \leq \bar{x} \leq x_m$ and $x_{n-1} \leq x \leq x_n$ for some $1 \leq m \leq n \leq N$, (41.11) implies

$$Y(x) - Y(\bar{x}) = f(x_{m-1}, Y_{m-1})(x_m - \bar{x}) + \sum_{k=m+1}^{n-1} f(x_{k-1}, Y_{k-1})\Delta x_k$$
$$+ f(x_{n-1}, Y_{n-1})(x - x_{n-1})$$

if $m < n$ and

$$Y(x) - Y(\bar{x}) = f(x_{n-1}, Y_{n-1})(x - \bar{x})$$

if $m = n$. Since $\Delta x^{(n)} < \delta$, $x - \bar{x} < \delta$ implies that $x_{m-1}, \cdots, x_{n-1}$ are within 2δ of $\bar{x}$. Therefore, (41.14) implies that

$$(f(\bar{x}, \bar{y}) - \epsilon)(x_m - \bar{x})$$
$$+ \sum_{k=m+1}^{n-1} (f(\bar{x}, \bar{y}) - \epsilon)\Delta x_k + (f(\bar{x}, \bar{y}) - \epsilon)(x - x_{n-1})$$
$$\leq Y(x) - Y(\bar{x})$$
$$\leq (f(\bar{x}, \bar{y}) + \epsilon)(x_m - \bar{x})$$
$$+ \sum_{k=m+1}^{n-1} (f(\bar{x}, \bar{y}) + \epsilon)\Delta x_k + (f(\bar{x}, \bar{y}) + \epsilon)(x - x_{n-1}).$$

Simplifying,

$$(f(\bar{x}, \bar{y}) - \epsilon)(x - \bar{x}) \leq Y(x) - Y(\bar{x}) \leq (f(\bar{x}, \bar{y}) + \epsilon)(x - \bar{x}). \qquad (41.15)$$

A similar result holds if $x < \bar{x}$ (see Problem 41.17) and we get (41.13).
Passing to the limit in (41.13) as $n \to \infty$ shows that for any $\epsilon > 0$,

$$\left| \frac{y(x) - y(\bar{x})}{x - \bar{x}} - f(\bar{x}, y(\bar{x})) \right| < \epsilon$$

for $|x - \bar{x}| < \delta$ in $[a, b]$.[3] This shows that $y(x)$ is a solution of (41.1).

41.4   Uniqueness and Continuous Dependence on Initial Data

Now we know that $\mathcal{F}$ contains a subsequence of Euler approximations that converges to some solution of (41.1). We next show that (41.1) must have a unique solution. In fact, we know this already from Chapter 40, but we give a different proof here using what is called a Gronwall argument.

---

[3]With a suitable interpretation if $\bar{x} = a$ or b.

Suppose that we have two solutions $y(x)$ and $z(x)$ of $y' = f(x, y)$ for $a \leq x \leq b$ with $y(a) = y_a$ and $z(a) = z_a$ and whose graphs are contained in $\mathfrak{R}$. We obtain an estimate on $|z(x) - y(x)|$ in terms of the difference between the data $|z_a - y_a|$. Subtracting the integral forms of the differential equations for z and y,

$$z(x) = z_a + \int_a^x f(s, z(s)) \, ds \text{ and } y(x) = y_a + \int_a^x f(s, y(s)) \, ds,$$

we get

$$z(x) - y(x) = (z_a - y_a) + \int_a^x (f(s, z(s)) - f(s, y(s))) \, ds.$$

Since the graphs of z and y are contained in $\mathfrak{R}$ for $a \leq x \leq b$, we can use the Lipschitz continuity of f to obtain

$$|z(x) - y(x)| \leq |z_a - y_a| + L \int_a^x |z(s) - y(s)| \, ds.$$

We define the continuous function $u(x) = |z(x) - y(x)|$ with $u_a = |z_a - y_a|$, so that

$$u(x) \leq u_a + \int_a^x u(s) \, ds. \tag{41.16}$$

We assume that $u_a \neq 0$ and leave $u_a = 0$ to Problem 41.18. If we set

$$v(x) = L \int_a^x u(s) \, ds,$$

then v is differentiable on $[a, b]$, and by the Fundamental Theorem 34.1, $v'(x) = Lu(x)$. Hence, (41.16) implies

$$\frac{dv}{dx} \leq Lu_a + Lv, \quad a \leq x \leq b.$$

We separate variables,

$$\frac{dv}{u_a + v} \leq L \, dx, \quad a \leq x \leq b,$$

and integrate from a to x, using $v(a) = 0$, to get

$$\log(u_a + v(x)) - \log(u_a) \leq L(x - a).$$

Some easy algebra gives

$$v(x) \leq u_a e^{L(x-a)} - u_a.$$

Putting this into (41.16) yields $u(x) \leq u_a e^{L(x-a)}$ for $a \leq x \leq b$. We summarize this last result as a theorem, named after Gronwall.[4]

Theorem 41.3 Gronwall's Lemma *Suppose that $u_a \geq 0$ and $u(x)$ is a nonnegative continuous function $a \leq x \leq b$ that satisfies*

$$u(x) \leq u_a + \int_a^x u(s)\,ds, \quad a \leq x \leq b.$$

Then

$$u(x) \leq u_a e^{L(x-a)}, \quad a \leq x \leq b. \tag{41.17}$$

Gronwall's Lemma implies that the solution of (41.1) depends continuously on the initial data. In particular if $z_a = y_a$, we conclude that $z(x) = y(x)$ for $a \leq x \leq b$. In other words, if (41.1) has a solution whose graph lies in $\Re$, then that solution is unique. We give this result as a separate theorem.

Theorem 41.4 Uniqueness for an Ordinary Differential Equation
Suppose that $f(s, y)$ is continuous in x and uniformly Lipschitz continuous in y for (x, y) in the rectangle $\Re$ given by $a \leq x \leq b$ and $|y - y_a| \leq M$. There can be at most one solution of (41.1) on $a \leq x \leq b$ whose graph is contained in $\Re$.

41.5   More on the Convergence of Euler's Method

Now we have shown that a family of Euler approximations corresponding to a sequence of meshes with decreasing step sizes contains a subsequence that converges uniformly to the unique solution of (41.1). We want to show that in fact the entire family of Euler approximations converges to the solution.

Given $\epsilon > 0$, we argue that at most a finite number of Euler approximations $\{Y\}$ in $\mathcal{F}$ can have graphs that lie outside the region enclosed by $x = a$ on the left, $x = b$ on the right, $y(x) - \epsilon$ below, and $y(x) + \epsilon$ above. See Fig. 41.9. Indeed, suppose an infinite number of the Euler approximations, do lie outside this region. Denote these approximations by $\{Y^{(n)}\}$ and the corresponding step sizes by $\{\Delta x^{(n)}\}$. Since $\Delta x^{(n)} \to 0$, we can use

---

[4]Hakon Tomi Grönvall, or Thomas Hakon Gronwall, (1877–1932) was born in Sweden, educated in Sweden and Germany, and worked and died in the United States. He was a powerful mathematician as well as a good physical chemist and civil engineer. In mathematics, he made important contributions to algebra, differential equations, mathematical physics, number theory, and real and complex analysis. Substantial parts of his career were spent as a consulting mathematician for engineers and chemists. A colorful personality; there is a legend (told to the author by Lars Wahlbin) that Gronwall did not die in 1932. Instead, he quit the mathematical rat race, made a lot of money in the stock market, bought an island in the South Pacific, and retired there happily.

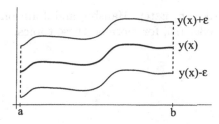

FIGURE 41.9. The region bounded by $x = a$ on the left, $x = b$ on the right, $y(x) - \epsilon$ below, and $y(x) + \epsilon$ above.

the argument above to show that there is a subsequence that converges to a solution of (41.1), and that solution cannot equal $y(x)$ by assumption. But this is impossible by Theorem 41.4.

We summarize as a theorem that we credit to the first three people to prove versions of the result.[5]

Theorem 41.5 Existence Result of Cauchy, Lipschitz, and Peano
Suppose that $f(x,y)$ is continuous in x and uniformly Lipschitz continuous in y for all (x,y) in the rectangle $\bar{\Re}$ described by $a \leq x \leq B$ and $-M + y_a \leq y \leq y_a + M$. Then there is a b with $a < b \leq B$ such that (40.1) has a unique solution for $a \leq x \leq b$. Moreover, any sequence of forward Euler approximations corresponding to a sequence of meshes whose maximum step sizes tend to zero converge to this solution.

41.6   Unanswered Questions

Theorem 41.5 implies that under the right assumptions, the forward Euler method can be used to approximate the solution to any desired accuracy. But it does not give any indication of the accuracy of a specific approximation. This gives rise to a slew of questions. How quickly does the forward Euler approximation approach the true solution as the mesh size decreases? How can we estimate the error of a particular approximation in order to decide whether it is sufficiently accurate or not? If we have an approximation that is not sufficiently accurate, how should we refine the mesh, i.e., decrease the mesh steps, in order to improve the accuracy.

---

[5]Cauchy gave the first general existence result for a first order nonlinear differential equation $y' = f(x,y)$. He proved that Euler's method, which had been described earlier by Euler, converges under the assumption that f is differentiable both with respect to x and y. Lipschitz proved the same result under the weaker assumption that f is continuous with respect to x and Lipschitz continuous with respect to y. Peano proved the most general result by assuming that f was merely continuous in (x,y), though Peano's result does not give uniqueness (see Problem 41.22).

See Braun [4], Eriksson, Estep, Hansbo, and Johnson [10], Henrici [13], and Isaacson and Keller [15] for more on these topics.

Chapter 41 Problems

In problems 41.1–41.5, you are asked to compute forward Euler approximations on uniform meshes for specific initial value problems. An efficient way to do these problems is to write a program in MATLAB $^{\text{©}}$, for example, that accepts the nonlinearity f, the interval $[a, b]$, the initial value y_a, and the number of mesh points, and computes and plots the corresponding approximation.

41.1. Compute the forward Euler approximations for the problem $y' = x + y$ for $0 \leq x \leq 1$ using both initial values $y_a = 0$ and $y_a = 1$ and uniform meshes with $N = 5, 10, 20, 40$. Plot the results in all cases.

41.2. Compute the forward Euler approximations for the problem $y' = y^2$ for $0 \leq x \leq .9$ using the initial value $y_a = 1$ and uniform meshes with $N = 40, 80, 160, 320$. Plot the results in all cases. Compare to the expected behavior of the true solution.

41.3. Compute the forward Euler approximations for the problem $y' = 1 - y^3$ for $0 \leq x \leq 1$ using both initial values $y_a = 0$ and $y_a = 1$ and uniform meshes with $N = 5, 10, 20, 40$. Plot the results in all cases. What is the error of the approximations when $y_a = 1$?

41.4. Compute the forward Euler approximations for the problem $y' = y^{1/2}$ for $0 \leq x \leq 5$ using the initial values $y_a = 0$ and $y_a = .0001$ and uniform meshes with $N = 20, 40, 80, 160$. Plot the results in all cases. We could think of the second initial condition as "zero" plus some experimental error. Which solutions of this problem are found by the forward Euler approximations?

41.5. (a) Compute the forward Euler approximations for the problem $y' = 2xy$ for $1 \leq x \leq 2$ using the initial value $y_a = e$ and uniform meshes with $N = 10, 20, 40, 80$. Plot the results in all cases and compare to the true solution $y(x) = e^{x^2}$. (b) This solution also satisfies $y' = 2y\sqrt{\log(y)}$ and $y(1) = e$. Repeat the computations for this new problem and compare the accuracy of the forward Euler approximations to those obtained for the first problem. (c) Compare the two different problems if we solve them on $[0, 1]$ with $y_a = 1$. Do you anticipate difficulties with one of the problems?

41.6. The **backward Euler method** is constructed using the right-hand endpoint on each subinterval. The nodal values $\{Y_n\}_{n=0}^{N}$ are determined by a set of nonlinear equations. First, we set $Y_0 = y_a$. The equation for Y_1 is

$$Y_1 = Y_0 + f(x_1, Y_1)\Delta x_1,$$

and in general for $1 \leq n \leq N$,

$$Y_n = Y_0 + \sum_{k=1}^{n-1} f(x_k, Y_k)\Delta x_k + f(x_n, Y_n)\Delta x_n.$$

By using the Contraction Mapping Principle Theorem 15.1, prove the equation for Y_n has a unique solution for sufficiently small time step Δx_n. Describe the set of valid initial iterates. *Hint:* You will need to use the continuity assumption on f.

41.7. Provided (x_{n-1}, Y_{n-1}) is inside $\Re$, prove the local solution $\tilde{y}$ of

$$\begin{cases} \tilde{y}'(x) = f(x, \tilde{y}(x)), & x_{n-1} \le x \le x_n, \\ \tilde{y}(x_{n-1}) = Y_{n-1}, \end{cases}$$

exists for all sufficiently small Δx_n. *Hint:* Use Theorem 40.2 or Theorem 41.5.

Problems 41.8–41.14 have to do with equicontinuity and Arzela's Theorem 41.1.

41.8. Decide if the following families are equicontinuous or not, and give a reason for your answer.

(a) $\left\{ \sin\left(x + \dfrac{1}{n}\right) \right\}$ on $[0, \pi]$ (b) $\{x^n\}$ on $[0, 1]$

(c) $\left\{ \dfrac{nx}{1 + nx^2} \right\}$ on $[0, 1]$ (d) $\{x^2 + \sin(n)x\}$ on $[0, 1]$.

41.9. Prove that a finite set of uniformly continuous functions is equicontinuous.

41.10. Suppose that $\{g_n\}$ is a family that is uniformly Lipschitz continuous on $[a, b]$ in the sense that there is a constant L such that $|g_n(x_2) - g_n(x_1)| \le L|x_2 - x_1|$ for $a \le x_1, x_2 \le b$ and all n. Show that $\{g_n\}$ is equicontinuous.

41.11. Suppose g is continuous on $\mathbf{R}$. Suppose the family of functions $\{g_n\}$ on $[0, 1]$, with $g_n(x) = g(nx)$, is equicontinuous. What can you say about g?

41.12. Suppose that $\{g_n\}$ is equicontinuous on $[a, b]$ and converges pointwise on $[a, b]$. Prove $\{g_n\}$ converges uniformly on $[a, b]$.

41.13. Construct the first three bands b_1, b_2, b_3 used in the proof of Arzela's Theorem 41.1 for the family $\{x^2 + x/n\}$ on $[0, 1]$. *Hint:* Use the fact that $x^2 \le x^2 + x/n \le x^2 + x$ for $0 \le x \le 1$.

41.14. Suppose that $\{g_n\}$ is a uniformly bounded family of continuous functions on $[a, b]$. Define

$$G_n(x) = \int_a^x g_n(s)\, ds \text{ for } a \le x \le b.$$

Prove there is a subsequence of $\{G_n\}$ that converges uniformly on $[a, b]$.

Problems 41.15–41.21 have to do with the proof of Theorem 41.5.

41.15. Verify (41.9).

41.16. Verify that (41.13) holds when $\bar{x} = a$, $\bar{x} = b$, $\bar{y} = y_a - M$, and/or $\bar{y} = y_a + M$.

41.17. Verify (41.17) when $x < \bar{x}$.

41.18. Prove (41.17) holds when $u_a = 0$. *Hint:* Using the same v as in the proof of (41.17) when $u_a > 0$, multiply $v' - Lv \leq 0$ by the integrating factor e^{-Lt}. Show that

$$\frac{d}{dt}\left(e^{-Lt}v\right) = \left(v' - Lv\right)e^{-Lt}.$$

Then use this fact.

41.19. State and prove a more general Gronwall Lemma for a nonnegative continuous function u satisfying

$$u(x) \leq u_a + \int_a^x g(s)u(s)\,ds, \quad a \leq x \leq b,$$

where $u_a \geq 0$ and $g(x)$ is a positive continuous function on $[a, b]$.

41.20. As part of proving Theorem 41.5, we show that the Euler approximation Y_n approximately solves the differential equation in the sense of (41.13) and then pass to the limit to show that limit y does solve the differential equation. As an alternative, prove that the limit y solves the integral equation (41.2). *Hint:* Show that

$$\left|\int_a^x f(s, y(s))\,ds - Y(x)\right|$$

can be made uniformly small for $a \leq x \leq b$.

41.21. What goes wrong in the proof of Theorem 41.5 if we drop the assumption that the sequence of meshes corresponding to the family of forward Euler approximations $\mathcal{F}$ have their maximum step sizes tend to zero?

Peano's proof that the forward Euler method converges to the solution uses weaker continuity assumptions on f than we assume in Theorem 41.5, and his version of the theorem gives convergence even when there is not a unique solution. However, his proof requires defining continuity for functions of several variables and using some properties of such functions. The uniform Lipschitz assumption in Theorem 41.5 is actually sufficient to give a constructive proof that does not resort to Arzela's theorem. In the next two problems, we ask you to prove Peano's original result and to give a constructive proof that the Euler method converges.

41.22. Peano's version of Theorem 41.5 holds for functions $f(x, y)$ that are merely continuous. $f(x, y)$ is continuous at (x_1, z_1) if for any $\epsilon > 0$ there is a $\delta > 0$ such that $|f(x_2, z_2) - f(x_1, z_1)| < \varepsilon$ for any (x_2, z_2) with $|x_2 - x_1| < \delta$ and $|z_2 - z_1| < \delta$. It is possible to prove that a function that is continuous on a closed, bounded rectangle $\bar{\Re}$ is uniformly continuous and uniformly bounded on $\bar{\Re}$, with the obvious definitions of these notions.[6]

Assume that $f(x, y)$ is continuous on a rectangle $\bar{\Re}$ defined by $a \leq x \leq B$ and $|y - y_a| \leq M$. Prove that there is a b with $a < b \leq B$ such that any family of forward Euler approximations defined on a sequence of meshes on $[a, b]$ whose mesh sizes tend to zero converges to a solution of (40.1). This result does not give uniqueness!

Hint: Define b as in the proof of Theorem 41.5, where $\tilde{M}$ is a bound on $f(x, y)$ on $\bar{\Re}$. Follow the same arguments with appropriate modifications using the uniform continuity and boundedness of f on $\Re$.

41.23. Modify the analysis of the rectangle rule for integration in Chapters 25 and 34 to give a constructive proof of Theorem 41.5.

---

[6]This result is not really harder to prove than the analog for one dimension discussed in Chapter 32. However, it requires some geometric notions that we have decided to avoid in this book. Otherwise, we could have proved Peano's version, which is just a modification of the proof of Theorem 41.5 given above.

A Conclusion or an Introduction?

It did not seem right to finish with a conclusion since this book is only an introduction to the vast subject of analysis. Instead, we finish by discussing where to go after reading this book.

The sources listed in the References are a good starting point for further explorations. The books by Courant and John [6] and Lay [17] discuss real analysis at roughly the same level as this book. However, Courant and John focus more on calculus, while Lay adopts a more abstract approach. The calculus text by Bers [3] also contains some rigorous analysis. The book by Rudin [19] is the classic introductory text in real analysis. It is more abstract than this book and goes further in some aspects of analysis. It is the next book on analysis to read after this book. Thomson, Bruckner, and Bruckner [20] cover much of the same material as Rudin but in a more modern style. The author consulted all of these books frequently while writing this text.

Complex analysis, which is the analysis of functions of complex numbers, is one of the most beautiful areas in mathematics. It should be studied right after real analysis. A standard text is Ahlfors [1], which indeed is a classic book in mathematics.

Grinstead and Snell [12] is a favorite introduction to probability.

Two general textbooks on numerical analysis are Atkinson [2] and Isaacson and Keller [15]. These books also offer material on general analysis not often discussed in standard analysis texts. These two books discuss Newton's method and the solution of ordinary differential equations, but these subjects are sufficiently complicated to warrant their own books. Dennis and Schnabel [9] is a great source for learning about the solution of nonlinear equations. The textbook by Braun [4] and the classic book by Henrici [13] discuss many interesting aspects of differential equations and their numerical solution. For a more modern perspective, the reader might consult Eriksson, Estep, Hansbo, and Johnson [10], which explains how many engineers look at the numerical solution of differential equations.

As far as the history of mathematics is concerned, there are some good sources [5, 7, 11, 16, 18] listed in the references. All students of mathematics should eventually read the classic books by Kline [16]. Mathematicians should also read Davis and Hersh [8] when they reach that point when they begin to question what they do and why.

It is reasonable to guess that those who have found their way through most of this book are likely to have more adventures and discoveries in analysis awaiting them. Whether these lie in mathematics, science, or engineering, whether future investigations consist of proofs or computations, or whether the goal is to model the physical world or to seek out mathematical truths, remember that it is *all* analysis.

References

[1] L. AHLFORS, *Complex Analysis*, McGraw-Hill Book Company, New York, 1979.

[2] K. ATKINSON, *An Introduction to Numerical Analysis*, John Wiley and Sons, New York, 1989.

[3] L. BERS, *Calculus*, Holt, Rinehart, and Winston, New York, 1976.

[4] M. BRAUN, *Differential Equations and their Applications*, Springer-Verlag, New York, 1984.

[5] R. COOKE, *The History of Mathematics. A Brief Course*, John Wiley and Sons, New York, 1997.

[6] R. COURANT AND F. JOHN, *Introduction to Calculus and Analysis*, vol. 1, Springer-Verlag, New York, 1989.

[7] R. COURANT AND H. ROBBINS, *What is Mathematics?*, Oxford University Press, New York, 1969.

[8] P. DAVIS AND R. HERSH, *The Mathematical Experience*, Houghton Mifflin, New York, 1998.

[9] J. DENNIS AND R. SCHNABEL, *Numerical Methods for Unconstrained Optimization and Nonlinear Equations*, Prentice-Hall, New Jersey, 1983.

[10] K. ERIKSSON, D. ESTEP, P. HANSBO, AND C. JOHNSON, *Computational Differential Equations*, Cambridge University Press, New York, 1996.

[11] I. GRATTAN-GUINESS, *The Norton History of the Mathematical Sciences*, W.W. Norton and Company, New York, 1997.

[12] C. GRINSTEAD AND J. SNELL, *Introduction to Probability*, American Mathematical Society, Providence, 1991.

[13] P. HENRICI, *Discrete Variable Methods in Ordinary Differential Equations*, John Wiley and Sons, New York, 1962.

[14] E. HILLE, *Thomas Hakon Gronwall - In Memoriam*, Bulletin of the American Mathematical Society, 38 (1932), pp. 775–786.

[15] E. ISAACSON AND H. KELLER, *Analysis of Numerical Methods*, John Wiley and Sons, New York, 1966.

[16] M. KLINE, *Mathematical Thought from Ancient to Modern Times*, vol. I, II, III, Oxford University Press, New York, 1972.

[17] S. LAY, *Analysis with an Introduction to Proof*, Prentice Hall, New Jersey, 2001.

[18] J. O'CONNOR AND E. ROBERTSON, *The MacTutor History of Mathematics Archive*, School of Mathematics and Statistics, University of Saint Andrews, Scotland, 2001. http://www-groups.dcs.st-and.ac.uk/~history/.

[19] W. RUDIN, *Principles of Mathematical Analysis*, McGraw–Hill Book Company, New York, 1976.

[20] B. THOMSON, J. BRUCKNER, AND A. BRUCKNER, *Elementary Real Analysis*, Prentice Hall, New Jersey, 2001.

[21] T. YPMA, *Historical development of the Newton-Raphson method*, SIAM Review, 37 (1995), pp. 531–551.

Index

Undergraduate Texts in Mathematics

(continued from page ii)

Undergraduate Texts in Mathematics